TRUCK COMPANY OPERATIONS

2ND EDITION

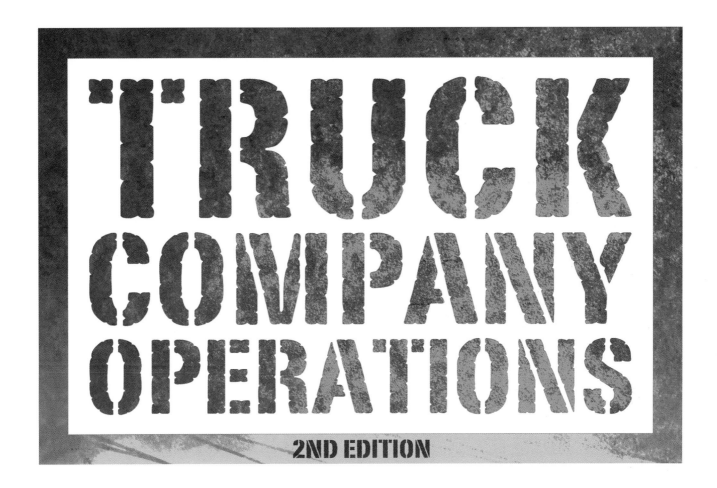

TRUCK COMPANY OPERATIONS

2ND EDITION

John Mittendorf

Fire Engineering

Copyright © 2011 by
PennWell Corporation
1421 South Sheridan Road
Tulsa, Oklahoma 74112-6600 USA

800.752.9764
+1.918.831.9421
sales@pennwell.com
www.Fire EngineeringBooks.com
www.pennwellbooks.com
www.pennwell.com

Marketing Coordinator: Jane Green
National Account Executive: Barbara McGee
Director: Mary McGee
Managing Editor: Jerry Naylis
Production Manager: Sheila Brock
Production Editor: Tony Quinn
Book Designer: Susan E. Ormston
Cover Designer: Karla Pfeifer

Illustrations by Paul Bunch and Jim Coburn

Library of Congress Cataloging-in-Publication Data

Mittendorf, John, 1940-
  Truck company operations / John Mittendorf. -- 2nd ed.
      p. cm.
  Includes bibliographical references and index.
  ISBN 978-1-59370-218-2
  1. Fire extinction. 2. Fire engines. 3. Command and control at fires. 4. Fires. I. Title.
  TH9146.M58 2010
  628.9'259--dc22

                                                                    2010017173

Printed in the United States of America

4 5   15 14

This book is dedicated to the memory of Captain Lane Kemper of the Los Angeles City Fire Department. He was a trusted friend, mentor, and forcible entry guru. If it were not for cancer, Captain Kemper would still be devising ways to enter and exit buildings and making the fireground safer for firefighters

And most of all, the day that I gave my life to Jesus Christ was the most important event in my life and ultimately allowed me to write this book.

Contents

Acknowledgments

A special thanks goes to the organizations and individuals who were willing to provide resources and information for the first edition of *Truck Company Operations*. For the second edition, I'd like to thank the following departments and individuals for sharing their expertise and time: Bend (OR) Fire Department, Layton City (UT) Fire Department, Overland Park (KS) Fire Department, Surprise (AZ) Fire Department, Consumes (CA) Fire Department, Midvale (UT) Fire Department, Peoria (AZ) Fire Department, Lynchburgh (VA) Fire Department, Phoenix (AZ) Fire Department, Fire District 3 (OR) Fire Department, Roseville (CA) Fire Department, Medford (OR) Fire Department, Willmar (WI) Fire Department, Glendale (AZ) Fire Department, Green Township Fire & EMS (OH) Fire Department, and Chief Dave Dodson, Chief Dennis Compton, Chief Vincent Dunn, Chief Steven H. Higgs, Chief Guy Keirn, Chief Mark Nichols, Chief Douglas J. Witsken, Erin C. Sawall, Kurt Zingheim, Rob Cravens, Kevin Trost, Allen Maier, Mike Gagliano, Rick Haas, Josh Roten, Randy Frassetto, Joel D. Andrus, Brain Maier, William Gustin, John J. Lewis, the numerous photographers who have provided some of the photos, and Paul Bunch for creating the majority of the graphics. There are probably other individuals and departments that I have forgotten, but I will have to plead old age and a fading memory.

Finally, and most importantly, I would like to acknowledge the various instructors who have spent a great deal of time preparing and delivering training programs on a national level, which, as a result, have significantly increased the safety and operational effectiveness of the fire service. Instructors such as Chief Vincent Dunn (*Collapse of Burning Buildings*), Francis Brannigan (*Building Construction for the Fire Service*), Chief Alan Brunacini (*Fire Command* and *Essentials of Fire Department Customer Service*), Chief John Norman (*Fire Officer's Handbook of Tactics* and *Fire Department Special Operations*), Chief John "Skip" Coleman (*Incident Management for the Fire Service*), Chief John J. Salka Jr. (*Get Out Alive*), and a host of others who have willingly devoted their time and talents to share their expertise. However, it is worth mentioning that along with these training programs, there have recently been three unique training programs that have not only captured the attention of the fire service, but have been long overdue from the perspective of changing the status quo. These programs are "The Art of Reading Smoke" by Chief Dave Dodson, "Air Management for the Fire Service" by the Seattle Boys, and lastly, "Pride and Ownership" by Chief

Rick Lasky. Not surprisingly, these three training programs have been given to standing room only crowds at the Fire Department Instructor Conference and Exhibition (FDIC), which is an indicator of the relevance of these subjects. Of particular note, "Pride and Ownership" holds a special place in my heart, as several of the subjects covered in Chief Lasky's book are passion, the desire to achieve a goal, and love of the job. It was primarily these factors that allowed me to become a firefighter for the Los Angeles City Fire Department (LAFD), because the first time I applied for the department I was not accepted and came very close to choosing another profession. Please allow me to relate the background of this true story as it may inspire someone to focus on their love of the job, desire, and passion to pursue a goal.

When I applied for the LAFD in 1961, I sailed through the written, interview, physical agility, and psychological exams until I was examined for the last portion, the medical exam. At that time, candidates had to be a minimum of 5 feet 8 inches tall and I measured 5 feet 7-3/4 inches tall, so I was promptly disqualified for being too short. Later, when I spoke with my father, who was a captain in the LAFD, he recommended that we see a chiropractor and determine if there were any options. After a few minutes with the chiropractor, he stated that people are taller in the morning than in the evening, and it was possible to stretch a person 1 to 2 inches. With this encouraging information, I promptly set about to see how far I could be stretched. To accomplish this, a winch from a boat trailer was attached to one end of a sheet of plywood and two stirrups were attached to the other end. This allowed a person to lie on the plywood with feet anchored by the stirrups with a padded strap placed under the chin with some tension applied by the boat winch. With this configuration, I went to sleep for the night, and the next morning measured almost 5-feet 9-inches. Armed with the encouraging possibility of being taller (for a short period of time, as I have been shrinking ever since), I filed a protest with the city to be re-measured. On the evening prior to the day of the re-measurement, I again slept on the plywood stretching machine but with increased tension by the winch. On the morning of the re-measurement, my father put the plywood in the back of his pickup truck (with the tailgate down), took and extra turn on the winch handle, and drove 60 miles to downtown Los Angeles for the re-measure. Obviously, there were more than a few curious glances from passing motorists on the freeway to Los Angeles. After arriving at the location where re-measurements were done, I quickly got out of the back of the pickup truck and gently ran into the building to be re-measured. Interestingly, the measurement device had two buttons on the floor (so a candidate could not raise up slightly) and the top portion of the device was brought forcefully down on the top of your head to achieve an accurate measurement. Thankfully, I measured 5-feet 9-inches, passing the medical examination, and I promptly joined the LAFD shortly thereafter. Is this story true? Absolutely—no one could make it up! Did persistence, desire, and a love for the job result in a lasting career with the best profession on earth? Absolutely!

In conclusion, an important note of thanks to Bobby Halton and Diane Feldman of Fire Engineering for all they do for the fire service, and Jerry Naylis who persisted in the development of this book.

Truck Company Basics

Introduction

As the population of this country began to increase from the early days of the first settlers, the number of buildings used for residential and commercial applications also increased. Wood and labor were plentiful and cost effective, so the number of wooden buildings along with the density of buildings in populated areas rapidly increased, which soon resulted in a stark realization that wooden buildings readily burn. Additionally, it became apparent that there was not an organized method to extinguish the growing frequency of fires and reduce the loss from these fires.

In its infancy, fire suppression efforts consisted of bucket brigades. These had a minimal affect on structural fires, but did illuminate the need for more effective solutions for the mitigation of fires. For this reason, early volunteer fire brigades were formed with Ben Franklin being known for his early efforts is this area. As these initial volunteer brigades became more organized and experienced in the various stages of fire suppression operations, the buildings were also being built larger and higher than single-story residential structures, and as a result, fires in multi-story buildings became more difficult to reach and extinguish. Along with the increasing organization and developing professionalism of the fire service came the knowledge that fire suppression often consisted of two basic components—the ability to reach a fire and then put it out. This realization led to the use of ladders to reach fires above the first floor in buildings and ultimately to the use of specific companies that carried ladders and tools instead of carrying hoses and having to pump water. This diversity of fire apparatus led to terms such as pumper, hose companies (which are still used by some modern fire departments), and ladder companies.

The first attempts at using dedicated ladder companies resulted in human-drawn wagons with wood ladders up to 20 feet in length. With the advent of horse-drawn apparatus, more ladders were carried, and lengths increased to 35 feet. The introduction of gasoline-powered apparatus saw ladder length further increase to 50 feet. During this time, pompier-type ladders were used that enabled fireground personnel to ascend the exterior of multi-story buildings by attaching this type of ladder to exterior windows. This operation was instrumental in using the nomenclature of *hook and ladder* companies that denoted apparatus that carried ladders. As hook and ladder companies became more integral in fire suppression operations, it became apparent that fireground operations involved more facets than the application of water and the raising of ladders. Search, forcible entry, overhaul, and salvage operations were often necessary to ensure the complete extinguishing of fire, saving trapped occupants, and reducing the cost of loss from fire and its by-products. In addition to these growing facets of fireground operations, the addition of hydraulically operated aerial ladders and specialized tools and equipment evolved into a specialized company with personnel responsible for specific tasks other than the actual extinguishment of fire.

Today, this facet of the fire service is known as truck companies (some are still known as hook and ladder companies). The advancements of modern technology has produced apparatus with aerial ladders capable of extending well over 100 feet, platforms with 1,000 pound load ratings and the capabilities of flowing 2,000 gallons per minute (gpm), booms that can be articulated over walls and other similar obstacles, and a vast array of tools and equipment for specialized fireground operations.

With all of the modern advances in technology that fire suppression personnel currently enjoy, one element has remained constant during the evolution of the fire service—water puts out fires, not ladders and other types of tools and equipment. However, when overviewing a typical fireground operation involving a structure fire, it is a fact that engine companies extinguish fires, but truck companies determine *how* a fire will be extinguished. This concept is readily apparent when we consider the definition of a truck company as follows—a truck company is primarily staffed and equipped to perform multiple specialized fireground operations. These operations are search, ventilation, forcible entry/exit, ladders (aerial device or portable), salvage, overhaul, utilities, checking for extension of fire, lights, and elevated master streams. Notice the preceding operations do not focus on the application of water, but are based on logistical operations (or support functions) that are necessary to ensure water reaches the seat of a fire in a timely manner and the safety of attack personnel is maximized. Today, it is widely recognized that the success of fireground operations is largely proportional to the ability of a fire department's truck company accomplishing its roles in a timely and safe manner. However, not

all fire departments are equipped with apparatus as well as staffing whose primary focus is logistical operations because fully staffed departments and fully equipped truck companies are not generally found in rural or small town settings. In this case, another definition for a truck company should also be considered—in the absence of specialized apparatus and staffing whose primary focus is logistical operations, other companies must then be assigned these tasks. As opposed to the first definition of a truck company, this definition places the emphasis on companies (which are normally engine companies) that must be equipped and trained to provide logistical support in addition to the application of water.

An effective engine company can quickly become ineffective if it is unable to gain access to a fire by a blocked entry into a structure or advance a hose line due to a lack of visibility, high heat conditions, or a sudden flashover. A staffing level of three or four personnel on an initial engine company cannot simultaneously complete the following basic operations that are normally required at a routine fire in a single-story, single-family dwelling:

- Force entry into a building as well as provide a means for safe exit for interior personnel.

- Ventilate the products of combustion to provide a safe environment for occupants and suppression personnel.

- Conduct a primary search for trapped occupants and affect a rescue, if necessary.

- Place portable ladders, if necessary.

- Advance a hose line.

- Advance a backup line.

- Apply water to extinguish a fire.

- Eliminate potentially dangerous utilities.

The mitigation of structure fires takes a coordinated effort between companies that initiate an aggressive attack on the seat of a fire and companies that provide the necessary logistical support to ensure a fire is extinguished in a timely manner with an emphasis on safety for any civilian occupants and/or fireground personnel. Unfortunately, this is a concept that some chief officers and city managers do not understand and/or do not implement due to other priorities that are perceived as more important. As an example, when a fire department budget is reduced, what is the first portion of the department that often receives a cut? Staffing and/or the training section. This is more than an interesting dilemma due to the fact that it takes people to complete necessary fireground tasks, and the most important non-emergency function accomplished by a fire department is training! Every book that has been written on fireground operations has always emphasized the need for safe and timely truck (logistical) operations. Although this emphasis is well deserved, it is even more relevant today than yesterday for at least five basic reasons that are routinely encountered on the modern fireground:

- The presence of lightweight construction is an implicit constant in addition to new types of alternative building materials and methods appearing on a frequent basis. These factors have significantly reduced the amount of time before buildings or portions of modern buildings collapse, posing a significant danger to fireground personnel engaged in interior or exterior suppression operations.

- The increased need for fortified and new security measures has resulted in the need for innovative operations to force entry into residential and commercial structures by fireground personnel. The universal use of plastics in some building materials and the contents within residential and commercial buildings has resulted in fires that burn at a higher heat release rate, reducing the time for safe extinguishment of structure fires. Additionally, in combination with the use of plastics is the common fireground hazard known as flashover which is capable of killing exposed human life within three to five seconds.

- The replacement of conventional materials (for example, cellulose-based materials) with synthetic compounds (for example, petrochemical) has resulted in a wide range of new and deadly fire gases, particularly the pronounced presence of hydrogen cyanide which is 33 times more deadly than carbon monoxide.

- The development of double- or triple-pane windows has created fires that are fuel rich and oxygen dependent, leading to fires that are contained inside a structure longer, building high heat levels due to a lack of natural ventilation.

- The fire service experiences less fires today as compared to yesterday. Just as sprinklers, smoke detectors, and other similar factors have collectively reduced the number of fires, the experience level of personnel has also been reduced. Therefore, when a modern firefighter arrives at a typical one-room fire in a single-family dwelling (which NFPA indicates is 60 to 70% of our fires), the excitement level can be abnormally high and create an environment that can result in personnel forgetting a few basics, thereby reducing fireground safety and increasing the potential of injury or death.

Additionally, virtually every fireground officer will support the need for at least six basic operations at most structure fires:

- Initial attack line
- Backup line
- Ventilation
- Forcible entry/exit
- Search
- Ladders (depending on the height or particular hazard of the building)

Four of the six previous fireground priorities are typically considered logistical operations. Therefore, to ensure the rapid extinguishment of modern structure fires as well as ensure the safety of trapped occupants and/or personnel engaged in suppression operations requires the ability to prioritize and quickly provide logistical operations prior to or in conjunction with fireground suppression operations regardless of the designation of a company.

Company Definitions

From a simplistic viewpoint, the application of water on a fire in concert with the logistical functions necessary to mitigate a structural fire can be accomplished by engine companies, truck companies, multi-functional companies (such as quint apparatus), specialized companies (such as rescue, hazmat, etc.), or any combination of these companies. Suppression companies are commonly classified or defined as follows:

Engine companies. Triple; consists of hose, water, and pump.

Truck companies. Truck; consists of an aerial device (for this book, the term aerial device will include aerial ladders and platforms), a full compliment of ground ladders and appropriate tools and equipment.

Quint companies. Quint; consists of an aerial device, pump, water, hose, and ground ladders.

Specialized companies. Specialized companies, such as heavy rescue, hazmat, rescue ambulances, ladder tenders, and other similar types of companies; often used to supply additional staffing and equipment at structural incidents.

Operational Considerations

When formulating or operating as a truck company, how are operational priorities perceived and how do they assist in developing operational guidelines and/or SOPs? Consider several factors, such as the perceived importance of being assigned to a truck company, the assignment of personnel to a truck company, and training.

Perceived priority

From a simple perspective, the phrase *perceived priority* means that an assignment to a truck company should be considered a promotion instead of a demotion. This perception (or lack of this perception) starts at the top command level of a fire department.

Personnel assignment

The assignment of personnel to a truck company is a key consideration that affects company operational effectiveness. The first priority should be the officer because this person sets the standard for the company, establishes fireground priorities at an incident, is responsible for developing appropriate expertise in company personnel, and ultimately determines the overall success of the company. The second priority is the qualifications of the driver-operator. As delineated in chapter 2, the safe operation of an apparatus responding to an incident, aerial device operations if required, and driving back from an incident is of utmost importance for the safety of the personnel assigned to the apparatus and operational effectiveness at an incident. The third priority is the remaining personnel assigned to a truck company. Three key considerations are a desire to be assigned to a truck company, fireground experience, and fireground maturity. To summarize the third priority, fireground maturity is the ability to prioritize and accomplish fireground priorities without supervision.

Using the aforementioned overview of personnel assignment as a starting point, let's focus on experience and training.

Officer experience. Generally speaking, the assignment of an officer to an engine company is simplified by the primary focus, which is normally attack operations. This statement is not meant to diminish the importance of these companies, but it is a fact that the primary equipment carried by engine companies includes a pump, water, hose, and related equipment. However, when considering a truck company, it readily becomes apparent the primary focus is more diverse, and in some cases can be considerably more varied. Search, ventilation, forcible entry/exit, ladders, utilities, salvage, and other operations can be a simultaneous primary consideration, and the priority of these operations can be different for each incident. Therefore, a prime ingredient to consider for a prospective truck company officer is fireground maturity that can enable the ability to efficiently prioritize fireground needs with the available company staffing. (Remember, fireground priorities normally exceed available staffing on a typical truck company.) This concept is summarized as follows:

- Engine companies have a narrower focus than truck companies, and provide an excellent opportunity for new officers to develop fireground experience.

- Truck companies often encounter incidents that demand more operations than company staffing can accomplish. This requires fireground experience that can enable the ability to prioritize necessary operations.

Consider all options before assigning a new fireground officer to a truck company. If a person without a truck company background is promoted to the position of company officer and is initially assigned to a truck company, the chance of diminished effectiveness of that particular truck company is high.

A quint demands an officer who is familiar with engine and truck company operations, is able to size up the many and varied fireground needs, prioritizes those needs, and then utilize limited personnel (remember that the average staffing level for quints is three people) to accomplish those priorities in an effective sequence. It is vitally important that a quint officer understands that the first priority for a quint *may* not be to stretch an attack line. For these reasons, assigning a new or inexperienced fire officer to a quint can be a risky decision.

Driver-operator experience. Most fire departments operate with the position of apparatus operator (or driver) as a promotional, non-promotional, or rotational position. Although these assignments can operate successfully, there can be a subtle yet marked difference among these positions. Consider the position of the person who is responsible for the driving and operation of a truck company from three basic perspectives.

- A person who is assigned to a particular apparatus on a regular basis will tend to develop a more intimate relationship with the apparatus than personnel who are rotated. This can result in increased pride (as the apparatus belongs to a regular assigned person) and efficiency (the rotated person can easily adapt the idea that the next operator will take care of maintenance and other similar considerations). This has been personally witnessed on numerous occasions.

- Rotated personnel can experience a reduction in experience and sharpness because their responsibilities are divided and not consistent.

- When considering the abilities and expertise necessary for operation of a typical truck apparatus and aerial device, a promotional position of apparatus operator, driver, or chauffeur with appropriate compensation is a more logical and professional approach as opposed to a non-promotional position.

Personnel experience (firefighters). The same considerations that apply to the officer and driver also apply to truck company personnel. As fireground needs often exceed company staffing, it is also necessary for company personnel to be able to prioritize fireground operations on initial arrival and throughout an incident if required. Remember, it is often necessary for truck company personnel to set and potentially modify priorities in the absence of the company officer because it is often impossible for a company officer to be with all company personnel throughout an entire incident. This concept can be summarized as follows:

- Truck company firefighters need to possess the ability to work safely and independent of a hose line.

- Some firefighters make excellent engine company members but less than average truck company members. Because the reverse is also true, assign personnel based on qualifications, and do not use a truck company as an afterthought for some personnel.

Training

Of all the attributes that collectively combine to develop an effective truck company, training (along with attitude) is one of the most essential elements. It is also one of the most variable elements. Remember that training is the most important non-emergency function in the fire service. Everything that is done (direct traffic as apparatus are backed into the fire station, start a power saw, and even including the way firefighters eat at the kitchen table) is a result of some type of training. The same perspective also applies to the expertise of truck company personnel and is a direct reflection of their attitude and the leadership qualities of a company commander. After 30 years in the fire service, allow me to relate two stories involving perspective and attitude that are still clearly remembered, starting with attitude.

As the company commander of truck 29, I was sitting in my office at about 7:00 p.m. doing some paperwork because we had just finished with dinner. It was during the summer months and there were about two hours of daylight left. Several firefighters walked into my office and confidently said, "Hey, Captain, there are still several hours of daylight left. How about we take the truck out and spot the aerial to some buildings in our district?" Fundamentally, that clarified three important attributes about these firefighters. Their attitude was "let's take some time and sharpen our fireground skills;" training could be fun and a rewarding experience; and training was more important than television.

The second story involves perspective and my wife. As the commander of the in-service training section, I worked administrative duty Monday thru Friday. Saturdays were used for home maintenance, and on Sundays my wife and I went to church, which involved driving thru several towns. Every Sunday at about 8:30 a.m. we drove past a large fire station with four bay doors that were normally open. This particular fire station housed an engine company, a truck company, a chief, and a rescue ambulance. It did not take long before I noticed that the various apparatus were clearly visible at 8:30 a.m. (driving to church) and again at about 11:30 a.m. as we again drove by the fire station (returning home). I always looked forward to driving by this fire station to see if the apparatus were there on our Sunday morning trip (and yes, they normally were). At about the 10th month while driving by the fire station, my wife said, "When do you suppose they go out and drill?" Although I was not familiar with the personnel in that fire station, I could easily presume that the personnel assigned to that particular truck company did not place a high priority on training in their district.

In summary, it seems there is always enough time to maintain proficiency in high-tech operations such as paramedic certification, high-angle rescue, rapid intervention teams (RIT), and so on because this *stuff* is important to the overall qualifications of the modern firefighter. However, the real question is this: Is adequate time devoted to the basics, such as proper ladder placement, forcible entry techniques, familiarity with search operations

under challenging circumstances, the proper use of a power saw, efficient use of the SCBA bottle on your back, and so on? It is my firm belief that training options for a truck company are endless, and can be relevant and fun.

Quints

Consider training as applied to quint apparatus, which is composed of an aerial device, water, hose, ground ladders, and a pump. This equipment gives a quint the flexibility to become an engine company or a truck company, depending on the needs of an incident. However, and this is the big however, this concept can be severely limited by the absence of a properly trained crew and weak leadership. In fact, training (officer and crew) is the cornerstone of the effectiveness (or ineffectiveness) of a quint. But, you are probably thinking, "No problem, I can train my officers and crews." Really? Want an example of ineffective training? Look at apparatus placement when you go to your next structure fire. Is there a need for improvement? Normally, there is lots of room for improvement.

So, the simple solution of training may not be the total answer. Although proper training is possible if the proper commitment is made and executed, many quint officers and crews do not understand the quint concept. So, what is the quint concept? Basically, if a quint has the potential to perform engine and/or truck company operations, a quint officer must determine the correct operation (engine or truck) and operate accordingly. That means a quint may not be an engine first, and truck second. Unfortunately, when a common firefighter is an initial responder into a structure fire, what is their first love? Is it water on the fire, ground ladders, search, ventilation, forcible entry, utilities, salvage? Answer . . . : It's first water every time, and someone else can handle the other stuff! Therefore, a quint officer must be able to determine the appropriate need(s), prioritize those needs based on available staffing, and have a crew that is willing and able to perform those priorities without wanting to immediately initiate an initial attack line.

Consider two more points associated with quints—officer experience and specialization. A quint demands an officer who is able to size up the many and varied fireground needs, prioritize those needs, and then utilize limited personnel (remember, the staffing on most quints is three people) to accomplish those priorities in an effective sequence. That takes fireground experience and an understanding of engine and truck company operations. Another consideration is specialization. It is imperative that a quint officer *equally* train a quint crew in engine and truck company operations (or other operations the quint is capable of performing). One advantage of separate functional companies (engine, truck, hazmat, heavy rescue, etc.) is the tendency of those companies to specialize in their particular responsibility. If responsibilities are combined, *cross training* becomes essential for company effectiveness. Cross training also means that later arriving crews are capable of using a quint apparatus for another operation, if necessary. As an example, if a quint is first-in to a structure fire and initiates the initial attack line, the next company should be able to use the quint's aerial device and ventilation equipment for a vertical ventilation operation (if that was the next priority).

To summarize quint operations, allow me to make the following three observations:

- The prime reason why quint apparatus are often ineffective as a multi-purpose company is a lack of strong and knowledgeable officers.

- Put an engine company officer on a quint and the possibility of the quint becoming an expensive engine company is high.

- Combine engine and truck company operations into a quint, and you will most likely see your engine company's proficiency remain the same, and truck company expertise slowly diminish.

Staffing Considerations

When considering appropriate staffing for a truck company, monetary constraints normally have the most impact on the number of personnel assigned to a truck company. The national average is three persons; however, insufficient staffing levels (for example, three persons) will minimize and/or reduce effectiveness at numerous types of incidents. Conversely, staffing levels should be formulated to enhance the capability of a truck company and the concept of accountability, which is based on the principle of safety. This is why NFPA 1710 (Section 3-2.2.2) calls for a minimum staffing level of four persons for a truck company assigned to a paid fire department. Additionally, other sources also call for a minimum staffing level of four persons for truck companies (see Training Resources at the end of this chapter). Remember that weak training plus weak staffing equals a disaster in waiting.

Consider staffing levels for a truck company and a quint company. Additionally, assume the minimum requirements that should be accomplished at a routine fire in a single-story, single-family dwelling are:

- An initial attack line

- A backup line

- Ventilation

- Forcible entry/exit

- A search (or the ability to ascertain that there are no interior victims)

These five considerations require a minimum of eight to ten personnel. Obviously, other considerations such as ladders, utilities, a RIT team, salvage operations, and other necessary operations would require additional personnel. Too, if a truck company is first-in to a structure fire, the concept of 2-in/2-out should be followed as mandated by OSHA. That means that the first-on-scene truck company to a structure fire cannot conduct interior operations with less than four persons (unless there is a *known* rescue).

Truck company

The minimum staffing level should be four persons with an optimal staffing level consisting of six members. For additional information on recommended minimal truck company staffing levels, see Training Resources at the end of this chapter.

Quint company

The minimum staffing level for a quint should be six persons with optimal staffing consisting of eight members. If these recommendations seem abnormally high, remember that a quint was hypo-thetically designed to simultaneously provide engine and truck company operations (particularly when a quint is first-in to a structure fire). Additionally, most incident management system (IMS) models require the first arriving company to also establish command functions. Obviously, these considerations can take a staffing level greater than three persons (again, the national average). Although the previous recommended staffing level is a rarity due to fiscal constraints, why buy a quint and limit its capabilities?

Assignments

When responding to a structure fire, a truck company needs to quickly size up the incident and prioritize fireground concerns as necessary. Fire-ground responsibilities and ensuing assignments to personnel will vary from incident to incident and are commonly determined from four basic viewpoints:

- Primary and secondary responsibilities
- Inside-outside operations
- Pre-designated assignments
- Scene arrival

Primary and secondary responsibilities are often used as a starting point to categorize logistical respon-sibilities from a broad viewpoint; however, the following lists are not in a particular order because each inci-dent determines the order of fireground priorities.

Primary and secondary responsibilities

Fireground logistical operations can be catego-rized from two basic viewpoints:

- **Primary:** These operations are normally the highest priority at structural incidents and are forcible entry-forcible exit, ventilation, search, and ladders (aerial and portable).
- **Secondary:** These operations are normally a secondary priority at structural incidents and are utilities, salvage, overhaul, and master streams.

Inside-outside operations

- **Inside responsibilities:** These operations include search (and rescue, if necessary), check for extension of fire, forcible entry/exit, and fire attack if a quint is utilized. Responsibilities such as lighting, salvage, and overhaul are normally post-fire considerations.
- **Outside responsibilities:** These operations include ventilation (horizontal, vertical, pressurized), utilities (electrical, gas, water), ladders (aerial device or portable), and forcible entry/exit.

Although forcible entry is often considered an exterior (or outside) operation, for this discussion forcible entry is considered as an operation consist-ing of two procedures: allowing personnel to enter a structure, and allowing personnel to get out of a structure. This normally requires forcing the initial entry point and then forcing additional exit points on the exterior of a structure as necessary. If forc-ible entry is assigned solely to interior personnel (who are normally responsible for forcible entry and search), the initial entry point is normally forced and then search operations are initiated. This is a prime reason why forcing additional exit points are often overlooked or significantly delayed. Remem-ber that virtually all structures that are entered by suppression personnel do not offer multiple entry or exit openings unless they are provided by exterior fireground personnel. Additionally, carrying typi-cal forcible entry tools (irons, lock pullers, and so on) into a structure while conducting a search does not normally simplify an interior search operation. However, there are buildings that do require internal forcible entry operations (tenements, apartments, commercials, and so on), and conversely there are structures that internal forcible entry is normally not a concern (such as the typical residential home). Therefore, forcible entry/exit can be part of inside and outside responsibilities because this assign-ment should be dependent on the type of structure that is encountered. If forcible entry is assigned to inside responsibilities, it is imperative that if exte-rior forcible exit is necessary, it be quickly assigned to another company (personnel).

The preceding viewpoints can allow personnel to quickly size up an incident and then categorize priorities into two simple considerations that consist of operations inside and/or outside a structure.

Pre-designated assignments

Basic operational duties can be pre-determined as a routine operation. This can be done before an incident through training and crew development. The five basic and most often performed truck company assignments are ladders (which can be initially implemented by company personnel and may consist of the implementation of portable ground ladders and/or an aerial device), forcible entry and exit, search, ventilation operations, and utilities.

These basic assignments are often assigned by seating position (also known as riding assignments or riding positions) and can enhance standardized procedures, safety, teamwork, preplanning, continuity, and the selection of tools and equipment as follows.

Standardized procedures: Assume a typical staffing level of three (the national average) and each shift begins with the following pre-determined duties for a structure fire (figure. 1–1). Keep in mind that these assignments may differ depending on apparatus configuration.

- Officer (O): Forcible entry/exit, search (with firefighter #1), and placement of jack pads on officer side (if necessary)
- Driver (D): Aerial device placement (if necessary) and exterior ventilation
- Firefighter #1: Search (with officer), and placement of jack pads on driver side of apparatus (if necessary)

If a search is not necessary, the officer should assist with ventilation, and firefighter #1 should assist with forcible entry/exit and utilities. Additionally, notice that firefighter #1 is seated behind the driver. This allows a timely placement of jack pads (if necessary) by firefighter #1 on the driver's side of the apparatus and by the officer on the officer's side of the apparatus. This enables the driver to quickly complete any duties in the cab and then prepare to raise the aerial device when exiting the cab of the apparatus.

Members can be initially utilized for portable ladder operations, if necessary, because ladder operations can often be an initial priority. (Remember that when personnel have been assigned other responsibilities, effective and timely ladder operations are often overlooked.) Although a company officer may modify operational tasks as each incident indicates, this company now has a starting point and a basic standard operating procedure (SOP) for each shift. This concept can be modified to fit any staffing level or apparatus configuration on a truck company. The mark of a good truck company is its ability to prioritize and perform multiple tasks simultaneously.

Assume a staffing level is four, and each shift begins with the following pre-determined duties for a structure fire (figure 1–2).

- Officer (O): Exterior forcible entry/exit and utilities
- Driver (D): Aerial device placement and exterior ventilation

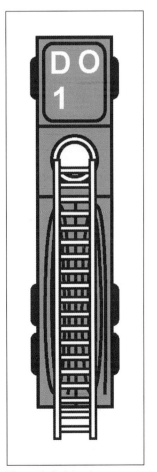

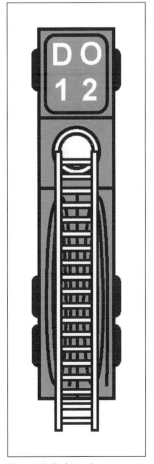

Figure 1–1. A truck company with a staffing of three

Figure 1–2. A truck company with a staffing of four

- Firefighter #1: Search (with firefighter #2) and placement of jack pads on driver side of apparatus (if necessary)

- Firefighter #2: Search (with firefighter #1) and placement of jack pads on officer side of apparatus (if necessary)

If a search is not necessary, firefighter #1 is responsible for assisting with ventilation, and firefighter #2 is responsible for assisting with forcible entry/exit and utilities.

Again, members can be initially utilized for portable ladder operations as necessary if ladder operations are an initial priority. The staffing level of this company now has a starting point and a basic SOP for each shift.

Safety. The ability to know what other members are doing can enhance fireground safety because members know what is currently being accomplished, what priorities need to be accomplished next, and who will be accomplishing those priorities. This augments fireground safety because fireground priorities can be accomplished sooner, resulting in a earlier mitigation of an incident and improved safety considerations for fireground personnel and occupants.

Teamwork. Each member is aware of their responsibility and the responsibility of other members. This can enhance setting fireground priorities because each member must be able to perform tasks without the direct supervision of an officer.

Pre-planning. Each member of the company in figure 1–2 can focus on specific operational procedures and techniques that may be necessary from two perspectives. First, each member is responsible for determining the presence and readiness of the tools and equipment they may be using. Second, consider what each member will be evaluating as they approach a structure fire. Instead of wondering what tasks may be assigned by the officer, firefighters #1 and #2 are focusing on the priority of an interior search and, if necessary, where potential victims may be, where to enter the building, and other similar considerations. If a search is not necessary, firefighter #1 will assist the officer with forcible entry/exit and utilities, and firefighter #2 will assist the driver with ventilation operations. (This concept applies to figure 1–1.)

Continuity. If the officer on duty for a particular shift is not the regularly assigned officer, the officer is more likely to be in command of a company that knows their individual responsibilities and operates as a team. This can be beneficial to the effectiveness of the company while working with a visiting officer. In this case, a substitute officer must determine what SOPs (if any) are in place and which member is responsible for a specific task. Additionally, if company members are aware of the responsibilities to be accomplished by other members, the process of determining what tasks need to be performed as individual tasks are completed is simplified.

Some departments that utilize this concept post these responsibilities above each seat in the apparatus. When a person is riding in a particular seat, the expected responsibilities should be known. This can be particularly advantageous for companies that rotate personnel and/or volunteer companies that do not operate with specific personnel assigned to an apparatus. As an example, a tag with forcible entry/exit and utilities denotes the person in this seat is responsible for these responsibilities. Whether these duties (or riding assignments) are assigned on a daily, weekly, or other basis is dictated by personnel, staffing levels, apparatus training needs, and goals. Does the concept of predetermined duties strip a company officer of the ability to designate specific duties at an incident? No. The concept of predetermined duties offers a starting point for company personnel but still allows an officer to reassign duties as an incident dictates. As an example, assume a truck company that operates with pre-designated assignments arrives on-scene to a structure fire, and incident conditions dictate an extended roof ventilation operation. In this case, the company officer can quickly assign all members to ventilation operations on the roof.

Tools and equipment. The concept of pre-designated assignments can also be expanded to the proper selection of tools and equipment for an intended operation, and can be denoted as general inside/outside operations and/or on riding assignment tags. Either viewpoint sets the starting point (for a firefighter who occupies a particular seat) for fireground tasks and related tools and equipment. As an example, consider some common fireground tasks and associated tools and/or equipment. It is assumed that each person will be equipped with the following minimum equipment—appropriate personal protective equipment, SCBA with mask that will be worn when encountering a contaminated atmosphere (and the ability to monitor the air supply), a personal alert safety system (PASS) that is activated, a pickhead axe with scabbard (delineated in the ventilation chapters), a radio, flashlight, search rope (200 feet by 3/8 or 5/16 inch), wire cutters (remember that dropped/hung ceilings are suspended by thin wire), and door chocks.

Common tools and equipment used by *inside* teams include:

- Thermal imaging camera
- Six-foot pike pole (although a pike pole can be beneficial in checking for extension of fire, they can be cumbersome for search operations), hook
- Search rope or search line with bag
- Irons and/or rabbit tool, if necessary
- Officer tool (commonly referred to as an "A" tool)
- An inside ladder (12 or 14 foot), if necessary, to allow attack lines to access fire in overhead areas such as attics, etc.

The tool and equipment list for inside operations is focused on two primary areas—a timely search, forcible entry (depending on the incident), and the ability to check for extension of fire for attack personnel. Lengthy tool and equipment lists look impressive on paper, but must be carried by personnel, often in challenging and minimal space conditions. Additionally, as staffing increases, so does the ability to perform more operations with a larger selection of tools and equipment. Truck companies with minimal staffing levels must judiciously choose the basic tools and equipment that will allow the completion of an intended task but also allow the freedom to complete the task in a timely and safe manner.

Common tools and equipment needed by the *outside* team include:

- Ladders (aerial and/or portable ground ladders)
- Irons (halligan-flathead axe) and/or hydraulic forcible entry tool; if forcible entry is necessary inside a structure, the search team would need to carry these tools.

- Through-the-lock tools
- Rotary saw with multi-use blade
- Chain saw with carbide chain
- Blowers
- Six-foot pike pole, rubbish hook, and other types of hooks
- Utility shut-off tools
- Thermal imaging cameras (can be beneficial for exterior operations; more commonly used inside a structure to find interior fire and potential victims)

Common tools and equipment used by riding or seat assignments for a staffing level of three (refer to figure 1–1) include:

- Driver: Ladders (aerial and/or portable), blowers, and chain saw with carbide chain, pike pole/hook, and/or rubbish hook
- Officer: "A" tool, thermal imaging camera, and a pike pole/hook
- Firefighter #1: Irons and/or hydraulic forcible entry tool, through-the-lock tools, rotary saw with multi-use blade

Common tools and equipment used by riding or seat assignments for a staffing level of four (refer to figure 1–2) include:

- Driver: Ladders (aerial and/or portable), chain saw with carbide chain, pike pole/hook and/or rubbish hook
- Officer: Irons and/or hydraulic forcible entry tool, rotary saw with multi-use blade, utility shut-off tools
- Firefighter #1: Thermal imaging camera
- Firefighter #2: Six-foot pike pole or hook and inside ladder (to be initially left at entry point to structure)

Remember that this staffing level allows for two two-person teams, depending on incident priorities:

Scene arrival

In the absence of inside-outside operations and pre-designated assignments, assignments can be delegated when arriving at an incident. Although this method is utilized by many departments, it can have several drawbacks.

Pre-planning. If company personnel are not responsible for particular assignments, it is still necessary to ensure that all tools and equipment on an apparatus are present and ready for immediate use. Although this should be a responsibility of all personnel (or at least the driver) assigned to an apparatus, this basic responsibility is often overlooked. This is a primary reason why gasoline-powered equipment is sometimes found with near-empty gas tanks and tools are not in their normal location because company members are not responsible for any specific equipment.

Size-up. As a truck company approaches an incident, the absence of inside-outside or pre-designated assignments can often hinder the size-up process. Without specific assignments, personnel do not begin a task-related size-up until the company officer delegates assignments. As an example, assume a truck company is approaching a residential structure fire. As the incident comes into view of the company members, their focus will likely consist of general observations of the incident. A specific size-up of the incident and related assignments will normally begin after the company officer assigns specific responsibilities. This can reduce the available time to formulate appropriate actions and enhance additional pressure on company personnel, as opposed to personnel using the concept of inside-outside or pre-designated assignments that can allow personnel to begin a specific size-up before stopping in front of an incident.

Assignment Applications

Let's apply the inside-outside responsibilities, pre-designated assignments, and scene arrival operations to several typical fireground scenarios, remembering that:

- The priorities for each incident will vary and determine the assignment of available staffing.

- It may be necessary for all company personnel to be assigned to a single task (for example, extensive roof ventilation operations).

- If a quint is utilized with a staffing of three, two interior persons are required to implement an attack line, if necessary, but only if a minimum of two additional personnel are present and/or with the knowledge there is a known search and/or rescue.

Assume a truck company with a staffing level of three (figure 1–1) is the first truck on-scene to the scenario in figure 1–3, which consists of a single-story residential structure attached to a two-story residential structure that has a well-involved second-story, attic and roof.

Figure 1–3. A two-story structure with attached one-story structure as viewed from an initial size-up

Additionally, it appears the single-story structure is vacant and the status of the two-story is undetermined. Size up this incident from the following viewpoints.

Scene arrival

Upon arrival, the driver and firefighter would likely be looking at the fire involving the upper portion of the two-story structure, thinking about possible extension of fire into the single-story structure, and wondering what tasks they will be assigned by the officer.

Pre-designated assignments

As the incident becomes visible to truck personnel, the driver would consider raising a portable ladder to the roof of the single-story structure, and determine if ventilation is necessary in the first floor of the two-story structure and single-story structure. The officer would notice that forcible entry does not appear to be a concern for the single-story structure (or if it is, it should be relatively simple); however, the need for forcible entry into the first floor of the two-story structure will need to be evaluated. The firefighter will also evaluate search concerns for the single-story structure and the first floor of the two-story structure. Basically, the preceding initial size-ups can be accomplished before the truck stops at the incident.

Inside/outside responsibilities

As the incident becomes visible to truck personnel, inside responsibilities will consist of evaluating the need for a search in the single-story structure as well as the first floor of the two-story structure, and a check for extension of fire into the single-story structure and first floor of the two-story structure. Outside responsibilities will consist of evaluating forcible entry and ventilation for the first floor of the two structures, a portable ladder to the roof of the single-story structure, and control of utilities. As these general responsibilities are evaluated, specific assignments can then be determined by pre-designated assignments (riding positions).

Scene arrival

Next, assume a truck company with a staffing level of four (figure 1–2) is the first truck on-scene to the scenario in figure 1–4, which consists of a vacant single-story residential structure that has plywood over the exterior doors and windows. Additionally, it appears there is a significant amount of fire showing from the rear portion of the structure, and the initial engine company on-scene is trying to force entry through the front door.

Upon arrival, the driver and two firefighters would likely look at the fire involving the rear portion of the structure. They might think about possible extension of fire into the front portion of the structure, the barricaded doors and windows, a company attempting entry through the front door, and wonder what tasks they will be assigned by the officer.

Figure 1–4. A vacant single-story residential structure as viewed from an initial size-up (LAFD file photo)

Pre-designated assignments

As the incident becomes visible to truck personnel, the officer would quickly notice that the windows and doors have been barricaded. The structure is secure, which would significantly minimize search considerations. However, easy entry into and out of the structure is not an option unless the barricades are removed. There will be fire extending through the attic from the rear to front of structure. The officer and driver must evaluate forcible entry and forcible exit (quickly removing the door and window barricades) and ventilation, and firefighter #1 and #2 would evaluate an interior operation consisting of pulling the ceilings to determine the extension of fire, allowing the attack company to extinguish the extending overhead fire, and consider the potential need for a primary search. Basically, the preceding initial size-ups can be accomplished before the truck stops at the incident.

Inside/outside responsibilities

As the incident becomes visible to truck personnel, inside responsibilities will consist of evaluating a primary search in the structure and checking for extension of fire in the attic. Outside responsibilities will consist of evaluating forcible entry, forcible exit, ventilation for interior operations, and control of utilities. As these general responsibilities are evaluated, specific assignments can be determined by pre-designated assignments (riding positions).

Apparatus Response Criteria

Single-family dwellings require at least two engines, one truck, and one chief. This provides an engine for fire attack, an engine for back-up and/or exposures, a truck for support operations (ventilation, search, forcible entry/exit operations, utilities, and so on), and a chief for command. Some departments (that are capable) are dispatching an additional company for RIT operations. This results in an initial dispatch of three engines, one truck, and one chief.

Always compare the size, construction style, and type of single-family dwellings in a district in relation to the appropriate response criteria because some single-family dwellings require additional companies for successful mitigation. As an example, a two-story single-family dwelling consisting of 5,000 square feet, 10-foot ceilings, and a tile roof will quickly exhaust the capabilities of two engines and one truck. In this scenario, an additional truck is mandatory, and an additional engine may also be necessary. From a practical perspective, some single-family dwellings should be considered as a moderate-sized commercial occupancy, based on what there is to burn or fuel load.

Structures *larger* than a single-family dwelling require at least three engines, two trucks, and one chief. This provides an engine for fire attack, an engine for back-up or additional fire attack, an engine for exposures, a truck for ventilation operations and forcible entry/exit operations, a truck for additional support functions (ladders, utilities, search, and so on), and a chief for command. If an additional company is dispatched for RIT, the initial assignment would increase by one company.

The recommended average companies has long been one truck per two engine companies. For additional information on this ratio, see Training Resources at the end of this chapter.

Miscellaneous Considerations

When engine and truck company apparatus are quartered in the same fire station, consider responding the truck company first or ahead of the engine company for optimum access to the incident; enhanced spotting requirements (truck companies don't have

the flexibility of engine companies); and improved truck operations preceding or at least simultaneously occurring with engine company operations to get crews into action sooner.

If the officer on a truck company is a higher rank than engine officers, other considerations can also be enhanced, such as the transfer of command. As detailed in chapter 2, the first officer on-scene is the incident commander. In the absence of a chief officer and as the number of on-scene companies increase, however, the chance of freelancing will also increase. Assume several engine companies arrive at an incident prior to the arrival of a truck company. The position of the incident commander is initially assumed by the first arriving engine officer and is maintained with the arrival of additional engine companies. When the first truck company arrives on-scene, however, the position of incident commander can automatically transfer to the truck company officer and is maintained until the arrival of a chief officer. This progressive transfer of command will automatically increase the effectiveness of the command structure for an incident, and minimize incident freelancing.

Additionally, if a truck company is a later-arriving company to an incident, the automatic assumption of command normally allows the truck officer to view and analyze the incident from a broad perspective because the on-scene companies are often committed to a narrow perspective or area of responsibility.

Ten Commandments of Truck Company Operations

It is an accepted fact that fireground operations consist of two specific viewpoints—fire attack and logistical operations—yet these viewpoints are interrelated from the perspective of safety and a timely mitigation of an incident. Although water extinguishes fires, logistical operations determine how a fire is extinguished and ensures a fire stays extinguished with a minimal loss to property and personal belongings. Therefore, the importance of truck company operations (logistical operations) cannot be overstated. With these thoughts in mind, let's review basic operational considerations as applied to a truck company by first looking at five basic rules of fireground safety and then examining the ten commandments of truck company operations.

Five basic rules of fireground safety

Most fireground problems and/or actions that are prime contributors to fireground inefficiency, injury, or death are commonly the result of fireground personnel violating basic safety principles. As an example, fireground problems can often be fittingly divided into mistakes and dumb mistakes. Fireground mistakes can be defined as "you knowingly violate a basic safety principle and get away with it!" Although every firefighter is familiar with this definition, the end result is often placed under the heading of "no harm, no foul." This means that if there are no resultant injuries, there are generally no changes to SOPs. Conversely, dumb fireground mistakes can be defined as "you knowingly violate a basic safety principle and it bites you." Obviously, these types of mistakes can result in serious consequences. These simple yet relevant definitions underscore the need for every firefighter to be fully aware of the basics, and practice them on a continual basis. To summarize these two basic viewpoints, let's review five basic rules of fireground safety that can equally apply to every firefighter, regardless of rank or company assignment.

Falling debris always has the right-of-way. This simple yet basic rule has not changed since the advent of the fire service. However, it has never been more applicable than today because of alternative building materials (and methods) that are commonly used in typical residential and commercial buildings. As a result, these materials do not offer the structural integrity of the building materials of yesterday. Therefore, it is vitally important that modern firefighters are continually aware of where they are in relation to a fire. As an example, the fire in figure 1–5 has extended into the fascia on the front of the building and presents a significant collapse hazard to the personnel with hose lines who are within the collapse zone.

When this perspective is applied to modern lightweight truss construction, firefighters should re-evaluate their intended operation and location when fire is known or suspected above *or* below. If it is not safe enough to stand on it, it also is not safe enough to stand below it!

Figure 1–5. These firefighters are within the collapse zone of the fascia. (LAFD file photo)

Equipment on your apparatus in the street is unusable on the fireground. Practically applied to fireground operations, a firefighter should always have a versatile tool at hand. A classic example is a pickhead axe and a scabbard. The scabbard allows an axe to be easily carried yet leaves both hands free to accomplish other operations and tasks. A pickhead axe is one of the most versatile tools in the fire service, yet they are often found mounted on fire apparatus in front of an incident when they should be ready for immediate use by fireground personnel. Always try to maximize your fireground flexibility. Empty-handed or ill-equipped firefighters often have to go back to an apparatus and retrieve a tool to complete a specific task, or suddenly wish they had a tool to save their lives!

If the fire is in range, so are you. The closer fireground personnel are to a fire, the closer they are to a potential structural collapse. Although this rule applies to any type of construction, it noticeably applies to lightweight construction from three perspectives. First, a noteworthy number of modern residential and commercial buildings are constructed with a fascia on at least the front portion of the building. If fire has extended into a fascia, this style of construction suddenly becomes a potential collapse hazard over the entry and exit point (normally the front door) of the structure as illustrated in figure 1–6. Where do most firefighters enter a structure for suppression operations? The front door!

Figure 1–6. Fascias can readily collapse over the front portion of a structure, creating entrance and exit problems.

Next, it is a known fact that lightweight trusses *can* collapse in as little as five to seven minutes when exposed to fire. If fire is overhead (attic, interstitial space, and so on) of advancing firefighters, they can be in the wrong place at the wrong time unless the fire is quickly extinguished. This is a good reason why the two basic tools that should be carried by initial entry personnel should be a hose line and a pike pole or hook. Finally, truck company firefighters who have been in the fire service for a number of years readily admit that when applied to roof ventilation operations, the old axiom of "if a conventional roof feels strong, it probably is" was used as a basic roof safety rule. This rule does not apply to roof ventilation operations on lightweight roofs (metal or wood) if the fire is underneath your location. This basic fact means that roof ventilation is not a feasible operation over fire when it is exposing lightweight structural members.

What you see may not be what you get. Modern building construction is very adept at making new buildings look like old buildings. Why? Because older buildings normally required more time to construct because of the size and shape of building materials and the fact that older buildings utilized more decorative materials and/or styles to enhance the exterior look of a building. Today, these methods and techniques are not cost effective. Several examples of how the exterior appearance of modern buildings can be deceiving are:

- Brick veneer attached to the outside of a building in order to give the building a substantial masonry look
- Exposed two-by-six inch rafter tails that are attached to two-by-four inch rafters within the building
- Foam cornices that are attached to the exterior of a building with adhesives and then covered with plaster type materials

The general theme of modern building construction is to construct a building as fast and cheap as possible, which ultimately saves construction costs but also dramatically changes what a modern building really is behind an attractive exterior façade.

Your fireground time is slowly being minimized. Modern building materials and the widespread use of plastics in residential and commercial buildings have changed the way modern fires burn. Lightweight truss construction has been around since 1960, so there is no excuse for a firefighter to be unaware of the fast failure rate of lightweight trusses when they are exposed to fire or high heat. Now it appears the building industry is preparing to adopt the widespread use of glued lightweight trusses that will continue to erode fireground time.

Additionally, the use of plastics (or petrochemical-based compounds) are readily found in residential and commercial buildings. This has resulted in fires that burn hotter and faster than the fires of the 1960s and 1970s. Additionally, modern buildings can hold heat longer due to energy-efficient windows, which can lead to flashover occurring about the time suppression personnel arrive on scene. As the building industry continues to adopt and use building materials that are smaller in size, use geometry instead of mass for strength, use adhesives instead of nails, and use oil-based synthetic compounds for furnishings, the fireground will continue to offer less time before structural collapse and/or flashover conditions. This can result in insufficient time to safely extinguish a routine structural fire and/or the ability of interior personnel to exit a structure in a timely manner.

Ten Commandments of Truck Company Operations

Using the preceding fireground safety rules as a foundation, let's review ten common fireground considerations from a truck company perspective. These can be called the ten commandments of truck company operations. Although the following is not all-inclusive, it focuses on firefighter safety and hopefully provides a starting point for thought and discussion. Remember that fireground logistical considerations, from reduced staffing levels and increased constraints (such as two-in/two-out, RIT teams, and so on), are slowly being diluted.

1. Don't forget your primary mission

Structural fireground operations consist of two basic procedures—fire attack and logistical operations; however, virtually all firefighters joined the fire service so they could put the wet stuff on the red stuff. This fact becomes amusing when an engine company can beat another engine company into that company's district and put first water on the other company's fire (which promotes unsafe speeds and compromised teamwork). The best attack company on this planet is worthless if they cannot get to a fire and/or extinguish a fire in a safe and timely manner.

This perspective is the primary responsibility of logistical operations that are designed to ensure that ladders, forcible entry, forcible exit, and ventilation operations allow an attack company to put the wet stuff on the red stuff in a safe and timely manner. This basic concept is the focal point of truck company operations and implies that the *first* priority of truck company personnel is to support attack operations and not stretch an initial or additional hose line (although this may be necessary in some instances).

2. It's the basics before arriving on-scene

Without a doubt, company efficiency and firefighter safety begins long before a truck company responds to an incident. The need to pre-plan a district cannot be over-emphasized. Although the term pre-planning can apply to numerous viewpoints, it specifically applies to successfully operating within your fireground office with a truck company. The definition of your fireground office is the sum total of the buildings, related hazards, and the ability to effectively use your truck apparatus within the various constraints of your district. Every member should review their apparatus at the start of each tour of duty, and that includes the cleanliness and readiness of tools and equipment, particularly those items that each member is primarily responsible for. That is one reason why pre-designated assignments (or riding positions) can be beneficial. One specific pre-check to consider is the readiness of your SCBA bottle. Is it full, clean, and ready for immediate use? Volunteer or smaller paid companies that do not have a daily or regular change (or rotation) of personnel should have an SOP for the constant review of the readiness of their apparatus and equipment.

Company personnel should also be familiar with the various types of buildings and related hazards within their district. Forcible entry problems, minimal space constraints for portable ground ladders, floor plans that defy recognition from the exterior of a building, considerations within a structure (such as overcrowding conditions, poor exit possibilities for occupants, and so on) and a host of other considerations should not be classified under the heading of a surprise when operating on the fireground. Remember that most truck apparatus are significantly heavier and larger than other types of apparatus. Therefore, common response routes should be evaluated for applicable hazards, and any spotting constraints to buildings should be evaluated for any alternative options that can be utilized at an incident. As an example, the three-story multi-unit residential structure in figure 1–7 has no aerial device access from the rear and both sides of the structure. Additionally, the front of the structure is blocked by heavy vegetation and is serviced by a narrow one-lane street. It should be determined before an incident how a large truck apparatus will operate at this scene, not during an incident because this structure can present a sizeable life hazard during a fire.

Fig. 1–7. This structure presents significant access problems for a truck company and must be pre-planned before an incident instead of during an incident.

incident is not expected by civilians and/or motorists, and the addition of cell phones and modern vehicles with advanced soundproofing and superior sound systems do not enhance the ability of civilian drivers to recognize and make allowances for responding emergency apparatus (figure 1–8). Additionally, the excitement of operating modern, large fire apparatus with lights and sirens attempting to clear traffic does not make allowances for a fire apparatus driver who is exceeding a reasonable speed and safe operation of a truck apparatus during a response. The old adage of "you cannot put the fire out if you are not on-scene" is still as true today as it was yesterday.

Know your assignment and job. Truck company priorities can be considerably more diverse than engine company responsibilities; therefore, each truck company member must be a fireground expert in the numerous logistical tasks that vary at each incident. Maintaining expertise in EMS, RIT, confined-space rescue, hazmat, and other similar operations is an important component of the training priorities of a company officer. The need to ensure competency in the *basics*, however, should be at the top of the list of training priorities because fireground injuries and deaths are often the result of violating the basics. This is why it is vitally important for a truck company officer to take the necessary time to ensure that company members are *grounded* in the basics when applied to building construction and logistical responsibilities such as ventilation, search, utilities, ladders, and so on.

Respond with vigilance. Webster's dictionary defines vigilance as *alert, watchful, on guard*. This definition is particularly applicable when responding to an emergency incident but is also applicable when returning to quarters. Remember that the third highest cause of firefighter injuries and deaths is accidents with fire apparatus. You cannot mitigate an incident until you safely arrive on-scene and properly place and/or spot your apparatus. Therefore, the responsibility for the delivery and safe operation of apparatus and personnel is the sole responsibility of the officer and driver. It is common knowledge that an apparatus responding to an

Fig. 1–8. Driving fire apparatus can be challenging, particularly when responding to an incident. (LAFD file photo)

3. Determine available fireground time

The ability to determine fireground time is a key consideration in determining the amount of time a firefighter has to accomplish an intended task; this equally applies to interior and exterior operations. The ability to determine available fireground time is derived from a combination of three factors.

Understanding building construction. Interestingly, building construction progressed from a heavy style of construction until about 1935, to a moderate style of construction from 1935 to 1960, and to a lightweight style of construction from 1960 until several years ago. Although this progression of style and size of structural members has undergone numerous changes, from heavy timber to the common lightweight truss of today, it potentially is about to take a giant leap backwards with lightweight trusses that use glue to hold them together instead of nails or gang-nail plates. Stated from another viewpoint, modern buildings are constantly reducing fireground time to every firefighter in this country.

Many factors must be considered when conducting a size-up at a structure fire, yet it is imperative that at least two basic questions are considered: Is fire exposing conventional or lightweight construction, and what is the resultant time before structural collapse? The following baseline has proved fairly accurate in determining how much time the two basic types of construction—lightweight or conventional—will allow before collapse: if fire is exposing lightweight truss construction for more than five minutes, re-evaluate your intended operation; and if fire is exposing conventional construction for more than about 20 minutes, re-evaluate your intended operation.

Reading the environment. In concert with reading the type or method of building construction, reading the environment from outside and inside a building can provide a useful guideline to how much time you have to conduct your intended operations. When arriving on-scene, initially evaluate the environment from the exterior of the incident. Remember that heavier and/or darker smoke results in minimal or no interior visibility, which dramatically slows your interior operations and can also present conditions that are more favorable to a flashover. Additionally, when evaluating the environment, be sure to think about the fire from the perspective of confinement versus extension. If extension is a reality, what is the degree of extension and where is it going?

Personnel engaged in interior operations must also read the interior environment with the intent of gauging the length of time until extinguishment and/or ventilation operations begin to improve the environment. Remember that truck company personnel are often conducting interior operations (for example, search) without the presence of a hoseline (which can be an effective source of protection). Under these circumstances, it is vitally important for truck company personnel to constantly monitor and/or evaluate the interior environment to which they are committed.

Evaluating the fireground clock. When the term *fireground clock* is mentioned, some of the initial factors that come to mind are: how advanced is the fire, how long did it take someone to become aware of the fire and report it, what was the dispatch time, and what was the length of time for response? Although these factors are certainly applicable to fireground operations, there is another fireground clock that is also applicable. That clock is defined as once personnel arrive on scene, how long will it take to make a visible impact on the fire? Notice this definition does not focus on the extinguishment of fire, but rather the time it will take personnel to exit the apparatus, make sure their complete PPE is ready, force entry into the building, complete ventilation operations, conduct a search, and so on until the fire begins to be extinguished. This definition is significantly different from the first fireground clock definition that focused on the time elapsed prior to companies arriving on-scene. Obviously, the time it will take to make a visible impact on a fire varies from incident to incident, but you should have a general idea of how much time is available to accomplish an intended operation.

As an example, assume the truck to which you are assigned is the first truck (along with an engine company) to stop in front of a recently constructed two-story residential structure. This structure is attached to a single-story residential structure (figure 1–9) that is typical of some residential structures in your district.

Figure 1–9. Fire exposing lightweight trusses can result in a lack of sufficient time for interior operations.

Additionally, assume your initial size-up indicates that fire visible from the second floor portion of the two-story structure has extended into the attic and resulted in heavy smoke, and some fire is visible from several attic vents as illustrated. Because you are familiar with this type of structure, you feel that the fire in the attic is consuming lightweight truss construction. You are also aware that fire is not static, but dynamic (it keeps burning until you put it out). Applying our second fireground clock definition, you assume that it will take you approximately four minutes to get off your apparatus, get your PPE ready to go, and accomplish basic logistical considerations of forcible entry/exit and ventilation. Additionally, you assume it will take your partner engine company four minutes to get off their apparatus, get their PPE ready to go, and stretch an initial attack line into the structure.

At this point, three considerations should command your undivided attention—the four minutes to start attack and logistical operations just spotted the fire an additional four minutes; the four minutes given to this fire and/or to conduct interior logistical operations also means four minutes has elapsed without mitigating the fire (which will continue to burn and weaken the structure); and the national average for the collapse of lightweight trusses when they are exposed to fire is around five to seven minutes. Therefore, does this simple scenario give you adequate time to enter the structure and extinguish the fire before collapse of the lightweight truss construction? Probably not. Additionally, remember that if fire is showing upon your arrival at a structure fire, you should consider how long the fire was burning before your arrival and add that to your fireground clock implementation time.

4. Prioritize fireground considerations

This commandment is potentially the most important consideration for truck company personnel because each incident is unique and always requires evaluating and prioritizing fireground operations. Although fireground priorities are developed during an initial size-up and during an incident (continuing size-up), there is a subtle factor that still influences the process of prioritizing fireground operations that begins in the initial portion of a firefighter's career. Firefighters are commonly taught from their first

days in a training academy that several popular acronyms such as RECEO VS, LOUVERS, LOVERS U, and AL-VES-SCOUP can be used as a baseline for determining fireground priorities. Whereas fireground acronyms can be a starting point and also used as a recall technique, they often do not adequately address modern fireground priorities.

As an example, let's look at the most popular acronym RECEO, which stands for rescue, exposures, confinement, extinguishment, and overhaul. To dissect this acronym and apply it to modern fireground priorities, a basic question should be asked—what is the most important fireground priority? A common answer is search or rescue (which RECEO indicates is a first priority) and without a doubt, search is an important fireground priority, particularly when it is known there is someone trapped in a burning building. Fire attack and ventilation operations are another common answer. An interesting viewpoint to this question was in the "Roundtable" section in the March 2008 issue of *Fire Engineering* magazine. This Roundtable asked, "what is the primary function to be performed on the fireground?" Of 28 responses, the majority answered "search" with "ventilation" a close second in priority.

Despite the fact that each incident determines fireground priorities, there is one priority that is a constant at every incident, and that priority is firefighter safety. There are no one-way tickets in the fire service because every apparatus that responds to an incident should focus on two basic priorities—returning to quarters with the same number of personnel that responded to the incident, and not adding an additional problem to an existing problem! Interestingly, RECEO and the other popular acronyms do not address firefighter safety (go back to RECEO or LOUVERS and see if you can find firefighter safety listed), although some say that search addresses firefighter safety. However, most firefighters view search as being conducted inside a building for trapped or possibly trapped occupants.

So, if firefighter safety is the top fireground priority, what is the second most important fireground priority? Some will quickly answer "search," "ventilation," or "fire attack." Based on operations that are most often conducted at structural incidents, forcible entry and/or ventilation should be the next most important fireground priorities. Let's look at forcible entry first. Before a search is ever conducted and water is put on a fire, forcible entry is normally required at most incidents before interior operations are initiated. Therefore, if forcible entry is a common requirement before most interior operations, forcible exit should be the next most important fireground priority based on the principle of "if it is important to get into a building, it should be as important if not more important to be able to get out of a building." Interestingly, a common rule that is applied to ladders for above-ground operations is "a minimum of two ladders should be raised." This basic safety rule should also be applied to forcible exit to provide interior personnel multiple means of egress (two means of egress at the very least).

Next, let's take a quick look at ventilation. Based on NFPA statistics that indicate flashover is one of the top three fireground problems encountered on the modern fireground, and that fires today burn significantly faster and hotter than the fires of yesterday, ventilation should (depending on conditions) be a requirement to improve the interior environment of a structure *before* the implementation of interior personnel. Ventilation may consist of horizontal, vertical, positive pressure, or any combination of these ventilation operations. Therefore, in the interest of firefighter safety, ventilation should be considered a priority equal with forcible entry and forcible exit (both of which are primarily designed to enhance the safety of firefighters, and can also assist trapped occupants).

When firefighter safety has been maximized by ensuring firefighters can enter and exit a structure in a timely manner, and/or the environment has been improved so visibility has been enhanced and the potential of flashover has been minimized, then search, fire attack, and other interior operations are next on the priority list. To summarize the preceding thoughts on forcible entry, forcible exit, and ventilation operations, if current fireground operations focused on elevating the priority of forcible entry, forcible exit, and ventilation operations prior to interior operations, firefighter injuries and deaths would be significantly reduced. What makes this statement so providential is the fact that a high percentage of fireground operations currently place search and fire attack operations ahead of taking the appropriate time for logistical priorities to ensure the safety of fireground personnel.

For a moment, let's step out on thin ice and ask a pertinent question: Has search been overemphasized in this country at the expense of firefighter safety? The answer is yes when search is placed ahead of firefighter safety. To carry this question one step further, let's consider an additional perspective: The words *avoidable* and *needless* should be applied to nearly all injuries or deaths that occur to firefighters engaged in search operations for persons who might be in a structure, and prior to ensuring the ability to get out of the structure in a timely manner and/or minimizing the potential of a flashover. Notice this perspective focused on the word *might* instead of the words *are known to be* because there should be a significant difference between the two when establishing fireground priorities. Even when victims are known to be inside a structure, we must also be able to recognize conditions that do not support saveable life. Risks should be reduced to the absolute minimum before searching. Is it a good use of resources to initially deploy a truck to an interior search, compared to using that truck for necessary ventilation that allows an engine company to get in to perform extinguishment and a basic primary search? To summarize the preceding thoughts on creating priorities that are based on firefighter safety, the following fireground priority list is recommended as a long overdue replacement for RECEO and other similar acronyms:

- Firefighter safety
- Forcible entry/forcible exit/ventilation
- Fire attack and/or search
- Other necessary priorities

The order of forcible entry, forcible exit, ventilation, search, and fire attack would obviously be determined by conditions encountered at each incident.

Notice that the preceding recommended order of priorities starts with ensuring the safety of interior personnel, and does not initially include a RIT team. A RIT team should not be implemented at the expense of the basic staffing necessary to mitigate a fire. The basics are defined as an attack line, backup line, ventilation, search, and forcible entry/exit. In July of 2008, Vincent Dunn, retired deputy chief, Fire Department of New York, authored an excellent article titled "Does Aggressive Firefighting Cause Firefighters To Become Caught And Trapped?"

which should be read by every firefighter. In the article, Chief Dunn made the following comments regarding RIT teams:

Has the fire service oversold the abilities of the rapid intervention team concept? I believe it has. A revealing study of the effectiveness of a rapid intervention team in reaching a victim was conducted by the Phoenix Fire Department. Arizona State University conducted tests with Phoenix firefighters to determine how long it would take a rapid intervention team to find and rescue a trapped firefighter. The tests revealed it would take on average 18 to 27 minutes to find and rescue a firefighter who transmits a Mayday. (A Mayday is the accepted national standard that refers to sending a distress call by a firefighter who is in need of being rescued as quickly as possible.)

Interestingly, the Los Angeles Fire Department conducted similar tests and found the rescue times to be virtually identical. Do these tests mean that RIT teams should not be implemented at structure fires? The answer is no. However, these tests indicate that RIT teams may not be as effective as some articles and current emphasis suggest. A RIT team should not be implemented at the expense of the basic staffing necessary to mitigate a fire, and specifically RIT teams should not be implemented from an initial truck company until that truck company has finished with the essential logistical priorities. It is important to note that RIT teams are an adjunct to fireground operations; they are not primarily responsible for the extinguishment of fires and/or logistical operations.

Using the previous discussion on developing fireground priorities, let's apply a general priority list to initial truck company operations as follows:

1. Firefighter safety
2. Ladders (Generally, the best time to raise ladders, if necessary, is when a truck first

arrives on scene as this is likely the only time that all truck personnel are together.)

3a. Forcible entry, forcible exit
3b. Ventilation (The priority of these is determined by each incident.)
4. Other logistical operations as necessary

When this priority list is applied to a common truck company, it should be easy to see that three of the top five initial priorities (forcible entry/exit, ladders, and ventilation) are designed to maximize the safety of fireground personnel who are committed to interior operations. Notice that search is not initially listed. If conditions were to warrant the identical need for ventilation, forcible entry/exit, and a search, what operation would you do first? If a structure fire needs to be ventilated and/or the building is not easy to exit from the interior, attempting a search without the completion of these considerations has the potential to make you a national statistic. In summary, forcible entry/exit and ventilation are two tasks that most contribute to a truck company's primary mission, that is, making the building/environment safer for firefighters operating inside.

5. Become a "Tool Time" expert

Truck companies carry a wide range of tools and equipment that allow company personnel the ability to perform numerous tasks such as forcible entry, search, ventilation, control of utilities, salvage, overhaul, ladders, confined space rescue, vehicular extrication, and so on. Although it can be debated that some engine companies can also be responsible for most of these same tasks, it is a fact that engine companies are normally focused on the implementation of hoselines. However, the term focus becomes more diversified when it is applied to truck operations and often with minimal staffing constraints. Therefore, it is imperative that truck company personnel become proficient (and maintain that proficiency) in all aspects of the abilities of their particular truck company. This challenge becomes appreciably more important when the diversity of tasks for which truck firefighters are responsible can often result in the application of an old adage that says "firefighters master many tools, but have they forgotten the basics?"

Here's a great truck company drill that focuses on the proficiency of the basics perspective. Pick a typical fireground operation (for example, forcible entry, search, ventilation, and so on). Then, have each member of the company select a tool and explain how the tool would be used for the selected fireground operation. Repeat this exercise as necessary until all appropriate tools and equipment have been removed from the apparatus. This drill can be significant from three perspectives: The firefighters basically conducted the drill; the firefighters have reinforced where the tools and equipment are located; and the firefighters have also reinforced how the tools and equipment are used.

6. Become a mental general contractor

Buildings are put together by construction crews and, conversely, buildings can come apart when exposed to the destructive affects of fire. Knowing how a building is constructed can assist in determining how it can fail. The following examples illustrate why it is important for every firefighter to study building construction and for seasoned firefighters (familiar with building construction) to take the time to tutor younger firefighters, particularly during overhaul operations and fire prevention inspections.

Attic fire. If fire is within the attic of lightweight truss construction, the trusses that are exposed to fire can suddenly and totally collapse into a building. This type of collapse can present a severe danger to interior personnel. Conversely, a fire within the attic of conventional construction (stick-frame construction) can often result in the jack rafters, rafters, and roofing material collapsing onto the ceiling joists, which can prevent the collapsing material from falling into the building (figure 1–10). Obviously this can be a significant safety benefit to interior personnel.

Newer wood frame. Newer wood frame buildings use platform construction, which resists the vertical extension of fire within the walls and into an attic. Conversely, older wood frame buildings used balloon-frame construction, which allows the rapid upward spread of fire in exposed walls and into an attic.

Moldings. When crown, base, window, and door moldings are exposed to fire, they must be removed during overhaul operations to check the gap behind the moldings for any remaining fire.

Figure 1–10. Collapsing material on ceiling joists can result in minimal danger to interior personnel.

Wiring. Older structures with a two-wire electrical service is an indicator of encountering knob-and-tube wiring, which is an ungrounded system that can present an electrical hazard to unsuspecting firefighters.

Cornices. Newer buildings with plastered cornices are normally attached with an adhesive. This type of construction does not present a suitable location to place an aerial device.

Unreinforced masonry. Unreinforced masonry construction can readily collapse outward when exposed to fire.

7. Use your protective equipment defensively, not offensively

Without a doubt, the personal protective equipment (PPE) of today is far superior when compared to the protective equipment of yesterday. Since the advent of hoods and high-tech materials, modern protective equipment does a remarkable job of protecting firefighters in a hazardous environment. However, this protection comes at a price that must be understood by every firefighter who wears PPE. To analyze this statement, let's briefly compare the PPE of yesterday to the PPE of today. Yesterday, PPE generally consisted of a helmet, breathing apparatus, gloves, turnout coat and pants, and boots. This left the wrists, neck, and ears unprotected. However, it was common practice for firefighters to use these unprotected areas to feel the heat and provide a warning when heat levels increased to either noteworthy or unbearable levels. Today, modern PPE consists of the same elements plus wristlets and hoods that fully encapsulates the firefighter wearing the PPE (figure 1–11).

Figure 1–11. Modern personal protective equipment can fully encapsulate a firefighter and mask the environment around a firefighter.

Although the addition of wristlets and hoods may not seem like a large addition to the overall PPE package, it has resulted in several drawbacks. Firefighters are now fully encapsulated, which makes it more difficult for the human body to ventilate itself. And the capability of modern PPE to shield a firefighter from heat has resulted in masking the environment around a firefighter, thus making it more difficult to monitor the external environment that surrounds personnel. Interestingly, when a firefighter begins to detect the temperature of an environment as warm or *very warm* but still feels able to operate within that environment, the temperature is around 400°F to 450°F. Although this temperature does not seem

high, remember that if this is the temperature at the floor, the temperature at or near the ceiling of a typical single-family dwelling is likely 800°F to 1,000°F, which is the approximate flammability range of the modern thermal layer.

Modern PPE was primarily designed as a defensive weapon, not a proximity suit. Although it can enable a firefighter an extra measure of safety while committed to a contaminated atmosphere, it is relatively easy (without proper training) to use modern PPE as an offensive weapon (instead of a defensive weapon). This allows a firefighter to advance into a hazardous atmosphere further and faster than ever before, and at a time when buildings are more prone to collapse and environments are flashing over with more frequency. Therefore, how do you use your PPE, and how do you monitor the environment around you?

8. Continually evaluate your fireground environment

All firefighters are aware of the benefits of conducting an initial size-up when a structural incident comes into view. Considerations such as the type of building, amount and extension of fire, evaluating fireground priorities, and numerous other factors begin to set the stage for the mitigation of the incident. Conversely, another size-up that can be potentially more important is one defined as a continuing size-up that continues until the incident has been mitigated. The focal point for this commandment is that interior personnel must continually evaluate and monitor their environment until the fire has been extinguished and/or the environment has been adequately ventilated.

However, the capability of PPE to shield firefighters from their external environment has resulted in the need for all firefighters operating in a hazardous environment to constantly monitor the status of their environment. This is particularly important for truck company personnel operating within a structure because they may not have a source of protection (the availability of a charged hose line) and/or they may be operating alone. (See the Developing Indicator section in chapter 5, Reading Smoke, for more information on this subject.)

9. When in doubt, ventilate

Of all the fireground operations that deserve attention at a structure fire, ventilation operations are a priority more often today than at structural fireground incidents of yesterday. This is an obvious statement from the perspective that what is burning *today* is significantly different than what was burning *yesterday*. The fires of yesterday were primarily composed of conventional materials (cellulose-based materials) that burned slower and cooler than fires of today. Fires of today routinely involve synthetic materials (petrochemical-based compounds) that burn hotter, faster, and can produce a denser smoke than the fires of yesterday. These factors cannot only more severely inhibit the visibility of interior personnel, but can result in an increase of flashovers that can instantly produce a fatal environment to human life.

When considering operations from the perspective of quickly developing fireground priorities, remember that forcible exit and ventilation collectively combine to significantly increase the safety of interior suppression personnel at most structure fires. This is not to imply that every structure fire requires holes cut in roofs, windows broken, and positive pressure blowers implemented. As with other fireground tasks, ventilation must be coordinated with attack and/or search operations. However, when ventilation is properly executed, four major benefits normally result: visibility improves, interior temperatures are reduced, flashover conditions are minimized, and dangerous fire gases decrease while the percentage of oxygen increases. These four benefits are worth considering during the initial size-up at any structure fire encountered by suppression personnel—and as a general rule, the earlier ventilation is started, the better!

10. Ensure the viability of your escape route inside, outside, and topside

Throughout this book, the terms *forcible entry* and *forcible exit* are synonymously used from the perspective of safety. If it is important to get into a structure, it should be as important, if not more so, to get out of a building. Unfortunately, firefighters routinely force entry into a structure but do not succeed in providing alternative, multiple, or timely exits points for interior personnel. Let's look at three forcible exit considerations.

Inside. A fundamental rule in wildland firefighting operations states that personnel operating on a hill shall never let a fire fishhook below them. This same rule can also be applied to structural firefighting operations and can be stated as interior personnel shall never let a fire above them (for example, an attic) get between them and their exit point. From a simplistic perspective, always keep your primary entry and exit point available to enhance your ability to exit a structure when necessary. In some cases, this dictates that personnel pull ceilings to verify the condition of an attic when fire is known or suspected to be above personnel. If fire collapses a ceiling between personnel and their entry/exit point, exit options will be compromised.

Topside. When fireground personnel are assigned to roof ventilation operations, what is the minimum number of ladders that should be raised to the roof? Two. Although every firefighter is aware of the right answer, it is more than interesting to look at pictures of fireground operations from around the country and see that a single ladder is the common answer (figure 1–12).

Figure 1–12. A single ladder to a roof does not provide multiple ways of egress. (LAFD file photo)

So, let's apply the tenth commandment of ensuring the viability of your escape route and raise at least two ladders to a roof when operating above ground. The phrase *at least* means that if the number of personnel exceeds two, then the number of ladders raised should also exceed two. As an example, if four personnel are on a roof, there are two ladders to the roof, and it is suddenly necessary to exit the roof, what do firefighters #3 and #4 do while firefighters #1 and #2 are exiting the roof via the two ladders? Always take the time to enhance the safety of your above ground operations.

Outside. Ensuring the viability of your escape route options from the interior of a building means to provide at least two ways to get out of a building. This is normally accomplished from the exterior of a building by forcing additional doors (other than the initial entry door), appropriate windows (and this means opening the entire window with all glass removed), and any other considerations that could obstruct or slow down exiting the building. Several examples of obstructing considerations are security bars and plywood-OSB panels over doors and windows on vacant or abandoned buildings (figure 1–13).

Figure 1–13. Obstructions over windows and doors should be removed for the safety of interior personnel.

In these cases, obstructing devices shall be removed prior to or during the initial phases of interior operations. Maximize your ability to exit a building.

Training Resources

The following resources can provide additional information and/or another viewpoint for the subjects discussed in this chapter.

- Truck Company Functions, Roundtable, *Fire Engineering*, March 2008.
- Dedicated Company Assignments, Paul Urbano, *Fire Engineering*, March 2008.
- Truck Company Riding Positions, Scott Lyons, *Fire Engineering*, April 2004.
- Truck Company Dos And Don'ts, Tom Murray, *Fire Engineering*, April 2002.
- Important Tactical Decisions, Jack Cottet, IFSTA, Winter 1988.
- Suburban Ladder Company Operations, Armand Guzzi, Jr., *Fire Engineering*, March 2008.
- How Good Is Your Truck Company?, Robert Hanley, *Fire Command*, December 1976.
- Quint Apparatus, Roundtable, *Fire Engineering*, March 2003.
- Setting Up A New Ladder Company, Mark Kuhar, *Fire Engineering*, December 2001.
- Fire Attack! Command Decisions And Company Operations, Warren Kimball, NFPA.
- Managing Fire Services, John Bryan and Raymond Picard, Institute for Training in Municipal Administration and International City Management Association.
- *Fireground Tactics*, Emanuel Fried, chapter 4.
- *The Fire Chief's Handbook, Fourth Edition*, Fire Engineering, chapter 15.
- *Fire Fighting Principles and Practices*, William Clark, Fire Engineering, chapter 10.
- *Modern Suburban Fire Fighting*, Dick Sylva, Fire Engineering, chapter 11.
- Weak Staffing, Letters To The Editor, *Fire Engineering*, February 2003.
- Organizing The Truck Company, Theodore Jarboe, *Fire Command*, May 1984.

Certification, Response, and Placement Considerations

Introduction

It normally takes a coordinated effort between the application of water and support functions to result in a safe, timely, and successful fireground suppression operation. With a common need for forcible entry, ventilation, search, portable ladders, and operations that may require the use of an elevated aerial device, the effective placement of truck apparatus in combination with the placement of engine companies should be a routine occurrence. However, when observing everyday fireground operations, it seems that the placement of truck companies could be classified as:

- An afterthought instead of an initial consideration

- An apparatus that is being saved for an upcoming parade

- An apparatus that arrives at an incident after the prime spots have been taken by several engine companies and then is parked down the street from the incident, used only as a resource for tools

- An urban exposure protection tool

- An apparatus that is capable of extending an aerial device over 100 feet in the horizontal position and lifting 500 to 1,000 pounds, yet is incapable of maneuvering into proper operational position due to the size of the apparatus

- An apparatus with a 100-foot aerial device that has been forced to spot 104 feet away from an objective

Although there are other common examples of ineffective truck apparatus placement, the preceding examples summarize operations that do not fully utilize the capabilities of a truck, and are often commonplace. Therefore, before a truck company can be effectively utilized, it must first be safely driven to an incident by a qualified operator and then strategically positioned to use its capabilities to their fullest. With these thoughts in mind, let's examine the following six basic considerations that can dramatically affect the operational capabilities of a truck company.

Qualified Operator

Before any truck apparatus responds to an incident, the initial requirements for safe, timely, and competent operation of an apparatus is a qualified operator with the right perspective. Notice the two key words—*qualified* and *perspective*. The focus of the word *qualified* is the difference between certified and qualified. A driver can be certified, but may not have the experience and practical qualifications to competently operate an apparatus with an aerial device. The second word, *perspective* (or attitude), is one of the most important attributes of an apparatus operator because it takes a qualified operator who can also display the appropriate amount of courtesy and patience to safely drive to and return from an incident.

The importance of this statement is underscored by National Fire Protection Association (NFPA) statistics that indicate the third highest cause of death and injury to firefighters is responding to an incident and driving back to a fire station. Stated from another perspective, a United States Fire Administration firefighter fatality report indicates that since 1984, apparatus collisions have accounted for about 25% of all firefighter fatalities. Although most apparatus operators are fully aware of this statistic, an inordinate number of fire service vehicular accidents continues to be a problem. Not surprisingly, if fire apparatus are involved in an accident and representatives from the legal profession become involved, some of the items expected to be requested for review are training records, the department's driver certification standards, and the department's driver *recertification* standards. In some cases, a program that maintains the proficiency of an operator can be as important as the initial certification process.

As a starting point, departments should be aware of two NFPA standards that outline minimum standards for driving apparatus and establishing responsible driver training programs. These NFPA standards are:

- NFPA 1002, Standard on Fire Apparatus Driver/Operator Professional Qualifications
- NFPA 1451, Standard for a Fire Service Vehicle Operations Training Program

As an example, the Los Angeles Fire Department employs a comprehensive driver training program that is used for an initial certification and an annual recertification program for apparatus operators and persons who may be required to drive fire apparatus. The program consists of four distinct portions:

- Phase 1 consists of classroom study and review.
- Phase 2 consists of a driver training course and driving rodeos.
- Reviews class 1 and 2 drivers license requirements and testing procedures.
- Training video tapes are used to enhance specific points/criteria

The program is scheduled once a year by a driver training coordinator for members who are currently certified. All written and driving scores are reviewed by company commanders. Anyone with a below-average score is given a "prescription for improvement," which consists of a personalized program to raise the skill level of that particular individual. Newly appointed members (to the rank of firefighter) are required to successfully complete both Phase 1 and 2 of the program to qualify for their initial Class 2 driver's license (as required by California). The complete driver training program, including the specifications for a driving rodeo, can be reviewed on the Los Angeles Fire Department training web site at http://lafdtraining.org/ists/books/bk08v1107.pdf. Another driver training program of interest was developed by the Albuquerque Fire Department and detailed in the article "Increasing Your Driver Safety Awareness" which can be found in Training Resources at the end of this chapter.

Response Criteria

It is the evening rush hour. Traffic is heavy and congested. Long lines of cars with windows rolled up and loud stereo radios are carrying tired and impatient people home. Suddenly, a fire engine with siren screaming and lights flashing enters our scenario. Its goal is a successful emergency response with minimal delay. As the fire engine charges through the dusk, the surrounding cars appear to be almost a blur. Suddenly, however, one of them makes itself known by swerving abruptly towards the path of the oncoming fire engine. How the driver of this fire engine reacts in this circumstance affects not only those persons whom the driver has just met, but many others during the response to an incident. The driver can assist motorists on their journey home, can cause delays, or can be responsible for an accident that may cause injury or even tragedy.

The driver also has an important degree of control over the future of the personnel on the fire engine. Is our driver mature and skilled? Does the driver know the rules of the road? Is the driver aware of the responsibility for the welfare of personnel on the fire apparatus and civilians? The answers to these questions are answered each time a driver operates fire department apparatus. It is the responsibility of all drivers (regular or acting) to operate fire department apparatus in a prudent, timely, and safe manner. State laws have given the fire service the right of way when driving to an emergency under most circumstances, but a good driver never demands it. A good driver drives defensively, keeping aware of changing conditions, and is aware that at times that motorists and pedestrians may not see apparatus or hear warning devices due to car radios, iPods, and cell phones. For these reasons, motorists and pedestrians don't always yield the right of way, nor do they have a reason to expect the sudden appearance of an emergency fire apparatus. Additionally, a good driver keeps the apparatus under control at all times and never takes anything for granted.

In the fire service, an accident caused by irresponsible driving has many repercussions. Lives and property that could have been saved may be lost. Firefighters, civilian drivers, and innocent bystanders may be injured or killed. A fire service driver (and responsible fire department) may be held accountable for the unsafe operation of fire department vehicles. Lawsuits and/or criminal prosecution could be initiated as the result of negligence. However, the smooth, efficient, and careful operation of apparatus will result in a response that is prudent, timely, and safe. Let's restate the last sentence from a different perspective. When operating fire department apparatus, drivers must exercise sound judgment coupled with sensible well-timed operations designed to exactly meet the needs of the occasion that will result in a response that is free from danger, damage, and injury.

Drivers must realize that the safety of fire department personnel, civilian drivers, and the people at an incident hinge on the safe operation of fire apparatus. Fire department personnel cannot perform their duty as public servants if they are unable to get to the location of an emergency in a timely and safe manner. Therefore, let's consider the basic skills needed to safely drive fire department apparatus during emergency and non-emergency responses.

Preparation

The preparation for an emergency response begins long before the receipt of an alarm. The knowledge and readiness of your apparatus, basic driving skills, and familiarity of your district (appropriate response routes, water supply, target hazards, access challenges, and so on) must be known to efficiently and safely operate apparatus during an emergency response. Additionally, the readiness of an apparatus and equipment is an excellent indicator of the level of pride of a driver.

Specifically, the knowledge and readiness of an apparatus begins at the start of each tour of duty because there is no excuse for compartment doors that are not properly secured and inadvertently open during a response and other similar surprises. Making certain you and your apparatus are ready begins with a careful check of at least the following:

- Check your personal protective equipment to be certain it is complete and ready for use This includes checking your SCBA bottle to ensure it is fully charged and your facepiece is clean and ready for immediate use.

- Check the fuel and oil levels; if necessary, start the engine and check the air pressure gauge. In all cases, check the emergency lights for proper operation. Checking the siren inside a fire station is not recommended.

- Check the air brakes and slack adjusters.

- Check the apparatus by doing a complete walk-around that consists of opening and securely closing each compartment door. While each door is open, check the compartment contents for the appropriate inventory, cleanliness of equipment, and proper fuel levels.

- Double-check the power tools. As an example, make sure the proper blades are mounted. If necessary, start each power tool.

- Check under the apparatus for any leaks that would not be classified as normal.

After receiving an alarm, verify the location of the alarm and determine the best route to the incident (based on time of day, weather conditions, and so on). As a regular or back-up driver, it is essential that you are *totally* familiar with your district, streets, intersections, traffic patterns, and potential routes of responding apparatus from other companies. Take the necessary time to learn your district, and consistently practice good defensive driving skills. Study your intersections and streets that offer the best routes of response as if you were pre-fire planning a structure. Give special attention to blind, uncontrolled intersections and changing traffic patterns. There is no substitute for experience and knowledge of your district. Additionally and of equal importance, be thoroughly familiar with the general attitude and driving habits of the public that you will encounter, and consider how other drivers react to emergency apparatus during a response. (At times, you probably wonder if your emergency lights and siren are operating.)

Always exercise additional caution when driving in an unfamiliar district or with different apparatus. Remember that apparatus built prior to 1991 may not meet recent NFPA standards and will likely have fewer safety features. The time involved in getting to an incident is seldom reduced during the actual response. Time is primarily saved by getting to your apparatus quickly and knowing your district without having to study a map to determine response routes or incident location. Know where you are going before leaving the fire station.

Responding

After you have verified the incident location and determined the best route to the incident, consider the issue of wearing personal protective equipment (PPE) while driving. Since the advent of modern protective equipment, there has been a debate regarding drivers wearing personal protective equipment while driving apparatus emergency to an incident. Interestingly, it seems as though modern apparatus and personal protective equipment has increased in size, yet the actual space available for operators to drive apparatus has decreased. Although this should be an SOP that each department should thoroughly research, it is easy to see that personal protective equipment consisting of turnout pants, coat, and boots does not enhance the maneuverability and comfort level of a driver in a confined space. For this reason, it is recommended that depending on the size of a driving compartment, most drivers are better able to safely operate modern fire apparatus while not dressed in personal protective equipment. See Training Resources for the Driving Apparatus in Bunker Gear Roundtable.

Now, get into the driver's seat, fasten your seat belt, and start the engine. Check the dash gauges (oil pressure first and air pressure second) and turn on emergency, head, and running lights. Verify that everyone is seated and buckled in safely. Regarding the use of seat belts, two basic rules should apply: all personnel are using their seat belts correctly, and personnel do not put on personal protective equipment without being secured by seat belts while responding to an incident.

As you proceed out of quarters, check for vehicle and pedestrian traffic. Consider that you are not expected, so proceed cautiously. When it is clear and/or traffic has yielded, proceed on your response. Now is a good time to consider the following three items.

Familiarity with apparatus. If drivers were asked "are you familiar with the driving requirements of your particular apparatus," most drivers would quickly respond with a strong "of course!" However, it is doubtful that few drivers have ever tested the difference and capabilities of standard air brakes and newer electronic braking systems such as anti-lock brakes. Therefore, drivers must have a clear understanding of their apparatus and its limitations. A driver should not exclusively depend on built-in safety devices to stay out of trouble. As an example, when evaluating braking considerations, remember the following facts:

- Total stopping distance is a combination of perception and reaction distances.

- Air brake systems take longer to stop than hydraulic brake systems. Remember that air brakes have an inherent lag time of .5 to 1 second longer compared to hydraulic brakes.

- Apparatus with ABS brakes require a different technique than apparatus without ABS brakes. The advantage of ABS brakes is they keep the tires from sliding and skidding, and as a result, a driver can maintain steering control. Remember that sliding tires lose steering control. Therefore, to quickly stop with ABS brakes, floor the brake pedal and stay there to let the ABS system do the work as opposed to non-ABS brake systems that require a driver to feather the brakes to keep them from locking up and skidding.

- Truck tires do not stop as well as passenger car tires because of their increased hardness (which makes them more slippery) for increased wear and load-carrying capacity.

- Auxiliary braking devices such as retarders and engine brakes can significantly increase braking capability.

- Kinetic energy can be a deadly enemy. The larger and heavier an apparatus, the more kinetic energy an apparatus has and the more distance will be required to stop before hitting something.

- The old motto of "drive as if your life depends on it because it does" still applies today as much as yesterday.

Speed. When responding, never drive over your head or faster than you can safely operate the vehicle. Going faster is not as important as being able to appropriately stop. The relevant racing motto of "to finish first, you must first finish" is applicable when driving emergency and is translated as "to fight fire and save lives, you must first safely get there." Speed is the largest single factor in total stopping distance. If speed is doubled, perception-reaction distance is doubled and braking distance is about four times as great. Consider the following distances on dry roads with good traction.

Speed	Perception/Reaction Time	Braking Distance	Total Stopping Distance
15 mph	17 feet	30 feet	47 feet
30 mph	33 feet	115 feet	148 feet
60 mph	66 feet	466 feet	532 feet

Defensive driver considerations. A defensive driver makes allowances for personal deficiencies, for the lack of skill and knowledge on the part of the other driver, and recognizes an inability to control the unpredictable actions of other drivers, pedestrians, or road and weather conditions. If necessary, a defensive driver will give up the right of way and make necessary concessions to avoid a collision. Expect the other driver to act in one of two ways—predictably or unpredictably. If another driver does the predictable, most accidents are prevented. However, a defensive driver will try to predict the unpredictable.

Seven steps of visual awareness

In a moving apparatus, a driver does not always have time to see everything along the road or to think about what actions to take in a tight spot. An aid to enhancing driving skills is to utilize seven visual steps that let a driver read the traffic picture in one quick glance and correctly make automatic decisions. All drivers utilize some of the visual steps that we will consider, but most drivers take so long on each step or are distracted that they often fail to finish all seven steps. Practice these steps until they are automatic:

1. Know your blind spots: When sitting in the driver's seat, be totally familiar with the areas around your apparatus that you can and cannot see. As an example, figure 2–1 depicts the blind spots as viewed from the drivers seat of a 1999 Pierce Seagrave quint, 100-foot rear-mount platform.

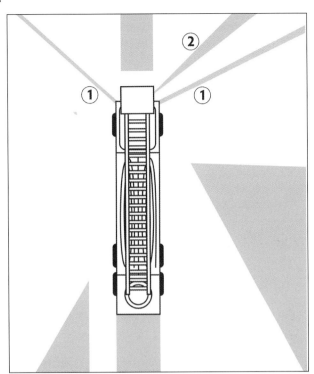

Figure 2–1. Being familiar with the blind spots for an apparatus will enable a driver to anticipate blind areas that may be hiding objects that can contribute to an accident.

Of particular interest with this apparatus:

- When viewing the area in front of the windshield from the driver's seat, the platform overhang blocks viewing ahead much further than one-half city block. This eliminates or minimizes viewing and anticipating traffic lights, intersections, and so on.

- The small shaded areas that traverse outward from the front corners of the apparatus (1) are the areas blocked by the A pillars.

- The mobile data communication (MDC) screen mounted on top of the dash by the officer blocked the drivers view to the front-right portion of the apparatus (2).

- Remember that all areas that can be *easily* viewed and are hidden from a driver will vary, depending on each type and make of apparatus.

The knowledge of a specific apparatus allows a driver to anticipate blind areas that may be hiding objects (vehicles, personnel, and objects) that can contribute to accidents. Remember that blind spots will dramatically change when the area around an apparatus is viewed from rear-view mirrors that are flat, equipped with bubble mirrors, and so on. Therefore, it is strongly recommended that every driver of any apparatus take the time to be familiar with the blind spots of their particular apparatus.

2. Look ahead: When responding, how far do you look ahead? Your objective should be to set your eyes on where you will be, not where you are now. This translates into one to two blocks in the city and a quarter to half mile on highways. Always keep your eyes moving and not fixed on a central point. At night, watch beyond your headlights. If poor visibility forces you to look low to find your direction or keeps you from seeing beyond your headlights, the look-ahead step will pull your speed down to a safe level. Additionally, you will have steering problems unless the driver's seat lets you sit erect with your line of vision at least two inches above the steering wheel.

3. Get the total picture: Move your eyes up and down and side to side to help determine (or see) the total picture. In the city, this picture is at least sidewalk to sidewalk and one to two blocks ahead. The total picture contains the following factors that can affect your judgment and resultant actions: stationary and moving objects (parked vehicles, pedestrians, and so on); warning devices (signals, stop signs, and so on); and the ability to predict actions or events (a car with its backup or turn indicator lights on, and so on).

4. Constantly move your eyes: Constantly keep your eyes moving. This includes glancing near and far ahead, to the sides of the apparatus, and into the rear view mirrors. This habit is restful to your eyes, develops the total picture, and keeps you alert.

5. Maximize your perimeter space: A key consideration of emergency driving is to maximize the space around your vehicle and leave yourself an out at all times. This allows space to avoid an impending accident. Potential accidents are reduced by avoidance through maneuvering. This requires a place to maneuver to or in. Remember, with no maneuvering space, you need extra space to stop.

6. Allow for sufficient following distance: Whenever you are following other emergency apparatus, consider the following factors that affect your total stopping distance: weight of apparatus, road conditions, and perception/reaction time plus brake lag is 1.9 seconds under ideal conditions. At 30 mph, this equals over 100 feet before your apparatus begins to stop. Always allow enough distance between you and other apparatus to safely stop.

7. Make sure they know your intended actions: Never assume other drivers know what you will do, and don't count on traffic laws (or traffic preemptive devices such as opticoms) to help you. You must communicate with other drivers and pedestrians to ensure they know your intended actions. This can be accomplished by eye to eye contact, warning lights-siren-horn, turn indicators and hand signals, positioning your apparatus, and speed of apparatus. Slowing down gives motorists and pedestrians more time to react.

As we continue on our response, let's consider eight additional factors that will affect your efficiency and safety.

Apparatus position

The position of apparatus on a road is important when driving emergency, as civilian drivers often appear to be oblivious to the response of emergency apparatus from any direction. When responding on a roadway without a raised median, consider placing the apparatus close to or on the center of the roadway (figure 2–2) if possible.

Figure 2–2. When responding on a roadway without a raised median, placing the apparatus close to or on the center of the roadway can assist in clearing traffic.

This can result in traffic being more visible to the driver and apparatus more easily seen by oncoming traffic. It can influence oncoming drivers to pull over where there will be additional room for maneuvering in heavy traffic and at intersections, and decreases the chance of rear end collisions with civilian vehicles that suddenly stop.

When responding with multiple apparatus, and depending on the configuration of a roadway or highway, consider the offset formation. This formation consists of the lead apparatus traveling as close to the center of the road as possible, the second apparatus on the left or right side of the lead apparatus, and a third apparatus (if present) behind the second apparatus and on the opposite side of the lead apparatus. This results in improved visibility of all apparatus to oncoming traffic, increases the responding apparatus drivers visibility of traffic conditions ahead, influences oncoming drivers to pull over, decreases the chance of apparatus colliding with one another in sudden stop situations, and leaves the driver of each apparatus an out.

Passing vehicles

When passing other vehicles, pass on the left if possible. This may require crossing a double-yellow line(s) into oncoming traffic lanes. If the street you are responding on has a raised center divider and all lanes are blocked, or if a vehicle in the left lane has committed to a turn, it may be advantageous to pass on the right. Reduce your speed as necessary, and make certain that all other drivers you are passing are aware that you are passing them on the left or right side. Remember, most states require cars to pull over to the right and stop when approached by vehicles responding emergency. Additionally, always be alert to vehicles entering from side streets, driveways, or roundabouts.

Intersections

Historically, most accidents involving emergency apparatus occur in intersections. As an example, in 2007 33% of all civilian vehicular accidents occurred in intersections, 25% of vehicular deaths occurred in intersections, and 69,000 deaths occurred in intersections. When approaching an intersection, a defensive driver is prepared to stop by removing a foot from the throttle and placing it over the brake pedal, even if the signal is green. By positioning a foot in this manner, you are prepared for the unexpected and you reduce reaction time and total braking distance. By watching well ahead, you can anticipate a green light turning red. If it is green, common sense dictates it will change to yellow or red. You can also take note of a "Walk-Don't-Walk" pedestrian signal. If it is flashing red or a low number, you can anticipate an impending signal change.

When entering an intersection and cross traffic is yielding to you, consider every open lane as a potential hazard. This is especially true of lanes blocked from view by a truck or bus. Also, don't overlook left turn lanes or gutter lanes because there may be a driver not yielding the right of way or in a position where the driver's view of emergency apparatus can be obstructed. Approach these open lanes with caution, and be prepared to stop to avoid a potential collision. A policy designed to reduce accidents that occur in intersections states that single-unit companies shall stop at all red lights, stop signs, and intersections with obstructed-impaired vision, look both ways, and when safe, proceed with caution; and the first apparatus in a convoy shall momentarily stop at all red lights, stop signs, and blind intersections where vision is obstructed or impaired. The remaining apparatus shall slow down, look both ways, and when safe, proceed through the intersection with caution.

One-way streets, curves, and hills

When responding to an emergency on a one-way street, drive as close to the left as possible. Driving in this manner should cause other vehicles to pull to their right and stop. In some instances, drivers will pull to the nearest curb requiring an apparatus driver to pass on the right. If this happens, pass with extreme caution. Driving around curves can present challenging circumstances. When a vehicle rounds a curve, centrifugal forces tends to force the apparatus outward. For this reason, it is more advantageous to round a curve at a constant speed or to be accelerating out of the curve rather than braking. This ensures the rear wheels will oppose the centrifugal forces pushing the vehicle to the outside of the curve and will provide a controlled and safer turn.

Additionally, consider the height (or center of gravity) of the apparatus (for example, a 100-foot rear-mounted platform, and so on) as apparatus will lean to the outside of a curve). In terrain with rolling hills or dips, avoid crossing over the center line. Oncoming vehicles may not be seen until it is too late to avoid a collision. If you are responding in an area with steep hills, sound your horn as you near the top of a hill. This can warn any drivers on the other side of the crest of your approach.

Road conditions

When driving in inclement weather conditions, the key to a safe response is to slow down according to the severity of weather conditions. Visibility can be impaired during inclement weather, and traction of apparatus tires can be severely reduced on wet roads. A conscientious driver is also aware that different road compounds influence stopping distances. As an example, a wet concrete roadway will require approximately 35% to 40% more stopping distance than dry concrete roads. Asphalt also presents special considerations when it is wet. Asphalt, when first laid down, is oiled and then cooled. This oil eventually rises to the surface and, when wet, combines with the water on the road surface. This combination makes the roadway extremely slick (particularly during a first rain), and extra caution must be exercised when driving on this type of wet pavement.

Regardless of what road compound is used, the combination of dirt, fine rubber dust, oil dropped by thousands of vehicles, and finely divided water particles of fog or rain can make any roadway slick as ice until enough rain has fallen to wash it off, particularly on corners. After a good, hard rain has cleaned it off, traction will improve, but any roadway is still dangerous when it is wet. When driving in fog, reduce apparatus speed. Never assume that the road ahead is clear. Keep your headlights on low beam. If high beams are used, the light is reflected back at a driver, making visibility even worse. Also, avoid sudden stops and lane changes, remembering that during quick maneuvers the higher center of gravity of your apparatus can suddenly become very evident.

When stopping, tapping the brake pedal a few times will warn other drivers behind you that you are slowing down or stopping. Remember, being a superior driver on any roadway is a matter of good attitude and judgment. A safe vehicle and good physical condition of a driver also plays an important part in safe driving, but driver attitude is the key.

Sirens

Let's briefly review the capabilities of sirens when utilized on emergency responses, particularly when few civilian drivers with windows in the up position and stereos in the loud mode can hear a siren over 100 feet away. Sirens on emergency apparatus are directional (particularly electrical sirens) with greater intensity straight ahead as compared to either side or rear of your apparatus and consist of a frequency that increases or decreases in intensity as heard by an observer. As a siren approaches an observer, the frequency will appear to be high. As the siren is near an observer the frequency will be normal, and as a siren moves away from an observer the frequency will appear to be lower. This is known as the Doppler effect. Additionally, tall buildings, hills, winding roads, blind intersections, or heavy brush can reduce siren audibility by as much as two thirds. To increase siren audibility in clearing traffic, use the full fluctuating/high-low tones. Do not ride the siren button; it will produce a high, constant sound that will reduce the siren's effectiveness. Due to noise levels in an apparatus cab area and the possible use of noise suppressing ear protectors, personnel may not always hear the warning devices of other responding apparatus.

Remember that other emergency vehicles may be responding from other routes and directions. This fact dictates that drivers must use extra caution at commonly used intersections. Additionally, some drivers consider the air horn an effective audible device (which it is) in warning motorists of approaching responding apparatus. When using an air horn as a warning device, remember an air horn may not be considered a legal emergency vehicle warning device by itself (depending on your area) and should not be used in place of a siren.

Additional factors to consider when responding with emergency warning devices to unknown incidents are as follows:

- When apparatus are dispatched to an incident, all apparatus respond Code 3 (emergency with appropriate warning devices active).

- When the initial company arrives on-scene and gives a size-up that indicates the incident is an emergency in progress, all responding apparatus continue with all warning devices active.

- When the initial company arrives on-scene and gives a size-up that indicates the incident is a potential false alarm (for example, nothing showing) but will investigate, responding apparatus continue to the incident but in a non-emergency mode (emergency warning devices are not operational). However, if the initial company determines the incident is an actual emergency, the responding companies can immediately respond Code 3. If the incident is determined to be a false alarm, the responding apparatus return to quarters. NFPA statistics indicate that if nothing is showing on arrival, the incident is a false alarm over 90% of the time.

See Training Resources for the Roundtable article "Response To Structure Fires."

Backing up

Any discussion involving the operation of fire service apparatus should involve the possibility of having to back up an apparatus during a response, backing up for positioning at an incident, or backing into quarters after a response. A fundamental question to consider is, "Does your department have an Standard Operating Policy (SOP) on backing apparatus?" If your department does not have an SOP on backing apparatus, it should. Remember that approximately 30% of fire service vehicular accidents occur while moving backwards.

If your department has an SOP on backing apparatus, it must be followed whenever an apparatus is moved backwards, regardless of its location and the need for quickly repositioning the apparatus. When backing an apparatus:

- Prior to backing the apparatus, the driver should quickly walk around the apparatus, if possible.

- Always have a person at the rear of the apparatus who can be seen by the driver.

- The driver should not solely rely on backup cameras, backup alarms, and so on.

- The driver and person at the rear of the apparatus should agree on and use predetermined hand signals. As an example, the airline industry uses a set of predetermined hand signals (which appears to be a standard, regardless of the airline) to safely move and park aircraft.

- If the person at the rear of the apparatus is not visible to the driver, the apparatus should not be moving.

- The driver should be responsible for the area in front of the apparatus. The person at the rear of the apparatus is responsible for the area at the rear of the apparatus, and that includes overhead obstructions.

Tillering

Because many fire departments employ tractor-drawn (or tillered) apparatus that involves a measure of expertise in driving the drawn trailer, let's briefly look at this operation. Interestingly, the word tiller has an interesting history. Tiller is commonly used in reference to the lever that turns the rudder of a boat from side to side. The person who steers the tiller is referred to as a tillerman. The seat the tiller-man occupies is called a tiller seat or tiller bucket.

Safe and proper tillering is not just a matter of climbing into a tiller bucket and steering the trailer. Steering correctly and in a safe manner depends on a tillerman's ability to properly react to the movement of the truck. The following is a summary of good tiller techniques:

- When traveling forward, place the hands at the 9 and 3 o'clock position.

- When the tractor makes a turn, the tillerman must turn the steering wheel in the opposite direction of the turn. This will cause the trailer to swing away from the turn.

- When backing up, place both hands at the bottom of the steering wheel and push the steering wheel in the direction you want to go.

- Oversteering is defined as allowing a trailer to move well past the inline position after a turn is completed. To prevent this situation, begin straightening the tiller wheels before the tractor has completed the turn.

- Overtillering is defined as tillermen turning the steering wheel unnecessarily. Unless necessary, the trailer belongs in an inline position behind the tractor.

- Overhang is the portion of the trailer that extends from the center of the trailer wheels (pivot point) to the rear of the trailer. Because the rear end of the trailer will swing on the pivot as the truck turns, it is important to allow enough room for the trailer overhang to clear a turn on both sides and rear of a trailer.

- Every tillerman should be familiar with the blind spots that are created from the tillerman's vantage point.

- Before mounting the tiller bucket, always check the direction the tiller wheels are pointed. If the wheels are in a position other than straight, be prepared to compensate as the truck begins to move forward.

- A good tillerman steers only as necessary.

A complete lesson plan on tillering including a general design for a field test course can be obtained at http://lafdtraining.org/ists/books/bk08v1107.pdf, Module #9.

Approaching the Incident

As you approach the location of an incident, SLOW DOWN! This simple basic maneuver can assist the driver, company officer, and other members of a company by:

- Giving more time to find a specific address

- Giving the company commander additional time to size-up an incident and select the best spot for the apparatus

- Giving company personnel time for a personal size-up of an incident and what needs to be done

- Enabling company personnel to determine if an aerial device will be needed.

As you approach the incident, make sure your head lights are on low beam. This keeps your lights from blinding other drivers approaching from the opposite direction, and also does not blind personnel that are working in front of your apparatus.

Multi-Company Placement

Now that the apparatus has been driven to an incident in a safe and timely manner, it will be necessary to position the apparatus to effectively utilize its capabilities to the best advantage. However, before we consider basic rules of multi-company placement at fireground incidents, let's briefly look at positioning apparatus on roadways, particularly limited access highways because most fire departments may not be exempt from positioning apparatus within the flow of civilian drivers. Although this section focuses on roadways and limited access highways, some of the principles also apply to positioning apparatus in city surroundings. The primary difference between these two locations are higher speeds and heavy trucks on limited access highways, and slower speeds, minimal heavy trucks, and more congestion in city surroundings. In all cases, it is dangerous to be working in or near traffic regardless of the location of any roadway. More firefighters and EMS personnel have been killed or injured on limited access highways than any other type of roadway system in the United States. The following considerations will overview common guidelines:

- Initially, place apparatus between the incident (consisting of work area, patients, and responders) and flow of traffic by using your apparatus as a visible large traffic barrier.

- Block two lanes, if possible (one lane for the incident, and one lane for your work area).

- Prior to dismounting the apparatus, the driver and/or officer should remind personnel to dismount the apparatus on the side away from the flow of traffic.

- Apparatus should be in a blocking position before personnel dismount the apparatus. The only exception to this is either the driver or officer, depending on which person must exit the traffic flow side. This should be done with great care and respect for errant civilian drivers who are likely looking at the incident and not the fire department apparatus committed at the incident. Consider that when a driver looks away from a given direction, they will inadvertently turn the steering wheel in the same direction as they are now looking.

The Federal Manual of Uniform Traffic Control Devices (MUTCD), published by the Federal Highway Administration, provides guidelines for the safety of personnel when operating in or near moving traffic. Specifically, Section 6B.01, Fundamental Principles of Traffic Control outlines steps that must be taken by responding agencies to protect pedestrians, victims, and first responders during an incident. Of particular interest are the following two guidelines (see the MUTCD web site www.mutcd.fhwa.dot.gov/index.htm for specifics):

1. The four stages of temporary traffic control are (1) the advance warning stage, (2) the transition stage, (3) the activity stage, and (4) the termination stage (figure 2–3).

The advance warning stage warns motorists of an impending hazard. The transition stage is designed to re-route traffic around the incident. The activity stage includes a buffer zone and the actual location of the incident. The termination stage is designed to re-route traffic back to their original lanes.

2. The duration of an incident is divided into three time constraints with specific guidelines for each duration of 1 to 30 minutes (minor), 31 minutes to 2 hours (intermediate), and an incident that exceeds 2 hours (major). As can be anticipated, as the length of an incident increases, the guidelines and regulations become more stringent.

Now consider the basic principles of multi-company operations, or how truck companies can operate within a framework of multiple companies at an incident.

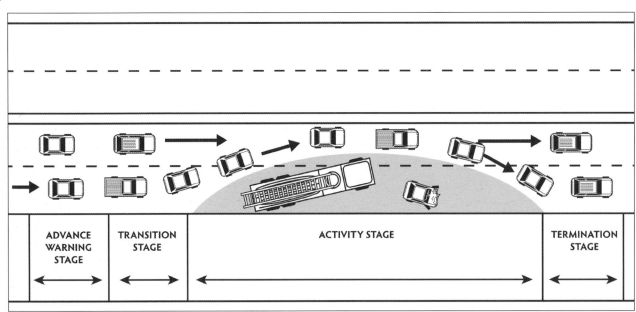

Fig. 2–3. The four stages of temporary traffic control are the advance stage, transition stage, activity stage, and the termination stage.

Note that because engine companies are normally the first apparatus to arrive at an incident and begin to form the foundation for the balance of an incident, the following section is also directed at engine company officers.

Incident Commander Responsibilities at the Fireground

The first-in officer is always the incident commander, and is the person who develops the initial plan for the incident. The goal of an incident commander is to achieve maximum effectiveness from all resources working together, including engines and trucks. A lack of initial direction usually results in independent action (or freelancing) by additional companies who do not consider the overall goal of a team.

Effective communications

Simply stated, communication provides a connection among an incident commander, on-scene personnel, and responding resources. When an

incident commander decides on a plan of direction, the commander should communicate that plan as clearly as possible. Fireground communications should not be top secret. For example, assume the first-in company to a working fire in the rear of a single-story, single-family dwelling is a typical engine company. The officer conducts a quick mental size-up and then gives the following radio size-up: *"Dispatch from Engine 1, we are on-scene at 1420 Elm Street with fire visible from the rear portion of a single-story, single-family dwelling. Engine 1 is taking an attack line inside the structure via the front door."*

In this size-up, the officer has described incident conditions and initial suppression actions. However, without specific directions, the next arriving company can easily spot next to Engine 1 and decide what they feel is the best action to take (this is called freelancing). If this happens, it is the combined fault of the officer on Engine 1 for ineffective communications, and the officer of the second-in company for not asking for specific directions prior to becoming involved in the incident.

In contrast, let's modify the initial size-up as follows: *"Dispatch from Engine 1, we are on-scene at 1420 Elm Street with fire visible from the rear portion of a single-story, single-family dwelling. Engine 1 will be command and I am taking an attack line inside the structure. Additionally, I want the next arriving company to provide ventilation operations for my company."* In this size-up, the initial officer has clarified that Engine 1 is in command of this incident, the actions of Engine 1, and the preferred actions of the next arriving company.

This simple approach has eliminated any fireground freelancing between the first two arriving companies and has both companies working together. If the initial officer does not adhere to the principle of assigning specific tasks to responding companies (when necessary), additional arriving companies should ask for specific instructions prior to arriving at an incident. This will force the initial officer to communicate what actions are necessary by later arriving companies. As an example, assume you are the officer of the second-in arriving company (which for this discussion is a truck company). Without any specific assigned task, try the following radio message: *"Engine 1 from Truck 2, we are several blocks away. How can we help you?"*

An initial size-up should include the address (location) of incident, observed conditions, initial action being taken, instructions to incoming resources (if appropriate), and request for additional resources (if appropriate). To simplify the initial size-up process, let's combine the preceding considerations into an initial size-up pre-plan as follows:

"DISPATCH FROM ___1___, I AM ON-SCENE AT ___2___, I HAVE A ___3___ _4-5-6-7___."

1. Your designation. You should be familiar with this one.

2. Location of incident. This ensures all resources are responding to the same address, and in combination with the previous consideration, gets you talking without contemplating what to say as the incident burns down around you.

3. Accurate description. Describe what you see, not what you suppose. This allows incoming resources to start their mental size-up. An ineffective initial size-up is illustrated by the following example: *"Dispatch from E-1, we are on-scene and have smoke showing."* Is there a difference between food on the stove and a well-involved structure with exposures

because both of these examples have smoke showing? Of course. So, describe what you see and be relatively descriptive. If you are the initial company and have insufficient information for an initial size-up, try the following: *"Dispatch from E-1, we are on-scene at 123 Highland Parkway. I am unsure of the extent of this incident. I will give an additional size-up when I see what I have."* When speaking on the radio, don't let your mouth fill in the blanks your eyes don't see. If you have additional resources breathing down your neck and you are unsure about incident needs, additional resource actions, and placement, tell the responding additional resources to stage back from the incident until you have a chance to evaluate the needs of the incident. Don't let the speed and excitement of incoming resources force you into a corner, or, enhance fireground freelancing.

4. (optional) Your actions. If necessary, describe your initial actions.

5. (optional) Resource placement and actions. If necessary, give incoming resources specific assignments (and that includes the initial truck company).

6. (optional) Additional resources. If you need (or think you will need) additional resources, ask. Be proactive, not reactive.

7. (optional) Modify your original size-up when you have seen more. Update as necessary.

This simple initial size-up pre-plan can be used for any type of incident, so let's consider a simple example. Assume you are E-1 responding to a reported structure fire at 120 South Grant Avenue, and the initial assignment consists of two engines and one truck company. As the initial company approaching the incident, you observe a two-story, single-family dwelling with fire showing from a window on the second floor. Using the preceding initial size-up example, your size-up would be as follows: *"Dispatch from Engine 1 (1), we are on-scene at 120 South Grant Avenue (2), and have a two-story, single-family dwelling with fire showing from a window on the second floor* **(3)**. *E-1 will be taking an attack line inside the front door* **(4)**. *I want the next engine company to provide me with a back-up line, and I want the first truck company to perform ventilation operations* **(5)**. *Additionally, I want another engine company to respond to this incident* **(6)**." Remember, the initial size-up is your opportunity to ensure that additional resources can follow in the footprints that you left in the sand.

Efficient apparatus placement

After the first-in officer assumes command of an incident, conducts a size-up, and determines a course of action, it is time to issue specific instructions to company personnel at an incident as well as incoming companies, if necessary. Let's combine this example with the preceding examples and apply them to figure 2–4. This incident was a four-story apartment complex with a fire in one unit on the fourth floor.

Figure. 2–4. Proper spotting techniques can result in the correct placement of companies at an incident.

Notice the engine companies (in front of the structure) have left the front of the structure open for the later arriving truck company to provide quick access to the fourth floor and/or roof. The apparatus placement in this incident did not happen by accident, so let's analyze two different size-ups that can minimize or maximize the effectiveness of this fireground operation.

Engine 1 arrives on-scene, stops in front of the apartment complex, and gives the following size-up: *"Dispatch from Engine 1, we are on-scene at 320 Grant Avenue, with fire showing from an apartment on the fourth floor of a four-story apartment complex. Engine 1 is taking an attack line to the fourth floor."* In this scenario, the first-in officer has enhanced the possibility of the next arriving engine company also spotting to the front of this structure, and in combination with the congested street, possibly preventing the first-in truck company from using an aerial device for quick access to the fourth floor and/or roof. Consider this well-known question: "How many apparatus/vehicles does it take to prevent a truck company from securing an effective spot?" The answer? One.

Now consider the next size-up, which was the actual size-up and resulted in the apparatus placement as illustrated: *"Dispatch from Engine 1, we are on-scene at 320 Grant Avenue with fire showing from an apartment on the fourth floor of a four-story apartment complex. Engine 1 will be command and is taking an attack line to the fourth floor. Additionally, I want the next in engine company to provide a back-up line on the fourth floor, and the first-in truck company to ladder the roof in the front of the building, provide ventilation operations, and check for extension in the attic."*

Notice the first-in engine company did not spot directly in front of the building, and the first-in officer also communicated that Engine 1 is in command of this incident, gave specific directions for the second-in engine company, and relayed specific directions for the first-in truck company.

It is a fact that the proper placement of apparatus will provide the opportunity for initial companies to use their abilities in the most effective manner. Therefore, if the placement of initial companies (which are normally engine companies) is correct, other apparatus (such as truck companies, heavy rescues, and so on) placement can also be correct. The first-in officer must assume the responsibility for considering *both* engine and truck company responsibilities (figure 2–5). If the first-in truck cannot properly spot to a building due to poor engine company placement, it is normally the fault of the first-in officer. If the placement of the initial assignment at an incident is not correct, it is impossible to call a time-out, regroup, and start over again.

Figure 2–5. The first-in officer must assume the responsibility for considering initial engine and truck company placement and responsibilities.

The placement of engine companies normally offers more flexibility than the placement of truck companies. This simple but often forgotten principle is based on the fact that it is possible to pull another length of hose, but impossible to stretch a ladder. If a truck company with a 75-foot aerial device is needed to rescue a trapped occupant on the fourth floor of a four-story building and is forced to spot 78 feet from the fourth floor by one engine company or chief officer, what has been accomplished? Depending on the needs of an incident, it may be necessary for initial engine companies to give the best spot to an incoming truck company and pull additional hose to reach an objective.

Additionally, the proper placement of companies will maximize fireground capabilities, whereas the improper placement of companies will minimize capabilities and often create an unsatisfactory condition that is unnecessarily tolerated for the balance of an incident.

The initial companies responding to an incident (specifically the first two companies) should consider approaching an incident from opposite directions. This will enable the responding companies to see more sides of a building, and develop a primary and alternative source of water. Too, the second

company will not drive over a supply line from the initial company in the street. Opposing companies can easily park in line, leaving more of the street open for other apparatus, such as truck companies.

In figure 2–6, Engine 11 was the first-in company, and Engine 15 (partially hidden by the police car) was second-in. Engine 11 has a source of water from a hydrant to the right of the photo, and observed the right side (D) and front (A) of the building before stopping. Engine 15 has a source of water from the left side of the photo (alternative supply), observed the *left side* (B) and front of the building (A) before stopping, did not have to drive over or around the supply line for Engine 11, and parked in-line to Engine 11, leaving half of the street open for other companies, such as a truck company.

Figure 2–6. The initial companies to an incident should consider approaching from opposing directions. (LAFD file photo)

Apparatus arriving after the initial company (or companies) should stay back in an uncommitted position unless a specific task has been assigned. Simply stated, responding companies that do not have an assigned task should stop at least a block away from an incident. Adherence to this principle will minimize incident congestion and maximize scene access and spotting opportunities for additional companies (if necessary). Let's look at two different scenarios to expand on this concept.

Nothing showing. When nothing is showing as initial companies approach an incident, the emphasis should focus on initial companies that are necessary to determine the status of the incident. As an example, let's assume the one engine and one truck in figure 2–7 responded to a reported fire in apartment 411 in the four-story condominium as pictured.

Figure 2–7. Apparatus arriving after the initial company should stay back in an uncommitted position unless a specific task has been assigned.

Utilizing the principles we have been discussing:

- The initial engine company would spot for possible future operations, and conducting an initial investigation.

- Additional engine companies spotting at the nearest intersection. This will enhance room in front of this incident, the ability to bring a source of water (if necessary), and allow uncommitted companies to quickly access an appropriate location for the incident.

- The truck (which was the second-in company) has spotted to the corner of the four-story condominium, is available for above-ground operations (if necessary), and has left the street open for the placement of the next-in engine (if necessary).

Working incident. If a working incident is encountered, the assignment of specific tasks by the incident commander should be able to dictate proper apparatus placement. If specific tasks have not been delegated, responding companies should ask the incident commander for an assignment prior to arriving on scene. As an example, assume you are the officer on a truck company that will be the first arriving truck company to a structure fire, the first-in engine officer

has only assigned responsibilities to the first two arriving engine companies, and has neglected to assign duties to your first arriving truck company. As you approach the incident, ask the IC the following question: "Engine 1 from Truck 1, we are several blocks away, how can we help you?"

In summary, additional companies other than the initial company responding to an incident should not drive into an incident and look for something to do without specific instructions. Adherence to this principle will eliminate a noteworthy percentage of fireground freelancing. Remember that strategic and tactical options are minimized if fireground congestion eliminates or detracts from proper apparatus placement.

Apparatus Placement Priorities

Because truck company placement can be enhanced or limited by numerous factors, let's consider key placement considerations. Assuming a responding truck company can effectively spot at an incident, placement is normally dictated by three factors—incident priorities, initial priorities, and functional capabilities.

Incident priorities

Does the immediate and future incident priorities in concert with the type of apparatus being utilized require an elevated master stream appliance? Or require access to the upper floors or roof of a multistory building? Does it operate as an initial engine company (if a quint), rescue a trapped occupant, or is it simply used as a resource for tools and equipment? Remember, it is the responsibility of the first arriving officer to consider *both* engine and truck company responsibilities, and ensure that the proper placement of apparatus will provide the opportunity to use their capabilities in the most effective manner. For simplicity, most incidents can be seen as either incidents beyond your immediate control, and incidents within your immediate control.

Incidents beyond your immediate control may happen only once or twice in your career and can provide an instantaneous wake-up call. A classic example occurred in Fresno, California, early on a Saturday morning when a Lear jet crashed into one of the major downtown intersections, killing all occupants on the aircraft and igniting several blocks of commercial and residential occupancies. If you would have been first-in, what would have been your course of action (other than wondering why you were first-in)? The point is, this incident was way beyond the control of the initial company regardless of the type of company. Therefore, if an incident significantly exceeds the capability of an initial company, the primary focus is to begin to initiate scene stability, stage incoming resources back from the incident until initial priorities are determined, conduct a quick size-up to determine the extent and priorities of the incident, and ensure additional resources are assigned the proper responsibility to start mitigation of the incident.

Most incidents that are encountered on a frequent basis, however, can normally be classified as either within the control of initial resources or somewhat within the control of initial resources (needing some additional assistance). A simple method to use in quickly determining initial priorities is the *Rule of 4* approach, which stands for access, ventilation, search, water. To expand on this concept, assume you are a company officer on a truck company and are responding to a reported structure fire. As you approach the incident, start your size-up by considering the factors discussed in the following sections.

Building type. You need to determine if this is a commercial or residential building. From a basic viewpoint, a residential building will have a high life hazard from about 9 p.m. to 7 a.m. Most fatalities occur in residential structures because the occupants (at night) are non-ambulatory. Few fatalities occur in commercial buildings because the occupants are ambulatory and are probably waiting outside the building when the first apparatus pulls up. Additionally, determine if the commercial building is vacant or occupied; if it is past normal business hours or over a weekend; and so on.

Smoke or flame. Determine if anything is showing. NFPA statistics indicate that if nothing is showing, it is normally a false alarm. If something is showing, determine to what extent.

Initial priorities

Initial priorities can be easily separated into four main concerns (Rule of 4): access, ventilation, search (or rescue), water:

- **Access:** The best engine company in the world is useless unless they can put the wet stuff on the red stuff. This may involve forcible entry, ground ladders, or possibly an aerial device. As an example, look at figure 2–8. This was a well-developed fire on the sixth floor of an eight-story commercial. The time of the incident was 2 a.m., and the windows were dark, indicating the building was closed and/or vacant. Applying the priority of access to this incident was easy. Before an attack was initiated on the fire, it was necessary to first put suppression personnel up to the fire via an aerial device as pictured. An additional example are fires in residential occupancies that require the front door to be forced prior to initiating interior operations.

Figure 2–8. In this incident, access to the fire was the first priority. (LAFD file photo)

- **Ventilation:** When a structure obviously needs to be ventilated, this operation should be quickly initiated to allow the safe interior access (by attack and/or search teams). This minimizes the possible chance of a flashover occurring.

- **Search (or rescue):** If it is known that a search or a rescue is necessary, these operations can trump other fireground priorities. As an example, look at the multistory hotel in figure 2–9. By looking closely, five victims are visible in the windows. In this incident, what is the first priority? Rescue.

Figure 2–9. When search or rescue operations are necessary, these operations can quickly become a first priority. (LAFD file photo)

- **Water.** Some incidents require elevated master streams (from a truck company perspective), so the implementation of master streams becomes the priority.

Functional capabilities

When truck and/or quint companies respond to an incident, priorities in concert with the functional capability of on-scene apparatus determine the following placement:

Truck company first on-scene. If nothing is showing and an aerial device will not be necessary, spot the truck away from the building to allow the first-arriving engine to spot for potential use of hose lines.

If fire is showing and an aerial device is necessary, spot for appropriate aerial device use. The first-arriving engine company can pull additional hose if necessary. (Definition: *the address of the building belongs to a truck company.*)

Quint first on-scene. Although the initial placement of an engine or truck company (single-purpose apparatus) can be relatively uncomplicated, the initial placement of first-in quint apparatus (which is a multi-purpose apparatus) can be challenging. When quint apparatus are first on-scene and nothing is showing (meaning an aerial device is not initially necessary), spot and operate as an initial engine company. If an aerial device is necessary *and later arriving apparatus have an aerial device*, spot as an initial engine company, but leave room for later-arriving aerial device apparatus (depending on the length of time it will take for later-arriving aerial devices to arrive). If the estimated time of arrival (ETA) of a later arriving aerial device is unacceptable, then the initial quint must spot so the aerial device can be utilized.

If later arriving apparatus *do not have an aerial device* and an aerial device is not initially necessary, spot as an initial engine company. If an aerial device is necessary, spot so the aerial device can be utilized. This concept places the primary focus on positioning the quint to use the aerial device, which may necessitate positioning the apparatus out of a prime position for potential use of hose lines.

If fire is showing, and later-arriving companies *have an aerial device*, spot as an initial engine company, leaving appropriate room for the company with the aerial device if aerial device operations are necessary (depending on the ETA of later-arriving aerial device or devices).

If later-arriving apparatus *do not have an aerial device* and an aerial device is not necessary, spot as an initial engine company. If an aerial device is necessary, spot as a truck company, but operate as an initial engine company. This concept allows a quint to provide an initial attack line for the fire and an aerial device that can be used by later arriving companies.

As an example, assume a working fire in the interior of the structure in figure 2–10, and fire attack in concert with roof ventilation is necessary. The first-arriving company (quint 3606) has been positioned so the aerial device can be used.

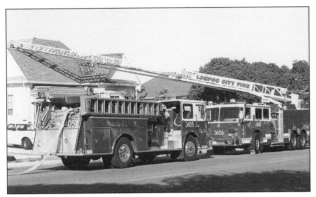

Figure 2–10. The first arriving company (quint 3606) has spotted as a truck company but will operate as an initial engine company.

With a staffing level of three, the officer can make a quick size-up by looking at the sides/back of the building (or other appropriate size-up measures). While the officer is conducting an initial size-up, the driver can quickly raise the aerial ladder to the roof. While the officer and driver are busy with the aforementioned procedures, the firefighter can advance an attack line to the front door of the building. At this point the driver can then provide proper water pressure to the officer and firefighter who will advance the initial attack line. The next arriving company (3405) can use the aerial device for roof ventilation operations.

Operational Considerations

The actual placement of an aerial device to a specific objective is dependent upon numerous factors, which are primarily based on the expertise of the operator and the conditions that are encountered at each incident. Therefore, it is imperative that every driver/operator understand the capabilities (from a positive and negative viewpoint) of their particular apparatus, and are also familiar with the

constraints of their district (this means pre-planning and practice). Before we consider general operational considerations, let's look at two specific considerations that must be initially considered when an aerial device will be raised—maneuverability and reaching an objective.

Maneuverability

This often overlooked factor is based on the principle that unless an aerial device is able to be positioned in the appropriate location, effectiveness is compromised or negated. Remember that increased aerial device capabilities have also resulted in larger apparatus. As a result, maneuverability (turning radius) is dependent on different types of apparatus (for example, tillered apparatus, single-chassis apparatus with a single rear-axle, large single-chassis apparatus with tandem rear-axles, apparatus with steerable rear wheels, and so on), and the space necessary to spot modern aerial apparatus has also increased. As an example, the quint in figure 2–11 has utilized the entire width of this street to spot for aerial platform use. Considerations such as this example may require that larger aerial device apparatus respond ahead of other apparatus, and/or later arriving companies respond from the opposite direction (as illustrated), or engine companies allow a truck to spot first.

Figure 2–11. Large apparatus can utilize the width of a street when spotting for aerial device operations.

Additionally, to utilize the full capabilities of an aerial apparatus, an operator must be familiar with apparatus placement in challenging conditions such as overhead obstructions (trees and shrub); civilian vehicles that hinder prime spotting locations; the need to safely operate an aerial device while the apparatus is "short jacked"; the ability to safely operate an aerial device while near its safe operating limits; and if an aerial apparatus can be moved while the aerial device is raised, and if so, what are its operational parameters allowed by the manufacturer. These considerations are prime reasons why an aerial operator should consistently practice spotting, maneuvering, and raising an aerial device in challenging conditions and not just parking lots and wide streets.

Reaching an objective

Several key factors that affect the capability of an aerial device to effectively reach an objective are height of objective, set back, length of aerial device, presence of hazards, and apparatus constraints.

Height of objective and set-back. As the height of an objective and the set-back from an objective increases, so does the required length of an aerial device. Set-back is defined as the cumulative distance from the objective to the curb plus any additional distance an aerial apparatus must spot from the curb (see figure 2–12).

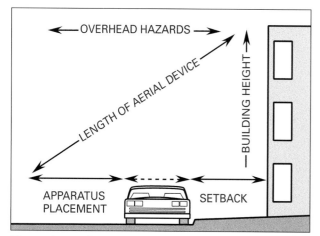

Figure 2–12. Set-back, height of the objective, length of aerial device, and the presence of hazards will affect the ability of an aerial device to reach an objective.

Although evaluating objective and set-back takes practice, general guidelines are as follows:

- Seventy-five-foot aerial devices can normally reach the roof of a four-story building with a moderate set-back. A moderate set-back consists of a sidewalk and parked vehicle with a minimal distance from the curb to the objective. However, if the set-back significantly increases (parkway, lawn, other apparatus, civilian vehicles, and so on), and/or a five- to six-story building is encountered, a 75-foot aerial device will not be adequate for roof access.

- One-hundred-foot (or over) aerial devices can normally reach the roofs of six- to eight-story buildings depending on the amount of set-back encountered.

- Aerial devices over 100 feet will provide additional reach depending on the additional length over 100 feet.

Hazards. In all cases, the presence of hazards such as power lines, civilian vehicles (the term *vehicles* can also encompass fire department vehicles as well as chief officer and police officer vehicles), and vegetation must also be considered. Any of these considerations may negate the safe operation of any aerial device.

Apparatus constraints. Depending on the type of apparatus, the way an aerial device is raised can reduce the reach of the aerial device. As an example, if an aerial device is raised perpendicular to an apparatus, the length of reach is not compromised. Conversely, if an aerial device is raised in line over the length of an apparatus (such as a rear mount raising the aerial device over the cab), the horizontal reach of the aerial device will be reduced by the length of the apparatus.

An aerial device is limited in its reach by the length of the aerial device and the positioning of the apparatus. Therefore, if an aerial device must be raised to an objective and is in close proximity to another company, position the aerial device inside of the other apparatus if maximum aerial reach is necessary, or outside if maximum aerial reach is unnecessary as in figure 2–13.

Figure 2–13. Spot an aerial device outside of other apparatus if maximum reach is unnecessary.

the retracted length of a ladder tower/tower ladder is the boom/aerial plus the platform plus any exterior appendages/fittings.

The *target* (or *scrub*) area is commonly defined as the area that can be reached by an aerial device (figure 2–14). However, if it is necessary to place an aerial device close to an objective, a telescoping device can be more effective in reaching a larger accessible area than an articulating device. Regardless of the type of aerial device, placing the device too close to an objective can limit the accessible area and minimize flexibility.

General Considerations

In combination with the preceding factors, let's review general operational considerations.

Understand definitions

As a general definition, *aerial device* is defined as an aerial ladder, tower ladder, or ladder tower. *Aerial ladder* is defined as an aerial ladder only, *tower ladder* is defined as a platform attached to a telescoping boom with a sideless escape ladder, and a *ladder tower* is defined as a platform attached to the end of an aerial ladder that can also be used for climbing.

Understand aerial devices

An aerial device is often only considered for use on structures of three or more floors. However, aerial devices can be raised by one person, and are the longest, strongest, widest, and heaviest ladder in our ladder arsenal. Therefore, aerial devices can often be effectively utilized for structures of less than three stories (and that includes some large single-story structures).

If an aerial device is spotted closer than the retracted length of the aerial ladder or ladder platform, the ladder platform may not be able to reach an objective that is perpendicular to the apparatus. As an example, assume the retracted length of an aerial ladder is 35 feet. The apparatus should be spotted more than 35 feet away from an objective if an unobstructed 180-degree sweep of the objective is desired. Additionally, remember that

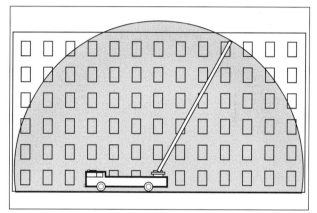

Figure 2–14. The target or scrub area is the area that can be reached by an aerial device.

Spotting techniques

Proper spotting techniques should still be adhered to (even for false alarms) to reinforce and refresh SOPs. Stated from another perspective, correct practice in non-emergency or false alarm-type incidents allows personnel to focus on objectives during emergency incidents, as the basics should be automatic.

Slow down

Although previously mentioned, the importance of slowing down when approaching an incident needs repeating. This simple consideration allows more time to conduct a size-up and determine appropriate apparatus placement based on incident factors. Remember that when an aerial apparatus is stopped, it rarely has the capability to reposition (if necessary) during the later stages of an incident. So, slow down and maximize fireground flexibility.

Placement

When considering the placement of the initial aerial apparatus to a building, common recommendations are to place the apparatus in front of a building for two basic reasons: for a specific requirement such as a known or visible rescue, firefighter access to a specific location, and so on; and to position the apparatus in case it will be needed during the incident. Obviously this second viewpoint can result in an aerial apparatus not being in the proper location to use an aerial device as an incident develops, and it can unnecessarily add to the congestion in front of a building. Therefore, unless an aerial device is needed in front of a building for a specific reason, the corners of a building can offer certain advantages, such as reaching multiple sides of a building (figure 2–15). Always remember to maximize fireground flexibility.

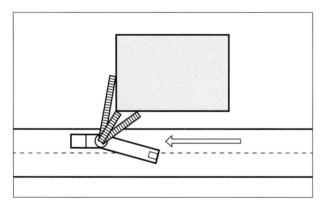

Figure 2-15. When spotted to a corner, an aerial device will be able to reach two sides of a building.

Placing an aerial device at corners. The corners of a building are the strong portion of a building, and are normally out of potential collapse zones. Placing an aerial device to a corner will normally not place the aerial device over horizontal openings, such as windows. In figure 2–16, notice fire extending from the windows on the front of the building, yet the aerial ladder still offers a safe means of exit from the roof.

Figure 2-16. Placing an aerial device to a corner will not normally place the device over horizontal openings, which can become an opening for extension of fire. (LAFD file photo)

Spotting to a far corner. By driving past a building and spotting to the far corner, company personnel will see three sides of a building (figure 2–17) and leave the front of the building open for other companies (such as engine companies, incident commanders, etc.). If the far corner is not appropriate, consider the near corner (the corner that is first passed by the apparatus) as an alternative spotting location.

Spotting to a corner also opens the front of a building for placement of other companies such as engine companies, chief officers, and so on. If necessary, an aerial apparatus could be moved to other locations as incident needs develop, as opposed to an aerial apparatus that is parked in front of a building and blocked by other apparatus.

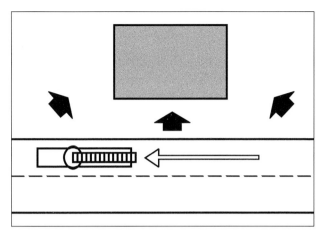

Figure 2–17. By driving past a building and spotting to the far corner, company personnel can see three sides of a building.

An additional consideration is spotting the apparatus to a corner that would also allow using the aerial device as a potential defensive-exposure protection tool. However, proper personnel access/egress routes should take first priority.

Approach from opposite directions

If two aerial trucks are responding to the same building, it can be advantageous to have both trucks approach the incident from opposite directions. This can result in more sides of the building being seen by the truck officers, not having to drive by initial engine company congestion in the front of the building, the trucks placed at the opposing corners of the building, and so on.

Reposition considerations

Generally, an aerial apparatus can pull forward to reposition if necessary. However, they can rarely back up to reposition.

Apparatus arrival

The earlier an aerial apparatus arrives at the scene of a structure fire, the better. Normally, an aerial apparatus should be no later than the second apparatus to arrive at the front of the fire building. If a truck and engine are housed in the same station and simultaneously respond to the same structure fire, the truck should respond first.

A successful beginning to fireground operations is dependent on the proper placement of the first-arriving apparatus from two basic perspectives. As a starting point for common residential structure fires, the minimum response should be two engines and one aerial device (even for simple residential structure fires). Ideally, the aerial apparatus should be the second apparatus to arrive, which in some cases, means the second-arriving engine must wait until the aerial device is properly positioned, if necessary. Also, unless an aerial apparatus is positioned in an appropriate location, its effectiveness is compromised or nullified. Increased aerial device capabilities have correspondingly resulted in larger apparatus. As a result, vehicle maneuverability has been compromised, and the space required to spot modern aerial apparatus has markedly increased. Depending on the incident, space constraints, and type of aerial apparatus, aerial apparatus may need to respond ahead of other apparatus and/or later-arriving companies may be forced to respond from the opposite direction (refer to figure 2–11).

Ladder and platform placement

Correct placement will also square a ladder and platform to an objective. Generally, aerial apparatus should be positioned as perpendicular to an objective as possible. This will allow both beams of an aerial device (either ladder or ladder tower) to simultaneously contact an objective (if the ladder happens to sag downward) and minimize potential ladder torsional stress. Some departments paint a mark on the turntable and the non-moveable portion of the apparatus by the turntable when the aerial is perpendicular to the apparatus. Therefore, when the marks are aligned, the aerial will be perpendicular to the apparatus and objective. This type of reference will be accurate only if the building and apparatus are parallel to each other (which is often the case).

However, due to the configuration of platforms (normally entered or exited from the front-corners, and master stream appliances are commonly mounted in the center-front portion of a platform as in figure 2–18), the placement of platforms to an objective is dependent on the intended use.

Depending on the sweep and desired angle of a master stream into a building, a platform is normally

Figure 2–18. Master stream appliances are commonly mounted in the center front portion of a platform, and exit-entry is accomplished at the corners of a platform.

Figure 2–19. A platform can be angled into a window with the bottom portion of the window level with the floor of the platform.

centered or squared to an objective and placed just below a window or bottom portion of a doorway. This will allow the stream to be angled upwards towards a ceiling, or horizontal if maximum reach is desired.

If a rescue is necessary from a window, a platform can either be positioned just below the window to facilitate ingress into the platform, or at an angle to the window and with the floor of the platform level with the bottom portion of the window to facilitate ingress into the platform (figure 2–19). This requires the turntable to be spotted just past or just before the objective.

However, if the openable portions open outward, be sure to provide enough space to open the doors. If a platform is positioned just below a window sill for a rescue, the presence of the platform controls and other stationary objects may complicate this operation.

If a rescue is necessary from a roof (or other flat area), the front-corner of a platform should be positioned at the appropriate angle to the victim so the openable portions of the platform will be in the best location to the intended victim for easy ingress into the platform.

When placing an aerial ladder to an objective, it should be positioned as perpendicular to the objective as possible. This will allow both base rails to equally rest on the objective if the ladder sags into the objective. In this configuration an aerial ladder is considered a supported ladder. However, if an aerial ladder is placed at an angle to an objective,

only one base rail will rest on the objective. In this configuration an aerial ladder is considered an unsupported ladder. It is also inherently weak due to torsional stress that can be created by ascending or descending personnel who will weight the unsupported base rail. This can cause the following considerations:

- Force the unsupported base rail downward, torsionally twisting the aerial ladder and potentially causing failure of the aerial ladder. The greater the angle of the aerial ladder to (or away from) the objective in concert with a given weight on an aerial ladder, the more critical this consideration becomes.

- An aerial ladder can slide against an objective towards the unsupported base rail creating a hazardous condition for personnel and the aerial ladder.

It's important to understand that an unsupported base rail is one reason why some manufacturers recommend operating their ladders in an unsupported condition, as some aerial ladders are stronger in the unsupported position as compared to supporting one base rail. When comparing an unsupported aerial ladder (both base rails are unsupported) versus an unsupported base rail, access on or off an unsupported aerial ladder is not perfect; however, the chance of structural damage to the ladder is minimal compared to personnel on a ladder with an unsupported base rail.

Figure 2–20. The center line of this aerial device is illustrated by the firefighter.

Correctly centering a ladder or platform to an objective can be accomplished by aligning the center line of a turntable with the center of an objective. The center line of a turntable is the hinge point of the bottom portion of an aerial device (firefighter illustrating in figure 2–20). In reality, the focus of this consideration requires spotting a turntable to an objective, not the apparatus.

Unfortunately, the increased size of modern apparatus and enclosed cabs have made this task more difficult for apparatus drivers. Because it can be challenging for a driver to consistently and correctly position an aerial device to various types of objectives, teamwork and training can enhance this three-step process:

1. Stop the apparatus short of the objective.

2. Use a company member to dismount and quickly draw an imaginary center line from the objective to a spot on the sidewalk or ground, and stand on that spot.

3. As the apparatus slowly drives by the member, the member motions the driver to stop when the turntable center line intersects the imaginary line (or the members position).

This process only takes a few seconds and results in a correct spot. Don't let an inconvenience (poor spot) become a problem or a compromise for the duration of an incident.

Spotting for aerial devices

When spotting for aerial device operations, the following minimum considerations will affect stability.

Angle of inclination. As the angle between an aerial device and the ground increases, so does the strength of an aerial device because maximum strength is obtained in the vertical position. In addition to strength, consider proper climbing angle. The preferred climbing angle for ladders (aerial and ground) is normally 70 to 80 degrees. However, it is easier to climb an aerial device positioned at 50 to 60 degrees.

Apparatus limitations. It is mandatory for apparatus operators to be thoroughly familiar with the strengths and limitations of their regularly assigned apparatus and apparatus they may be required to operate (for example, reserve, an apparatus assigned to another company, and so on). Older apparatus, particularly apparatus built before 1991, do not posses the same operational strengths and limitations as newer apparatus.

As a general guide, refer to figure 2–21. Apparatus achieve minimum stability if an aerial device is positioned 90 degrees to the apparatus (A), and as the aerial device is moved towards an in-line

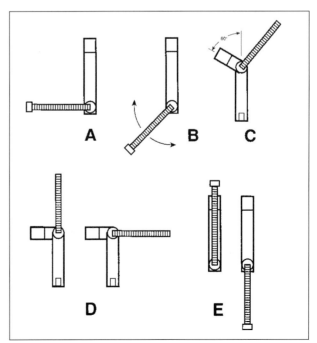

Figure 2–21. Depending on the type of aerial device and required operation, always consider how a spot will affect apparatus stability.

position, stability is increased (B). Maximum stability is achieved when an aerial device is positioned in-line with the apparatus (E). If the in-line position is not possible, tillered apparatus can enhance stability if the apparatus is positioned to an objective with a 60 degree outside jackknife (C), or 90 degree outside jackknife with the aerial device in line with the cab or trailer (D).

Extension. As the extension of an aerial device is increased, strength is decreased.

Support. It is advantageous for a ladder to be perpendicular to an objective because if the ladder settles into the objective, both beams will rest on the objective eliminating the possibility of the ladder twisting if a vertical-force weight is applied to the ladder. This is why manufacturers prefer an unsupported ladder over a ladder that is supported by only one beam.

Ground surface. When positioning apparatus for aerial device operations, always consider the surface that will support the apparatus. There are differences among concrete, asphalt in the winter as opposed to the summer, ice and snow as opposed to dry dirt, dirt as opposed to mud (remember that when operating master stream appliances, dirt often turns into mud), street manhole covers, and other similar considerations. Additionally, consider that roadways are *crowned* to force water to drain to the outside of the roadway. If spotting next to a curb, the apparatus will be tilted at an angle (figure 2-22), and will tend to favor or place more weight on the inside ground jack(s). For this reason, place inside ground jacks on the strongest surface possible (for

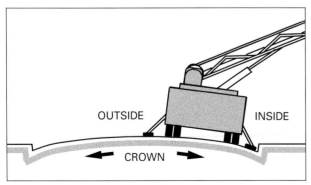

Figure 2-22. The crown of a roadway can tilt an apparatus and place additional weight on the inside ground jack(s).

example, concrete curbs can offer a superior footing surface as compared to an asphalt roadway in the summer heat).

Ground jacks

Ground jacks are often referred to as stabilizers or outriggers, and provide support for an aerial apparatus when an aerial device is in use. Ground jacks should be used when the footing is solid. If solid footing is not available, it may be best to depend upon the wheels of the apparatus for support, and not to depend upon ground jacks that may suddenly shift. Any movement of an aerial device caused by a shift when the jacks are on solid ground will be gradual and is an indication that the load or extension of the aerial device should be limited. An aerial operator can easily monitor the weight transfer from the outboard jacks (see figure 2–13) to the inboard jacks by periodically glancing (from the turntable) at the outboard jacks. If it is visually apparent during an aerial operation that the outboard jacks have loosened or are even slightly lifted off the ground, it should be evident that only the inboard jacks are now supporting the apparatus, and as a result, it may be necessary to limit any further extension and/or reducing the angle of inclination. Additionally, any movement of the aerial device caused by sudden sinking of the ground jacks when they are placed on ground that is not solid may cause the resultant magnified force at the end of the aerial device to cause a serious problem. On uneven terrain, it is generally best to set the low-side jack first. On level ground, set the off-side jack first. Jacks should normally be set to the point where they touch the ground with some pressure. If the ground has some irregularities such as gravel or a thin layer of dirt, force the jacks down to firm footing. In most cases, jack pads can dramatically increase the surface area of a ground jack and maximize support of the ground jack. However, particular attention should be focused on using jack pads on notable hills for obvious reasons. Some manufacturers restrict the use of jack pads on hills that exceed 6%. Always be sure to follow the instructions in your specific manufacturers operational procedures regarding ground jack and jack pad operations.

If a truck company is staffed with an officer, driver, and firefighter, the firefighter should sit on the driver's side of the apparatus. When the apparatus stops to position for aerial operations, the officer can exit the apparatus and place the officer-side jack pad(s) for placement of the ground jacks. Simultaneously, the firefighter can also exit the apparatus from the driver side and place the opposing-side jack pad(s) for placement of the ground jacks. This operation can be dramatically simplified by welding small-angle brackets on the jack pads (figure 2–23). This allows the pads to be quickly inserted onto the jacks. As a result, the jack pads will always be with a jack regardless of the distance from the apparatus. Additionally, always put the "handhold" on the jack pads towards the apparatus to eliminate the potential of tripping by fireground personnel.

Figure 2–23. Brackets affixed to jack pads can allow the pads to be quickly inserted onto the jacks. As a result, the pads will always be with the jacks regardless of their final positioning.

Short jack. It cannot be emphasized enough that aerial operators are familiar with the limitations imposed by the manufacturer of their apparatus when it is necessary to "short jack" the ground jacks. Normally, short jack means the jacks (typically on one side) have not been able to be fully extended outward from the apparatus due to some type of obstruction or hazard. As an example, some safety devices prohibit rotating an aerial device over the side that is short jacked.

Chock blocks. Chock blocks should be placed as soon as an apparatus is in position for aerial device operations. On level terrain, chock the back and

front of the dual wheels (or front wheels if the duals have been raised off the ground as per manufacturer recommendations) on the side the aerial device will be raised. On an incline, chock the down-grade side of the dual wheels (again, the front wheels if the duals have been raised off the ground). Chock blocks should always be placed before the ground jacks are set. Be aware that because of the advent of maxi-brakes on apparatus, chock blocks have changed from primarily preventing apparatus from inadvertently moving to distributing load carrying capacity across more tire surface area. Again, always be sure to follow the instructions in your specific manufacturer's operational procedures.

General safety

When spotting to a structure with the target area blocked to aerial device access (for example, electrical wires, trees, vehicles, and so on), spot next to a curb and raise the aerial device under the obstruction and alongside the structure. Although this spot may not be in the desired location, it can provide access (with an unsupported aerial device) to a potion of a building as a secondary option. Remember that placing apparatus under electrical wires is not a recommended option.

Electrical wires in the immediate vicinity of an aerial device pose an extreme hazard to safe operations. Preventing contact with the wires is the only practical safeguard against this hazard as there is no substitute for an alert operator who makes absolutely certain the apparatus does not provide a ground for electricity. Always consider all wires as charged wires, even though they may not be charged. If it is necessary to place an aerial device in the vicinity of electrical wires, keep at least 15 feet away from the wires. This is about 5 feet more than most training materials advocate. However, an additional 5 feet not only allows an extra measure of safety, it can also allow room from the potential of arcing and for the aerial device (or personnel on the aerial device) to inadvertently move and still maintain a safe distance from the wires. On a typical city street with power poles on one side of the street that feed structures on both sides of the street, remember the wires will be higher on the side with the poles and normally a different elevation across the street from the poles.

Under normal conditions when ground jacks are placed on concrete or blacktop, they may not provide a ground for an electrically charged apparatus. An exception is when jacks are placed on water, a metal railroad rail, or other similar materials. Any contact with electrically charged wires will cause an apparatus and aerial device to become electrically charged. Under these circumstances, any person who is grounded and comes in contact with the apparatus can be electrocuted. If an aerial device inadvertently makes electrical contact, the operator should rotate or retract the aerial device out of the electrical contact. If personnel are on the aerial device, they should keep their hands and feet on the rubber covered rungs until the aerial device is moved to a safe location and/or the electrical circuit is eliminated. If it becomes necessary to get off the apparatus while it is electrically charged, personnel should jump clear and not make simultaneous contact with the ground and apparatus. Personnel on the ground must keep clear of electrically charged apparatus.

The following are general safety tips:

- If an aerial device has been used as an access tool, it should never be moved unless approved by the personnel who originally used it.

- When climbing an aerial ladder, it is normally best to look forward or up when ascending, and downward (so you know where your feet are going) when descending particularly when carrying objects such as tools, equipment, etc.

- If a hoseline is extended from the end of an aerial ladder or platform, that aerial device will not be able to move as long as it is being used as an elevated standpipe.

- The more stuff that is mounted inside a platform, the less usable space there is, particularly when affecting a rescue from a building. One of the worst examples is to mount a stokes basket in or on a platform (how often do you use stokes at incidents?). Remember that aerial devices are most often used for firefighter access, not elevated master streams or rescue of civilians. One tip that can pay big dividends in cold areas is to use a piece of wood to stand on in platforms to insulate your feet from a cold

metal floor of a platform, particularly in extended operations.

- Alleys are poor locations for aerial devices because they are historically narrow, overhead electrical wires are often present, and modern aerial apparatus are large. Most alleyways may not allow fully extending all ground jacks on some apparatus.

- When considering the space necessary to spot an aerial device, a tower ladder generally needs more space to place the jacks and stabilizers than an apparatus with an aerial ladder.

- Tower ladders can be more effective than aerial ladders when placed in front of a building. Although both devices can provide access to a building and rescue operations for trapped civilians, a tower ladder can be more easily used for defensive operations if an offensive operation is suddenly switched to a defensive operation due to the flexibility of its capability to quickly direct heavy streams to specific areas.

- Platforms (attached to telescoping booms) that are placed on roofs should be slightly raised before retracting the boom.

- The master stream appliance on platforms should be carried with the appliance swiveled back (if possible) so there are minimal protrusions from the front of the platform. This will assist in placing the platform closer to an objective for rescue, placing personnel, and so on.

- Aerial ladders with master stream appliances that can be pinned back one section should carry the appliance in the *rescue position* (appliance stays back one section when the ladder is extended). The primary reason for this consideration is that aerial ladders are most often used for access/egress of fireground personnel, not rescue operations.

- When an aerial ladder is changed from being able to extend with the master stream appliance at the end of the ladder to the rescue position, extreme care must be taken to ensure the method of pinning has been completed as per the recommendations of the manufacturer. When operating pinnable waterways, be aware that their use is controlled by mechanical (for example, electrical or hydraulic) or manual means. The mechanical pinnable waterways are relatively consistent in their operation. However, the manual pinnable waterways depend on the proper insertion of a metal pin in the correct location, or rotating a handle attached to a cam for proper engagement. If this vital step is not performed in the appropriate manner, the monitor/nozzle mount and the last section of the waterway pipe can disconnect from the fly section and become an airborne missile. For a closer look at an incident that involved this consideration, see Training Resources, The Importance of Training Highlighted in Aerial Waterway Fatality.

Figure 2–24. The addition of accessories at the end of this aerial ladder has significantly increased its overall width.

- When placing an aerial ladder to a window or fire escape balcony, place the top rung flush or even with the sill or railing. Although the tip of the ladder will not be visible inside the building, the ladder will not block or reduce the size of the window opening. Depending on the type of window, some window openings are not much larger than the width of modern aerial ladders. This is a reason why accessories (ladder controls, spotlights, and so on) at the end of an aerial ladder should be kept to a minimum because they tend to increase the overall width of an aerial ladder. Another reason why accessories can be detrimental is their added width to the end of an aerial ladder can block the view of the tip of the ladder from an operator on the turntable, forcing the operator to leave the control pedestal to obtain a better view of the aerial tip as it nears the objective. As an example, figure 2–24 illustrates the ladder tip of a 105-foot rear-mount aerial ladder. Notice the extensions (spot lights, remote ladder controls, and so on) on either side of the ladder that significantly increase the overall width of the end of this ladder. This width will increase the difficulty of placing this ladder in a window and placing the ladder tip at an angle to an objective.

- Another item worthy of debate is the status of an aerial operator when a aerial device is in a substantial position (for example, to the roof of a building), locked, and being used as an access and/or egress device. This debate takes on even more importance when many departments are operating with minimal staffing yet basic fireground operations still need to be accomplished. Although an aerial operator must stay on a turntable when a aerial device is in motion (or may be in motion), there are times when an aerial device is positioned on a building, locked into position, and used as a stationary access-egress ladder. In these cases, if a department can afford to have an aerial operator stand on a turntable while other important fireground operations are completed, there is no debate. However, in cases of minimal staffing, the importance of completing important fireground operations near the apparatus-aerial device by the aerial operator should be (at the least) considered.

- There are two basic modes of operating aerial devices from a turntable when considering the use of the raise/rotate/extend controls. One mode is to use the controls individually, and the other mode is to use the controls simultaneously. Applying these two modes to operating an aerial device would be as follows:

For the individual operation, the operator would first raise the aerial from the bed to the appropriate height, rotate the aerial towards the objective, extend the aerial to or over the objective, and then lower to the appropriate position/location. This operation is the simplest to master and is beneficial to operators who are occasional operators.

For a simultaneous operation, the operator would first begin to raise the aerial. As the aerial clears the bed of the apparatus, the operator rotates the aerial towards the objective while the aerial is still being raised to the appropriate height. When the aerial reaches the appropriate height, the operator begins to extend the aerial towards the objective while finishing with the rotation of the turntable. As the turntable reaches the desired rotation, and while the aerial is being extended, the operator can lower the aerial to the desired position/location. Although this operation takes practice and a knowledge of the operational capabilities of an aerial, this method is normally faster than the individual method and definitely presents a more professional appearance. If this last statement is in doubt, consider how operators of heavy machinery (for example, front loaders, backhoes, and so on) operate their controls and apparatus. In summary, practice.

- There is a tendency to raise an aerial device too high (sometimes close to vertical) when the aerial is raised out of the bed of the apparatus and toward an objective. Operators should practice raising the aerial out of the bed of the apparatus a distance that will allow the aerial to be rotated towards the objective and, when extended, will only be a short distance above the objective. This ability minimizes multiple lowering operations to lower the aerial close to an objective, and also allows an operator the ability to more easily gauge the distance of the ladder over an objective.

- If it is necessary to use an aerial device for an above-ground rescue operation from a building (such as windows), evaluate the following considerations:

  - If possible, position the aerial device on the windward side of the person(s) to be rescued. This will minimize the presence of smoke during the rescue operation.

  - Bring the aerial device *down* to the person(s) to be rescued, not up to the person(s).

    If an aerial device is moved downward toward a victim, the victim cannot jump towards the aerial device before it is properly positioned.

  - If multiple persons are encountered, persons trapped in windows contaminated with smoke and heat are normally a higher priority than persons trapped in windows with no visible heat or smoke, and persons closer to the fire area are normally a higher priority than persons away from a fire area.

- If an aerial device will not be used at an incident, try to park the apparatus in close proximity to the incident. This enhances apparatus security, access to tools and equipment, and can be used to block a street to provide a safer work area at an incident.

- Later-arriving apparatus should not park directly behind aerial apparatus due to the potential of blocking access to the ground ladders. Leave at least 30 feet and enough room alongside truck apparatus to allow space to open compartment doors and remove equipment. Tillered apparatus can slightly position the trailer outwards from the tractor, allowing the ground ladders to be easily removed. This also allows closer placement to the truck apparatus for later-arriving companies (figure 2–25).

- The strength of aerial devices primarily depends on design rather than upon the strength of materials and does not allow for side stress. Aerial devices are much stronger with a load applied perpendicular to the rungs than when a load is applied in a sideways direction. The rungs of an aerial device may be considered tilted when they are not parallel with the plane of the earth's surface. If an aerial device is being operated on level ground, the rungs will remain parallel with the earth's surface when raised in any position. However, when operating on an incline with the apparatus parallel to a curb, the rungs will be tilted when the aerial device is raised to any position not in-line with the apparatus. The tilt will increase as the aerial device is moved from an in-line position to a 90-degree angle from the in-line position with the apparatus (the tilt with a 45-degree angle will be one-half as great as the tilt at a 90-degree angle). If the rungs of an aerial device are tilted, a portion of the weight of the aerial device and any load acts sideways on the aerial device. This is caused by a shift in the center of gravity from the center of the rungs towards the low side of the tilt. The greater the tilt (in concert with any load), the greater the shift in the center of gravity and the greater the force that will be applied sideways on the aerial device.

This side stress is immediately apparent to an operator who can experience increasing difficulty of operation as the sections of the ladder will tend to bind against each other as the aerial ladder is extended or retracted.

- When approaching an objective from an uphill or downhill direction on excessive street grades, position the aerial device to the objective from the downhill position to maximize the angle of inclination (figure 2–26) because aerial devices offer maximum strength in the vertical position.

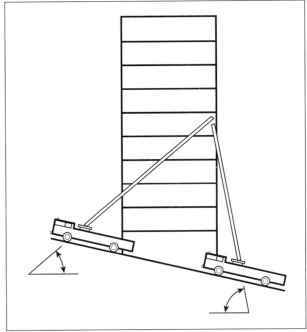

Figure 2–26. When approaching an objective from an uphill or downhill direction on excessive grades, position the aerial device to the objective from the downhill position to maximize the angle of inclination.

Figure 2–25. Tillered apparatus can position the trailer slightly outwards from the tractor to enhance the removal of ground ladders and closer placement of later arriving companies.

To accomplish this if traveling downhill, pass the structure and spot back to the objective. If traveling uphill, stop and spot before the objective. An exception is when maximum reach to the objective is necessary. In this case, the additional height gained by spotting from an uphill position may add essential length to the target area of the aerial device. Additionally, if a truck is stopped when traveling in an uphill direction, releasing the ladder locks to remove rear loaded ground ladders can allow the bedded ground ladders to slide onto the street.

- Although platforms are superior to aerial ladders in delivering personnel, tools, and equipment to an above-ground location, aerial ladders can also be used for this operation if two conditions are met—if aerial ladder personnel are properly trained to perform this operation, and if the fly section of the ladder is extended outward a distance (usually 6 to 8 feet) from the main section prior to personnel mounting the fly section and being raised to an above-ground location. The reason the fly section is extended is that this operation eliminates the *crossing rungs* or *shifting rungs* syndrome and allows personnel to stand on the end portion of the fly section rungs (and/or the foldable steps that are often provided by manufacturers) without having moving rungs from the section underneath the personnel. As an example, figure 2–27 illustrates an aerial after it had been extended about 10 feet (this distance varies for each manufacturer) and clearly shows the rungs of the last 10 feet of the fly section do not have crossing/shifting rungs. Additionally, notice this manufacturer has also installed solid plates below the rungs around the side mounted steps, so even if the ladder were not extended, personnel on this portion of the ladder would have solid material underneath the rungs they would be standing on. Contrary to popular opinion, this can be a safe operation if it is performed as previously delineated and if the rated tip load of the aerial ladder is sufficient.

- Breaking windows with an aerial ladder can be a controversial subject, however there may be times when this option is the only practical and timely solution to venting the interior of a hazardous environment. Obviously, the heavier the metal sash of a window, the more force will be necessary to break the window; therefore, knowledge of the type of window to be broken is necessary. In most cases, if an aerial ladder is projected through the upper portion of a window and then allowed to move downward, breaking the window, the weight of the ladder will be more powerful than trying to extend an aerial ladder through an entire window as this operation relies on the strength of the aerial extension motor. If this is the only practical means of venting

an area that needs to be vented to ensure firefighter safety, then the potential of some damage to the end section of an aerial ladder must be weighed against the issue of firefighter safety.

Figure 2-27. By extending an aerial ladder so the fly section rungs are beyond the rungs of the preceding section, the dilemma of crossing rungs can be avoided.

- The sophistication of modern aerial devices has allowed the capability to provide master streams that can be elevated above a fire, and for taking the place of hoselines in strip malls, taxpayers, elevated windows, doors, and so on. For this discussion, let's define elevated master streams as an aerial device that is capable of delivering 600 gallons per minute (1½-inch nozzle) or more. When utilizing elevated master streams, consider the following:
  - Interior personnel within the affected building should *not* be used because this operation is a defensive operation.
  - There are two modes for elevated master streams: short duration, which is designed

to quickly knock down a large body of fire; and extended duration, which implies that you will chase the building to the ground. The key to effectively using elevated master streams is to operate only long enough to knock down the main body of fire. Complete extinguishment with handlines if appropriate and structural stability is deemed safe.

- Elevated master streams from aerial ladders cannot be generally directed above the horizontal. However, these streams can be directed upward from platforms (for example, tower ladders) that are located above ground or from ground level.

- Portable monitors cannot be easily moved across the front of a building, but aerial master streams can.

- When elevated master streams are directed through openings such as windows, the stream is normally directed in a horizontal manner if maximum reach is desired, or at an upward angle so it will deflect off ceilings and walls to reach shielded areas or for greater dispersion of the stream.

- Master streams can be used in a defensive and offensive mode in specific conditions. As an example, a master stream is directed into a building for a defensive operation while a master stream is simultaneously directed into an adjoining building for an offensive operation (for example, strip malls/taxpayers).

- Evaluate directing an elevated master stream through holes in roofs. If a large portion or an entire roof has collapsed, elevated master streams can be effectively utilized. However, if fire has vented through a small portion of a roof, directing a master stream through this type of opening can negate vertical ventilation and drive the fire back into a building, enhancing interior horizontal extension.

- If elevated master streams are directed through windows and other similar openings from close range, remember that the close-up capability also increases the stream's destructive power which can be formidable. As an example, high-pressure, large-volume streams can easily fracture old chimneys, false facades, unreinforced masonry construction, and so on.

- A key point of consideration regarding the duration of use of elevated master streams is that water weighs 8.35 pounds per gallon. If a 1¾-inch tip is used, it will flow 800 gallons per minute or over 3 tons of water per minute. If the duration of the stream is 30 minutes, that equates to over 100 tons of water for each 30 minutes. Obviously, at the conclusion of this type of operation, there is the possibility of significant water accumulation and a potential of structural collapse, particularly during overhaul operations.

- When reviewing fireground pictures of defensive operations, it often appears that the closer apparatus can get to a building, the better! When defensive operations are utilized, the building becomes a prime candidate for collapse. Common collapse zones are the sides and ends of a building, and collapsing walls can collapse at least the height of a wall, and in some cases, twice the height of the wall (for example, unreinforced masonry construction). Therefore, defensive operations should focus additional emphasis on safety of personnel and apparatus by placing them in non-collapse zones and out of the smoke (upwind or windward side of a building). Why subject personnel to smoke from a building that will burn to the ground?

- Always evaluate the distance of an aerial device from a building when elevated master streams are being used. A common mistake is to assume that the collapse area of a wall would be outward and downward, similar to the dark triangular shaded area as illustrated in figure 2–28 (this is normally the case).

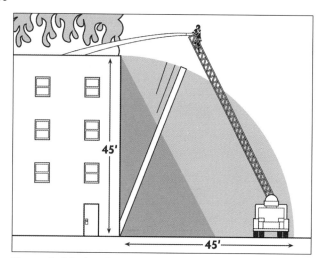

Figure 2-28. A collapsing wall can collapse outward and downward, or outward in an arc the height of the wall.

However, if a wall collapses outward in an arc, as illustrated by the lightly shaded arc (such as concrete tilt-up walls), it can strike the aerial device and/or apparatus. Always allow enough room for this potential.

- Firefighters that are operating an elevated master stream are normally in a position to observe the progress of the fire, the effectiveness of the elevated master stream, potential collapse problems, and other similar factors that an incident commander may need to know.

- Another area of debate is placing a firefighter on an aerial device to direct elevated master streams. Although a firefighter in this position is able to more effectively direct the stream and view the incident from a different perspective than fireground personnel, the assumption is that this operation can be inordinately dangerous and should not be attempted. As an example, see Using Master Streams From Straight Stick Aerials, Roundtable in Training Resources. However, this operation can be safely utilized if two considerations are met—if personnel are properly trained to operate elevated master streams, and the positioning of the aerial device is correct. If an aerial device is placed near or over a building

for elevated master streams, the aerial device and elevated personnel have been positioned in a potentially dangerous position. As illustrated in figure 2–29 (2), if the roof were to suddenly collapse and the resultant vertical extension of fire were to dramatically increase, the aerial device and personnel would be within the area of the vertically extending fire.

- However, if the aerial device and personnel are placed *away* from the building as illustrated in figure 2–29 (1), the aerial device and personnel are outside of any area of fire extension and resultant danger. Due to the reach of elevated master streams, they can be just as effective away from a building as they are on top or too near a building. This is another reason why straight streams should be used in elevated master streams instead of fog or spray streams.

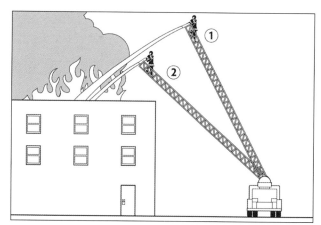

Figure 2-29. Elevated master streams can be safely operated by personnel on an aerial device if they are placed away from the hazard as illustrated by (1).

- Consider not shutting down an elevated master stream at the delivery point (nozzle) but rather at the base of the turntable, pump, or other similar location. If the supply hose/pipe is not properly bled, cold weather and retracting the aerial device can cause noteworthy problems because water freezes and is also a noncompressible liquid.

Aerial and Platform Placement

Prior to spotting apparatus for aerial operations, a driver must evaluate a combination of apparatus and platform or aerial ladder placement considerations that are based on SOPs, incident considerations, apparatus capabilities and limitations, and the expertise of the driver-operator.

After an apparatus has been appropriately spotted, assume it is necessary to place an aerial device to the roof of a four-story building with a flat roof surrounded by a 3-foot parapet wall. The operator must now make an important decision as the aerial device is raised to the roof. Specifically, how will the aerial device be placed to the parapet wall/roof (objective)? Let's consider some placement considerations for a platform and an aerial ladder.

Platform

Platform over the objective. If an operator decides to extend the platform over the objective and near a corner, the platform should be extended over the parapet wall and lowered as close to the parapet wall as possible. These two considerations will place the platform near the corner of the roof (strong area of construction), allow personnel to descend from the platform to the strongest portion of the roof, minimize the distance from the platform to the roof, and simplify access/egress to and from the platform.

However, this operation has resulted in an interesting dichotomy. Although the platform has been lowered as close to the parapet wall as possible, roof access from the platform may require a ladder between the roof and platform due to the ladder inclination angle, height of the parapet wall above the roof, and the distance between the parapet wall and base rails of the ladder. The distance between an objective and base rails of a ladder are determined by the expertise of an operator and the presence of a waterway below the base rails. If a waterway is not located below the base rails of a ladder, the ladder-platform can be placed as close to the parapet wall (or other objectives) as conditions allow. However, sufficient distance between an exposed waterway and an objective must be allowed to eliminate inadvertent damage to the waterway from potential movement of the ladder if being climbed by personnel and/or platform operations (figure 2–30).

Fig. 2-30. When a platform is placed above an objective to clear an exposed waterway, access to and from the platform can be compromised.

However, this consideration will increase the distance between the ladder/platform and the objective, potentially decreasing egress simplicity from the platform to the roof (as illustrated). This is one reason why a 10- or 12-foot ladder is attached to the aerial directly behind a platform.

Platform to the parapet/building. An alternative method is to place the platform to the parapet wall. By placing the center line of the turntable to the junction of the two walls, the platform can be positioned just above the parapet wall (figure 2–31) reducing the distance from the parapet wall to the platform.

Figure 2–31. This platform has been positioned just above the parapet wall to reduce the distance from the platform to the parapet wall.

The closer the platform is placed to the parapet wall, the more access and egress will be enhanced. In this illustration, personnel can exit the platform by either stepping from the platform onto the parapet and then to the roof, or, depending on the distance from the platform to the roof, from the platform to the roof. This option is not as practical or safe as placing the centerline of the turntable just outside the corner of the building. By doing so, the platform can be positioned next to the side of the building with the floor of the platform level with the top of the parapet wall (figure 2–32).

Figure 2–32. By placing the center line of a turntable outside the corner of a building, the platform can be positioned next to a building with the floor level with the top of the parapet wall.

Figure 2–33. A platform can also be positioned next to the front of a building with the floor of the platform even with the top of a building as illustrated.

Another variation is to place the center line of the turntable to the front wall by the corner and then position the platform to the building as illustrated in figure 2–33.

If a platform is being positioned by an operator who is on the turntable, it is easier to position a platform to the side of a building as compared to straight into a building (figure 2–33), because determining the proper distance between the platform and objective is easier when the platform is viewed from the side as compared to the rear. These two considerations can allow personnel to exit the platform and reach the roof without the use of a ladder between the platform and roof (depending on the height of the parapet wall above the roof), and reduce the possibility of an exposed waterway inadvertently striking the building. Remember that most platforms project forward from an aerial ladder and can allow placement without exposed waterway considerations if the platform is positioned next to an objective.

Articulating booms have the ability to articulate the boom over a vertical wall or parapet (fig. 2-34) and then lower the platform near the surface of a roof, simplifying the exposed waterway dilemma and enhancing egress-access to and from the platform.

Fig. 2-34. Articulating booms can articulate over an objective and then lower the platform near the surface of a roof.

When purchasing a platform, several factors can simplify and enhance the effectiveness of placement. Keep the pre-plumbed waterway close to the bottom of the base rails or rungs, particularly the fly section of the ladder. This allows placing the platform closer to an objective and can prevent inadvertent damage to a waterway if the ladder is placed too close to an objective.

Also, keep the platform simple. Ensure the platform is easy to leave and return to when personnel are carrying tools and equipment and wearing full protective equipment. Make sure it is easy to work from (for example, during ventilation operations), and not just able to lift 750 pounds at the tip while flowing 1,000 gallons per minute in any position.

There is a popular misconception that platforms are easier to work from than aerial ladders. As a blanket statement, this is a myth based on the fact that all platforms are not created identical in size, shape, and ability to offer easy access to and from the platform. (This also holds true for aerial ladders.)

Aerial ladder

Assume an aerial apparatus is proceeding past the front of a building with a flat roof. A 3-foot parapet wall has spotted to the far corner, and the operator raises and extends the aerial ladder over the parapet wall by the corner of the building. At this point, how far does the operator extend the aerial ladder over the parapet wall, and how far is the ladder positioned off the parapet wall (or how far is an aerial ladder positioned off an objective)?

Aerial extension. When extending an aerial ladder (or ground ladder) over an objective, forget the principle of three rungs or three feet over the objective, and focus on extending the aerial ladder over an objective far enough to be easily seen from the roof. This is essential when an aerial ladder is your primary escape route, and it is dark and/or smoky as graphically illustrated in figure 2–35. (Notice the tip of the ladder above the smoke.)

In dark and/or smoky conditions, a fluorescent tip and strobe light can enhance ladder identification.

Figure 2–35. Always consider extending an aerial ladder far enough that it can be easily seen, particularly if the ladder is a primary egress route. (LAFD file photo)

Aerial with a waterway/monitor. Similar to the platform section on waterways below the base rails of a ladder/platform, if an aerial ladder has a waterway that is located below the base rails of the ladder, the ladder must also be positioned far enough above an objective to keep the waterway from striking the objective during positioning as well as while personnel are climbing the ladder (because the ladder will move vertically or bounce while being climbed). This vertical movement is dependent on the extension, strength of an aerial ladder, and number of personnel on the ladder, but must be taken into placement considerations to allow enough vertical movement of the ladder without damaging an exposed waterway.

However, this can keep an aerial ladder above an objective and make access and egress difficult to impossible. This condition can be corrected by the following:

- Purchase an aerial ladder with a waterway above the base rails and/or with a pinnable waterway. A pinnable waterway allows the ladder to be extended without the waterway and monitor extending with the fly section. This enables closer placement of an aerial ladder to an objective, and enhances the ability of personnel to access or work from an aerial ladder (figure 2–36). This is why aerial ladders with low or sloping hand rails at the tip area are easier to work from than aerial ladders with high hand rails.

Figure 2–36. This aerial ladder with a pinned waterway has placed the waterway/monitor behind the fly section and allowed the tip of the fly section and personnel to be placed close to the work area.

- If an aerial ladder has a waterway below the base rails and/or a monitor at the outer end of the fly section (that is not able to be pinned back one section), easy access to and from an aerial ladder may be obtained by placing the center line of the turntable just outside the corner of the building and placing the extended aerial ladder next to the building at the top of the parapet wall (figure 2–37), or placing the turntable center line to the front wall near the corner and placing the extended aerial ladder just into the top of the building (figure 2–38).

- If an aerial ladder is extended into an objective, be sure the monitor is behind (or tilted backwards) the tip of the fly section as it approaches the objective (figure 2–38) to keep from damaging the monitor. Remember, it is easier to gauge the distance (from the turntable) between an aerial ladder and the side of an objective (similar to figure 2–37) as compared to extending an aerial ladder straight into an objective (figure 2–38). When evaluating the aforementioned conditions, placing an aerial ladder to the side of an objective (if possible) can be preferable to placing an aerial ladder straight into an objective due to ease of placement, access to and from the aerial ladder, and the potential of not striking the objective with the ladder/monitor.

Figure 2–37. If an aerial ladder with an exposed waterway at the fly section is placed alongside a building, access to and from the aerial can be enhanced.

Figure 2–38. When extending an aerial ladder into an objective, keep the monitor behind or tilted backwards from the tip of the fly to keep from damaging the monitor.

- When spotting with a pinnable waterway, be sure to spot far enough away from close objectives to allow extending the ladder with the waterway pinned behind the fly. This will allow placing the fly as close as possible over the objective without concern about the waterway/monitor striking the objective.

- Always remember that when an aerial ladder (or platform) is positioned off an objective, it is considered to be in the unsupported position. Some manufacturers reduce the load capacity of an aerial ladder when in the unsupported position.

On or off the objective. Let's assume an aerial ladder has a fly section that is not limited by a waterway/monitor below the base rails and has been extended over an objective the appropriate distance. As the aerial ladder is lowered into the objective, can the aerial ladder be placed on the objective? Most fire service personnel would answer this question as follows: "Aerial ladders must be positioned about

4 to 6 inches, 6 to 8 inches, or 8 to 10 inches above an objective for the following reasons:

- An aerial ladder placed on an objective can aid in building collapse.

- An aerial ladder of truss construction placed on an objective will reverse the design criteria of the truss configuration and weaken the ladder.

- The manufacturer requires it."

Although each of these three considerations has some merit, an aerial ladder that is positioned above an objective has a couple of noteworthy disadvantages. When being climbed by personnel, an aerial ladder can move or bounce vertically. This does not result in a stable climbing ladder for personnel and can result in an aerial ladder striking the objective. Also, depending on a combination of inclination angle, extension of an aerial ladder, number of personnel on an aerial ladder, type of aerial ladder, and distance from the bottom rails of an aerial ladder to the objective, an aerial ladder can settle and rest on the objective while being climbed by personnel. Although this will increase the stability of the aerial ladder for personnel, an aerial ladder can also suddenly lift off an objective as the last person exits the fly section, potentially creating a hazard to the unsuspecting person.

When the preceding two considerations are an operational concern, they can be minimized or eliminated by just placing an aerial ladder above an objective. Interestingly, there is no common standard for how far an aerial ladder should be placed above an objective other than various perspectives such as 2 to 4 inches, 4 to 6 inches, 8 to 10 inches, or an aerial ladder should not be placed or forced down on an objective. The reasoning behind these perspectives is based on the simple fact that modern aerials can be hydraulically raised and lowered, and as a result, can be forced downward by an operator, potentially creating a significant hazard to the aerial ladder and objective. This is the primary reason why all manufacturers recommend that aerials should not be placed on an objective. However, this viewpoint is worth mentioning from the perspective that when an aerial ladder is placed above an objective (for sake of discussion, let's say 4 inches) and climbed by personnel, it can settle downward and rest on the objective. Because the aerial has not been forced downward on the objective by an operator, this is normally an acceptable procedure. However, what about potential building collapse, weakening of the truss configuration, and manufacturers recommendations? Using the baseline of an aerial ladder that has lightly settled downward on an objective, consider these concerns.

Building collapse. If an unreinforced masonry brick parapet that has been severely exposed to fire and/or visible cracks are encountered, the resting of an aerial ladder on this type of objective is a consideration for obvious reasons. However, if an aerial ladder has lightly settled on a stable objective, the hydraulic system of the apparatus supports the majority of the aerial ladder weight, not the objective.

Truss construction. Because most aerial ladders are constructed in a truss configuration, the hand rails of the ladder are in tension and the base rails are in compression. If a truss ladder is supported on an objective, is the tension and compression configuration reversed, weakening the ladder? This is dependent on the force of the ladder on an objective. If a truss aerial ladder is forced against an objective with little regard for design criteria, the ladder can be easily damaged and can result in serious consequences and/or ladder failure. However, if the ladder settles on an objective, the tension/compression configuration is minimally reversed, resulting in a supported aerial ladder. It is interesting to note that some manufacturers rate their aerials at a higher load carrying capacity for personnel in the supported configuration as opposed to the unsupported configuration.

Aerial ladder and platform summary

In summary, how should an aerial ladder or platform be placed or positioned to an objective? If an exposed waterway is a factor, the aerial ladder or platform should be placed so any movement of the device does not damage the waterway. And if an aerial ladder has a pinnable waterway/monitor that has been pinned so it stays back one section when the ladder is extended, and in the absence of a manufacturer's recommendation, and if the ladder is extended over an objective, consider placing the ladder close to the objective. For this discussion,

the definition of *close to the objective* is approximately 2 to 4 inches, which will accomplish the following objectives.

If the manufacturer of your apparatus has not recommended a specific distance to place a ladder above an objective, a distance of several inches will have met the manufacturer's recommendation of not initially placing an aerial device on an objective. Also, if an aerial ladder is initially placed several inches above an objective, it will lightly settle into an objective when climbed by personnel. Interestingly, most manufacturers will accept the concept of an aerial ladder settling into an objective. As previously mentioned, this procedure has the ability to minimize movement of an aerial ladder while supporting personnel and ultimately improve their safety. For this reason, apparatus operators should practice placing an aerial ladder as close to an objective as appropriate.

Aerial ladder manufacturer recommendations. The concept of positioning an aerial ladder above an objective certainly addresses the issue of liability and unqualified apparatus operators. Yet, as we have previously discussed, aerial ladder operations can also be safely enhanced if an aerial ladder is correctly placed just above an objective, when appropriate. It is imperative that all personnel who operate aerial ladders and elevating platforms be intimately familiar with the strength, weaknesses, capabilities, and manufacturer's recommendations for their particular apparatus/aerial device. Although this may sound simplistic, it is interesting that a significant percentage of aerial ladder failures are the result of operator error, not ladder failure. The implementation of NFPA 1904 in 1991 requires all new aerial ladders and elevating platforms to have or exceed a rated capacity of 250 pounds supported at the tip, at full extension, and be able to operate in any position while carrying this load. However, NFPA 1904 should raise concerns that there are still a large number of aerial device apparatus in use that were purchased before NFPA 1904 and may not possess modern strength and operational capabilities. If an operator is not thoroughly familiar with the operational considerations and differences of newer aerial ladders and platforms versus older aerial ladders, and platforms versus the capabilities of various manufacturers' aerial ladders and platforms, a potential disaster awaits when applying the same operational considerations to all apparatus or when operating a newer aerial/platform today, and an older aerial/platform tomorrow. Consider that the greatest number of aerial devices in service were manufactured before 1991 and the adoption of NFPA 1904. As an example, consider the following three general classifications for aerial ladders and platforms:

First generation. Aerial devices purchased prior to 1980 generally do not posses the operational capabilities, hydraulic system sophistication, and metallurgical strength of newer aerial devices. Aerial ladders with no rated capacity were common. These aerial ladders normally use hydraulic pressure on one side of the hoist-lower piston, can be raised under hydraulic pressure, are lowered by gravity, and may also have small outrigger systems.

Second generation. Aerial devices purchased between 1980 and 1992 were developed as a result of regulatory concerns over structural and stability safety factors by ANSI, OSHA, and so on. These aerial devices have improved hydraulic systems, strength, K-bracing, and improved outrigger systems and operational capabilities as compared to first generation aerial devices. These aerial devices use hydraulic fluid on both sides of the hoist-lower pistons and are raised or lowered under hydraulic pressure. This means that these aerial devices can easily be hydraulically forced against an objective, with an expected negative result.

Third generation. Aerial devices purchased after 1992 to the present feature dramatically improved hydraulic systems, metallurgical strength factors, enhanced outrigger systems (improved stability), and significantly improved operational characteristics as compared to first and second generation aerial devices. These aerial devices can also be raised or lowered under hydraulic pressure.

Training Resources

The following resources can provide additional information and/or another viewpoint for the subjects discussed in this chapter:

- The Importance of Training Highlighted in Aerial Waterway Fatality, William C. Peters, Supplement to *Fire Engineering*, June 2008 .

- Communications: The First Five Minutes, John T. Carlin, *Fire Engineering*, January 2003.

- Setting The Stage With the First-Due Radio Report, James Mason, *Fire Engineering*, January 2007.

- Driving Apparatus In Bunker Gear, Roundtable, *Fire Engineering*, August 2005.

- Response To Fires, Roundtable, *Fire Engineering*, February 2008.

- The Qualities Of An Effective Engineer-Driver, Steve Prziborowski, *Fire Engineering*, February 2008.

- Increasing Your Drivers' Awareness, Paul Dow, *Fire Engineering*, February 2007.

- Tips For Driving More Safely, Steven M. DeLisi, *Fire Engineering*, November 2002.

- Drive To Survive, Chris Daly, *Fire Engineering*, February 2006.

- Manual on Uniform Traffic Control Devices, published by the Federal Highway Administration. www.mutcd.fhwa.dot.gov/index.htm

- For information on the LAFD Driver Training Program, see the training website at http://LAFDTraining.org/ists/books/bk08v1107.pdf.

- Tillering, LAFD Driver's Training Module, #9 Lesson Plan. http://LAFDTraining.org/ists/books/bk08v1107.pdf.

- Tiller Skills, Ted Corporandy, American Fire Journal, January 2001.

- Operating Safely On Roadways And Limited-Access Highways, *Fire Engineering*, February 2007.

- The Contributing Causes To Aerial Failures, Ralph Craven and James T. Steffens, *Fire Engineering* July 1996.

- Tiller-Mounted Ladder Operations, Al Hom, *Fire Engineering*, October 2004.

- Emergency Operations In Or Near Traffic, Scott Thompson, *Fire Engineering*, May 2007.

- Using Master Streams From Straight-Stick Aerials, Roundtable, *Fire Engineering*, April 2001.

- Master Stream Safety, Vincent Dunn, *Fire Engineering*, June 1991.

- Aerial Operations For Low Rise Buildings, Drew Smith and David Traiforos, *Fire Engineering*, October 2006.

- Strategies For Safer Driving, Steven M. DeLisi, *Fire Engineering*, February 2009.

- Reducing Firefighter Vehicle Crash Fatalities, Kristina Peterson, Harlan Amandus, James T. Wassell, *Fire Engineering*, June 2009.

Apparatus

Introduction

When contemplating the purchase of a designated truck company, one of the primary considerations will always center around purchasing an apparatus that is properly configured and equipped, and able to respond to a wide variety of incidents. Interestingly, departments that purchase truck apparatus normally fall into two categories:

- Purchasing a truck apparatus for the first time.
- Replacing or upgrading an existing truck apparatus.

Both of these categories should require an immense amount of study, comparisons with other types of apparatus, what type of apparatus will best fit the needs of a particular area, and fall within appropriate financial constraints. To determine what type of apparatus will most appropriately fit the needs of a particular department, an apparatus committee (composed of personnel within a department) is often created to determine what type of apparatus will be the best choice within the aforementioned parameters of needs and financial constraints. Unfortunately, this process can be more difficult than it needs to be from the following four perspectives.

After a truck committee makes their recommendations, the chief of a department circumvents this process and purchases an apparatus that was desired from the beginning. Obviously, this condition can create feelings categorized as *less than favorable* and as often happens, a department inherits an apparatus that does not maximize the needs of a department or an apparatus that is bigger and flashier than the apparatus of a neighboring department. It is important to remember that the chief will not have to use the apparatus in day-to-day operations and will probably not be aware of any resulting compromises or reduced operational effectiveness as the result of a questionable decision.

Also, not having access to various types and capabilities of available truck apparatus, a department will often purchase an apparatus based on the recommendation of a manufacturer. Although apparatus manufacturers can build excellent apparatus, they likely are not familiar with the specific needs of a particular area, and do not use the apparatus in *everyday* operations. Additionally, be very careful when purchasing an apparatus from a manufacturer with a model that has been returned, not accepted, or a pre-built model at a *reduced price.*

When purchased, a truck apparatus can be expected to last around 20 years, although sometimes longer. If the purchase project is not properly researched, a department can be forced to live with an apparatus that might not have been what a truck committee was expecting.

Finally, it is common practice for some departments to add on to another department's order. Remember that what may be appropriate for one department may not be the appropriate apparatus for another department.

Without a doubt, the style, configuration, and capabilities of truck apparatus has dramatically changed since the introduction of truck companies into the fire service (figure 3–1).

This chapter highlights various observations, types of apparatus, equipment inventories, and a few interesting points that have been observed from surveying numerous departments. Let's begin by looking at some general considerations that are not listed in any particular order of importance, but should be evaluated before purchasing a truck apparatus. It should be noted that the detailed specifications of apparatus will be left to a manufacturer because they are best qualified to expand on the specifications of their particular apparatus.

Figure 3–1. Modern truck apparatus have dramatically changed from the truck apparatus of yesterday. (LAFD file photo)

General Considerations

As a starting point, consider becoming familiar with the current edition of NFPA 1901 *Standard for Automotive Fire Apparatus* (available from NFPA) and

the publication *Fire Apparatus Purchasing Handbook* (available from PennWell). And if a truck apparatus is being replaced, its past history should be evaluated to determine what it was primarily used for and whether it met its intended purpose and, if not, what changes need to be made or considered for a new apparatus.

Understand the apparatus

One of the initial decisions is whether an apparatus will have a water tank, pump, and hose (quint) or be classified as an aerial fire apparatus. A quint is defined as "a fire apparatus with a permanently mounted fire pump, a water tank, a hose storage area, an aerial ladder or an elevating platform with a permanently mounted waterway, and a complement of ground ladders." An aerial fire apparatus is defined as "a vehicle equipped with an aerial ladder, an elevating platform, an aerial ladder platform, or a water tower designed and equipped to support firefighting and rescue operations by positioning personnel, handling materials, providing continuous egress, or discharging water at positions elevated from the ground."

Aerial ladders are generally more compact in size than platforms and can be a better choice if limited space is a concern. Additionally, the tip of an aerial ladder is smaller than a platform and is more easily placed between trees, overhead electric wires, balcony railings, closely spaced buildings, and other similar considerations.

Platforms can provide a larger working area for personnel than aerial ladders, and with a higher weight-carrying capacity. Additionally, ladder climbing and associated hazards can be eliminated, and platform mounted deck guns can safely deliver flows up to 2,000 gallons per minute.

Platforms can be purchased in three basic configurations, the first being the tower ladder. This configuration consists of a platform attached to an extendable/retractable boom. A sideless escape ladder is attached to the side of the boom and is considered not suitable for routine climbing due to the lack of side rails.

Next, the ladder tower configuration consists of a platform attached to the end of an aerial ladder. The ladder can be used for routine climbing. The ladder tower is considerably more popular than the tower ladder. Finally, the articulating boom configuration is unique from the perspective that a platform is attached to the end of an articulating boom. Although the concept is relatively new, it has gained popularity in departments that can use the flexibility of an articulating boom.

The higher a cab the more negative impact it will have on elevated operations over the front of an apparatus. Although some cabs are configured with a channel in the top portion of a cab to keep the overall height of an aerial device low, the aerial device must still be raised to clear the cab before initiating elevated operations.

A pre-plumbed waterway and monitor can add $20,000 to the overall cost of an apparatus and can increase the overall length of an apparatus by up to 3 feet.

If considering a single rear axle as opposed to tandem rear axles, remember that tandem rear axles are a superior configuration for supporting weight, providing enhanced braking capabilities and increasing stability. Although tandem axle configurations can add about 2 to 4 feet to the overall length of a chassis, it often does not compromise the turning radius of an apparatus. If there is a choice, tandem axles are normally a superior configuration, particularly for quints with a 75-foot aerial device. One-hundred-foot aerial ladders and platforms are normally equipped with tandem axles.

Investigate the feasibility of air ride suspensions because these systems can offer a superior ride and enhance apparatus longevity.

The appropriate apparatus that will best serve the needs of your district cannot be determined until the needs of your district are evaluated. As an example, always evaluate the population base, determine how fast is the community expanding and what community expansion plans are in the works, determine if the response area will be primarily residential or a combination of residential and commercial, if there is a noteworthy industrial area, if there are multi-story buildings, what the height of the tallest buildings are, if this number will remain static or taller buildings will be constructed, and so on.

An additional consideration is the affordable style of housing that is currently common in many areas of this country. As an example, condominium and townhome complexes can employ narrow streets, small cul-de-sacs, and challenging access considerations to these developments that will impact your decision on the best apparatus for your community.

A truck apparatus can be purchased from three basic perspectives—buy the biggest with all of the extras so you will have it if you ever need it; purchase a truck apparatus to be the centerpiece for the next parade; or determine what the truck will be used for the majority of the time (about 80%) and buy accordingly. Obviously, only one of these options is the most practical particularly when some truck apparatus cost over $800,000 and most responses in the modern fire service consist of grass, garbage, automobile fires, and EMS runs. You can be the judge on which is the most important for you.

Gross vehicle weight

Three ratings are necessary to determine the gross vehicle weight (GVW) of an apparatus: axle, springs, and tires. If the rating of one of these factors is less than the other two (in other words, does not fit the appropriate ratio), the GVW rating automatically falls back to a lesser rating. Each component must match accordingly.

Also be aware of the gross axle weight rating (GAWR) of the axles (particularly the rear) in concert with the anticipated *loaded* weight of the apparatus. As an example, if the GAWR of a single-axle quint is 31,000 pounds, this figure can be approached or exceeded as additional equipment finds its way into empty spaces.

Ladders

Until 1991, aerial ladders could be purchased with an unrated tip load. However, after 1991, aerial ladders were required by NFPA to have a minimum of 250-pound tip load at any angle between 0 degrees and the maximum angle of inclination, and at full extension (and platforms are required to have a minimum capacity of 750 pounds).

Although these requirements have dramatically increased the safety factor of aerial ladders and platforms, it has also resulted in a significant increase in room (or width) for outriggers to be properly placed to support the increased capability of an aerial device. This translates into the need for more maneuverability and space requirements for modern aerial devices (ladders and platforms).

If purchasing an aerial ladder, remember that modern aerial ladders can be equipped with a wide array of extras that can enhance flexibility but can also be a significant detriment to an aerial operator. Items such as a monitor, lighting, electrical outlets, breathable air outlets, tip controls, and other similar goodies can block the view of the end of an aerial ladder and can make placement (when viewed from the turntable) more difficult and sometimes impossible unless viewing the ladder tip from a location other than the turntable. This consideration would not fall under the classification of user friendly. Conversely, if purchasing a platform, the previous array of extras can often be mounted within or near a platform-basket without sacrificing the working area of personnel within a platform basket.

Should you choose a steel ladder or aluminum ladder? Both of these ladders meet NFPA standards, so there should be no major differences. However, there are several considerations:

- Aluminum will not need to be repainted.

- Aluminum will not rust although it can oxidize.

- Aluminum can be lighter.

Before purchasing an aerial ladder (steel or aluminum) with a pre-piped waterway and monitor, ensure the monitor is pinnable (will stay back one section when not being used). An un-pinned monitor can dramatically limit the operational capability of an aerial ladder.

If an aerial ladder will be used for roof ventilation operations, the height of the side rails should be considered. The lower the height of the side rails, the easier it will be for personnel to lean over the rails with power saws to ventilate a roof. Conversely, the higher the rails, the more difficult the operation will be.

The type of truck apparatus and overall length will affect the space that will be available for nesting ground ladders. As an example, apparatus with a water tank and pump will often have less horizontal space to nest ground ladders than a similar apparatus without a pump and water tank. Also, insufficient space to nest the length of ground ladders can require shorter ground ladders with additional sections (three or four sections as opposed to two or three sections). As the sections of ground ladders are increased, so is their width and the space that will be required to nest the ladders.

Hands-on demonstration

The importance of being able to have a prospective manufacturer demonstrate a potential truck apparatus in your district cannot be emphasized enough. If this is possible, remember three key considerations. First, pre-determine a typical route through your district and take the apparatus for a drive-thru to see how it performs, particularly in any congested areas and/or buildings with challenging considerations for aerial devices. Next, figure that fires are fought against burning buildings and not spacious paved parking lots, so consider evaluating a truck from a practical perspective in your fireground office and not a static display with ideal conditions. Finally, evaluate the ease of operation because modern safety standards have mandated the use of items such as micro-switches, interlocks, automatic leveling devices, and other assorted safety devices. From a simple perspective, how user friendly will the apparatus be, particularly for a volunteer department.

Drive your response district and make a note of any potential obstructions. When behind the wheel, ensure the new apparatus is able to make any sharp turns and easily traverse any dips or grades when traversing from one street to another. Evaluate the angle of departure (which is the angle from the bottom of the rear-most wheel to the rear underside of the tailboard). Additionally, the new apparatus should be able to travel under overhead power lines, low bridges, and other similar considerations. Be absolutely certain you know the overall height and length of the new apparatus to ensure it will fit into your fire station. Drive the apparatus in and out of the station before making any purchase decision!

During your drive-thru, evaluate the angle of a driveway ramp if the apron is not level with the apparatus floor in the fire station. There are two angles to consider: the angle of the apparatus leaving the station, and the angle for returning. As an example, when the truck leaves the station and the front wheels begin to traverse the down slope of the apron, the rear of the apparatus will raise upwards. This can affect the size of a fire station opening for apparatus.

Determine if there are any areas, such as narrow streets, that can affect the space necessary to position the outrigger jacks. Be sure to note if the apparatus can be short jacked and, if so, whether it can maintain full capabilities on the working side of the apparatus.

If there is an increase in weight, or if this is a first-time truck, make sure the weight will not be a factor on a fire station floor. Some floors are over basements and/or may not have been designed for the weight you are about to purchase.

Always try the operational control levers (raise, rotate, extend) to determine their ease of use. The length and spacing of the levers can be an advantage or a detriment. As an example, short levers can be difficult to feather and levers placed close together can be difficult to operate when wearing gloves

Consider where the apparatus will be serviced and maintained because this can be a major factor when choosing a manufacturer. And be sure to verify who will be responsible for any warranty work.

Compartment space

The amount and type of equipment carried on a truck company can affect how you will operate at an incident. Therefore, a key consideration is how much compartment space the new truck will have. Consider making an inventory of the equipment to be carried, and evaluate how it will be carried in the available compartment space of the new apparatus. Remember, there is a big difference between large doors that open to compartments that are only 24 inches deep and compartments that are significantly deeper (transverse).

Various hosebeds designs are available for quint-style apparatus. Consider the ability to load and access hose when necessary, particularly under emergency conditions. NFPA requires a minimum of 40 cubic feet of enclosed compartment space. Remember, this requirement is a minimum or a starting point. The point is to know how much enclosed compartment space will be required to store your equipment.

General compartment storage space examples are:

- 75-foot aerial quint, 175 cubic feet
- 75-foot aerial non-quint, 225 cubic feet
- Rear-mount, 100-foot platform quint, 206 cubic feet

- Rear-mount, 100-foot platform, non-quint, 350 cubic feet
- Mid-mount, 100-foot platform quint, 220 cubic feet
- Mid-mount, 100-foot platform non-quint, 350 cubic feet
- Tractor-drawn, 100-foot aerial non-quint, 400 to 600-plus cubic feet

When opened, rollup-type compartment doors retract out of the way and enable better access to compartment storage. They are significantly lighter in weight than hinged-type doors and can make a difference when applied to weight-sensitive apparatus. They also do not cause damage to an apparatus body if an apparatus is inadvertently driven out of a fire station with a door that is left open.

Hinged-type compartment doors are traditional, reliable, and easily repaired. When horizontally hinged doors are opened, they can provide cover for inclement conditions. However, they are significantly heavier than their roll-up counterparts and can significantly damage an apparatus body if left open when an apparatus is driven out of a fire station. If cost is an object, know that there is little or no cost difference between roll up and hinged type compartment doors.

Measure the dimensions of any compartments that will contain any specialized or large equipment. What's important is the dimension of a compartment opening when a door is fully open because equipment will need to be placed and removed from the compartment. Ensure this process will be simple and not a wrestling match.

If compartment depth is appropriate, tip-down trays and pullout boards can increase mounting surfaces for equipment and can aid in identifying missing tools and equipment.

Equipment

Heavy equipment should be mounted low for ease of use and safety to personnel. Also consider that equipment placed on top of apparatus can be difficult to reach, remove, and can reduce the target area of an aerial device.

Equipment that is frequently used should be placed so it is easily accessible. Always label portable equipment because this can simplify recognition at incidents.

NFPA allows a minimum of 2,500 pounds for equipment on aerial devices. From a practical perspective, this figure can often approach 5,000 pounds. If possible, it can be very beneficial to weigh the equipment that will be placed on the new apparatus because you may be surprised at the total weight. It is better to plan ahead than suddenly discover it will be necessary to compromise for 20 years. This can be a major consideration for quints with single rear axles.

Lighting systems that are available from both sides of an apparatus can be a significant benefit at incidents that need auxiliary lighting. This can reduce the need to reposition apparatus for lighting capabilities.

Final considerations

Mutual automatic aid agreements can affect your considerations. If there are no trucks available from other jurisdictions, then your purchase will be the star of Showtime! Conversely, if truck apparatus are available from other jurisdictions (considering their availability and response time frame constraints), the parameters for your truck apparatus might be less stringent. However, always consider the availability of other truck apparatus (and their configuration) to your incidents.

Apparatus Considerations

In this section, let's look at some common advantages and disadvantages of aerial fire apparatus. As a quick reminder, NFPA defines aerial fire apparatus as "an apparatus with a permanently mounted, power-operated elevating device," and divides them into four distinct categories: aerial ladders; aerial ladder platforms; telescoping aerial platforms; and articulating aerial platforms. However, for this section we will add four additional categories: mid mounts and rear mounts; tillered apparatus (or tractor drawn); quints; and ladder tenders. Ladder tenders are becoming more popular due to their lower initial and continuing cost of operation, and the routine type of responses most departments respond to on a frequent basis.

When comparing the advantages and disadvantages of specific apparatus, always consult a manufacturer to determine the apparatus capability and specifications, and how it can meet your specific needs and requirements. Additionally, always remember that NFPA 1901 is a minimum standard.

Table 3–1 through table 3–8 show the common advantages and disadvantages of rear-mounts, mid-mounts, platforms, articulating platforms, aerials ladders, quints, tillered apparatus, and ladder tenders.

Table 3–1. Rear-mounts, advantages and disadvantages

Advantages	Disadvantages
Most common type of aerial devices and can have more options than mid-mount aerial devices	Travel height is higher than mid-mounts
Can be less expensive than mid-mounts ($80-100,000)	Apparatus height can range from 11 feet to slightly over 12 feet
Can have better below grade operations than mid-mounts	Ride and handling attributes may not be as advanced as mid-mounts
Aerials are generally shorter in overall length than mid-mount aerials	Difficult to place aerial device on ground near apparatus
More flexibility with placement of pumps, foam systems, and amount of water carried	Normally the largest and tallest apparatus of aerial devices
Can have fewer restrictions on compartments and space for ground ladders that are mounted in the interior of the apparatus as compared to mid-mounts	Working area of an aerial device can be limited due to placement of turntable height from the ground
Ladder overhang is often less than the platform overhang for mid-mounts	Front overhang of platforms can be considerable and limit the visibility of a driver
Can offer a tighter turning radius than mid-mounts	Placement of platform is not easily accessible to personnel unless it can be first placed at ground level or accessed by a vertical walkway on the apparatus

Table 3–2. Mid-mounts, advantages and disadvantages

Advantages	Disadvantages
The aerial device can be more maneuverable than rear-mounts; as an example, a common operational range for a mid-mount aerial device is a negative 12 degrees to a positive 72 degrees, whereas a rear mount can be a negative 8 degrees to a positive 72 degrees	More expensive than rear-mount counterparts ($80-100,000) due to the complexity of their design
Can have better ride and handling characteristics than rear-mounts	For quint mid-mounts, placement of hose loads, water tanks, and space for equipment can be challenging
Improved visibility through the windshield as compared to rear-mounts	The lower profile of mid-mounts can reduce the amount of compartment space; can also limit the number of ground ladders and how they are nested
Shorter travel height than rear-mounts (10 feet to slightly over 11 feet); reduced height can be a significant factor depending on the size of fire station apparatus doors	Increased length as compared to a rear-mount
Shorter retracted length of an aerial ladder; composed of four or five telescoping ladder sections as opposed to the three telescoping ladder sections normally found on rear-mounts	Raised cabs can reduce operations over the front by over 30 degrees
The shorter retracted length can enable an operator to position the ladder tip or platform closer to the ground by the apparatus or lower on the face of an objective; shorter retracted length also can minimize the space required to swing and rotate the aerial or platform past trees, electric lines, buildings, and other similar considerations	Accessibility of a platform to personnel can be determined by it's location on the rear of the apparatus and/or accessible by a ladder on the rear of the apparatus
Easier to judge the position of the turntable center line when spotting to an objective	The rear overhang can be a detriment when the apparatus is turning
Increased target area as opposed to a rear-mount	Not as popular as rear-mounts

Table 3–3. Platforms, advantages and disadvantages

Advantages	Disadvantages
Minimum rating of 750 pounds (depending on the apparatus) as compared to a minimum rating of 250 pounds for aerial ladders; ratings of 1,000 pounds are common whereas aerial ladders are available in 250 pounds, 500 pounds, or 750 pounds	Initial cost is higher than an aerial ladder due to the increased size of the apparatus, complexity of the hydraulic system, and larger capacity rating
More working room (work station) within a basket as compared to the tip of aerial ladders; can be an advantage when considering rescue and elevated master stream operations	The overall size and weight of an apparatus will be larger than an apparatus with an aerial ladder
Articulating booms can reach over obstacles	The travel height (distance from the ground to the top most portion of the aerial device) will be higher than an aerial ladder
The bottom portion of platforms can be equipped with anchor points for lifting objects, which can be enhanced by the additional rating as compared to aerial ladders	Can be 7 to 8 feet longer than a comparable aerial ladder
Can provide some protection against radiated heat to personnel within a basket	Depending on capacity rating, the space required for the outriggers may be greater than an aerial ladder
A basket can safely accommodate more personnel than the tip of an aerial ladder	Can take more time to reach an objective than an aerial ladder
Can quickly place more personnel and equipment to elevated locations than an aerial ladder	A rear mount platform with a reach of approximately 100 feet can be an extremely large apparatus

Table 3–4. Articulating platforms, advantages and disadvantages

Advantages	Disadvantage
More maneuverable than platforms and aerial ladders	Enhanced capabilities requires a higher degree of expertise and training for an operator. This also includes maintaining an appropriate level of training and competence
Able to articulate over walls and parapets	Can be an expensive apparatus as compared to aerial ladders and platforms
The platform can sweep from side to side, a platform cannot	Due to the complexity of the articulating ladder/boom, maintenance and repairs can be more complicated
Can be easier to work from for roof operations as the ability to sweep the platform is an advantage as compared to a stationary platform	The configuration of the apparatus and articulating ladder/boom can affect the overall size of the apparatus and compartment space

Table 3–5. Aerial ladders, advantages and disadvantages

Advantages	Disadvantages
Compared to platforms, not as expensive; an aerial ladder can be $80,000 to $100,000 less than a comparable platform and $160,000 to $200,000 less than a comparable mid-mount	Operating from the tip of an aerial ladder for rescue and elevated heavy streams can be more challenging than from a platform
Lighter than platforms	The tip of an aerial ladder does not normally carry the same amount of equipment as a platform
Can have a faster implementation time than platforms	Some platforms can be effectively used to direct a master stream at ground level into some structures such as mini malls, taxpayers, etc.; not an easy option for aerial ladders
Can be easier to operate than a platform, particularly for the occasional operator	Depending on the tip load rating, ladders can be limited for lifting or hoisting heavy objects
Lower travel height than platforms	Master stream flows can be less than platforms (i.e., 1,250 gpm vs. 2,000 gpm)
Can be equipped with a bolt-on egress section for easy replacement; common angle for bolt-on egress sections is about 11 degrees, which can simplify egress and access for personnel	Many older aerial ladders are still in operation. An operator must know their limitations compared to newer aerial ladders

Table 3–6. Quints, advantages and disadvantages

Advantages	Disadvantages
The quint concept theoretically provides the ability to perform engine or truck operations	Most departments staff quints with three personnel; severely limits the capability of the basic concept of this apparatus; with three personnel, a quint is often either an engine or a truck and is normally used as an engine company
By combining an engine and truck into a single apparatus, staffing is normally reduced	Reduction in staffing can often result in the need for more companies to handle basic suppression priorities to achieve the desired results on the fireground
Equipped with pump, hose, water, aerial device, and ground ladders	Maneuverability can be more difficult as opposed to a standard engine company; this is dependent on the size of the quint
Self sufficient until other apparatus arrive	Reducing staffing levels with quints can be a short term advantage; not an advantage on the fireground
Advantageous for limited staffing constraints	A pump and associated plumbing can add 1,500 pounds, increase the wheelbase 20 to 30 inches, and reduce space that could have been used for other considerations
Multiple quints can increase the number of available aerial devices at an incident	A typical quint can have approximately 40% less compartment space for equipment as compared to a non-quint
Ability to initiate aerial device rescues as compared to an engine company	Due to limited nesting space, ground ladders may be forced to be configured with three sections instead of two sections; increases the necessary nesting width and can result in heavier ground ladders
Elevated master stream capability as compared to an engine company	The complexity of the quint apparatus can often result in increased maintenance costs as well as increased wear on components such as tires, brakes, fuel, etc.
One quint can cost less than one engine and one truck apparatus	The quint concept can often result in insufficient personnel for engine and truck operations; stated another way, insufficient personnel for attack and logistical operations
Good apparatus for smaller, rural, and volunteer departments	Quint apparatus are often used as a high-priced engine company. As a result, truck company experience and expertise can be slowly minimized or virtually eliminated

Table 3-7. Tillered apparatus (tractor drawn), advantages and disadvantages

Advantages	Disadvantages
Can be purchased with or without the capability of a quint. If the apparatus has the capability of a quint, the water tank, pump, and hose are located on the tractor resulting in the trailer containing the aerial device and equipment	Overall length can be longer than a comparable apparatus with a straight chassis.
Can have a lower purchase price as compared to single chassis apparatus ($80,000 to 200,000)	Requires two drivers. This can be a concern for smaller and/or volunteer departments.
The most maneuverable type of aerial device available, particularly when compared to their counterpart single chassis apparatus	Additional training and certification for the driver and tillerman
The ability to steer the trailer independent of the tractor in forward or reverse situations can be advantageous in congested areas, narrow streets, alleyways, and other similar considerations.	Normally only available with an aerial ladder
Superior compartment space as compared to a straight chassis apparatus (normally over twice the space) as compartments can be constructed in a transverse configuration	Not as popular as straight chassis apparatus
Greater selection of portable ground ladders. Ladders of 40 to 50 feet can normally only be carried on this type of apparatus	The capability of carrying more equipment requires a higher level of training for personnel assigned to this apparatus
Due to the increased compartment space, some departments combine the capabilities of a truck and a heavy rescue apparatus. This configuration can reduce the number of apparatus required to perform these two functions.	Tillermen often tend to "over tiller" the trailer

Table 3-8. Ladder tenders, advantages and disadvantages

Advantages	Disadvantages
Lower initial cost as compared to a traditional truck apparatus	Departments that use ladder tenders normally also have a truck with an aerial device. This requires two apparatus with associated costs.
Operational costs are significantly less than a traditional truck apparatus with an aerial device	If arriving on-scene and an aerial device is necessary, there will be a delay until one arrives.
Can be effectively used to handle responses that do not need an aerial device (which are most responses)	When an alarm is received at a station with a ladder tender and an apparatus with an aerial device, the appropriate officer must determine which apparatus to take based on initial information.

Recommendations

This author is frequently asked to supply an opinion on what type of apparatus to purchase, particularly for departments that are preparing to purchase their first truck apparatus. This is a difficult question to answer for three basic considerations: all municipalities are different; some departments are paid, some are volunteer, and the size and staffing available to staff truck apparatus can significantly vary; and fiscal restraints.

Having the opportunity to conduct truck company workshops in this country since 1983, I always ask why a department purchased a particular type of apparatus and also ask the available driver what its strengths and areas of needed improvement are.

I never ceased to be amazed at the number of truck apparatus purchased that are ill-suited for their intended function. What makes this statement significant is the fact that once an apparatus is purchased it can have a service life of over 20 years. As a side note, if you want to determine the realistic advantages and disadvantages of an apparatus, be sure to interview the driver of the apparatus, not the chief.

When surveying truck apparatus, three considerations have a significant influence on the purchase and operation of a truck apparatus. The cost for modern truck apparatus is rapidly escalating. As an example, the cost for a quint with a 100-foot aerial device is approaching one million dollars. Also, the majority of emergency responses can be classified as grass, garbage, emergency medical service (EMS) calls, and so on. Finally, aerial devices are primarily used for firefighter access and egress. Rescue and elevated master stream operations are few and far between at best.

With these thoughts in mind, why spend the big bucks for an apparatus that is rarely ever used for all of its potential capabilities, will not be easy to be maintained, and not be particularly user-friendly as opposed to a truck apparatus that is more than capable of performing its duties over 80% of its service life? As an example, a single-chassis quint with a 100-foot, rear-mounted platform is one of the largest and most expensive types of apparatus in the fire service. Most of these apparatus are overweight, not easy to drive (as compared to other types of truck apparatus), have minimal compartment space, and are very expensive to purchase and maintain.

Although some departments successfully use these apparatus for specific needs, a significant number of departments that were surveyed stated these trucks were purchased due to their superior capability to manage rescue and elevated master stream operations. However, the question that needs to be asked is *"How often do you rescue people and flow the big water from elevated heights?"* The likely answer is *"not very often."* Another pertinent question is *"How often will your new truck apparatus respond to the everyday routine calls, and what impact will that have on an apparatus that was not primarily designed to run the routine calls?"* The answer to this question is a major reason why ladder tender apparatus are gaining popularity.

Allow this author to step out on thin ice (if I haven't been there already) and offer some general suggestions to assist in your search for a truck apparatus that can operate most efficiently in and for your area of responsibility. I must confess that I am biased toward basing your apparatus purchase on what it will be used for most of the time and keeping the apparatus as simple (user friendly) as possible. This means that a single-chassis quint with a rear-mount 100-foot platform will not normally be at the top of the following recommendations.

Rural department, limited to one apparatus

Quint, 75-foot aerial ladder (unless a 100-foot aerial device is necessary), 250-pound tip load, tandem axles. A quint will be self contained; a 75-foot aerial ladder should reach most rural multi-story buildings; and the 75-foot aerial ladder will be easy to operate and result in a moderate size of apparatus. The tandem axles will provide enhanced stability, an increased ability to support the weight of the apparatus, and superior braking capabilities. The downside will be compromised compartment space.

Rural department, limited to one truck

If height of buildings is not a factor (residential), a ladder tender would be a good choice. This will provide an apparatus that is devoted to logistical operations with the compartment space to carry the necessary equipment. If no ladders are carried, basic ladder evolutions can be provided by ground ladders carried by an engine company or companies. However, the preference would be for the ladder tender to carry a few basic ladders. An additional consideration is the lack of an aerial device will not require a certified operator for an aerial device.

If height is a factor, a quint, 75-foot aerial ladder, 250-pound tip load, and tandem axles is a good option. This will provide the same considerations as listed under rural department, one apparatus, except a certified aerial ladder operator will be required.

Urban department, limited to one truck

If a quint is necessary, consider a tillered quint with a 100-foot aerial ladder and 250- or 500-pound tip load. Unless there is a specific need, remember that a 500-pound tip load aerial ladder will result in a larger and more expensive apparatus. This combination will provide an apparatus with a 100-foot aerial ladder capability, maneuverability, and notable compartment space because this apparatus often becomes a do-all support apparatus. Be sure to not overload the tractor as this can result in needing a tractor with tandem axles

If a quint is not necessary, a tillered 100-foot aerial ladder with 250- or 500-pound tip load. This truck will provide 100-foot aerial ladder capability, maneuverability, and notable compartment space.

Urban department, multiple trucks

If the capability of an aerial ladder and platform is desired, consider a mid-mount 100-foot platform and a tillered 100-foot aerial ladder with 250- or 500-pound tip load. This combination will provide the benefits of both apparatus. The mid-mount platform will have some benefits (as previously mentioned) over a rear-mount platform, and the tiller will feature maneuverability and notable compartment space.

Unless there is a specific need for a platform, consider tillered trucks with 100-foot aerial ladders with 250-or 500-pound tip load. Some departments specify only tillered apparatus for the primary advantages of maneuverability and compartment space.

In the debate between a rear- and mid-mount platform, or rear- and mid-mount ladder, unless there is a specific reason for a rear-mount, consider a mid-mount for its comparative lesser size but capability to perform the same basic functions as a rear-mount. However, remember there will be a significant cost increase for the mid-mount as compared to a comparable rear-mount.

Apparatus Configurations

Basic truck apparatus configurations are discussed in this section. Additionally, the inventory for the truck company in figure 3–2 is representative of what a typical truck company carries. All compartments are numbered from the front to the back of this apparatus, and primarily categorized as driver or officer compartments.

Figure 3–2. A rear-mount, 100-foot aerial ladder

Aerial ladder, rear mount 2006 Pierce 105-foot aerial 500-pound tip load

Front Bumper Compartment

- 2 jaws of life with 100-foot hoses each

Aerial Ladder

- 1 10-foot Fresno ladder
- 1 10-foot roof ladder
- 1 6-foot pike pole
- 1 pick head axe

Top of Apparatus

- 2 water rescue bags
- 1 wheelbarrow
- 1 decontamination bag
- 1 Carlson board
- 2 50-foot 1¾-inch hoses

Rear Compartment

- 2 6-foot pike poles
- 2 8-foot pike poles
- 2 10-foot pike poles
- 2 12-foot pike poles
- 1 10-foot rubbish hook
- 1 10-foot attic ladder
- 2 35-foot extension ladders
- 1 24-foot extension ladder
- 2 20-foot roof ladders
- 2 16-foot roof ladders

Officer Compartment #1
Compartment

- EMS equipment
- air bags
- 4 air bag bases
- 1 air bag box
- 2 water vacs with hose
- 1 200-foot cord reel

Slide out rack

- 3 lock cutters (16", 30", and 42")
- 2 long pick head axes
- 2 short pick head axes
- 2 4 lb sledges
- 2 carpenter saws
- 3 broom handles
- 3 squeegee handles
- 1 big-easy
- 1 hydrant valve tool
- 2 crow bars
- 4 pry bars

Officer Compartment #2

- 2 6-foot rubbish hooks
- 2 small scoop shovels
- 2 square point shovels
- 2 round point shovels
- 2 portable generators w/light head
- 2 portable lights (500W)
- 2 extension cords
- 2 light boxes
- 1 sump pump
- miscellaneous pig tails

Officer Compartment #3

- cribbing
- 2 circular saws
- 2 chain saws
- 4 extra metal blades
- 4 extra carbide blades

- 1 2-cycle fuel can
- 1 4-cycle fuel can
- 1 bar oil

Officer Compartment #4

- 1 Sawzall
- 1 air hammer kit
- 2 step chocks
- 1 ram anchor

Officer Compartment #5

- 2 long rescue struts
- 2 short rescue struts
- 1 bag rescue strut equipment
- 1 small hydraulic ram
- 1 medium hydraulic ram
- 1 large hydraulic ram
- spreader tips
- spreader arms
- 2 25-foot hydraulic lines
- 1 portable hydraulic unit
- 1 set extrication chains

Driver Compartment #1
Compartment

- rabbit tool (10,000 psi)
- 4 squeegee heads
- 2 broom heads
- 1 set irons with sledge
- 1 bucket grease sweep
- 1 Sked
- 1 half back
- 1 150-foot rescue setup
- 1 tool box (miscellaneous tools)
- rescue cache consisting of: 1 stokes pre rig, 1 cinch collar, 2 atrier, 1 throw bag, 1 spreader bar, 1 short rope, 1 wristlet, 9 long Prusik, 1 brake bar rack, 18 carabiners, 2 carabiners (large), 8 small Prusik, 1 aerial pulley, 2 300-foot rope bags, 4 rescue harness, 1 rescue rocket, 1 edge protector (small), 1 edge protector (large), 1 canvas mat,

2 load-releasing straps, 1 pick off strap, 1 edge roller, 2 gibbs ascenders, 1 edge roller, 2 4-inch pulleys, 2 single pulleys, tri-link, 1 rope grabber, 2 4-inch Prusik pulleys, 1 gathering plate, 1 swivel, 4 #8 plates with ears, 1 knot passing pulley, 9 orange webbing (20 feet), 12 blue webbing (15 feet), 9 yellow webbing (12 feet), and 9 green webbing (5 feet)

Slide out rack

- 1 lockout bag
- aerial tip parts
- 2 long pickhead axes
- 2 4-pound sledges
- 1 K tool
- 1 A tool

Driver Compartment #2

- 3 50-foot rope bags
- 2 come-alongs
- 1 bag of chains
- 1 little giant
- 1 stokes
- 1 6-foot rubbish hook
- 1 blower

Driver Compartment #3

- 1 leak kit
- 1 redwood plug kit
- 2 sets small slats
- 2 sets large slats
- 1 roll plastic sheeting
- 1 chimney mirror
- 4 carry-all
- 6 traffic vests
- 4 large salvage covers
- 4 small salvage covers
- 4 hall runners
- 1 case water

Driver Compartment #4

- 3 nail bags

- 1 roll red tape
- 1 roll yellow tape
- 4 cans marking paint (orange)
- 1 roll duct tape
- 1 residential sprinkler kit
- 1 plastic tool box (misc tools)

Driver Compartment #5

- 10 small traffic cones
- 1 ppv fan
- 1 can ozium
- 2 50-foot high pressure 5-inch hoses
- 2 traffic signs

This apparatus has several noteworthy features, including:

- A collision avoidance system for the aerial ladder
- Two Jaws (with 100 feet of hoses each) mounted within a box on the front bumper to facilitate working on extrications from the front portion of the apparatus
- Slide out racks next to the driver and officer portions of the cab that allow easy access to tools that are often used; the tray in figure 3–3 has forcible entry equipment mounted on both sides of the rack for easy access; also works as a quick indicator of missing equipment

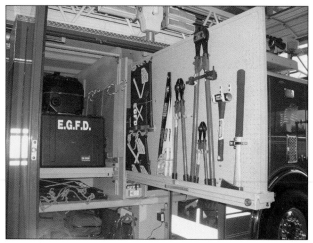

Figure 3–3. Slide-out racks offer easy access to tools and can be a quick indicator of missing tools.

Mid-mount 100-foot platform (quint)

Figure 3–4 shows a 2003 American LaFrance (LTI) quint with a 93–foot tower ladder 2,000 gallon-per-minute pump, and 300 gallon tank. It also has a 1,000-pound tip load.

Figure 3–4. A mid-mount 100-foot platform

Notice a lack of equipment/boxes on the top of the apparatus next to the aerial ladder. This affords a greater operational area and fewer considerations for the operator. Also, note the limited compartment depth. (See figure 3–5 where equipment has been mounted on vertical boards for easy recognition and storage.)

Figure 3–5. Vertical boards can also offer easy access to tools.

Each compartment is designated for a specific type of equipment (lighting, appliances, loss control, and so on). This department responds to numerous high-angle rescues, so appropriate rescue equipment is carried.

Tractor-drawn aerial ladder

Figure 3–6 shows a tractor-drawn, 100-foot aerial ladder. This is a 2007 American LaFrance (LTI) 100-foot aerial ladder with a 500-pound tip load.

Figure 3–6. A tractor-drawn 100-foot aerial ladder

This apparatus has several noteworthy features. Because the trailer affords large compartments, some of which are categorized as pass-through compartments, there are numerous multi-level compartments. Many of these can be pulled outward to allow easy access to equipment (figure 3–7). Each compartment is designated for a specific type of equipment, such as lighting, appliances, loss control, and so on. The compartment space is listed at 485 cubic feet.

Figure 3–7. Transverse compartments that incorporate significant storage space for equipment and can also integrate slide-out trays.

The bottom edge of the cab doors are higher than normal so the apparatus can park close to guard rails on roadways and still allow the cab doors to open.

This apparatus employs a portable monitor and large diameter hose that must be placed on the aerial instead of a pre-plumbed waterway. The reasoning for this configuration is that time is not a major factor in defensive operations, and the apparatus does not have to be placed out of service for maintenance on a pre-plumbed waterway.

A smaller turntable (84 inches instead of 96 inches) is used to provide more clearance between the turntable and cab. Strategic hooks were installed to allow tie off points for rope rescue anchors. A section of cord is placed between the two wheel chocks that allow both chocks to be simultaneously carried.

The ground ladders ride on Teflon slides for easy removal. Additionally, the balance point for each ladder has been marked so as a ladder is removed, a firefighter can pick a ladder up at the appropriate balance point. The only ladder that is mounted externally is a 24-foot extension ladder (R/R side of the trailer) that is primarily used for the single-story, single-family dwellings that are common in the area served by this truck.

A RIT bag consists of a large area search rope (200 feet) in a bag, a Murphy sled, a 1-hour bottle, mask, and pig tail for buddy breathing with 6 feet of hose.

Tractor drawn aerial quint

The tractor-drawn aerial quint is shown in figure 3–8. This is a 2007 KME, 100-foot aerial with a 500-pound tip load, a 300 gallon tank, and 1,500 gallon-per-minute pump.

This apparatus is similar to the preceding 100-foot tractor drawn aerial apparatus, the primary difference being the addition of a pump, water tank, and hose that are carried on the tractor (figure 3–9). The compartments and equipment that are carried are also similar.

A summary of this apparatus is supplied by the following article that appeared in the April-June 2008 issue of the Utah Fire Rescue Academy magazine *Straight Tip*, courtesy of and authored by Chief Stephen Higgs, Midvale Fire Department, Utah.

Figure 3–8. A tractor drawn 100-foot aerial ladder that incorporates the capability of a quint. (Photo courtesy of Stephen Higgs)

Figure 3–9. The pump, water tank, and hose are carried on the tractor. (Photo courtesy of Stephen Higgs)

Why a Tractor Drawn Quint?

"Having grown up in the fire service as a truck company purist, I was often critical of the quint concept. Over the past eight years as the chief of Midvale City Fire, I was forced to reconsider my position regarding the use of quints. Even though I have modified my position (is this a political flip-flop?), I still have what I consider to be legitimate concerns about quints. Aside from staffing and competing fire ground priorities, there are the following considerations:

- Quints are large and heavy.

- Based on their size and design they are not very maneuverable.

• Quints lack adequate storage space for ladders, hose and equipment.

"These are problems that are inherent with combining multiple functions in a single apparatus.

"Midvale has been operating a 75-foot rear-mount aerial quint for the past ten years. When it came time to look at replacing the aerial, we began a two-year quest to determine what type of apparatus to purchase. The first thing we did was create a list of the most restricted access areas in our city. We identified 12 locations where it was difficult or impossible to gain access with out current apparatus. The next step was to begin evaluating other apparatus configurations. We evaluated the following:

• Rear-mount 100-foot aerial

• Rear-mount 100-foot platform

• Mid-mount 100-foot platform

• Mid-mount 75-foot platform

• Rear-mount 75-foot aerial

• 100-foot tractor drawn aerial

"What we discovered through our evaluation was a reaffirmation of what we already knew or assumed. Due to the length of the straight frame apparatus, especially the rear mount and mid mount platforms, the apparatus could not negotiate turns in restricted areas without making multi-point turns. In many areas the apparatus could not make access, or took so long in the process that it would not have been feasible. Ladder nests were always the minimum NFPA compliment. The most extensive ladder compliment had only 127 feet of ground ladders. Hose beds were generally restricted in size and method of deploying the hose. The ability to carry additional 2½-inch hose was very limited. Loose tool storage was restricted, and in most cases was between 200 to 300 cubic feet of compartment space.

"At one point we considered the concept of a two-piece truck company. What is that? We considered purchasing a 75-foot mid-mount platform (non quint) and a small engine to respond in tandem (L.A. City task force concept). We would staff the platform with three or four firefighters and the engine with one firefighter. This solved some of our problems. The aerial apparatus could be kept at 40 feet in length or less for maneuverability, we would still have a unit with a pump, water tank and hose bed, and we would have adequate compartmentation between the two apparatus to carry all the equipment. This was our plan until we evaluated the tractor drawn quint concept.

"During our evaluation period, Park City Fire was gracious enough to bring their tractor drawn aerial (TDA) to Midvale and maneuver the apparatus through our 12 most restricted access points. The results were impressive. The tractor drawn aerial, even at 58 feet in length, was able to maneuver into areas that our current apparatus, including our engine, could not access. During our research we became aware of a tractor drawn aerial quint (TDAQ) that was being built for Los Angeles County Fire that was less than 56 feet in length. Our reasoning was this; if the 58 foot TDA could easily access and back out of most of our restricted areas, it stood to reason that an apparatus that was over two feet shorter would also be able to access these restricted areas.

"After an hour of examining a TDAQ, we were sold. The TDAQ was the answer to everything we were trying to accomplish, and we could do it all with one apparatus. Why were we so impressed? We were able to get superior maneuverability, 100 feet of aerial ladder, 1500 gallon-per-minute pump, 350 gallons of water, adequate hose storage with easy deployment and loading, over 200 feet of ground ladders, and more than 600 cubic feet of storage for loose equipment. It was a total package in one unit that provided for crew integrity. The icing on the cake came when we were able to add onto the L.A. County bid. With the add-on we were able to modify the specification to meet out needs. We eliminated the L.A. wooden ground ladders allowing us to add two more full transverse locker compartments. We also added a Telma drive line retarder. With the additions and deletions to the specification we were able to purchase the ladder for $681,000. We had anticipated spending as much as $950,000 to replace our ladder truck."

This apparatus has several noteworthy features. First, a pre-plumbed waterway was deleted and reduced the overall apparatus length by 3 to 4 feet. The apparatus now has an overall height of 11 feet 6 inches and a length of 55 feet 9 inches. All SCBA have been removed from the cab interior and mounted in compartments on both sides of the tractor and trailer for quick access and donning.

This was done to improve safety and allow time to properly size-up the incident prior to taking action.

In addition to the ladder nests and hose bed storage areas, the apparatus has over 630 cubic feet of storage space for loose equipment. There are seven 200 pound and two 500 pound slide-out trays in the transverse compartments.

The apparatus is equipped with a 5,000 kilowatt Honda generator that sends power to 12 mulitplex outlets from the aerial tip to a 200-foot cord reel. Charging bases for flashlights, radio batteries, and thermal imager are powered through the shore line as well as the generator.

Rear mount platform quint

Figure 3–10 shows a 100-foot, rear-mount platform quint. It is a 1991 LTI Simmons with a 100-foot platform, 750-pound tip load, 300 gallon tank, and 1,500 gallons-per-minute pump.

This apparatus has several noteworthy features. Although this apparatus was built in 1991 and will soon be replaced by a newer model, notice the low cab height, its relation to the ladder/platform turntable, and the lack of equipment compartments on the top portion of the apparatus. These factors maximize the operational capability of the aerial/platform and minimize operational considerations for the operator as the aerial is raised and rotated. Conversely, newer apparatus can be equipped with higher cabs and the presence of equipment on the top of the apparatus next to the aerial ladder. These factors can negatively impact the ease and range of operation of an aerial/platform.

A clever combination of a funnel and attached hose allows water to be quickly contained from broken sprinkler heads until the system can be turned off. This is an easy way to minimize water damage and cleanup operations.

Comparing a rear-mount platform to a mid-mount platform, notice the mid-mount platform in figure 3–4 is easier to access from the ground for personnel.

Articulated boom

An articulated boom is shown in figure 3–11. This is a 2004 E-One 114-foot Bronto Sky Lift with a 1,000-pound tip load.

The design and capability of the articulated boom allows the basket to be placed over a vertical parapet and/or wall and still place the basket on a roof, depending on the length of the articulated boom and height of the vertical obstruction.

This particular apparatus is not a quint, although a quint configuration is an option. This type of apparatus and resultant configuration has resulted in reduced compartment space. A quint configuration would have reduced the compartment space even further.

Figure 3–11. A 114-foot articulated boom/platform.

Quint

A quint is shown in figure 3–12. This is a 1991 E-1 75-foot aerial with a 1,250 gallons-per-minute pump, a 500-gallon tank, 500-pound tip load and 1,250 gallons-per-minute elevated stream.

Figure 3–10. A rear-mount 100-foot platform quint.

Figure 3–12. A 75-foot aerial ladder quint is one of the most popular aerial devices.

This apparatus configuration is arguably the most popular apparatus in use today. This particular apparatus is primarily configured as an engine company, although it does carry a basic assortment of truck company tools and equipment. It has a 75-foot aerial with a 500-pound rating.

Ladder tender

This particular type of apparatus is a departure from a traditional truck company in that it does not have an aerial device; it carries a full complement of truck equipment and is normally parked in a station that also has a truck apparatus with an aerial device. As previously mentioned, this concept has recently become popular for two basic reasons: The initial cost is significantly lower than the initial cost for modern truck apparatus with an aerial device, and the day-to-day operational cost is significantly lower for this type of apparatus.

Some ladder tenders carry few, if any, ground ladders. The thought behind this concept is that on-scene engine companies have basic ground ladders that can be used if necessary. Conversely, some ladder tenders carry a basic complement of ground ladders and are a more self-sufficient apparatus in varying conditions. Additionally, some ladder tenders are equipped with a small pump, water tank, and hose so their capability is enhanced.

The operational considerations vary but are somewhat similar for departments that were interviewed for this apparatus. Generally, there was no standard operational procedure for response priorities when determining which apparatus (ladder tender or aerial device) should respond to an alarm. This decision was normally the responsibility of the appropriate officer in the corresponding station. Obviously, a decision of this type is based on the type of response and/or the officer's knowledge of a district. If a ladder tender arrives at an incident and it is determined that an aerial device is necessary, a truck with an aerial device must be requested which can result in a varying amount of time until its arrival.

The ladder tender shown in figure 3–13 carries support equipment that is normally carried by a truck company in addition to ground ladders.

Figure 3–13. This ladder tender carries support tools and selected portable ground ladders.

Additionally, this apparatus also carries a public relations kit that is composed of the following items (figure 3–14):

- 2 smoke detectors
- after-the-fire brochures
- after-a-medical-emergency brochures
- 4 9-volt batteries
- pencils/erasers
- stack of book markers
- stack of coloring books
- restaurant coupons
- badge stickers
- 3 pass-it-on cards
- 3 ride-along tickets

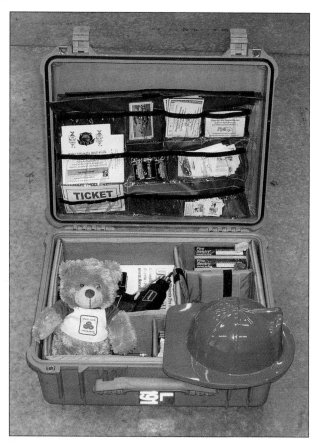

Figure 3–14. A cache of items that can enhance customer service at an incident

- stickers
- key chains
- senior brochures
- home safety brochures
- 3 citizen recognition—golden tickets
- 1 kids fire helmet
- 1 teddy bear

As a point of interest, each officer of all apparatus has access to a pro-card with a $200 limit that can be used to provide customer service at the officer's discretion, such as to purchase groceries, fix a flat tire, take a family and/or kids from an incident out for an ice cream cone, and so on. Giving cash is not an option.

Training Resources

The following resources can provide additional information and/or another viewpoint for the subjects discussed in this chapter.

- *Factory Inspections of New Fire Apparatus Equipment*, William C. Peters, Fire Engineering books and videos.

- 1901 *Standard for Automotive Fire Apparatus*, NFPA.

- Fire Apparatus Purchasing Handbook, William C. Peters, Fire Engineering books and videos.

- Brakes, Axle, Weight, Mark Berg, *The Minnesota Fire Chief*, November/December 1990.

- Primary Qualifying Questions, Selecting the Appropriate Model, American LaFrance Fire Apparatus, 2008.

- Apparatus Showcase, One of a Kind, Midvale's New Tractor Drawn Aerial Quint, Chief Stephen Higgs, *UFRA Straight Tip*, April/June 2008.

- The Sky's the Limit, Larry Safko, *National Fire & Rescue*, November/December, 2005.

- Replacing an Aerial Unit: Platform or Straight Stick?, William Peters, *Fire Engineering*, 1996.

- What are We Going to Call this Rig?, William Peters, *Fire Engineering*, October 2001.

- How to Avoid Big, Fat Fire Trucks: Become a Weight Watcher, William Peters, *Fire Engineering*, November 2002.

- To Buy (or Not To Buy a Quint), Jake Rixner, *Fire Engineering*, April 2001.

- Why Santa Clara Switched to Tillered Aerials, Jim Haney and Mason Weirhauser, *American Fire Journal*, April 1998.

- The 75-Foot Quint: Know What It Can Do, Bill Adams, *Fire Engineering*, February 2009.

- Apparatus Supplement to Fire Engineering, William C. Peters, *Fire Engineering*, June 2009.

Reading a Building

Introduction

When the term *size-up* is mentioned, what initially comes to mind? Some may say a size-up is one of the first actions taken by a first arriving officer to an incident; others would say a size-up determines a course of action in relation to a perceived risk, and so on. However, when we apply the term size-up to structure fires, several descriptions always seem to be mentioned, such as make a quick lap around the building to determine where and what the problem entails, decide on your course of action, and establish a mental picture to incoming resources.

Interestingly, most injuries and deaths in the fire service occur at structure fires. So, if fireground safety is a primary consideration, conducting a size-up that enhances your ability to mitigate the incident in a safe and timely manner is a cornerstone of efficient and safe fireground operations. Generally, the initial size-up lays the foundation for abatement strategy, and normally starts before arriving at an incident (pre-fire planning, training programs, and so on). Upon arriving at an incident, the initial size-up is expanded and fine tuned by the observable factors that are presented by the incident.

For this chapter, let's focus on the initial size-up as applied to reading a building. However, before we consider the physical aspects of a fireground structural size-up, let's consider two basic reasons why we should do one.

First, NFPA statistics indicate that injuries and deaths related to structural incidents are a result of three primary factors: the firefighter falling into it; it falling onto the firefighter; and the firefighter involved in a flashover (see figure 4–1).

Also, to use an analogy, when serving the ball in tennis, you are in control. You control the speed of the ball, its placement, and its spin. Similarly, prior to your involvement with a building that is on fire, the ability to read a building, evaluate its strong and weak points, and then determine your degree and place of involvement (interior vs. exterior) puts you in the drivers seat. To a large degree, after fireground personnel are committed inside or on a building, the building is often in control. So, take the necessary time and determine what specific factors about a particular building are important. Do not pass up the opportunity (figure 4–2) to take a few minutes and size-up your *growing problem*.

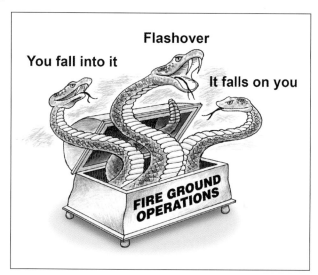

Figure 4-1. The three major fireground hazards are falling into it, it falling onto the firefighter, and flashover.

Figure 4-2. Do not pass up the opportunity to size-up a structure.

The ability to read a building in a timely and effective manner can significantly reduce at least two of the aforementioned NFPA statistics—you fall into it, and it falls on you. Flashover will be covered in chapter 5.

Although some older types of building construction are still in use, newer, more efficient, and more cost-effective construction methods are producing new buildings. Unfortunately, new construction methods are not usually designed to assist fire suppression operations. Considering the cost of labor, equipment, and building materials, it is not economically feasible to construct a structure

the same as the early 1900s. Today, heavy timbers have been replaced by two-by-fours and TGI trusses. Petrochemical-based compounds have replaced conventional interior materials, regardless of building type or size. As modern architects reduce the mass of a common structural member and change the chemical composition of building contents, we are losing one of our most valuable factors—*fireground time!*

Unfortunately, we are fighting structure fires the same way we did 40 to 50 years ago. Hose lines are taken inside an involved structure, and ventilation operations are initiated to improve the fireground environment. But are modern buildings the same as the buildings constructed during the 1920s and 1930s? They are not even close. Therefore, firefighters who can recognize and evaluate the strengths and hazards of buildings will increase their efficiency and safety. A working knowledge of building construction provides not only the necessary expertise to conduct a quick and accurate size-up of a structure, but also the foundation for effective, timely, and safe fireground operations.

When conducting a building size-up, undress the building in your mind (figure 4–3). Look past the exterior of a building and visualize what is inside (strengths and hazards) the building because what you initially see, may not be what you get.

Figure 4-3. Initially, always undress a building from the exterior.

Remember, when you first arrive on scene, every building tells a story. The question is, do you take the time necessary to give you a workable idea of what you want to accomplish during your intended operations, or do you sprint off of your apparatus to see who can be first inside the structure? Most often, the short amount of time required to quickly analyze a building that is under demolition (on fire) normally pays huge dividends (timely and safe operations) during suppression operations.

With the previous thoughts in mind, briefly consider two additional items that must be considered when reading a building: the venerable fireground clock, and how much time you have. Now, what does the term "fireground clock" mean to you? (Check out chapter 1 for more information about the fireground clock.) Some firefighters consider it the time a fire has been burning, dispatch time, time of response, type of materials that are burning, and other similar factors. Although these are important considerations, there is another consideration that is the most important. When you arrive on-scene and conduct your initial size up, ask yourself this question: *How long will it take me to make a visible impact on this fire.* Most often the answer to this question will answer the question of interior or exterior operations. Remember that fires are not static; they are dynamic. This means when you arrive on-scene, the fire will not wait until you put it out, but will continue to burn and weaken the building. So, the question is, how long will that take?

Until a fire is extinguished, it will continue to weaken a structure. Remember that during suppression operations, there are three factors that are working against you: fire weakens a building; gravity wants a building; and water weighs 8.35 pounds per gallon. So the perspective of fireground time (after you arrive on-scene) is a combination of the time it takes until you apply the first drop of water, and the additional time it takes for extinguishment.

Now let's introduce one more factor to the concept of fireground time: How much time the building will give you before it collapses. This perception is more than just a basic principle. It gives you the ability to determine your mode of operation (offensive or defensive), and if you are in an offensive mode, how long should you stay inside a building? With that thought in mind, it can be advantageous for you to develop a rough fireground baseline for the type

of construction in your area. From there, you can evaluate how much time you have before a structural collapse when fire is exposing (and weakening) old and/or structural members. See chapter 1 for a definitive look at developing a rough fireground baseline for conventional and lightweight construction.

To expand on the viewpoint of old and new construction, look at several examples that graphically display an escalating problem that has specifically targeted fireground operations. Figure 4–4 is a great example of common construction during the 1920s, 1930s, and 1940s.

Figure 4–4. Older construction was often significantly more substantial than modern construction.

The focal points in the photo are:

- Floor joists of rough cut 2-by-10-inch or 2-by-12-inch lumber; scissor bracing (or herringbone bridging) were used between the floor joists to keep the joists from twisting.

- The floor of the second story was 1-by-6-inch or 2-by-6-inch sheathing.

- Knob and tube wiring can be seen running between the joists.

Figure 4–5 is an example of construction that was popular from about the 1960s and is still commonly used today.

Figure 4–5. The modern lightweight truss has replaced the substantial construction of yesterday.

The focal points in this photo are:

- The floor joists of the second floor in this building under construction are composed of open web trusses. The size of the wood chords is 2 by 4 inches and separated by thin steel tubes.

- The steel tubes have been attached to the wood chord members by cutting slots in the chords and then inserting the tubes into the chords. This has effectively removed material from this connection point.

Figure 4–6. To save money, glue can be used to replace gang nail plates.

The construction in figure 4–6 is relatively new (at the time this is being written). History can be an excellent indicator of expectations.

The focal points in this photo are:

- The wooden trusses are constructed of pine, which is a relatively soft wood.

- The trusses are held together by glue, which breaks down when exposed to heat.

- The roof decking above the trusses is ½-inch oriented strand board (OSB).

These three examples graphically represent that the building industry has historically been reducing the size of structural members and fireground time. Additionally, figure 4–6 should be a chilling example that if glue (or other similar substances) is used to replace previous methods to secure structural members such as steel plates, 16 penny nails, and so on), it should be apparent that many modern buildings will not offer the same structural stability of older buildings when exposed to fire. In summary, although there are many other examples of building materials and methods that have collectively combined to reduce structural integrity and fireground time when a building is exposed to fire, the need to take the opportunity to read a building (to which you are about to commit) should be clearly evident because every building gives you an opportunity to read its story. To assist in being able to read a building, the balance of this chapter consists of looking at eight items of consideration: construction styles, roof styles, age, method, style, access and egress, name and utilities, and status.

Construction Styles

For simplicity, let's define construction styles as either conventional or lightweight.

Conventional

Conventional construction (2 by 6 inches or larger) derives its strength from dimensional width and height (mass; see figure 4–7), and generally employs independent members. The greater the span of a structural member, the larger it has to be to support a given load.

Additionally, conventional construction usually does not depend on the sum total of all its structural parts or members for its strength. This is demonstrated by mill-timber construction. Those structural members may be 8 by 8 inches or larger and will withstand a significant amount of fire before failing.

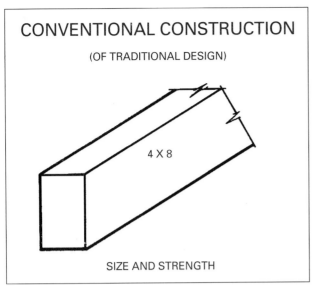

Figure 4–7. Conventional construction depends on the size of its members for strength.

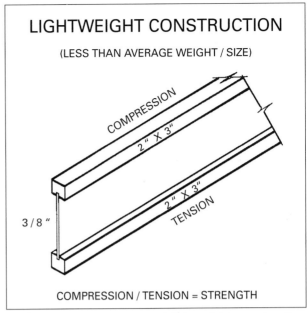

Figure 4-8. The strength of lightweight construction is obtained from small members being in tension and compression.

The larger the size of a structural member, the longer it will take to fail.

Lightweight

Unlike conventional construction, lightweight construction doesn't derive its strength from size. Typically, its strength is obtained from less mass being in compression and tension (figure 4–8). A single lightweight truss can span 70 feet and is composed of two-by-fours in compression and tension to form an integral unit. Although this structural member is strong, its components are relatively small, requiring less time for structural collapse when exposed to heat or fire.

Additionally, the strength of each member is dependent on the sum of the other members (this is often referred to as assembly-built). Thus, if one member fails, others may fail as well. The following is a comparison of some of the major differences between conventional and lightweight construction and what you can expect when they are exposed to fire:

- Conventional construction doesn't rely on the sum of its members for strength. A simple example is a gable roof in a dwelling. The ridge board and rafters form an integral unit; however, they are separate and distinct from the ceiling joists. When the attic is exposed to sufficient fire, the rafters and roof may only collapse onto the ceiling joists,

thereby preventing collapse onto firefighting personnel below. This condition can allow personnel to escape the structure in a timely manner.

- Lightweight construction is vastly different due to truss construction, which does depend on the sum of its members for strength. When a truss gable roof in a dwelling is exposed to sufficient fire, expect the exposed rafters (top chord of the truss), the roof decking, and the ceiling joists (bottom chord of the truss) to collapse as a unit into the structure, exposing firefighting personnel to falling, burning debris.

- Conventional construction can assist in partitioning fire which has the ability to limit the extension of fire. As an example, assume a two-story residence uses 2-by-6-inch joists between the first and second floor, and the joists run in an east-west direction. If fire extends from the first floor to the aforementioned joists, the fire can extend east to west but will have difficulty extending in a north to south direction.

- The size and configuration of structural members translates into a time factor that you must consider when deploying personnel and deciding between an offensive or defensive operation. An accurate estimate of the amount of time that a building can be considered structurally viable depends on the type of construction, how long the fire has been burning, the severity of the fire, perceived load on structural members/floors, extension, how long for extinguishment, available resources, and status of protective barriers (i.e., drywall, truss voids, etc).

Roof Styles

When reading a building, do you include the style of the roof in your size-up, and if so, what are you thinking about? Obviously, some factors are dependent on the type of roof construction in your particular area; however, West Coast roofs and East Coast roofs have a lot in common in construction methods and styles. The primary difference is the age of the roof. As an example, an older gable roof on the East Coast is basically the same as an older gable roof on the West Coast with the possible exception of a higher dead load on the East Coast roof because it can be expected to have more layers of roofing material (which can be a significant weight consideration), and a potential degradation of the wood due to termites, wood rot, and so on. In many cases, the type of roof is an excellent indicator of the age and type of construction in a building.

The six most common roof styles in the United States are gable, hip, flat, bridge truss, arch, and sawtooth (figure 4–9) and are overviewed in the following sections.

Gable roof

A gable roof has an A-frame configuration (figure 4–10). Conventional or ordinary construction consists of a ridge board plus rafters extending down to and across the outside walls.

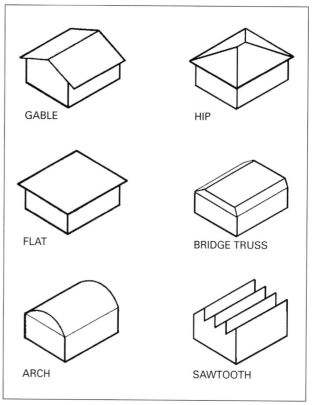

Figure 4–9. The six most common roof styles

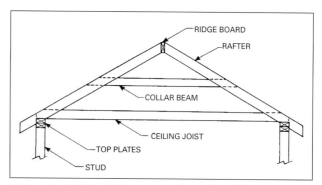

Figure 4–10. The A-frame configuration of a gable roof

The ridge board and rafters are usually 2 by 6 inches or larger. Rafters are usually 16 to 24 inches on-center. As detailed in the frame-wood portion of this chapter, rough-sawn 2-by-3- or 2-by-4-inch rafters spaced up to 36 inches on-center were also used for roof structural members. Additional support is provided by collar beams and ceiling joists. This roof is found in semi-flat to steep-pitch configurations.

Lightweight construction uses 2-by-3- or 2-by-4-inch wood trusses (sometimes metal) normally held together by metal gusset plate connectors (figure 4–11).

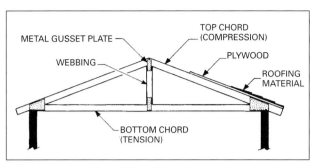

Figure 4-11. Lightweight wood trusses with metal gusset-plate connectors

Trusses share common features such as top chords, bottom chords, and webbing (supports between the top and bottom chords). Metal gusset plate connectors may vary in size, thickness, and depth of penetration. However, 18-gauge steel plates with prongs of ⅜-inch penetration are common. The bottom chord of the truss has replaced the 2-by-6-inch or larger ceiling joists found in conventional construction. Truss systems are enjoying widespread use in roof, floor, and rough window and door openings.

The common on-center spacing for trussed rafters is two feet. In dwellings, conventional or lightweight roof construction will be covered by a variety of products such as shakes, tile, or rock, and supported by space sheathing or ½-inch plywood or OSB type materials.

Strengths. Conventional construction uses ridge boards and rafters of 2 by 6 inches or larger. This type of construction will last longer than 2-by-4-inch trusses when exposed to fire. The strong areas of this roof are the ridge, hips, valleys, and the area where the rafters cross the outside walls.

Hazards. The use of lightweight 2-by-4-inch trusses with no ridge board is similar in external appearance and size to 2 by 6 inch or larger conventional construction. This similarity can easily mislead unsuspecting fire personnel. The presence of 2-by-3- or 2-by-4-inch trusses with metal gusset plate connectors or glue equals a short burning time and potential early failure rate. The trusses are under compression and tension; when the bottom chord or webbing fails due to fire damage (connector plates that become heated and pull out of the wood, or melting glue), the trusses will fail. The bottom chord of the truss has replaced the 2-by-6-inch or larger

ceiling joist of conventional construction. Expect collapse of portions of the roof or total collapse of the entire roof in a short period of time (depending on the extent of the fire). Newer roofs use 7/16 or ½-inch plywood or 7/16-inch OSB as decking instead of 1-by-4- or 1-by-6-inch space sheathing. Plywood and OSB will burn and fail at a faster rate than sheathing, and offers minimal resistance to fire. Particle board, chip board, and OSB are also currently used as roof decking to reduce building costs. These can be more hazardous than plywood. An important additional hazard is possible misidentification of lightweight construction. Identification can be enhanced by an intimate knowledge of a district and pre-fire planning. Additionally, the presence of 2-by-4-inch rafter tails under the eaves are an excellent indicator of lightweight truss construction. However, if the eaves are soffited, the rafter tails will not be visible. Be aware that trusses may be modified on commercial buildings to look like 2-by-6-inch rafters on the front or sides of a building, but in the rear they are commonly not modified. Therefore, the rear of a building can often allow you to see the real size of rafters used in a building.

Hip roof

The hip roof is similar to the gable roof, but notice the lack of the A-frame configuration (figure 4–12). The ends of the roof terminate in a hip configuration.

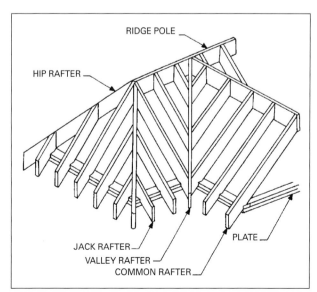

Figure 4-12. The hip roof resembles the gable roof but lacks A-frame construction.

Conventional or ordinary construction consists of a ridge pole plus hip rafters from the ridge pole down to and across the corners at the outside walls. Valley rafters are used where two rooflines join. Jack and common rafters complete the structural members. The ridge pole and rafters are usually 2 by 6 inches or larger. Rafters are usually 16 to 24 inches on-center, similar to the gable roof. Rough-sawn 2-by-3- or 2-by-4-inch rafters 36 inches on-center were also used in older wood-frame structures. In lightweight construction, the construction methods are similar to those used for gable roofs. Various degrees of pitch are characteristic of this style of roof.

Strengths: The ridge pole, valley rafters, hip rafters, and the places where the rafters cross the outside walls are the areas of strength. In conventional construction, the ridges and rafters are 2 by 6 inches or larger.

Hazards: The hazards of hip roofs are similar to those of gable roofs. The presence of 2-by-3- or 2-by-4-inch trusses for rafters will produce results similar to those of lightweight trusses in gable roofs when exposed to fire. Although the on-center spacing of the trusses is reduced, trussed gable and hip roofs can be covered with tile or other such materials. This enhances the collapse potential due to the increased dead load on the roof. Similar to gable roofs, reading the rafter tails (if exposed) is an excellent indicator of the size of the rafters.

Sawtooth roof

Sawtooth roofs are used to yield additional light and ventilation to buildings such as manufacturing-type occupancies (figure 4–13). They are commonly constructed with rafters of 2 by 8 inches or larger, and use wood and/or metal supports for bracing.

The sloping portion is covered with 1-by-6-inch sheathing (½-inch plywood or OSB in newer roofs) and composition roofing material. This type of roof is basically constructed the same today as it was during the 1930s and 1940s with the exception of the roof decking.

Strengths. Sawtooth roofs are generally well constructed. When exposed to fire, the early collapse of main structural members shouldn't be of primary concern. Additionally, this type of roof is easy to

ventilate—simply open the hinged panes of glass. The strong portions are at the perimeter of the building and the areas near the glass panels.

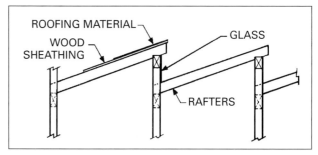

Figure 4–13. Sawtooth roofs provide light and ventilation to the interior of a structure.

Hazards. Consider the undersides of these roofs to be open or exposed to the interior of the structure. Newer sawtooth roofs are covered with ½-inch plywood or OSB decking which have little resistance to fire.

Older truss roofs

Normally, when the subject of fire suppression operations and older truss roofs are considered, the venerable bowstring roof immediately comes to mind as one of the most dangerous roofs a firefighter will encounter. Although this point of view has merit, the bowstring is not the only older truss roof (with similar hazards) that a firefighter will encounter. Interestingly, the following truss roofs are constructed with similar characteristics, the major difference is their external shape that can be easily read from the street.

Although the tied truss and bowstring roofs are both arched, there has been some difference of opinions regarding the nomenclature for these two roofs. Although the tied truss roof looks like a bow when viewed from the side, the tied truss nomenclature is derived from the fact that it is necessary to run a tie rod below each truss (due to the lack of a bottom chord and webbing) to keep the trusses from pushing the walls outward. The construction industry refers to this configuration as "tying the trusses" because these trusses rely on the tie rod for their stability. Conversely, although the bowstring also looks somewhat like a bow when viewed from the side, it is not necessary to tie the trusses for stability, hence the nomenclature of bowstring.

Bridge truss roof. These roofs are found on various types and sizes of commercial buildings primarily constructed during the 1920s, '30s, and '40s (figure 4–14). The wooden truss members are built from rough-cut 2-by-12-inch lumber. This usually constitutes a heavy grade of construction. Vertical metal tie rods may be used for additional support.

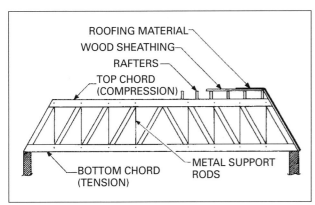

Figure 4–14. Bridge truss roofs are common on older commercial structures.

The rafters are 2 by 6 inches or larger, and are covered by 1-by-6-inch sheathing (diagonal or straight) and composition roofing material. Straight sheathing was used prior to the mid-1930s, and diagonal sheathing was favored afterward. Diagonal sheathing provides increased structural stability to a roof assembly compared with straight sheathing. Plywood and OSB decking can be found on top of straight sheathing in earthquake-prone areas.

Strengths. Bridge truss roofs are well constructed. When exposed to fire, early collapse of main structural members should not be of primary concern. This type of roof can fail predictably in sections depending on the type of fire. It is easily identified by its characteristic sloping sides and ends. The strong area is at the perimeter of the building.

Hazards. Its strength is dependent on the size of the lumber and the span of the trusses. The trusses are in compression and tension and can fail under severe fire conditions. The underside of this roof is usually exposed in warehouse-type structures.

Arch roof (bowstring). The bowstring is similar to the bridge truss roof (figure 4–15). This popular type of roof was constructed during the early to mid-1900s on both small and large commercial-type structures.

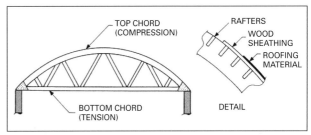

Figure 4–15. A bowstring truss is similar to the bridge truss except in visual appearance.

Usually large-size rough-cut (2 by 12 to 2 by 14 inches) wooden members comprise the arch trusses and related members. Some arch trusses have multiple members forming one truss arch. Rafters (2 by 6 inches or larger) are covered with 1-by-6-inch sheathing, straight or diagonal, and composition roofing material. Diagonal sheathing provides a stronger bond as each 1 by 6 crosses multiple members as opposed to straight sheathing.

Strength. Most roofs of this type are well constructed. When exposed to fire, early structural collapse of the arched trusses should not be of primary concern. Like the bridge truss roof, it usually fails in sections depending on the type of fire and the structural integrity of the roof. The strong area is at the perimeter of the building.

Hazards. The size of the lumber and the span of the arches determine the hazards. Most roofs of this type are well constructed. However, the underside is usually common to the interior of commercial/warehouse-type structures. The bottom chord of the arch trusses can be modified to allow storage in the attic area or ceilings (the tops of the bottom chords can be covered with sheathing or plywood) that make ventilation, fire control, and overhaul difficult (this can also apply to bridge and gable truss roofs). Additionally, lath and plaster can be attached to the bottom chords of the trusses, further increasing the weight that they must carry (and were not designed for).

Gable truss and parallel chord truss. The gable truss (A) and parallel chord truss (B) in figure 4–16 are similar with the primary difference being their characteristic shape. Both of these roofs were used on commercial buildings of moderate size and were constructed during the early to mid 1900s. Both of these roofs were constructed from a heavy grade of lumber (usually rough cut) and have characteristics similar to those of the bridge truss.

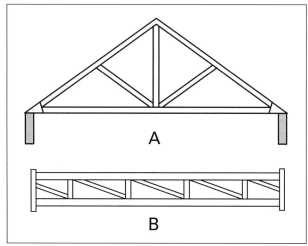

Figure 4–16. The gable truss (A) and parallel chord truss (B) are variations of older truss roofs.

Strengths. These roofs are well constructed and are a good example of older construction when lumber was plentiful. These roofs normally predictably fail in sections that are exposed to fire, while uninvolved sections do not. The gable truss is easily identified by its shape, and the parallel chord truss is common on older buildings with flat roofs. The strength of these roofs is the perimeter of the roof.

Hazards. The main hazard of these roofs is two fold. Although they are constructed of lumber that is significantly larger than modern trusses, they will last longer and thereby allow interior personnel to overcommit in suppression operations. Also, the underside of the gable truss is often exposed and vulnerable to fire. However, the parallel chord truss is normally not exposed to fire due to ceiling type materials on the underside of the truss, thus concealing fire within the trusses.

Lamella. A lamella is an egg-crate, geometric, or diamond-patterned roof that commonly has a higher

(or steeper) arch than the common bowstring roof (figures 4–17 and 4–18). It is constructed of 2-by-12-inch wood framing with steel plates and bolts at the framing junctions. (Note: Although the lamella roof is not a true truss roof, it is placed in this section because of its age, shape, and similarity with older truss roofs.)

The roof decking is 1-by-6-inch straight or diagonal sheathing with composition roofing material. This type of arch roof is supported by exterior buttresses or internal tie rods with turnbuckles. It is common on gymnasiums, recreational buildings, large supermarkets, and the like. This particular roof can often be recognized by its shape as it often has a higher hump type configuration.

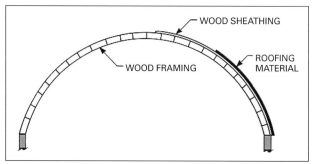

Figure 4–17. The lamella roof is common on structures that need large open interior areas without visible vertical supporting members.

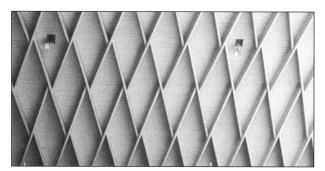

Figure 4–18. Diamond-patterned lamella roof

Strengths. Lamella roofs are solidly built, possessing good construction techniques and lumber. The strong area of this roof is at the perimeter of the building.

Hazards. Although these roofs offer some protection when exposed to fire, total roof collapse may occur if fire removes more than 20% of the roof structure. Total collapse of the roof can result from the domino effect.

Tied Truss. Although this roof is similar in external appearance to bowstring arch and lamella roofs (arched), it is significantly different internally because it uses metal tie rods to offer lateral support to the walls of the building (figure 4–19) in place of webbing and bottom chords. Tie rods (usually ⅝-inch) with turnbuckles are used below each arch member to ensure that the arches do not push the exterior walls outward.

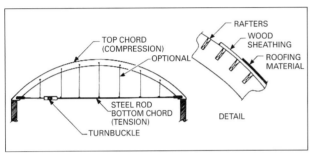

Figure 4–19. The tied truss roof uses metal tie rods to support the walls of a building.

The tie rods may pass through the exterior walls to outside plates, which can facilitate identification of this roof. Proper tie rod tension is maintained by turnbuckles. The top chords of the arch members may use laminated 2 by 12s or larger members, and 2-by-10-inch rafters are covered by 1-by-6-inch sheathing (straight or diagonal) and composition material.

Strengths. This type of roof uses a large size of lumber (2 by 12 inches or larger) and 1-by-6-inch sheathing as the roof decking. The strong area of this roof is at the perimeter of the building.

Hazards. The primary hazard of this roof is early failure of the metal tie rods and turnbuckles. The tie rods, which are in tension, provide lateral support for the exterior walls and prevent the arches, which are in compression, from pushing the exterior walls outward, thus thwarting collapse. There are basically three ways to form an arched roof—lamella, bowstring, and tied truss. The lamella and bowstring type roofs can be primarily wood (heavy timber) and can be quite durable when exposed to fire. The same cannot be said for the other form of an

arched roof (tied truss). Because this roof depends on the strength and security of its tie rods, failure may occur in sections that are exposed to fire. However, it may also be susceptible to total failure depending on the type of fire. Compared with the bridge, parallel, gable, bowstring, and lamella timber trussed roofs, the tied truss roof offers significantly less structural integrity during fire conditions.

In summary, let's examine the older truss roofs from several different viewpoints. In recent years, attention has been appropriately focused on timber truss roofs, particularly as a result of the fire in a bowstring truss roof located in Hackensack, New Jersey, that resulted in five firefighter fatalities. This incident focused on five principal hazards that are often ascribed to these types of roofs:

- Truss roof
- Easy to identify
- Weak roof
- Early failure rate
- Collapse without warning

Although most of these hazards can apply to any roof that is subjected to fire, let's consider several current popular perceptions.

Truss roof. Roof configurations that are composed of a top and bottom chord and separated by webbing are normally referred to as truss construction. However, this viewpoint doesn't take into account the *size* of the truss, which can directly affect the amount of time available before collapse. Compare the size of structural members in older heavy timber trusses and modern 2-by-4-inch trusses. Although the dead load (roofing materials, HVAC equipment, and so on) on the older trusses is greater, fireground experience has proven that timber trusses can offer significantly more time before failure than lightweight trusses. Also, the older truss roofs used steel plates and bolts at connection points instead of gang nail plates that penetrate wood members only ⅜-inch or use glue. These factors can affect the amount of time available to initiate or terminate appropriate operations.

Easy to identify. Although this statement applies to arched trussed roofs, it doesn't apply to all trussed roofs. Remember that all trusses are constructed according to the same principles and can fail in the same manner. Therefore, remember the following trussed roof shapes: bowstring and tied truss (arch), gable truss (triangular), bridge truss (pyramidal), and parallel chord truss (flat). Not all wood truss roofs exhibit the characteristic exterior hump or arch shape. Additionally, consider that the timber truss roof is not limited to long unsupported spans or particular types of buildings. The timber truss roof was one of the most popular roofs constructed until the 1950s and was used on a wide variety of commercial and industrial buildings. These include large warehouses, two-story office buildings, and even simple 20-by-40-foot commercial occupancies. Expect to encounter these roofs on many older commercial and industrial buildings, and know the different types of truss roofs in your area.

Weak roof and early failure rate, collapse without warning. The amount of time for failure cannot be reliably predicted for a roof. As a result, any roof can be dangerous and may collapse without warning during the early stages of a fire. Interestingly, history doesn't support the perception that timber truss roofs generally fail during the early stages of a fire. Although the definition of *early* is debatable, let's briefly consider the following noteworthy fires involving timber truss construction.

Waldbaum's Supermarket, New York City, August 2, 1978. Two factors contributed to the collapse of this bowstring arch: the double roof (rain roof) alteration, and the extent/severity of fire. The roof collapsed 32 minutes after the initial units arrived.

Hackensack Ford Dealership, Hackensack, New Jersey, July 1, 1988. Three factors contributed to the collapse of this bowstring arch: alterations that consisted of a heavy ceiling of cementious material on wire lath: auto parts storage in the attic, and the fact that the fire had burned for a significant length of time and was well advanced prior to detection. The roof collapsed 35 minutes after the initial units arrived.

Neither of these incidents supports the view that timber truss roofs fail quickly, even when they have been significantly altered from their original design. Fireground experience has indicated that timber truss roofs (except the tied truss) are capable of lasting over 30 minutes under heavy fire conditions.

The major hazard attributable to timber truss construction may not be the construction itself but a combination of the following factors.

Alterations that exceed the design criteria of the original roof. Unfortunately, these roofs are easily altered for numerous purposes. As previously mentioned, it is common practice to cover the top of the bottom chords with flooring that creates a significant storage space within the trusses. Additionally, it is also common to cover the underside of the bottom chords with materials to create a ceiling, particularly cementious materials in older applications. Consider the following three hazards when confronted by timber trusses with ceiling materials on the bottom chords of the truss members: ceilings add additional weight that the trusses may not have been designed to support; ceilings will hamper access and suppression efforts in the attic area (remember that ceilings can hide a well involved fire in the attic that may not be visible to interior personnel located below the ceiling); and ceilings can be indicative of storage, translating into additional weight on the truss assemblies.

The inherent size and strength of this construction can allow a fire to burn for a period of time in the upper portion of the trusses while personnel initiate extended interior attack and/or roof ventilation operations. The preceding hazards were all present at the Hackensack collapse.

Age. There is a significant difference between timber truss roofs located in the western and eastern states. Although these roofs can be constructed the same, the timber truss roofs on the East Coast are significantly older and have been subjected to harsh weather conditions, wood rot, termites, renovations, and other circumstances for a longer period of time than their West Coast counterparts. Not surprisingly, the majority of timber truss roof failures have occurred in the eastern states.

Depending on the size of the truss members, timber trusses can possess more strength than they are given credit for. The timber trusses in figure 4–20 span more than 90 feet with a significant portion of the truss members and all of the roof decking removed by fire, yet the trusses are still standing.

Figure 4-20. These timber trusses are still standing after significant portions of them have been removed by fire.

Operations. An aggressive attack on fires in timber truss roofs can be successful in some instances. Several factors are suggested considerations for suppression operations in timber truss roofs. First, and if possible, determine if the fire is only in the structure and not within the trusses, or if the fire is in the truss attic area and *involving* the trusses. If the fire is within the building but not involving the trusses, the collapse potential of the trusses is minimal.

Next, if the extent of fire is unknown, and/or the underside of the trusses have been covered by materials to form a ceiling and obscuring the truss attic area, these considerations should require the first company to briefly enter the structure, pull the ceiling, and evaluate the attic/trusses. If the fire is in the attic area and well involved, plan for a defensive operation. If a fire is in the attic area and not well advanced, or can be extinguished in an acceptable time period, an aggressive attack should be successful. If a fire is not in the attic area, an aggressive interior attack should also be successful.

Too, the presence of flammable stock in close proximity to trusses will dramatically enhance the extension of fire into a truss area, particularly if the trusses are exposed.

Also, the incident commander must carefully evaluate the type of construction, the degree of fire involvement, the approximate time the fire has been burning, and the resources available. As with other types of construction, a working knowledge of these variables can aid in deciding whether to initiate an offensive or defensive attack.

Avoid conducting an interior attack of long duration. If interior attack operations are initiated, suppression efforts should begin to control the fire. If the fire is not controlled and actually appears to be increasing, discontinue offensive suppression efforts.

Timely and accurate communications are necessary at any fire, particularly at one involving a timber truss roof and aggressive suppression operations. All personnel must be aware of the type of construction and the intended plan of operation. Interior and roof personnel must continually communicate their progress or problems to the incident commander.

Timber truss roofs can be among the most hazardous types of construction encountered by suppression personnel. However, they can also offer the strength and time necessary to conduct successful operations in and around them. Don't let the hazards of these roofs detract focus from the hazards inherent in all roofs. Remember that any roof can cause fatalities if the proper ratio of fire, time, and type of construction is present.

Flat roofs

Flat roofs have become popular for a wide variety of structures. Instead of constructing arch or sawtooth designs, just erect four walls and cover them with a flat roof. This is a cost-effective design, and becomes even more cost effective when lightweight materials are used. Although simple in appearance, flat roofs vary in construction methods and can hide modifications to the basic roof structure, often making it difficult to determine which type and/or method of construction has been used. The focus is to try and identify the type of flat roof based on a prior knowledge of a particular roof, the style of building, and the age of a building. Flat roofs can be categorized as follows.

Conventional flat roof. Wood rafters or joists of various sizes (2 by 6 inches and larger) are laid across the outside walls and may or may not have internal supports (figure 4–21).

The rafters may also be suspended by metal hangers. Some older roofs have bridging or scissor bracing between the joists to increase lateral stability (refer back to figure 4–4). The rafters are covered with 1-by-6-inch sheathing, plywood or OSB (in newer applications) and composition roofing material.

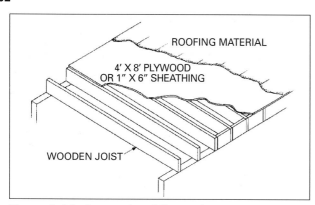

Figure 4-21. Conventional flat roof

½ inch) nailed to the I-beams. When 4-by-8-foot sheets used for decking are nailed to structural members (truss or conventional construction), a method called diaphragm nailing is used. Prior to nailing, the 4-by-8-foot sheets are placed so that the 8-foot dimension crosses the roof structural members and the 4-foot dimension parallels them. The sheets of plywood are then staggered every 4 feet like bricks in a masonry wall. Wooden I-beams are either hung from bearing walls by metal hangers or placed on top of the bearing walls. The common on-center spacing is 2 feet.

Strengths. Their susceptibility to fire is totally dependent on the size of the rafters, the on-center spacing, and the type of decking. Older roofs use rough-sawn lumber with numerous bridging supports and 1-by-6-inch sheathing as decking, which offers good resistance to fire. The area of strength for this roof is at the perimeter of the building.

Hazards. The degree of hazard is determined by the rafters' span, size, and on-center spacing and the presence of metal hangers used to suspend them. Roofs covered with plywood or OSB instead of sheathing present a significant problem. Plywood and other similar materials, typically of ⅜- to ½-inch thicknesses, offer minimal structural integrity under fire conditions. Plywood layers spread apart when they are exposed to heat and fire, and OSB burns rapidly; consequently, the decking itself may be burned out from below roofing materials without showing any signs of weakness above. All of these materials require specific ventilation techniques.

Flat Roof (Wooden I-Beam). These roof structural members (also known as TJI) consist of three main components: top chord, bottom chord, and stem (or closed web). Common two-by-fours are used as chords, but 2-by-3-inch chords are also used (figure 4–22).

Some chords may resemble plywood due to horizontal (longitudinal) laminations. This is a trade lamination process (referred to as micro-lam) that enables a cheap grade of lumber to be used for structural members. The stem is joined to the top and bottom chords by a continuous glued-edge joint and may be constructed from ⅜-inch plywood or OSB of the same thickness. This construction is very unstable until adequately braced with nailing blocks and 4-by-8-foot sheets of plywood/OSB decking (usually

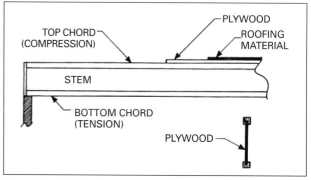

Figure 4-22. Wooden I-beam members consist of a top chord, bottom, and a stem.

Strengths. The strong area of this roof is at the perimeter, where the roof ties into the building.

Hazards. The principal structural hazards are twofold. The ⅜-inch stems and the 2-by-3- or 2-by-4-inch chords are in compression and tension. It will take little time for the ⅜-inch stem to burn, weaken, and cause the collapse of any truss chords and roof sufficiently undermined by fire. Buildings can be found with open and unprotected chords. Also, the glue used to form OSB used for stems can degrade with convection heat. The glue is hydrated—chemically wet. When the moisture is steamed off, the glue turns to dust. A common practice is to run heating and air conditioning ducts of various sizes through the stems. This removes a significant portion of the stem and gives fire horizontal access to adjacent I-beams, enhancing extension. Ventilation personnel must be aware of nailing blocks when cutting between and parallel to the top chords of the I-beams. Nailing blocks can also be found on open-web and metal gusset plate construction.

Flat roof (open-web construction). Open-web construction consists of top and bottom parallel wooden chords cross connected by steel-tube web members (figure 4–23). The top chord (supported) is under load and offers a bridging affect that causes it to be in compression. The bottom chord member (unsupported) is in tension.

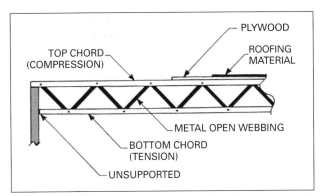

Figure 4–23. The top and bottom chords in open-web construction are connected by steel-tube web members.

Open-web construction is prefabricated at the factory before installation and is constructed either with parallel chords laid on edge or flat-laid chords. The steel-tube web members are prefabricated from one- to two-inch cold-rolled steel tubing. The ends are pressed flat into a semicircular shape with a hole punched through them. These flattened ends are inserted into slots in the chords and steel pins are driven through, completing the assembly. Spans of 70 feet are possible using a single two-by-four or two two-by-threes as top and bottom chord members. A single two-by-four up to 70 feet is made possible by joining different lengths in glued, mitered finger joints. The normal on-center spacing is 2 feet.

Strengths. The strong area of this roof is at the perimeter of the building, where the roof ties into the exterior walls.

Hazards. As with most lightweight construction, the hazards of this roof are numerous. It is basically constructed of two-by-threes or two-by-fours under compression and tension with decking of ½-inch plywood or OSB. These components offer minimal resistance to fire. The chord members are exposed in the interiors of some structures, increasing the exposure hazard to the roof. Expect a lack of fire stops in this construction. A noteworthy hazard

is the openness between the chords. Unlike conventional construction, the open space between the top and bottom chords will promote the lateral extension of fire, resulting in early collapse of the roof (this is actually horizontal balloon frame construction). Expect rapid failure due to the size of the lumber and its chord members in compression and tension.

Flat roof (metal gusset plate construction or glue). Trusses for roofs are constructed in a wide variety of methods, shapes, and styles. However, all of them share common features (figure 4–24). Lightweight wood trusses are predominantly composed of two-by-fours held together by metal gusset plate connectors.

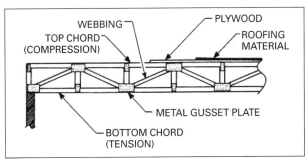

Figure 4–24. Metal gusset plate trusses share common features with other trusses.

Currently, trusses are being used that make use of glue instead of metal gusset plate connectors. Because this is relatively new, there seems to be some disagreement regarding the temperature at which the glue will melt. However, several non Underwriter's Lab (UL) tests have indicated that the glue will melt at between 250°F and 300°F. These types of trusses consist of top chords, bottom chords, and webbing (supports between the top and bottom chords). If metal gusset plate connectors are used, they vary in size, thickness, and depth of penetration. Eighteen-gauge steel plates with prongs of 3/8-inch penetration are common and are used in a wide variety of applications. You may encounter two-by-fours in spans of up to 80 feet. The decking is usually ½-inch plywood or OSB.

Strengths. The strong part of this roof is the area where the trusses cross the outside bearing walls (cantilever applications) or terminate on them.

Hazards. Extensive use of 2-by-4-inch trusses with metal gusset plate connectors equals short burning time and early failure of the roof. The trusses are under compression and tension. When the bottom chord or webbing fails (whether from connector plates that have pulled out or from deep char), one or more trusses can also give way. Rapid and total collapse is common if fire or high heat is exposing all truss members, metal gusset plate connectors, or glued joints. In addition to the metal gusset plates, it is easy to visualize that if trusses are encountered that are glued together, and the glue melts at a moderate temperature, collapse will occur relatively quickly, depending on the amount and extension of fire. Decking composed of ⅜- or ½-inch plywood or OSB is common. Plywood and OSB will burn and fail at a fast rate and offers little resistance to fire.

Flat roof (panelized). This roof may be found on wood, masonry, or concrete tilt-up slab buildings (figure 4–25). It is predominately common in western states and consists of four major components: beams (laminated wood or metal), purlins, 2-by-4-inch joists, and ½-inch plywood or OSB decking.

After the walls have been erected, the roof is begun with laminated beams spanning the length or width of the building. These beams vary in size, but 6 by 36 inches are common. They are supported at their ends by pilasters or saddles and may be bolted together to provide lengths well in excess of 100 feet. Wood or steel posts may provide additional support along the span. These beams are spaced 12 to 40 feet apart. Supported by these beams, wooden purlins are then installed with metal hangers on 8-foot centers. A common size for a purlin is 4 by 12 inches, with the length depending on the spacing of the beams. Metal gusset plate trusses are now being substituted for conventional purlins, resulting in substantial cost savings. Joists measuring 2 by 4 inches by 8 feet are then installed with metal hangers on 2-foot centers between the purlins, parallel to the beams. Sheets of ½-inch plywood or OSB are nailed over this framework. Composition roofing material covers the plywood or OSB decking. A three-layer insulation paper can be stapled to the underside of the roof between the beam and the purlin. It consists of tar-impregnated kraft paper covered on either side by aluminum foil.

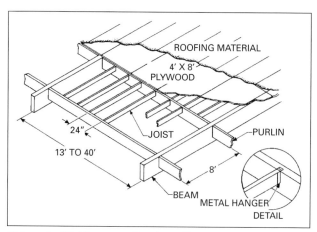

Figure 4–25. The panelized roof is common in western states.

Strengths. The strengths of this roof are its beams, the purlins, and in the regions along the building's perimeter.

Hazards. Four-inch hollow steel pipe can be found supporting the span of the beams. Expect weakening or collapse of these supports with subsequent failure of large portions of the roof under heavy fire conditions. Moderate to heavy fire intensities will quickly burn through the 2-by-4-inch joists and ½-inch plywood or OSB decking, which can result in vertical fire travel and a reduction in horizontal fire spread. When the insulation (kraft paper) is subjected to fire or sufficient heat, the foil covering will peel away from the middle layer of tar-impregnated paper. The paper will give off flammable gases that rise and build up between the insulation paper and the decking. When the ignition temperature of these gases is reached, they can flash, igniting the insulation paper and charring the surrounding wood. The burning insulation will fall away, contributing to extension. Fire is then able to expose the 2-by-4-inch joists and ½-inch plywood or OSB decking, which offer little resistance. If lightweight trusses are used for the purlins, expect additional and rapid roof failure if exposed to fire.

Flat roof (open-web bar joist). Open-web bar joist construction uses metal in a wide variety of buildings, large and small (figure 4–26). It is predominately common in the central and eastern states; however, it is now replacing panelized roofs in Western states due to a lower cost that is a result of an increasing scarcity of wood.

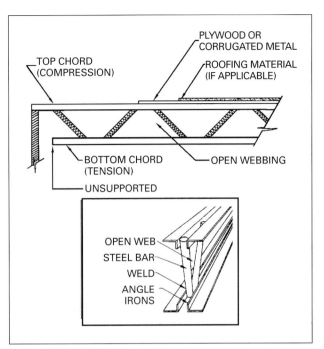

Figure 4-26. The open-web bar joist roof is common in central and eastern states, and is gaining popularity in Western states due to the scarcity and price of wood.

The top and bottom chords are usually made from 1/8-inch steel, and the web supports are solid ⅝-inch steel bar. Large buildings may have bar joists for girders spaced up to 45 feet apart. The joists are spaced at 8-foot intervals with corrugated metal or plywood decking. Corrugated metal is a common decking material for bar joist construction and may be bare or covered by alternating layers of tar and tar paper. These layers may also include a composition board or other type of material to provide insulation protection. Additionally, 4-by-8-foot sheets of ½-inch plywood or OSB with 2-by-4-inch joists are gaining popularity on the West coast. This roof may have composition covering the decking.

Strengths. The strong area of this roof is at the perimeter of the building.

Hazards. Metal exposed to fire or sufficient heat (800°F to 1,000°F for steel) will expand, twist, and possibly fail. The time necessary for roof collapse is of major concern when the entire roof is composed of metal. An additional hazard is travel of fire between the multiple layers of tar paper, insulation materials, and the corrugated metal decking, making extinguishment difficult. In large buildings, large spans are common that compounds the potential of collapse.

Flat roof (nonstructural lightweight concrete). A steel or wood substructure is covered by corrugated metal (Robertson decking). An air-entrained mixture of sand, cement, and occasionally pea gravel is pumped on top of the corrugated metal decking and a 4-by-4- or 6-by-6-inch wire mesh to a thickness of three or four inches. Composition roofing material makes up the final layer. These roofs provide additional insulating properties for buildings near airports, freeways, expressways, turnpikes, and the like (figure 4–27).

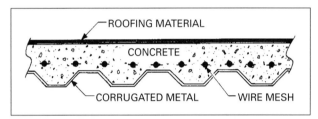

Figure 4-27. Nonstructural lightweight concrete roofs provide soundproofing as well as strength.

Strengths. Lightweight concrete roofs offer strong, hard surfaces. They are structurally sound and resilient to fire. The strong area of this roof is at the perimeter of the building. The remaining portions can also be strong, depending on the span of the roof.

Hazards. Although the roof is strong, an item of consideration is what holds the roof up. Common vertical structural members are hollow 4-inch steel pipes, wood members, and so on. Lightweight concrete roofs are difficult to penetrate without specific equipment. Masonry blades are ineffective. Use a rotary saw with a carbide-tipped wood blade or multi-use blade. A chain saw equipped with a carbide chain will also work. These and other methods will be more fully addressed in the chapter on ventilation.

The following three styles of roofs are not a specific style of roof; however, they have definite characteristics that should be considered.

Corrugated roof (flat or sloped)

Corrugated roofs (figure 4–28) are inexpensive and easy to erect, whether used in large or small applications. They consist of steel, aluminum, or fiberglass over a wood or metal substructure.

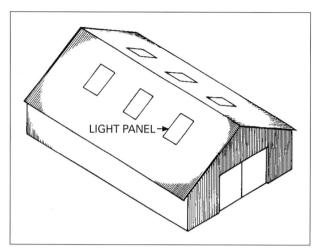

Figure 4-28. Although corrugated roofs aren't a specific style of roof, consider them extremely dangerous when exposed to fire.

Corrugated steel is usually of 18- to 20-gauge thickness. These roofs are normally found in flat (with varying pitches), gable, or hip configurations and are easy to identify.

Strengths. The strong areas of this roof are the ridge and those portions that cross the outside bearing walls.

Hazards. Corrugations may be steel, aluminum, or fiberglass. Expect rapid failure of these materials when they are exposed to heat or fire. Plastic or fiberglass light panels are common and will readily fail when exposed to heat or fire. Personnel must consider corrugated materials extremely hazardous for aboveground operations. Use a roof ladder or aerial device to distribute the weight load if it is necessary for personnel to be on this type of roof.

Rain roof

A rain roof (figure 4–29) is a roof that has been added on to an existing roof for the purposes of correcting a potential hazard from excessive weight, or to change the pitch of an existing roof so drainage from the roof will be improved.

As an example, a double bowstring roof results in a significant low point in the middle of a building where the two arches meet. This results in a bowl or low point that will collect snow and rain that can impose significant loads that can be detrimental. To correct this dilemma, a rain roof is constructed

over the existing roof (dotted line, figure 4–29A) that eliminates the low portion between the arches. A rain roof can also be constructed over an existing roof (usually a flat roof) that has suffered from improper drainage. This type of rain roof can yield a new pitch that can be tailored to correct any drainage problems (dotted line, figure 4–29B), give an existing roof a new look, and so on. Be aware that the new look from the rain roof can be a sloped flat roof, or other configurations such as a trussed gable roof that can hide the HVAC equipment (within the truss loft) on the original roof.

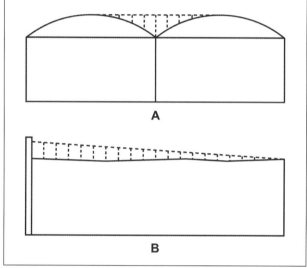

Figure 4-29. Rain roofs are normally a retrofit to an existing roof to enhance the drainage of water from an existing roof.

Strengths. In reality, this type of roof is a retrofit constructed over an existing roof. Therefore, the strengths are dependent on the type of construction that has been used. In any case, the strengths are the perimeter of the roof.

Hazards. The hazards of this construction are numerous and dependent on three basic factors. A rain roof is likely to not be detectable, thereby allowing fireground personnel to assume there is one roof instead of two. Also, it is obvious a rain roof will significantly increase the weight (dead load) on an existing roof that may or may not have been designed for the added weight. Finally, when roof personnel cut ventilation openings, unless it is obvious there is another roof below the rain roof, they will likely not observe any smoke/fire/heat from the opening, which can lead to a false opinion of actual conditions below the rain roof. Obviously, this can be a dangerous condition.

Inverted roof (raised roof)

An inverted roof might in some ways be similar to a rain roof, but there is a major difference. Normally, an inverted roof is constructed as a complete assembly, not a retrofit. A conventional roof is normally composed (going upwards) of ceiling materials (such as plasterboard), ceiling joists, rafters (2 by 6 inches or larger), decking, and roofing materials. An inverted roof is composed of (again, going upwards) ceiling materials, rafters (2 by 6 inches or larger), two-by-fours to support the decking, and roofing materials. This configuration uses the two-by-fours (or other similar members) to slope the roof for drainage.

Strengths. As opposed to the rain roof, this roof is constructed as an integral unit, so the dead loads should have been accounted for as opposed to a retrofit type of construction. The strengths are the perimeter of the roof.

Hazards. Compared to the rain roof, the hazards are somewhat simplified. However, because the roof is inverted, if fire is able to reach the attic space, it will not take long to weaken the top members (two-by-fours or similar members) because they are much smaller than the rafters supporting the ceiling. This can be a dangerous situation for roof personnel.

Attics and cocklofts

Before leaving this section on roofs, let's define three terms that are created by roofs and play an important part in fireground operations and priorities. These terms are cockloft, attic, and truss-loft, and are defined as follows:

Cocklofts are defined as a *small* space that is created when a roof is raised above the level of ceiling joists and rafters to provide a pitch for drainage. Cocklofts are common in older buildings (1800s and early 1900s) and are normally of conventional construction.

Attics are defined as a *large* space that is created by a steep pitched roof (arch, gable, and so on) for drainage and/or appearance. Depending on the type of construction, attics are normally large enough for storage, can be modified for additional living space, contain HVAC equipment, and so on.

Truss-lofts are defined as an attic space created by lightweight trusses.

Age

It is advantageous to determine the age of a building. From a simplistic viewpoint, it can give you a general idea of the type of construction and inherent hazards you may encounter and, in concert with the amount and extension of fire, gives you the ability to evaluate how much time is available to safely conduct your anticipated operations. Obviously, it can often be challenging to read the interior of some buildings, particularly those that have been remodeled or changed from the original construction and floor plan. However, attempting to read a building from the exterior is superior to being surprised or caught unaware by something that should have been anticipated. Buildings can be conveniently divided into four general classifications: pre-1935, 1935 to 1960, post-1960, and 2007. Let's begin reading the age of a building by considering buildings constructed prior to 1935.

Pre-1935

As delineated in the Methods section of this chapter, the earthquake in Long Beach, California, in 1935 initiated a new chapter in the construction of brick masonry buildings. Prior to 1935, these buildings used lime and sand for mortar, no rebar, and were referred to as unreinforced masonry construction (URM). Therefore, the combination of bricks, lime, and sand mortar were the primary structural members. However, not only did buildings constructed before 1935 feature URM (which is a prime candidate for collapse), they also generally featured heavy trusses, and a heavy style of construction (compared to today). As an example, the truss construction of pre-1935 buildings is generally categorized as truss construction, but it is significantly larger than the trusses of today. Therefore, the heavy truss of yesterday is capable of providing more fireground time (before collapse) than the modern 2-by-4-inch truss. Additionally, the average construction methods of pre-1935 buildings (residential and commercial) were larger than modern construction. As an example, a pre-1935 two-by-four was 2 inches by 4 inches, and lumber was often "rough cut" and full size. This is not true today because a modern two-by-four is 1½ inches by 3½ inches.

Remember that URM and trusses are not the only hazards that should be considered for these buildings. During this time frame, residential and commercial buildings could also contain balloon frame construction, roofs that have 60 to 100 years of roofing materials (which can easily be detrimental to the load capacity of a roof structure), brick noggin (in the late 1800s and early 1900s, brick was placed between wall studs to fill the gap and provide some sort of a barrier to fire), knob and tube wiring, multiple remodels, and the list goes on. Although these buildings can last longer in a structure fire than the buildings of today, their age may be their biggest detriment. This is a good reason to be familiar with the construction in your area.

Figure 4–30 is an example of a pre-1935 residential structure. Notice the following characteristics that are readily visible:

Figure 4–30. Older conventionally framed residential structures can be easy to read from the street.

- The foundation was made from field stones (stones common to an area and indicative of old construction).

- The exposed rafters are 2 by 6 inches.

- The construction style is referred to as *Craftsman*.

- There are two stories visible, and the attic may or may not be converted to an additional living area.

- The 1-by-4-inch stripping around the windows is indicative of hollow spaces on the left and right side of the windows (see chapter 10 on overhaul).

- Although not visible, this structure should contain balloon frame construction.

1935 to 1960

This general time frame was characterized by improved building codes, conventional building methods, and decent building materials. From a fireground perspective, this was the best time frame for building construction that had minimal surprises, and a standardized approach to building construction. As an example, URM was not used, lumber was still generally full size, lightweight trusses were still a gleam in the eyes of the building industry, and synthetic materials were not as common as today (although the interior contents can be modern which means synthetic). These factors combined to produce a fireground that was capable of an enhanced fireground time as compared to the modern fireground, and with minimal surprises. (However, don't think this is promoting a safe fireground!)

Figure 4–31 is an example of a 1935 to 1960 residential structure. Notice the following characteristics that are readily visible:

Figure 4–31. This residential occupancy is typical of the homes that were constructed between 1940 and 1960.

- This is a split-level home; first floor at grade level and upper level over the garage.

- It is wood frame construction with wood lap siding and some masonry stone on the exterior.

- It has a common gable roof with composition shingles (probable conventional construction).

- This is a relatively simple floor plan with a square footage of about 1,800 to 2,000 square feet.

Post-1960

From 1960 to present, although it is common to find strict building codes, you will also find light-weight trusses (the presence of lightweight trusses should not surprise any firefighter because they have been commonly used since the 1960s in residential construction as well as a growing number of commercial applications), alternative building methods, fascias on numerous buildings, and—last but not least—numerous synthetic materials in residential and commercial buildings. Alternative building methods and materials have yielded an interesting array of products (such as ABS plumbing, tin-can studs, foam and membrane roofs, vinyl siding, curtain construction, structural insulated panels, reflective board, and so on).

One of the most dangerous factors that has influenced the modern fireground is the widespread use of synthetic materials that has resulted in fires burning two to three times hotter and faster. In fact, what is a common fireground problem that was virtually nonexistent about 30 years ago? Flashover. So, when we summarize the age of a building (based on the history of building construction), we can safely assume that what a firefighter encounters today will be significantly different that what will be encountered 10 years from now—and guaranteed, it will not be to the benefit of firefighters.

Figure 4–32 is an example of a 1960 to present residential structure. Notice the following characteristics that are readily visible:

- This home was constructed in 2000. Interestingly, it is also called a Craftsman style of home. (Compare this with the Craftsman home in figure 4–30.)

- The construction is 2-by-4-inch truss.

- There are two levels and no visible basement.

- The 1-by-4-inch stripping around the windows are for decorative purposes only.

Figure 4–32. This Craftsman-style residence is typical of modern truss construction.

2007

Although the preceding time frames overview significant changes over the past 100 years (1900 to present), why consider the year 2007 as another change in the building industry, and how these changes will affect your fireground from the perspective of operations, priorities, and safety? Briefly, go back and look at figure 4–6 for a few moments and reflect on the prospect that wooden structural members can be glued together, and potentially use glue that melts at a relatively moderate temperature. It is easy to imagine what will happen when this type of construction is exposed to fire. Additionally, it is also easy to visualize the impact of fireground operations if this construction proves to be a cost-effective construction alternative (and it probably will). In summary, when you drive thru your district, look at your fireground office and analyze your buildings from an age and hazard/strength perspective.

Style

For size-up purposes, let's look at the most common styles of buildings. However, whether there are 10, 16, or 20 styles of buildings is not as important as what you are thinking about when confronted by a particular building style. What is also important is what a particular name of a building means to you from the perspective of hazards, fireground strategies/tactics, and so on. Let's overview the common types of buildings.

Mobile homes-manufactured housing

Because there are many mobile homes (in many areas, mobile home-type structures can also be known as manufactured housing) and mobile home parks (figure 4–33) across the country, these types of structures, although relatively simple, can present a formidable challenge when exposed to fire.

Figure 4–33. Manufactured housing structures will rapidly burn and are often located in close proximity to other similar structures.

From a simplistic viewpoint, mobile homes are a very lightweight style of construction that is constructed on a tow-able trailer, towed to the appropriate location, and mounted on a foundation of jacks, cinder blocks, and so on. Mobile homes can be relatively small (800 square feet) in a single unit, or larger (over 2,000 square feet) by joining multiple units together. Normally mobile homes are a single story structure and constructed of two-by-twos, two-by-threes, and other similar lightweight materials.

Fire. Due to the comparatively limited size of mobile homes, lightweight structural materials and the use of synthetics, fires in these structures burn fast and hot. If a fire company is on-scene fairly quickly and finds a fire that is not well developed, an aggressive mode can be accomplished (with a primary search if necessary). However, if a fire is well advanced, interior operations (attack and search) are doubtful. Additionally, the smoke from these fires can contain a high percentage of formaldehyde due to the use of plywood panels that use this chemical as a preservative.

Exposures. Some mobile home parks that have positioned the mobile homes in close proximity to each other have enhanced the ability of a well involved fire to easily spread to other mobile homes. In this case, it is important to prioritize the seat of the fire versus possible extension.

Ventilation. Due to the rapid spread of fire in these structures, any type of ventilation other than horizontal ventilation is doubtful. If a fire is not well advanced, positive pressure ventilation (PPV) can be timely and effective.

Self storage units

Structures dedicated to public storage are an interesting mix of length, often of minimal construction, and contain numerous units filled with an unknown quantity of storage that can have a wide range of possibilities. Although these structures can be found in buildings that have been converted to storage units, or new multi-story units, for the purpose of this discussion we will focus on the most common of these structures, the single-story version (figure 4–34).

Figure 4–34. Storage units are popular and can contain a wide variety of unknown contents, some of which can be very hazardous.

Construction can vary from masonry walls (concrete block or pre-formed concrete panels) to wood frame covered with corrugated metal or wood panels such as T-11 siding. What seems to be a common denominator is a cheap roof that has two characteristics: a roof made from lightweight trusses or two-by-fours used as the primary structural members and covered with plywood, OSB, and roofing materials, or corrugated metal (not a substantial type of construction); and the interior walls that are normally ½-inch sheetrock over wood or metal studs that separate the units and normally do not go to the roof, leaving an extension route for fire between the units. As a result, a simple one-unit fire can quickly spread to other units.

Access. Access to units can be a standard man-door, or varying sizes of metal overhead doors (depending on the size of the unit). Locks vary as the renter provides their own lock(s). Although the locks can be substantial, the hasps holding the locks may not be.

Contents. For a simplistic structure, the potential contents are where these buildings demand a high degree of respect. Because the renters can often have 24-hour access and the only key to a unit, the contents can easily consist of normal or standard articles as well as illegal substances, hazardous materials, ammunition, and other similar items of note. Expect the worst because these structures are capable of delivering noteworthy surprises.

Basements and cellars

Although basements and cellars are not a particular style of building, they can dramatically change the appearance and related hazards of any building. Therefore, they have been included as a noteworthy style of building.

Interestingly, some parts of this country have very few basements whereas some municipalities have an inordinate number of basements. Another item of note is that some areas (East Coast) have interior and exterior doors that provide multiple means of egress and access, and some areas (Salt Lake City area, for example) normally have only an interior door to the basement, providing a single means of access and egress. If you are familiar with basements, you are well aware that a fire in this portion of a structure can be one of the most difficult fires

a firefighter will ever encounter because a copious amount of storage is normally found in basements, and most importantly, fire, heat, and smoke normally rise upward whereas suppression personnel normally travel downward to access the fire.

Access. There is a major difference between one way to access a basement and multiple ways of access. A single access point can mean the same route for ventilation, fire spread, and firefighter access to a fire.

Contents and size. The contents and size of a basement often determines the severity of a fire. Because of this, there is a potential difference between a basement fire in a typical single family dwelling as compared to a commercial-type structure.

Extension routes. Obviously, normal extension routes are vertical, so any vertical channels (pipe chases, balloon frame construction, and so on) are potential routes for extension. Additionally, remember that if a single access route to a basement is used by suppression personnel, that same route can also feed oxygen to the fire and will also be used as a ventilation route by a fire.

Grade floor. From a simplistic viewpoint, the floor over a basement (normally the grade floor) is also the roof of the fire. So, until the fire in a basement is extinguished and/or the integrity of the floor is verified, remember where you are standing.

Definitions. A basement or cellar can have different definitions depending on several factors (figure 4–35). Some definitions summarize cellars and basements as either partially or completely below grade, but there is a difference between the two:

- A basement (A) is less than 50% below grade and will normally have large windows.
- A cellar (B) is over 50% below grade and will normally have small rectangular windows.
- A sub cellar is an underground level below a cellar.
- An above-ground cellar is underground, but can be entered from the exterior. This type of cellar is found under homes on hillsides, and is also known as a *walkout*.
- A crawl space is between the grade and the floor above, and is normally used for access

under a home. Normally, it is not tall enough to stand in and not a living space.

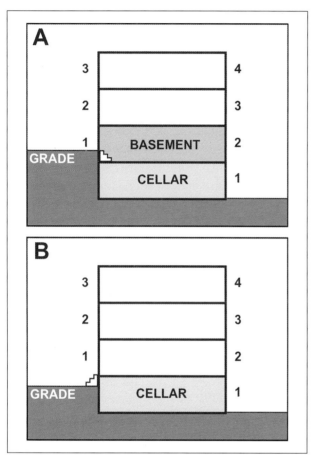

Figure 4–35. It can be important to recognize the difference between a basement and cellar.

Single family dwelling

NFPA statistics indicate that 70% of our fires are in single family dwellings (SFDs), and 70% of those fires are confined to one or two rooms. Although this type of structure is relatively simple, this is where we have most of our fires and problems. Interestingly, when responding to a SFD fire, most firefighters tend to think 1¾-inch lines and quick knockdown. However, not all SFDs are single story and 1,800 square feet. The main points of considerations are as follows.

Time of day. The most dangerous time of day for life safety (sleeping occupants) is from 10:00 p.m. to 6:00 a.m.

Construction. Is it conventional construction, or lightweight construction? Remember, if the rafter tails are exposed, the presence of 2-by-4-inch tails (figure 4–36) are an indicator of lightweight trusses.

Figure 4–36. If rafter tails are visible, they can give an indication of the type of construction.

Size and height. There is a major difference between a single story, 1800-square-foot SFD and a two or three story, 6,000-square-foot SFD (figure 4–37).

Figure 4–37. Based on what there is to burn, this is not an ordinary single-family dwelling.

The larger the square footage, the closer the SFD becomes to being a moderate-size commercial (considering what there is to burn and what it will take to put it out), and the additional floors will complicate access, extension of fire, and so on. As an example, compare the SFDs in figure 4–31 and figure 4–37.

In figure 4–31, the SFD is a common wood-frame, split-level home of approximately 1,800 square feet. Most of the structure is single story, access and egress is straightforward, the floor plan is uncomplicated, and a fire involving several rooms would be classified as routine by most fire departments.

Now compare figures 4–31 through figure 4–37. In many areas, this structure can also be classified as a SFD and commonly known as a Victorian-type structure. These structures were commonly built during the 1930s to 1940s and often incorporated large floor plans. This particular structure is probably around 6,000-plus square feet in size. A closer look reveals that the structure has been converted into a duplex (notice the two address numbers and entry doors in the front). In some areas of this country, SFDs can easily exceed 10,000 square feet and are primarily constructed with 2-by-4-inch lightweight trusses (two classic examples are some areas around Dallas, Texas, and Bucks County which is North of Philadelphia). Based on what there is to burn, are these types of SFDs really a SFD, or are they a commercial occupancy?

Converted attics. Attics that have been converted into living areas add another story to a SFD and are often difficult to access. Look for curtains, flower boxes, and/or air conditioners in attic windows as an indicator of a converted attic.

Windows. Windows can be an excellent indicator of the interior floor plan. As an example, smaller rectangular windows that are located further up a wall than normal windows (middle window, second floor, figure 4–32) are indicative of a kitchen or bathroom because these windows need to be above counter tops. If it is necessary to use a window for interior access (last option), choose the largest window to the appropriate area that may also be used as an exit opening.

Floor plan. Be familiar with typical floor plans in your area—it's that simple and that important.

Exit/entry points. Any time you enter a structure, ask yourself how many ways are there to get out if necessary. As an example, if you are the first truck company on-scene, the first company has stretched a line through the front door, and you are responsible for a primary search, is the front door enough entry/exit point(s) for you? I would hope not.

Attic vents. Attic vents in any structure can be an excellent indicator of fire extension (or a lack of fire extension) in the attic area, particularly if the attic area is over a fire.

Different building styles

The following styles of buildings fall into an interesting mix of styles, features, configurations, and nomenclature. As an example, a few short years ago, the nomenclature for apartments and condominiums was somewhat standard. However, a quick review of modern terminology reveals that these "standard" nomenclatures have dramatically changed, and as a result, terms such as urban village, mid-rise condominium apartments, flats, lofts, apartment homes, studios, townhomes, apartment condominiums, multiplexes, land-o-miniums (and so on) can form a bewildering array of terms that do not have a standard meaning, and can have a wide range of meanings depending on different areas. As a result, although the following terms may vary from the perceived norm, they focus on specific characteristics that should be considered when reading a building. Whether the following terms are old or new, familiar or unfamiliar, standard or non-standard, the primary focus is what a specific name means to you with respect to strengths and hazards. When you look at a specific building, what would be your designation? For an expanded look at specific building characteristics, see the Training Resources section at the end of this chapter.

Initially, consider two specific configurations that will affect some residential and commercial structures—the concept of garden and center-hallway floor plans.

Garden floor plans. The term *garden* comes from buildings with units that opened or faced a center area or garden area within or outside a complex. This idea originated from inhabited occupancies so the occupants could view and/or enjoy a garden environment. Even though the terminology may vary for these types of structures, the basic idea is that they do not have a center hallway, and each unit opens to the exterior of a structure (figure 4–38).

Figure 4–38. Units in garden apartment configurations open to the exterior of a building.

This configuration is common in residential and commercial occupancies. Common considerations are as follows.

Common attics. Most of these occupancies are composed of multiple units, thus common attics are the norm. The presence of division walls (which are supposed to partition a common attic) can be identified if they project above a roof. The major consideration with division walls is if are they intact, or if they have been *violated* by cable, electrical, plumbing personnel, and so on. The presence of attic vents can be an excellent indicator of extension of smoke and/or fire in a common attic.

Easy access. Because these occupancies feature units that open to the outside of a building and not a center hallway (which can also contain self-closing fire doors), grade floor units are easily accessible, and exterior stairs to floors above grade simplify access and egress to and from upper floors. This can enhance fireground operations by simplifying aerial device operations, and simplified access to specific units.

Simplified operations. As opposed to a center hallway structure, fire extending out of a unit or units will vent to the exterior of a structure. This enhances the implementation of resources and can minimize search operations.

Age. Basically, the age of a building is an excellent indicator of the type of construction—lightweight or conventional.

Center hallway floor plans. Center hallway occupancies can be easily found in commercial and residential configurations in virtually any municipality.

A simple description for this type of structure is "a hallway runs down the middle of a building with units on either side of the hallway, and each unit is primarily accessed from the hallway" (figure 4–39). The main points of consideration are as follows.

Figure 4–39. Units within a center hallway configuration open to a center hallway within a building.

Hallways. This configuration gives the most units per building when compared to garden-type occupancies because the hallways run down the middle of a building with occupancies or units on either side of the hallways.

However, during a fire, these same hallways can also become a channel for fire, heat, and smoke. Upon arrival, determining the status of the hallways is a key factor in determining fireground priorities and operations. As an example, if a fire is determined to be located in one of the units, and the hallways are clear, the fire is likely a room and contents fire. However, if the fire and/or smoke/heat have extended to the hallway(s), the fire is much more advanced and a major search will likely be necessary.

Age. The age of the building is a good indicator of conventional or lightweight construction. Again, if it was constructed during the last 30 years, what type of construction do you think it is? One note about lightweight and conventional construction—if you think it is lightweight and it turns out to be conventional, that's in your favor. Conversely, if you think it is conventional and it turns out to be lightweight, that is not in your favor.

Common attic. This is a standard unless equipped with division wall(s). Most all of the newer roofs and resultant attics are constructed from 2-by-4-inch lightweight trusses, and a lack of fire stops are common.

Lobby. Be careful with this one. Older, more expensive, more exclusive center hallway motels and hotels are configured so the upper floors can be accessed by stairs and/or elevator(s) in the lobby area. However, the newer budget motels/hotels (for example, Hilton Garden Inn, Holiday Inn, Comfort Inn, Fairfield Inn and Suites, and so on) are normally configured so the upper floors are accessed only by an elevator from the lobby area, and stairs that are located in the ends of the building. If fireground personnel need to access upper floors with fire suppression equipment, it will be necessary to use the stairs at the ends of the building. These buildings are often referred to as Holiday Inn Lites.

Now, let's look at ten specific types of buildings that may or may not use standard terminologies. Remember—terminology is not as important as what these buildings mean to you as applied to fireground operations.

Modular dwellings

These structures can be found almost anywhere in this country, and are commonly recognized by the following characteristics. Each module normally consists of two, four (fourplex), and sometimes more occupancies, but normally no more than eight separate occupancies. Often, the modules are grouped into clusters in a particular development (figure 4–40).

Figure 4–40. Buildings classified as modular are easy to recognize and are normally limited to about eight units.

Modular dwellings are normally two stories. Each occupancy within the module can be a single story, or two-story, and can be conventional construction,

but are normally lightweight construction. There are usually no hallways, and each occupancy has its own entrance doorway. They can be advertised as apartments or condominiums.

Common attics. Common attics are a standard in these structures unless a division wall is present. However, even if a division wall is present, consider it violated (cable, electrical, plumbing, and so on) until proven secure.

Construction. If this type of structure has been built in the last 30 years or so, it is likely of lightweight construction, even if it appears to have masonry walls. Know your fireground office!

Access. Access is relatively easy because first floor occupancies are accessed by an entrance door on the first (grade) floor, and second floor occupancies are accessed by stairs to a doorway or a landing that serves adjacent occupancies. The point is this: Know your secondary means of access/egress from the second floor. If you planned ahead, it has to be some type of ladder (portable or aerial device).

Walls. Walls are walls, right? Wrong! For this discussion, there are three basic types of walls: Division walls go from the floor through a roof; partition walls go from the floor to a ceiling only; and offset walls are similar to partition walls because they are also used to separate rooms, occupancies, and so on. However, the big difference is found in the location of the studs (offset) in the wall. This allows easy extension of fire.

Apartments, condominiums, and townhomes

Terms that describe apartments, condominiums, and townhomes can be confusing. They can often be used interchangeably, particularly apartments and condominiums. As an example, an apartment complex can be constructed and then rented. In some cases, however, if business does not meet expected projections, the complex can then be sold as condominiums (which can generate instant cash and reduce the constant challenge of renting vacant units).

The structure in figure 4–41 is an example of a building in a complex that was originally rented as apartments. The occupants were evicted, and the complex was then sold as condominiums.

Figure 4–41. Is this building an apartment, condominium, or townhome?

However, even if the name of these types of buildings are changed, the basic structure, floor plan, and inherent hazards remain the same. For simplicity, let's categorize these three structures as follows.

Apartments. Webster's dictionary defines apartment as "a room or suite of rooms to live in, one of a number in an apartment building." These buildings can be an individual structure or grouped into clusters, and can vary from several units in a single building to numerous units in large buildings. Additionally, structures classified as apartments are normally rented and not owned by the occupants. As a result, they can pose a higher risk of fire because the occupants do not own the building/unit as opposed to occupants who own their living quarters and are responsible for their property. As an example, the structure in figure 4–42 is an older six-story hotel that has been converted to an apartment building. One story is now commercial occupancies, and the other five stories are apartments.

The apartments are primarily accessible from a center hallway configuration. Common considerations are as follows.

Geographic location. Older apartment-type structures are often confined to one building, some of which are large buildings as in figure 4–42. However, it is common for newer apartment structures to be smaller, some of which are built on terraced or sloping property that will result in a different look from the front of the building as opposed to the rear (two stories from the front and three stories from the rear, for example). In this case, it is important to determine the location of a fire and be able

to designate the proper location to resources. Terraced buildings are notorious candidates for poor or non-existent access for placing apparatus to the sides and/or rear of these buildings.

Figure 4–42. This old masonry center hallway hotel has been converted into apartments.

Newer apartments (and condominiums) can also be found in clusters as in figure 4–43. The presence of these configurations mandates a previous size-up to determine the best method to locate the fire, and manually provide resources to buildings that are not easily accessible to parked apparatus.

Figure 4–43. Apartments, condominiums, and townhomes can often present challenging access considerations.

As an example, the buildings in figure 4–43 were over 200 feet from the nearest area available to spot apparatus. As a result, an aerial device would be unusable in figure 4–43.

Age. The age of a building is a good indicator of the type of construction (lightweight or conventional) and how that can affect fireground time. Compare figure 4–42 and figure 4–43. Obviously figure 4–42 is an older building composed of conventional and masonry construction, and the units in figure 4–43 are lightweight construction, lightweight trusses, and vinyl siding. Both of these types of construction dramatically affect fireground time.

Size. The size of a building can dramatically affect the key fireground factor of implementation time. As an example, let's apply implementation time to figure 4–40 and 42. In figure 4–42, the structure is six stories tall, a large building (from the concept of square footage), and is constructed from conventional and masonry materials. Although these materials will resist fire for longer periods of time when compared to modern construction, it is easy to see that a fire on the fifth floor that has extended into a center-hallway configuration will take a noteworthy period of time to access and extinguish in concert with a time consuming search. Conversely, figure 4–40 is a two-story structure of lightweight construction and of a much simpler configuration and size. Although the lightweight construction can fail in a short period of time, a fire on the first floor will be easy to access and extinguish along with a simplified search.

Configuration. Regardless of the type of building, apply the concept of garden apartment or center-hallway configurations. The large apartment building in figure 4–42 is constructed with a center hallway, and the apartments in figure 4–43 utilize the garden configuration that can simplify fireground operations when compared to center-hallway configurations. Additionally, remember that some older, large apartment/hotel-type occupancies were constructed with light and air shafts into interior apartments. If fire extended into a shaft from one of the apartments (particularly the lower ones), the shaft can act as a vertical chimney to upper apartments with windows common to the shaft. These shafts are normally not found on modern apartment type construction.

Condominiums. Webster's dictionary defines condominium as "one of the units in a multi-unit dwelling, each separately owned." Although the visual appearance of some condominiums and apartments can be similar, there is normally a difference in the ownership of these two structures. Apartments are normally rented and can be at the lower end of the cost scale (comparatively speaking); condominium units are normally owned by the occupants and tend to be at a higher part of the cost scale. As delineated in the previous section on apartments, the hazards with condominiums are similar. A review of the common considerations are as follows:

Configuration. Condominiums can be found in both garden-type apartment or center-hallway configurations. Depending on the type, condominiums have the ability to dramatically affect fireground priorities and operations. Remember that fire (or its by-products) that has extended to a hallway will create additional fireground problems when compared to a fire in a garden apartment configuration. Whether the building is an apartment or condominium configuration, the floor plan and associated hazards are likely the same. As an example, although the building in figure 4–44 could easily be described as a condominium complex, it was listed by real estate agents as "apartments for rent."

Figure 4–44. Apartments and condominiums can have similar floor plans.

Would a different designation change fireground hazards and/or priorities? The obvious answer is no. Additionally, be aware that in some cities, large, older warehouse structures have been gutted, remodeled, and sold as condominiums with the intent to use an existing structure, bring people back to the downtown areas of a city, and reduce travel time to these areas for employment. These types of structures are generally large, feature older construction, consist of center-hallway configurations, and have multiple inhabited units inside the structure.

Geographic location. In some areas, it is common for these types of buildings to have limited access to the sides and no access to the rear of the complex and/or particular building. In these cases, the only access will be from the front ("A" side) so apparatus placement and positioning should have been pre-planned.

Construction. Figure 4–44 is a graphic example of a building that has recently been constructed and, therefore, is a prime candidate for lightweight truss construction. Notice the potential of a common attic over the four units. If a fire on the third floor of the end unit, left side of photo, extends into the attic, it can travel to the end unit on the right side of the photo unless resources pull ceilings to get ahead of the fire.

Size. Compare the buildings in figures 4–44 and 4–45. Both of these buildings should be condominiums, but a fire in a second floor unit (figure 4–44) will be simple when compared to a fire in a fourth floor unit, middle of the complex in figure 4–45, and will be significantly more challenging.

Figure 4–45. Some condominium buildings can be large and present formidable fireground challenges.

Townhomes. Webster's dictionary defines townhome as "a two-story or three-story dwelling, a unit in a complex of such dwellings." Although there are numerous similarities when reading townhome-type structures, their hazards are similar to apartments and condominiums. However, there are normally two distinctions that can be inherent to townhome-type structures. They resemble a number of SFD structures that have been constructed to form one single building. Also, the units in a single building share common walls and a common attic. As an example, in figure 4–46, this building consists of four three-story dwelling units. This configuration can provide the maximum number of dwelling units per square foot of property, hence their popularity in this country.

Figure 4–46. Townhomes are attached residences and often share a common attic.

However, the presence of common walls and attics are a potential avenue for the extension of fire to uninvolved units within the same building.

Hazards. As delineated in the previous sections on apartments and condominiums, the hazards with townhomes are similar. However, the popularity and rapid construction time that has been made possible by lightweight construction has enhanced the proliferation of interesting alternative construction methods that were unknown a few years ago. Let's look at a prime example of why a fire department must (not should) be aware of new construction methods and styles in its district or area. Figure 4–47 appears to be a row of common townhome-type structures with each building containing four occupancies, each over a *walkout basement.*

Figure 4-47. Lightweight construction has made the apartment, condominium, and townhome-type buildings very popular.

This particular set of townhomes is located near Cincinnati, Ohio, and features the following items:

- There are three entry doors on the right side of the building, one entry door in the front, and one entry door in each garage (for a total of five). If you were the initial truck company arriving to a visible fire on the second floor in the middle of the building, which door would you choose for quick access to the second floor? An interesting nonvisible feature is that the primary access to the basement is the doorway within each garage!

- The interior construction is composed of glued trusses (see figure 4–6), some of which support 4-inch concrete flooring above the trusses. Not surprisingly, the common attic is also composed of glued trusses.

- Last but not least is an emerging construction method that is not unique to just these townhomes. In multiple occupancies where space is at a premium, the laundry room can be located in an area that is distant from an exterior wall. As a result, it is necessary to run the exhaust ducting from the dryer(s) to an exterior location, which is done by using a more powerful motor in the dryer to force hot air and lint through the ducting. In these structures, the ducting was over 150 feet long and traveled through various areas within the structure before exiting the building. Of note, a soffit is used to conceal the ducting in visible areas, and in many

areas the ducting was crushed behind the security strapping. In a few years, this condition will result in a fire in excessive lint that will be difficult to find.

Brownstones, railroad flats, row houses, and tenements

The following four buildings (brownstones, railroad flats, row houses, and tenements) all share a common potential fireground priority of search based on the flammability of the interior contents, narrow hallways, open vertical stairways, multiple floors, common attics/voids, and high-occupant load. Additionally, the previous section on definitions of basements and cellars is applied to these four buildings. For a more definitive look at these buildings, see Training Resources at the end of this chapter.

Brownstones. Structures referred to as brownstones are commonly found on the East Coast, particularly New York and New Jersey, and were constructed primarily during the early 1900s. Although a row of numerous brownstones can form what appears to be a large complex similar to tenements or row houses (figure 4–48), a distinctive feature of brownstones is a sunken court or patio area in the front portion of the structure that is several feet down from the sidewalk level. This configuration allowed light and access into the basement area and could also serve as a small outdoor patio area during appropriate weather.

Figure 4-48. Brownstones can be identified by the characteristic basement entrance under the stoop. (Photo courtesy of John J. Lewis)

Access into the basement from the patio area was easily accomplished from a doorway that was located under the front stairs (also known as a stoop) and can be seen in figure 4–48. Access to the rear of these buildings is historically poor to nonexistent. Common considerations are as follows.

Configuration. Brownstones were initially constructed as an inhabited occupancy that varied from two to four stories with a basement. Each individual occupancy is approximately 20 to 25 feet wide and 40 to 60 feet deep. As previously mentioned, the front entrance to the basement is through the door under the stoop. Originally, the basement was primarily used for the kitchen and dinning area with upper floors consisting of living and sleeping areas. Interior access to these areas from the basement is provided by an internal stairway from the basement to the next level. Exterior access to the level above the basement is provided by stairs (stoop) in the front of the building. In some cases, a cellar can be located below the basement and is often identified from the exterior by an old coal chute in the front sidewalk and/or a metal grate in the sunken court. Brownstones can also have common vertical shafts, dumbwaiters, and heating ducts from the cellar to upper floors. Common cocklofts are also common over adjoining occupancies.

Age. Older buildings (in this case, the early 1900s) can be depended upon to contain three noteworthy considerations:

- A heavy grade of conventional construction instead of lightweight members. Although the front portion of the building is often of masonry construction, the balance of the building is constructed of wood, often displaying wood siding on the side and back. The roof will be of substantial construction and will be covered by numerous layers of roofing materials.

- The older the building, the higher the chance of remodels, the use of alternative building materials (such as lightweight), and modified floor plans that can incorporate floor levels changing from a residential to commercial occupancy.

- Numerous concealed spaces and voids.

Railroad flats. The focus of railroad flats is not so much the style of building, but rather the interior configuration of the building. Although railroad flat-type occupancies can be a stand-alone building as in figure 4–49, their configuration can also be found in other buildings such as old and new tenements.

Figure 4–49. Railroad flat configurations are similar to two parallel box cars separated by a narrow hallway. (Photo courtesy of John J. Lewis)

Because the floor plan is unique, this configuration is summarized with the following considerations.

Configuration. The term railroad flat comes from the internal configuration of the apartments that resemble two parallel railroad cars that run from the front of the building to the back and are separated in the middle of the building by the stairs and a narrow hallway. Normally, the bedroom areas are toward the front of the building with the kitchen area toward the back of the building. There are two entrances to each flat. An entrance by the staircase landing enables access to a kitchen area, and an entrance at the other end of the flat enables access to a bedroom/living room. Normally, this entrance is blocked by furniture and other similar items, and the door can be expected to be secured by noteworthy locking devices.

Three additional items of consideration are the vertical stairs, stacked kitchen/bathrooms, and poor rear access. Because these buildings are multi-story, a vertical stairway serves the upper floors from grade level. This configuration is no different than a horizontal center hallway configuration; both of these configurations can willingly enhance the extension of heat, smoke, and fire to uninvolved areas. As an example, if a fire in a flat on the first floor in figure 4–49 extends into the hallway/stairs, it can quickly extend to the top floor trapping any occupants on the 2nd, 3rd, and 4th floors.

In multi-story inhabited buildings, bathroom and kitchen areas are normally placed one above the other (stacked) to minimize plumbing and other similar costs. This configuration will result in voids and pipe chases that can run from the bottom floor to the top floor. Additionally, it is common practice to run plumbing vents through a roof, which can provide an open pipe chase into a cockloft/attic area.

These buildings historically back up to a similar building on the opposing street; therefore, rear access can be compromised, and fireground personnel can encounter fences, dogs, and other similar obstacles. Apparatus access to the rear of these buildings can be difficult to nonexistent.

Age. The older the building, the older the construction. This has some advantages and disadvantages. The advantages are these buildings were primarily constructed of rough-cut wood construction, often with a masonry front. This will result in an increase of fireground time and stability as compared to modern lightweight construction and brick veneer fascias. The disadvantages can be numerous. Common cocklofts, cornices, vertical pipe chases, coal chutes, heating chases, and numerous voids collectively combine to enhance the extension of fire, and increase the difficulty of finding the extension of fire. Roof ventilation personnel will encounter roofs that commonly used 1-by-6-inch sheathing with numerous layers of roofing materials. Additionally, these buildings are prime candidates for renovation that can change the older construction to newer lightweight members and a different floor plan from the original. This is where a familiarity

of your fireground office is crucial to success on the fireground.

Row house. Structures that are known as row houses were originally constructed to meet the demand for affordable housing (figure 4–50). Similar to brownstones but lacking a front sunken courtyard that provided access to a basement, these structures were also constructed in the early 1900s from full-sized wood, and may have brick masonry on the front of the building.

Figure 4–50. Row houses were popular due to their ability to meet the demands for affordable housing. (Photo courtesy of John J. Lewis)

The size can vary from 20 to 30 feet wide to 35 to 65 feet deep and are commonly two to four stories high. The popularity of these structures often resulted in a single building that consisted of numerous occupancies that shared a common party wall; buildings that run an entire city block are not unusual. Common considerations are as follows.

Construction. Row houses are, as the name implies, a row of attached buildings that share common walls between each occupancy. During the late 1800s and early 1900s, this type of construction commonly utilized balloon-frame and braced-frame construction, brick or wood on the front of each building, common cornices, basements/cellars that were common between adjoining occupancies, and brick noggin that was nothing more than brick placed between the studs to provide a measure of protection against the extension of fire and add some strength to load bearing walls. However, it can be easily seen that the addition of brick will dramatically increase the dead load of a wall. Although constructed from full-size lumber, row houses are well known for their ability to enhance the extension of fire to adjoining occupancies.

Common attic. The common wall between occupancies is normally not a true fire wall, but from a definitive viewpoint, it is a party wall that leaves the cockloft common over the adjoining occupancies unfirestopped. This can be the case for the length of the entire building, which can often be a city block long. Obviously, a fire in a single occupancy that extends into the common cockloft area can endanger the entire building unless it is stopped. The roofs are flat with a slight slope for drainage considerations. They are commonly 1-by-6-inch sheathing with numerous layers of roofing materials that can slow roof ventilation operations. Interestingly, older row houses will often have smaller cocklofts than newer row houses.

Access. Similar to brownstones, the rear portions of these buildings commonly face the rear of an opposing building. Therefore, access can be difficult to impossible from the presence of fences, dogs, the lack of an alleyway, and other similar considerations. This condition can mandate that apparatus will have to use the front of the building for the only access area.

Tenements. Tenements were primarily constructed as common rental apartment buildings (figure 4–51) around the turn of the century. Therefore, tenements are of older construction of wood and/or masonry construction and often vary between 20 to 40 feet wide, 40 to 90 feet deep, and two to six floors in height with a common count of four to eight apartments per floor.

Basements/cellars are also common with a stairway from the basement to the interior of the building. A vertical stair/shaft then continues vertically above the stairway (from the basement/cellar) to serve the upper floors within the building. Common considerations are as follows.

Old law vs. new law. A major consideration with these buildings (and some other types of buildings) is a configuration known as old law, new law, or retrofit. Buildings that incorporated the configuration of old law incorporated the following characteristics:

Figure 4–51. Tenements were primarily constructed as common rental apartment buildings around the turn of the century. (Photo courtesy of John J. Lewis)

- An interior stairway from a basement/cellar to the interior of the building. The stairway may or may not be fitted with fire doors, and if so equipped, the doors may not be closed. This configuration can allow a fire in a basement to travel vertically upwards and into the building. Fire, heat, and smoke are then free to continue their upward extension route within the building via the vertical stair/shaft, trapping occupants within the upper portions of the building.

- Stairways and stairs from a basement/cellar to a penthouse/bulkhead are constructed from wood and will readily burn if exposed to fire.

- The floor joists separating a basement/cellar from the floor above are exposed to the basement/cellar area and vulnerable to fire in the basement.

Buildings that incorporate the *new law* or retrofit configurations are significantly different from the perspective of simplifying fireground operations and incorporate the following retrofit or modifications:

- Basement/cellars are not accessible to the interior of the building above a basement/cellar. This acts as a barrier to fire extending from the basement/cellar to the interior floors above the basement/cellar.

- Stairways are not primarily made of wood but of noncombustible materials.

- First floor joists are not exposed to the basement/cellar area.

Construction. Due to the age of these buildings (late 1800s for old law and early 1900s for new law) and the need for low-cost housing, full-size lumber was the norm with masonry construction. However, the following contribute to fireground challenges:

- Close proximity to other similar buildings
- Adjoining buildings
- Common attics/cocklofts
- A high probability of renovations that change the original floor plan
- The probable use of lightweight members for renovations
- Attached basements/cellars
- Light and air shafts
- Tin ceilings

Fire escapes. Fire escapes not only provide a means of egress for building occupants, they also provide a means of access/egress for fireground personnel. Due to the age of these buildings, fireground personnel should always try to determine the integrity of a fire escape before use. Normally, fire escapes that are attached to wood and masonry construction that is not un-reinforced masonry construction (URM) can be more substantial than fire escapes attached to URM because this type of masonry construction used lime and sand instead of Portland cement. Additionally, it is a good idea to lower the escape ladder on fire escapes to ground level to simplify occupant egress and negate the possibility it does not accidentally deploy on top of fireground personnel.

Stairways. As previously mentioned, the stairways in old-law-type buildings traverse from the basement/cellar to a bulkhead/penthouse on the roof are constructed of wood and a straight run configuration (one over the other), and are historically narrow. New law or retrofit stairways do not traverse from the basement/cellar to the roof, and primarily use noncombustible materials. However, the presence of an unfirestopped vertical stairway above the basement will enhance the travel of fire, heat, and smoke to uninvolved floors, trapping unsuspecting occupants and significantly increasing search considerations and priorities.

Configuration. These buildings can vary from a single, unattached, moderate-size building (figure 4–51) to numerous attached buildings, and to large buildings of H, E, U, T, and similar configurations. These buildings were originally constructed to serve as a community within a community and will have all of the aforementioned hazards in addition to a higher occupant load.

Taxpayer

This style of building is an example of two for the price of one. In the late 1800s and early 1900s, it became cost effective to build one building to serve two purposes—commercial and residential in one. A single two-story building was constructed on a single piece of property instead of two buildings on separate properties. Property owners also constructed these buildings to offset their property taxes. The first floor was commercial; the second floor was residential. It was common for the business proprietors to live above their place of business; therefore, these buildings can combine a noteworthy fire load of the commercial occupancies with the life hazard of residential occupancies.

In some cases, these buildings were three stories, consisting of first floor commercial occupancies and the second and third floor residential occupancies as shown in figure 4–52. Of interest, notice that some of the first floor occupancies have been modified and covered over to increase the size of adjoining occupancies.

Figure 4–52. Taxpayer buildings consisted of commercial occupancies on the first floor and residential occupancies one or two stories above.

Note the buildings to the left and right of the three-story taxpayer in the figure are two-story configurations. Interestingly, the modern version of this building is the venerable strip or mini-mall, which we will look at next. Both the two- and three-story versions of these styles of buildings share some common features and hazards, such as common attics and adjoining commercial occupancies. Common considerations for taxpayers are as follows.

Construction. These buildings were primarily constructed in the late 1800s to early 1900s; therefore, the construction will be conventional construction primarily composed of the following methods:

- Wood that was full size and often rough cut lumber

- Standard masonry construction (for the applicable time period) that was not typical of modern masonry veneer

- Standard un-reinforced masonry construction (URM)

Although URM is not considered a strong style of masonry construction when exposed to fire (see the following Methods section on masonry construction) it can be a superior style of construction as opposed to the modern style of brick veneer. Additionally, these buildings employed full-size lumber in conventional configurations, did not use lightweight trusses, used lath and plaster for walls and ceilings, and so on. Due to the age of these buildings, remember that they have likely been remodeled numerous times to reflect changing economic conditions and owners. This equally applies to the commercial occupancies on the first floor and the residential occupancies on the floor (or floors) above. Therefore, expect a mix of the old and new in these older buildings. The roofs will normally be flat sloped and consist of 1-by-6-inch sheathing and numerous layers of roofing materials. Roof ventilation personnel will find that vertical ventilation will be challenging and time consuming as compared to modern roofs. Tin ceilings, a heavy grade of floors, large display windows, dropped or suspended ceilings, and some cornice/facades were used, although the use of facades was significantly different than the modern use of facades/fascias.

Common attic. Although these buildings can be expected to have a common attic, it is not over the commercial occupancies on the first floor, which consist of floor joists supporting the upper floor(s). The common attic is above the top floor occupancies. Therefore, a fire in the first floor occupancies will not have an easy extension route to the attic when compared to a fire in the top floor occupancies.

Cellar. Depending on the geographical area, cellars can be found under these buildings in two basic configurations—a large cellar under the entire building that may or may not have been partitioned (fire is normally free to extend to uninvolved areas in this type of configuration), or separate cellars that will not enhance the extension of fire under this type of building. Conveyor belts, chutes, and other similar vertical passageways can be found in these types of buildings, which can spread the vertical extension of fire, heat, and smoke.

Occupants. The inhabitants/residents are usually non-ambulatory during the hours of 10:00 p.m. to 7:00 a.m. Conversely, the commercial occupancies pose the minimum hazard for occupants because they are ambulatory during business hours.

Access. The front of these buildings offers the best access to the interior by utilizing the doors and/or display windows. Due to the age of these buildings, expect strong doors and/or overhead metal doors. Access to the upper floor inhabited occupancies is normally limited to a minimum of two stairways (uniform building code requirement). Access from the rear can be challenging because an alley will normally be behind the building (which offers minimal space for apparatus and can be filled with dumpsters, trash cans, overhead wires, and so on). Each occupancy is normally accessed by a single substantial door with numerous locking devices.

Contents. These buildings have the potential to present a wide variety of hazards, including hazardous materials due to the diversity of occupancies within a single building.

Mini-malls

This type of structure should be on every fire-fighter's list of favorite structures because most municipalities have at least one. Mini-malls have been around for years, which means that although they can share common features with a taxpayer, there are some basic differences that are normally a result of the old versus the new. Two of the common configurations in these structures are a single-story structure, and the presence of fascias (figure 4–53). Common considerations are as follows.

Figure 4–53. Mini-malls normally share two common hazards: common attics and fascias.

Common attics. Similar to center hallway and garden apartment structures, common attics are normal because most mini-malls are nothing more than numerous occupancies/units in a rectangular, single-story building. This results in a common attic the length of the building. Again, the presence or absence of division walls will have a major impact on the length and hazard of a common attic in this type of structure (unless they have been breached). However, remember that division walls do not normally extend into a fascia, which allows an attic fire to extend past the division wall via the fascia. Additionally, look for the presence of attic vents. Truck companies on a roof normally have an excellent view of ventilators and other signs that indicate the condition of the common attic area.

Old vs. new. The older mini-mall type buildings (single story or taxpayers) were commonly conventional construction, and masonry exterior walls were also common. Newer mini-malls are (for the most part) lightweight construction (which is rapidly including glued trusses), and if masonry exterior walls are present, are probably brick veneer. Additionally, newer mini-malls are commonly equipped with some type of mansard, fascia, or other overhang on the front of the building.

Access. Normally, access is relatively easy. Each unit has an entrance in the front with display windows, and a door in the back. Often, the rear entrance is a metal-clad door in a metal frame, and interior security bars across the door are common. However, some mini-malls (particularly older ones) can have overhead rolling metal doors or scissor gates over the front entrances. An additional consideration is the presence of parking lots and/or areas behind these structures that can afford sufficient space to spot apparatus.

Fascias. Fascias are considered in more detail when we consider the method of construction in the next section, but for now, let's focus on two considerations. First, and potentially most important, most firefighters make initial entry into a structure through the front door. Because firefighters are creatures of habit, if you suspect or know that fire has extended to a fascia that is hanging over the front door you are about to enter (on any type of structure), you should consider an alternative means of access (such as the rear door). Secondly, most fascias do not have fire stops or sprinklers, and are common to the attic area. Therefore, fire in a structure that extends to the attic area can easily spread to the exterior fascia, which is nothing more than an exterior common attic capable of spreading fire around the exterior of the building.

Commercials

Not much can be said about commercial buildings (figure 4–54) other than they come in a wide variety of sizes, heights, floor plans, contents, and so on. Three primary differences among commercial buildings and residential occupancies are size (square footage), fire load, and the lack of a standard floor plan.

Figure 4-54. Commercial buildings come in a wide variety of configurations, styles, and sizes.

It is for this reason that an initial size-up should include the approximate size and type of business, if known. There is a significant difference between a single-story commercial building that is a 30-by-30-foot barber shop and a single-story commercial building that is 150-by-200 feet and contains a plating plant. Additionally, when considering life hazard, you don't often lose an occupant inside a commercial building as opposed to a residential occupancy, so search considerations in commercial buildings can be less important as opposed to residential buildings. Common considerations are as follows.

Age. The approximate age of a building can be an excellent indicator of the type of construction (conventional, lightweight, real masonry—not brick veneer on a metal or wood frame) and provide some insight into the approximate interior of the building. As an example, older commercials may have been renovated numerous times to accommodate various owners, thereby making a size-up of the floor plan difficult. Numerous renovations also increase the chance of substandard construction. Conversely, newer commercial buildings often have a standard floor plan that can often be recognized from the exterior of the building.

Attic. This one is simple. Check to see if the building has a common attic, an attic that has visible division walls, and if attic vents are visible. Attic vents can be an excellent indicator of extension of smoke/fire into an attic.

Size. This one is not so simple because commercial buildings come in an unlimited variety of sizes, which also includes height. Consider if the building

is part of other buildings, is a stand alone, is a single story or multiple story, and if multiple story, how many stories. Remember these points:

- The higher a fire from the ground level, the longer it will take to get your resources to the fire.
- The business name on the outside of the building is an indicator of what you should expect to find inside.
- The time of day often indicates if the building is vacant or occupied (also consider holidays and weekends).
- The larger the building, the more resources will be required to find and fix a problem.

Configuration. The name on the exterior of a commercial building is an excellent indicator of the interior of a building. Consider if the building is all commercial, or commercial on the first or lower floors and residential on the upper floor or floors. This will dramatically change your search considerations in concert with the time of day. Some of the many configurations to consider are office, retail outlet, and warehouse.

Roof location. This one is a sleeper. If a commercial building has parapet walls, a fascia/mansard, or if the roof is not visible from the ground, you should determine where the roof line is below the parapet wall, mansard, and so on, if you are going to the roof for roof operations. If the roofline is 5 feet or more below a parapet wall, you should consider how many ladders it will take to safely enable you to go from the ground to the roof. Usually two ladders will suffice—one from the ground to the top of the parapet wall, and one from the top of the parapet wall to the roof. If you were on a roof and the roof suddenly started to collapse, you may (or may not) be able to jump 5 feet with full PPE to reach the top of that parapet wall if a ladder was not there. Pre-plan your roof operation from the ground before you commit yourself. From the ground, consider any visible indicators, including the distance between top floor windows and the top of parapets, fascias and mansards, scuppers that indicate the roof line, and the visible presence of equipment (HVAC units, air conditioning units), turbine ventilators, and so on.

Institution

These are buildings that typically fall into this classification are hospitals, jails (or detention facilities), senior living care facilities (figure 4–55), and so on.

The focus on these types of occupancies is the presence of non-ambulatory occupants or occupants that will have difficulty exiting a building in case of an emergency. Another concern is the size of some of these structures because they can be large, multistory facilities. More common varieties of this style of building are referred to as assisted living, dependent, non-dependent, and so on. Many municipalities are experiencing an increase in senior care facilities. Common considerations are as follows.

Figure 4–55. Buildings classified as senior care living can come in a wide variety of configurations and care-levels.

Construction. Lightweight or conventional construction will give you an idea of fireground time, but more importantly, you need to determine if this type of building in your area of responsibility has sprinklers and/or smoke-fire detection systems. Occupancies equipped with sprinklers normally have minimal fires, but if you have retirement type facilities that are non-sprinklered (older-type buildings or areas where these considerations are not required), you have a potential major hazard in your district. If you have non-sprinklered buildings with a noteworthy occupant load, particularly non-ambulatory occupants, you should have more than a good idea of what your actions would be if this type of building has a fire or similar emergency.

Rescue (evacuation). With the preceding thoughts in mind, any fire, smell of smoke, or other type of emergency will not have a great effect on staff personnel; it's the non-ambulatory occupants you should be concerned about. Additionally, elderly occupants can panic at the slightest hint of a problem. If you doubt this concern, just try walking through one of these occupancies with your PPE on and see what happens. Additionally, it is not uncommon for staff personnel to not be acutely aware of their responsibilities in case of an emergency and be counted on to assist with meaningful assistance in evacuation if the need arises.

Size. The size of this type of occupancy will give you a good idea of the magnitude of a potential rescue, although size and type of care can be deceiving. As an example, the building in figure 4–55 is a single-story facility that can easily be ignored. Notice the sign on the lawn lists the facility as "Senior Attended Living." What this listing does not say (a few will) is that the facility specializes in Alzheimer's patients in all stages of this disease. Although some facilities can be one to four stories in height, the facilities that often slide under the radar are the single-family dwellings that have been converted to assisted-living facilities. From the street, these structures can look no different than a routine single-family dwelling, but the interior may be filled with senior residents that can often be classified as non-ambulatory. Additionally, this type of facility is often supervised by someone who is hired to make a call to appropriate emergency personnel in case of a problem. This individual is often ill equipped to assist in an emergency.

Configuration. It is important to know the various types of facilities in your district and their level of care. With this information, you can preplan for an emergency. (That also applies to responding fireground personnel.) As an example, you should know the difference between limited care, independent living, dependent living, assisted living, long-term care, and other terms that may apply to your area of responsibility. Remember that most if not all of these facilities are of the center hallway configuration because of security concerns.

Public assembly

Typical buildings that fall into this classification are restaurants (figure 4–56), theaters, night clubs, and so on. Common considerations are as follows.

Figure 4–56. Normally, the fire service does not lose occupants in public assembly buildings.

Ambulatory. Unlike institutional occupancies, most if not all occupants in these buildings are ambulatory (able to leave in case of an emergency). Although there can be large occupant loads, the fire service normally has few problems in these buildings unless there has been an explosion or other infrequent problem.

Exits. Normally, these types of buildings must meet strict building codes for exits that can serve the maximum occupant load.

Configuration. The business name on the outside of a building is an excellent indicator of what can be expected inside the building. As an example, restaurants come in a wide variety of sizes, floor plans, and occupant loads. Typically, floor plans can be challenging, and the fire load is often above average. The common fire problem is often found in the kitchen area.

Attic space. These buildings are excellent candidates for multiple attic spaces. As the type of occupancy changes, so can the ceiling height. As an example, assume you have a restaurant with a distance of 5 feet between the roof line and the windows. You probably have a single attic space. Conversely, assume there is 15 to 20 feet between the windows and roof line/parapet. You may have multiple attic spaces. An indicator of this is when interior personnel pull the ceilings and find no fire. At the same time, roof personnel open the roof in the same area and also find no fire, yet conditions indicate there is fire somewhere in the attic area. As a side note, the restaurant in figure 4–56 is the third different restaurant in this building (and the most successful).

Warehouse

We started size-up considerations of commercial occupancies with the following statement: Commercial buildings come in a wide variety of sizes, heights, floor plans, contents, and so on. This statement also applies to warehouse occupancies because they also come in a wide variety of sizes, heights, floor plans, contents, and so on. Common warehouse occupancy considerations are as follows.

Size. What comes to your mind when the term warehouse is mentioned? Probably a large type of building because most warehouse-type buildings are used to store materials and/or contain a manufacturing process. Therefore, we are talking about a type of building that can be noteworthy in size. There is a wide variance in warehouse sizes and floor plans, so an initial size-up should include the approximate size because there is a significant difference between a warehouse occupancy that is 30 by 30 feet and one that is 150 by 300 feet.

Contents. The business name on the outside of the building can give a clue to the expected contents and fire load inside the building. As an example, the business name on the building in figure 4–57 indicates this is a retail outlet that also serves as a warehouse to store materials associated with carpets.

If interior operations are anticipated, then consider the presence of rack-tiered storage that will divide the interior into aisle ways, and the possibility of collapsing rack-tiered storage onto or behind interior attack personnel. Additionally, bear in mind that ventilation operations in a building of this size can be challenging at best.

Figure 4–57. Warehouse buildings can often contain a considerable fire load.

Mezzanine. It is common practice for occupants of these buildings to construct a mezzanine over the office area (and probably not to code) and then use this handy out-of-sight area to store their stuff and junk. Unfortunately, if a fire occurs and the sprinkler system is activated, this area can quickly become overloaded as the storage absorbs water. This potential condition can suddenly become important to personnel entering through the office area to advance to the interior of the building. Conversely, mezzanines are not normally located over the rear portion (loading dock doors) of these buildings.

Access. The typical warehouse often has just a few ways into and out of the building, and, can have few to no windows. As an example, consider the typical concrete tilt-up warehouse with a single man-door for access to the office area in the front, and several loading dock doors in the rear. Additionally, most concrete tilt-ups have no windows. Therefore, the lack of openings in these buildings should dramatically change your thinking regarding access/egress openings and your ability to mount interior operations if conditions warrant. To summarize this consideration, in many cases the large overhead doors (such as loading dock doors) can be a superior means of access, egress, light, and ventilation as opposed to a single man-door that opens only to the office area (which also may be under a mezzanine).

Parapet walls. If personnel are assigned to the roof for ventilation operations and the roof line is not visible, then consider the potential distance of the roof line below the parapet wall. Remember, if the distance is 5 feet or over, then it will take two ladders for personnel to safely reach the roof (one ladder from the ground to the top of the parapet, and another ladder from the top of the parapet to the roof). Also, the presence of scuppers will visibly give an indication of the roof location behind a parapet wall. As an example, look at the parapet wall on the front of the building in figure 4-57 as opposed to the actual roof line on the right-side of the same building.

Built to burn. Imagine you have a building that is located in an industrial/commercial section of a municipality. An interior fire has made some progress before being identified and reported. The building has minimal openings to allow access/egress for fireground personnel; the interior of the building has minimal or no partition walls and a moderate to heavy fire load; and the building has a lightweight roof. What you have is a building that was built to burn and will likely result in a new parking lot.

High rise

Buildings that fall into this classification are normally considered to exceed 75 feet in height, as shown in figure 4–58.

Figure 4-58. High-rise buildings can present a formidable challenge if a fire is located above ground.

However, when contemplating fireground operations in any building that has multiple floors, two extremely important fireground rules should always be considered:

- Equipment on your apparatus in the street is unusable on the fireground. If this rule is specifically applied to high-rise buildings, equipment that is left on your apparatus parked in the street is unusable for above-ground operations.

- The higher the fire from grade level, the longer it will take your resources to reach a fire.

Common considerations for this type of building are as follows.

Height. The height of the building in concert with the appropriate amount of initial resources on the fire floor to affect extinguishment makes an above-ground fire challenging. At the start of this chapter, we briefly considered the definition of the fireground clock from the perspective of how long it will take you to make a visible impact on the fire. The height of a fire in a high-rise building will keep the fireground clock ticking while fireground personnel are advancing upward toward the fire floor. Therefore, a major consideration is the size and extent of fire when first units arrive, and the size and extent of the fire when the initial unit (or units) arrive on the fire floor. Therefore, an interesting perspective is what would happen if initial resources think the fire could be extinguished if initial resources were put on the fire floor and/or the floor above for extension containment) in concert with delaying implementing some of the initial IMS system as applied to high rise buildings? Remember, the IMS system does not flow any water or conduct ventilation operations.

Access. If a fire is above the reach of aerial devices, there are only two ways to access a fire above ground: stairways or elevators. Elevators are the fast and easy method, and stairways are slower but safer. There have been numerous deaths to initial fireground personnel due to elevators unexpectedly going to the fire floor. So, if you ever step into an elevator, and as the elevator is traveling upward you begin to hope the elevator does not go to the fire floor, you might be in the wrong place at the wrong time. A good rule of thumb (particularly for initial companies) is to never step into an elevator if the elevator can potentially go to the fire floor. Remember, even though it might be an apparent false alarm, and/or you are using your trick fire department elevator override key, you and your crew stand to lose (big time) if the elevator malfunctions and inadvertently goes to the fire floor.

HVAC. Heating, ventilation, air conditioning (HVAC) systems were designed to move air around the interior of high-rise and other multi-story buildings. Unfortunately, they can also move heat and smoke. If you suspect or know there is a fire, consider turning off the HVAC system until you have a better idea of conditions.

Hallways. Normally, high-rise buildings incorporate two types of hallways: a center hallway configuration with hallways that run from one end of a building to the other; and circuit hallways that run around the perimeter of a center core. In the latter example, the core contains stairways, elevators, pipe chases, and so on.

Curtain construction. This construction style usually results in a gap the approximate size of a fist between the exterior panels and the metal substructure. This gap can assist the HVAC system by allowing the upward movement of air within the building (or selected parts of a building). However, this gap can also allow for the upward travel of heat, smoke, and fire.

Method

Let's continue with the methods of construction and take a look at metal, concrete, frame wood, frame stucco, fascias, masonry, and curtain.

Metal

Metal buildings can be put into two basic types of categories: all-metal construction, and metal frame structures with a decorative exterior. The all-metal buildings can be constructed of corrugated steel, corrugated aluminum, and corrugated fiberglass (which is not a metal, but looks like metal from the street). All three of these materials look identical from a distance but will act differently in a fire. Corrugated steel will last a longer period of time than aluminum or fiberglass but, not surprisingly, all three of these materials can quickly fail when exposed to fire. A word of caution: be familiar with the differences of these materials.

Conversely, metal frame buildings will commonly use a metal frame structure and then finish the exterior with glass, marble, decorative aluminum panels, brick veneer, and other similar materials. A common example is a modern three-story office building that appears to be of brick construction (figure 4–59). It would be easy for an unsuspecting firefighter to assume that the building offers strength (brick veneer attached to an erector set does not equal strength) and a lack of vertical extension in the exterior walls (again, brick veneer attached to metal supports which are in turn attached to the metal frame structure leave plenty of room for vertical extension).

The metal frame-type buildings are currently popular because the exterior can be completed quickly and provide protection for construction crews that are working on the interior of the building. A significant number of metal frame buildings also employ curtain construction and metal studs (known as tin-can studs). It is common knowledge that metal begins to fail between 800°F and 1,000°F.

Figure 4–59. Modern buildings that appear to be of brick construction may be brick veneer attached to a metal frame.

Concrete

Concrete buildings can be found in two basic configurations: older concrete buildings and newer concrete tilt-up buildings. Older concrete buildings can be expected to have conventional construction, and in some cases, the older style of timber construction. This conventional (heavy) style of construction can be found in the roof and interior construction (floor joists, ceiling joists, and so on). Although the age of some of these buildings may have resulted in remodels that can detract from the strength of the original construction, these buildings can offer significant fireground time when compared to modern buildings, and early collapse in a fire is not the norm.

Conversely, newer concrete buildings (such as the concrete tilt-up shown in figure 4–60) have solid concrete exterior walls, but normally feature lightweight construction in the roof and interior construction.

Figure 4–60. Modern concrete tilt-up buildings normally have lightweight roofs, mezzanines, and a noteworthy fire load.

Although the exterior walls seldom collapse, early collapse of the roof and portions of interior construction can be expected when exposed to fire.

Mezzanines. Remember that concrete tilt-up-type buildings often have mezzanines that were likely constructed after the occupants took occupancy, possibly of substandard construction. These can support significant weight, and are normally located over the office area as well as the main entrance point into the building.

Occupancy. Modern concrete buildings are normally used for warehouse-type occupancies (concrete tilt-up) whereas older concrete buildings are commonly used for warehouse and residential occupancies. Today, some of the older concrete warehouse-type buildings have been converted to residential condominium-type buildings that are a significant alteration from the original building.

Wood frame

This method of construction may be the most common type of construction in the country. Every community has these types of structures because they have been built since the 1700s. Additionally, virtually every portion of wood construction is capable of burning with rapid fire spread. Common considerations are as follows.

Type and size. On your initial size-up, this building falls into the classification of either a residential or a commercial type of structure. If it is a residential, consider the following types—single story, single-family dwelling; multiple story, single-family dwelling; center hallway residential; garden apartment residential; tenement; row; railroad flat; and so on, all of which can be with or without a basement/cellar.

If it is a commercial, consider the following types—center hallway commercial; garden apartment commercial; commercial or residential, ordinary commercial/residential-type occupancy (figure 4–61); any of the previous examples with or without a basement/cellar.

Figure 4–61. Wood frame buildings can be residential, commercial, or a combination of both.

After you have identified the type of wood frame building, consider the approximate size of the building. As an example, it could be a typical older, single-story, single-family dwelling, or it could be a common two-story, wood frame house with a basement/cellar and the attic converted into an additional living area. This would total four stories (basement/cellar, first and second floor, and converted attic.) Remember, the typical Queen Anne or Victorian-type structures are normally referred to as homes, houses, single-family dwellings, and so on, but when the size is evaluated for fireground operations, they are really much larger than a routine house.

Age. The older the structure, the more renovations will be common with hidden spaces, substandard construction, and other similar types of variances. Too, the older and larger an inhabited structure, the greater the possibility of the structure being divided into multiple/separate living spaces for multiple families. A typical consideration that can have an effect on fireground time is the age of a roof. Most roofs are re-roofed approximately every 15 to 20 years, so it is easy to see that a structure that is 100 years old can have a significant dead load on the roof if the old roofing materials were not removed prior to installation of each new roof. Always be prepared for the presence of an inverted or rain roof.

Balloon, platform, and bungalow construction. If these structures were constructed prior to the 1940s or possibly the 1950s, they can be prime candidates for balloon frame construction, which features no fire blocking in the walls and a lack of plates at the top and bottom of vertical walls. Instead, a ledger is attached to the inside of a wall, which is used to support the floor joists/ceiling joists and can allow the rapid upward extension of fire. Conversely, platform construction that has been universally used since the 1950s employs fire blocking in the walls and a plate at the top of vertical walls. These two features will restrict the vertical extension of fire above a plate. Remember, the key to balloon frame construction is to open the walls above a fire and check for extension of fire—and the sooner the better.

Bungalow construction commonly uses rough-sawn 2-by-4-inch studs and rafters. An on-center spacing of 36 inches is not uncommon, and does not result in a substantial roof for ventilation personnel. The ridge, if present, is composed of a 1-by-6-inch ridge board. The lack of a ridge board results in the 2-by-4-inch rafters being butted together. This type of construction can be classified as an old type of lightweight construction. Age, termites, multiple layers of composition on the roof, and various types of remodeling can seriously affect the structural stability of these buildings. This applies to single-story and multi-story structures.

Utilities. Older wood frame structures (residential and some commercials) can be expected to have knob and tube wiring throughout the structure, unless the electrical has been updated. Before romex and conduit, wires were suspended on ceramic insulators and run through ceramic tubes when passing through walls. This wiring also used the typical fuse box on the service porch. An excellent indicator of this wiring is the age of the structure. This was used about the same time as balloon frame construction, a two-wire service into a structure, and 1-by-4-inch stripping around the windows. This was used to hide the open space on either side of double-hung windows.

As a side note, did the older wood frame building in figure 4–61 originally employ balloon frame construction and knob and tube wiring? Answer—most likely, which means the balloon frame construction is still present but the knob and tube wiring has *likely* been replaced or modified.

Frame stucco

Frame stucco buildings (figure 4–62) are relatively common, yet straightforward with few outstanding hazards other than some common considerations.

Construction. From a simplistic viewpoint and until recently, these buildings were constructed with wood structural members on the interior and stucco on the exterior. Although newer buildings of this type are similar to older buildings of the same construction, exterior decorative enhancements (corbels and such) are now made from foam, placed on the exterior, and covered with a plaster-like material or genuine plaster (figure 4–63). If fire extends to the aforementioned foam, it will give off a synthetic smoke that can be extremely hazardous to suppression personnel. Remember that modern

cornices may not be a viable support for the placement of aerial devices.

Figure 4–62. Frame stucco buildings can be relatively common with few outstanding hazards.

Figure 4–63. Newer cornices will not support the weight of aerial devices.

Age. If a building was constructed before the 1960s, the building is likely conventional construction with no outstanding hazards (other than old buildings that may have balloon construction and knob and tube wiring). If the building was constructed during the past 30-plus years, the frame stucco portion of the building is probably substantial; however, the roof may be of lightweight construction. Don't forget the newer types of construction such as structural insulated panels (SIP), tin-can studs, and other similar types of alternative construction.

Fascias. Newer frame stucco buildings (for example, strip malls) are prime candidates for fascias.

Fascias

For this discussion, let's put fascias, mansards, and overhangs in the same classification. This method of construction is common in most areas of this country (figure 4–64) because the Spanish style of architecture on some buildings is popular. Also, as the popularity of flat roofs has increased (the cheapest style of roof), so has the popularity of fascias that are used to dress the front and sides of a building, as well as hide equipment and machinery (HVAC systems) on the roof.

Figure 4–64. Fascias are commonly found on a wide variety of structures, creating an additional fireground hazard.

As an example, imagine what a McDonald's fast food restaurant would look like without the venerable fascia. Of all the modern building construction hazards, you should consider this method of construction as one of the most dangerous, yet easily identifiable.

Construction. Fascias come in two distinct varieties: an integral part of the construction, and an add-on method of construction. Quite often, the add-on varieties are attached to the exterior of a building as a retrofit. This may result in a partition between the attic space and fascia. However, the fascia has been added to the existing building, which can result in minimal strength in the attachment method. Conversely, fascias that are an integral part of the construction can have a stronger method of

support, but will often not be separated from the attic space by some type of partition or fire wall. Interestingly, an overhang that is part of lightweight trusses can be cantilevered one-third of the length of a truss. So, if a lightweight truss is 30 feet in length, the truss can be cantilevered 10 feet.

Additionally, when looking at the fascias in your district, ask yourself if they are equipped with sprinklers and/or fire partitions, and if they are separated from the attic. Naturally, the answer to these considerations is normally no. This means a fire that extends into an attic area can easily spread to a fascia, which then becomes an external common attic around the exterior of a building that is also capable of spreading fire around the exterior. Remember, if you encounter a fire that has started in the interior of a building with a fascia, you need to evaluate the possibility of fire extending to the attic and then to the fascia, which means you may need to check three distinct areas for fire. One last point under this consideration—when you encounter a fascia, remember to consider if it is lightweight or conventional construction. You know what the answer will be most of the time.

Entry. In some cases, humans are creatures of habit. As an example, you usually sit at the same place for lunch and dinner at the firehouse. So, when we apply this human characteristic to fireground operations, we can normally make two bold statements: Regardless of the size of a fire, fireground personnel will stretch a 1¾-inch line most of the time, and regardless of the location of a fire in a structure, fireground personnel will stretch the initial line through the front door. When we apply the last two statements to fireground operations that involve buildings with fascias, a potential problem can suddenly surface, courtesy of human habits—initial fireground personnel will likely enter a structure through the front door and underneath a fascia that is famous for its tendency to collapse when exposed to fire. So, if you suspect that a fascia on the front of a building and/or over your entry and exit point is (or might be) exposed to fire, consider entering through a back door.

If this scenario is worthy of consideration, consider sending initial personnel to the back of a structure (for initial entry) and ensuring they have access to a rotary saw with a metal cutting blade because the doors at the rear of these structures are normally metal clad doors in metal frames. Although this approach may involve more time, it can significantly increase fireground safety.

Extension/overhaul. If a fascia has been exposed to fire and must be overhauled, remember that gravity wants the fascia, fire weakens construction, and water weighs 8.35 pounds per gallon. This means there are three factors working against fireground personnel while overhauling a fascia (or any building for that matter) that has been exposed to fire. There are three basic ways to overhaul a fascia (figure 4–65):

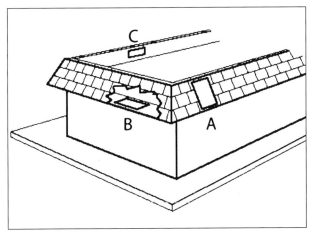

Figure 4-65. Access to the interior of a fascia is normally accomplished from three perspectives.

- Personnel can stand underneath a fascia and pull the bottom portion to access the interior (B). As the uniform building code requires compressed metal lath and plaster over a public walkway, this can be a difficult operation at best. Additionally, personnel will be standing in the potential collapse zone.
- If the fascia has a slope (or pitch) on the front portion, you can try to work from the exterior (ground ladder or aerial device) down into the fascia (A). This is difficult and dangerous from a ground ladder. If the fascia is flat on the front portion, you can try to pull the finish material from the front with a

tool such as a pike pole, but you will also be located in the potential collapse zone.

- If the roof is safe, personnel can access the interior of a fascia by quickly removing the back side of the fascia (which is normally ½-inch plywood, OSB, and so on), and conducting the overhaul from the roof (C). If the fascia were to collapse during this overhaul scenario, the collapsing material would fall away from personnel.

Consider the following four-point fascia size-up when confronted with this style of construction (figure 4–66):

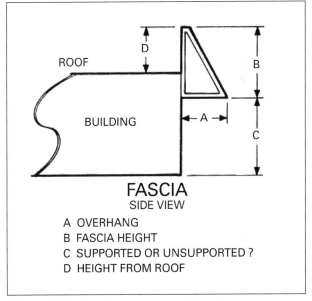

Figure 4-66. Fascia size-up considerations

Overhang (A). Overhang is the distance a fascia extends from a building. As the distance increases, so should your concern about structural integrity whenever a fascia is exposed to fire. As the size of a fascia increases, so does its complexity and the materials used in its construction (which can easily be lightweight construction). The size of a fascia can have a direct effect on the area, path, and travel of fire. Unless proven otherwise, expect any fascia to lack fire-stopping capabilities, sprinklers, and be open to the attic. If a fascia is exposed to fire, expect it to collapse outward *at least* the distance of the overhang.

Height and shape (B). The height and shape will affect structural stability, the quantity of building materials present, and the potential path of fire. Remember, a fascia is an external common attic that easily conceals and spreads the travel of fire around the exterior of a building. Keep such factors in mind during ladder operations. On buildings without fascias, the roof type (arch, sawtooth, and so on) can often be identified from the ground. This is helpful (and often necessary) when laddering the roof area. The style (curve) of a fascia and/or distance out from the building can hide the type of roof and/or also keep a ground ladder from reaching its top, further hindering access to a roof.

Supports (C). Some fascias are equipped with vertical supports (pilasters, columns, and so on) that serve as decorative additions to a fascia, but also serve as an external support. Obviously, these supports can enhance the strength and safety of a fascia, but unfortunately they cost additional money and are limited in their use. Conversely, most fascias are unsupported, which does not enhance structural stability when exposed to fire (see figure 4–53).

Height from the roofline (D). Fascias can conceal a roofline, so ladder operations can present an additional challenge. When laddering a fascia and the roofline cannot be seen, the height of the fascia above the roof should be determined. If this is not done, an additional trip back to the ground for another ladder may prove necessary. Additionally, personnel on a roof without a safe and easy means of egress to the top of a fascia (on the backside) are in trouble if the roof begins to collapse. If the distance from a roof to the top of a fascia or parapet wall exceeds about 5 feet, a ladder from the roof to the top of the fascia/parapet wall is necessary. Consider the following when determining the roofline of a building:

- Rafter tie plates: These indicate the location of the roof rafters which will designate the roofline.

- Windows: Rooflines run between the top floor windows and the top of a fascia/parapet.

- Equipment on the roof: Air conditioners and heating units that can be seen above a fascia indicate that the roof is in close proximity.

- Attic vents: Rooflines are between the attic vent and top of the fascia/parapet.

- Scuppers: A scupper is the actual level of a roof. The scupper in figure 4-64 is about 4 feet from the top of the fascia.

- Perimeter: Fascias are normally constructed on the front and sides of a building, not on the back of a building. If present on the back (for example, such as McDonald's, White Castle, and so on), they will have an access opening for roof access/egress.

Masonry

Although any type of construction that is exposed to fire can present a significant hazard, specific types of construction are capable (and willing) to pose unique hazards. Such is the case with un-reinforced masonry construction (URM). Fortunately, this method or type of construction is often easily recognizable and can be expected to present a specific hazard when exposed to fire—collapse. Let's look at some typical considerations.

Pre-1935. Prior to 1935, brick buildings were constructed with the following characteristics: mortar consisting only of sand and lime; a lack of rebar; exterior walls about 13 inches thick and tapering vertically upward (reducing in thickness); parapet walls (which can often be of a noteworthy height) around the perimeter of a roof; floor and roof joists that are "let" into a joist pocket into the exterior walls; and roof and floor joists that are *fire cut* (ends are cut at an angle) so they will pull away from a wall and fall into the interior of a building without collapsing the exterior walls.

Post-1935. As a result of the Long Beach, California, earthquake in 1933, the following revisions were implemented to new masonry construction: cement used in the mortar; steel rebar to be used; exterior walls at least 9 inches thick; and all joist and rafters anchored to exterior walls. This was normally accomplished by bolting a ledger board to a wall and attaching the joist/rafter to the ledger via a metal hanger.

Post-1959. After the Tehachapi, California, earthquake in 1959, many building codes (applied to masonry construction, and specifically URM construction) were further modified (including some retrofits) as follows: 4-to 6-inch concrete cap on top of parapets along public walkways and exits; parapet could be no more than 16 inches high (including the cap); and rafters and joists had to be anchored to exterior walls with a steel anchor bar or rod.

With this brief history of URM modifications, one would think these buildings have been adequately strengthened to withstand the effects of earthquakes and, for our discussion, the effects of fire. However, this is a false premise. Interestingly, some of the aforementioned modifications have significantly increased the hazard of collapse to fireground personnel. Let's look at identifying this type of construction, related hazards, and some operational considerations.

Identification. Identification of URM buildings is relatively easy. Look for the following trademarks (figure 4–67):

- **Rafter tie plates:** From the exterior, they can be in the form of a star, square plate (that when turned sideways will look like a diamond), straight bar, S-shaped rod, and so on. One advantage of visible tie plates is they are an indicator of the location of floor joists and roof rafters.

- **Deeply recessed windows:** In these buildings (and most masonry buildings), window frames are normally set to the inside of the exterior walls, exposing about 8 inches of wall.

- **Concrete bond beam cap on top of the exterior wall:** Bond beams may also be present over windows and between floors (for additional stability).

- **King row:** Within every 4 to 7 rows of bricks, one row will have been laid on end for additional strength.

- **Lintels over windows:** Windows will have arched or straight lintels over the windows. Normally, the lintels are arched.

- **Lime mortar:** The lime mortar between the bricks is often white, porous, and can often be removed by a finger or knife. Additionally, some older URM buildings have what appears to be sloppy workmanship that

has resulted in bricks that have not been uniformly laid and uneven mortar joints.

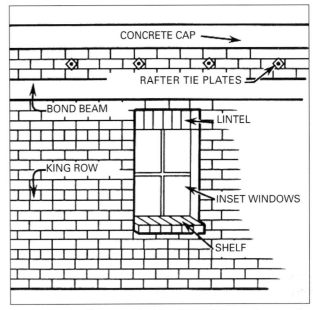

Figure 4–67. Characteristics of unreinforced masonry construction

In some URM buildings, the exterior walls have been plastered to give the building a newer look, or a higher-quality brick was used on the front of the building. In these cases, the sides of the building do not normally use the better brick, and the inset windows and/or lintels over the windows are still visible on plastered buildings.

Hazards. Originally, these buildings were designed so floor and roof joists would pull out from the walls during a fire (remember the fire cut?), thereby preventing wall collapse. When joists and rafters are tied to the exterior walls (remember the rafter tie plates?), collapse is more likely, and the collapse is normally outward a distance greater than the height of the wall. Another common hazard is the presence of arch type roofs on URM construction. If a collapsing arch truss is connected to hip rafters at the end of a building, the truss will push the hip rafter outward, also pushing the corresponding wall outward with considerable force. Additionally, if these buildings are old (built in the late 1800s and early 1900s), think about how many layers of roofing materials are on the roof, and then determine if it is possible that the roof is carrying a severe dead load. If so, this makes a collapse more likely.

Operational consideration. Operational considerations are simple if you consider that a floor and/or roof collapse will push the walls outward. With that thought in mind, remember the potential collapse zone may be outward a distance greater than the height of the wall. The front and rear walls are at the greatest risk and the side walls pose a secondary risk. Safe areas for resources would then be located at the corners of a building and away from the building at least the distance of the height of the walls (figure 4–68).

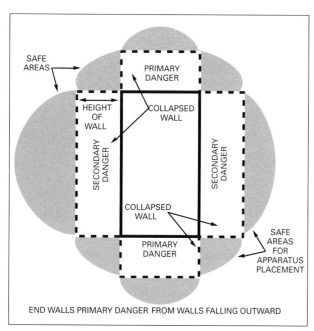

Figure 4–68. Unreinforced masonry construction collapse zones

Additionally, departments that conduct roof ventilation operations may find metal straps across the roof and underneath the roofing material if the building has been retrofitted for earthquakes. There is a significant difference between URM buildings and masonry buildings constructed after 1935. These masonry buildings (or post-1935) use cement for the mortar and rebar. Post-1935 masonry walls do not have a tendency to collapse, particularly cinder block walls. Additionally, these buildings look different as follows—a better quality of brick; no lintels over the windows (unless the architecture is made

to look older); windows do not have as large an offset; no king row (unless the architecture is made to look older); and no rafter tie plates.

Curtain

Curtain construction is a recent addition to the tool bag of the building construction industry that goes something like this: If you want to quickly build a nice looking, multi-story commercial building (such as a modern two-, three-, or four-story office building with glass exterior, or a high-rise building), erect steel beams, coat the beams with fire retardant materials, and attach/bolt exterior panels of glass, slate, concrete panels, brick veneer, or other similar materials to the steel beams. After the exterior is finished, two initial goals have been accomplished: The exterior has a finished look; and construction personnel can now finish the interior with some protection from the elements of weather. Although this method of construction is fast, it also provides a vertical passageway to fire and smoke. When the exterior panels are attached to the steel beams, a gap about the size of your fist can normally be found between the beams and exterior panels. Although this gap can assist the upward travel of air within the building and enhance the operation of the building's HVAC system, it can also vertically spread the travel of fire, smoke, and other dangerous by-products of fire. Although this gap has been sealed in some buildings, there are numerous buildings with this gap.

Vertical extension. Remember, there are five ways fire and its by-products can vertically spread in newer multi-story buildings: poke thru construction, lapping, radiated heat thru flooring, HVAC ducting, and the gap between the exterior panels and interior beams.

Vinyl siding and ABS plumbing

Vinyl siding and ABS plumbing are two subtle hazards that don't fit into any specific classification, but should be considered. Vinyl siding is becoming more popular as a replacement and retrofit to common wood lap siding. It is a plastic-based material and will easily melt and burn in a fire, spread fire faster than thicker conventional siding, and give off toxic fumes that are not associated with wood lap siding. In most cases, vinyl siding can be identified by its narrower horizontal dimension (7-inch width for most wood siding and 4-inch width for most vinyl siding). Vinyl siding is easily identified by a close visual inspection and by tapping on the vinyl with a hand.

ABS plastic is used in plumbing applications. If fire contacts this material, it suddenly becomes a wick through a structure and gives off great quantities of black stringy smoke. The first time I experienced this material in a residential fire, it was assumed that we had an arson fire due to multiple fires in remote locations. If you open walls and find this material, or if roof personnel see smoke issuing from plastic plumbing vents, you may have a fire in plastic plumbing that is spreading through the walls and attic spaces.

Access and Egress

When entering a burning building for any reason, consider that the building is under demolition. As a result, before forcing entry into a structure under demolition, consider not only how you are going to enter the building, but how you are going to get out. Additionally, entry and exit points can also be used by occupants. So, entry and exit points should be evaluated for occupant exit considerations and firefighter entry and exit options. Before we look at several types of buildings with noteworthy access and egress considerations, let's quickly review one basic access and egress fireground rule: The minimum number of openings available to interior fireground personnel is always two. And quite often, the more openings the better! If the modern fireground now features flashover as a common potential problem, interior personnel should evaluate the number of ways to exit a structure if the need suddenly arises. Remember: Personnel who have been caught in flashover-type conditions commonly verify that you have three to five seconds to exit a building if caught in a flashover! Do not limit yourself to one way in, and more importantly, one way out. Here is one more point for careful consideration: Most firefighters deaths occur as a result of running out of air and smoke inhalation, which

is why the new buzzword in the U.S. Fire Service is air management (see chapter 5 for more details on this subject). If an environment suddenly becomes untenable and/or you run out of air, your primary consideration will be your nearest exit option. Let's look at several examples of buildings and problems that can offer entry/exit challenges.

Vacant buildings

Vacant buildings, particularly buildings that are boarded up, are potentially one of the most dangerous buildings you will encounter. Although you can expect to find substandard construction, holes in the floors, and other similar problems, you are initially confronted with the fact that all doors and windows are covered up. Therefore, the more openings you create during suppression operations (and the sooner the better), the safer your interior operations become.

Concrete tilt-up

Quite often, concrete tilt-up-type commercial occupancies have a single man-door into an office area, one or two overhead doors in the rear of the building, and no windows (refer to figure 4–60). If fire conditions and resource space permits, go for the overhead doors because the openings will be significantly larger than a single man-door that may also allow entry under a mezzanine.

Security bars

Security bars are becoming more popular in numerous municipalities as the concern for security increases. Normally, security bars are placed over all windows (and doors) for maximum security. The more security bars that are removed, the better, particularly in the areas where interior personnel are expected to be operating. When considering doors and windows for access/egress openings, several basic thoughts should come to mind—which opening is normally used by occupants and/or firefighters, which opening is normally larger, which opening can be made larger with the least amount of effort, and which opening can be opened with the least amount of effort.

Windows

Windows are relatively simple because they can be easily broken to provide quick access into the interior of a building. However, there are several rules of engagement to remember:

- The size of a window can be an indicator of the floor plan of the building. As an example, large windows can indicate display areas in commercial buildings, and living/dining rooms in residential structures. Smaller windows that are higher up on a wall on residential structures can indicate a kitchen/bathroom because these windows are above a counter top.

- Window glass comes in several noteworthy varieties. See chapter 6 for more details on breaking glass windows.

- If a window must be used for access/egress, remember to remove the entire window (glass and any restricting frames) because you may need to use the window for egress.

- If windows in vacant buildings are restricted by plywood/OSB or other materials (or windows in inhabited buildings by security bars), remove as many obstructions as possible if fireground personnel will be entering the building. Remember: Make a building safe and easy to exit.

Doors

Without a doubt, doors normally provide several advantages for access/egress when compared to windows: they are larger—sometimes much larger—entry and exit points for occupants, and allow entry and exit lower than windows (visibility, lower heat levels, and so on). Similar to windows, when fireground personnel are committed to the interior of a structure, make the opening as large as possible, and ensure there are at least two ways to exit a structure. With these basic thoughts in mind, consider these examples:

- Assume you are the first truck company to a fire in a residential dwelling, and the first company has forced entry through the front door for an initial attack. Based on the concept of a minimum of two openings that allow ingress and egress, this forcible entry is not complete. As you size-up a structure, determine how easy egress will be after you are inside and if additional forcible entry is necessary.

- Assume you are the first company to a fire in a concrete tilt-up commercial building. All things being equal, would you force entry through the front man-door into the office area, or would you force entry through the rear overhead door that is used for loading and unloading materials? Based on the concept that a larger opening is better, the rear overhead door would be a superior opening. Additionally, if the front man-door into the office area is broken, glass will likely be on the ground where personnel are entering and exiting the building, creating a potential hazard. And if there is a mezzanine over the office area, personnel will be working under an additional potential hazard.

Name and Utilities

The business name on the front of a building can be an excellent indicator of what you can expect to find inside of a building. As an example, although residential buildings normally do not have a name on the front (unless they are a commercial residential building), every firefighter is aware that you can expect to find occupants inside; depending on the time of day, the number of occupants can significantly vary. However, when observing a commercial building, your concern should center on what is inside the building. The business name on the front is an excellent indicator of the contents. Additionally, any materials, stock, and/or storage outside a building can also be an indicator of what you can expect to find inside. As an example, look at figure 4–69, which is a building that contains three occupancies.

Figure 4-69. The name on the exterior of a building can be an indicator of what can be anticipated inside a building.

The first occupancy in this building is Dividing Line Youth Ministries. Likely a minimal hazard based on what there is to burn. The next is Weathermaker Heating & Air Conditioning. Probably a moderate hazard, based on what there is to burn. But the third business in this building is Pool Supplies. This business could be a maximum hazard based on what there is to burn and the added potential of hazardous materials.

Utilities should also be a consideration when entering a building. This is covered in more detail in chapters 12 and 13, but for this discussion, let's consider electrical and gas utilities as our main concerns. Always look at the size and number of the utilities that enter a building as an indicator of a potential interior hazard. If you are thinking about entering a structure where darkness and/or smoke has resulted in minimal or zero visibility in concert with heavy utilities entering the same structure, re-evaluate your priorities. The following are examples of electrical and gas utilities that are considered not within the norm:

- Duplex service (two wires) is an indicator of knob and tube wiring.

- Quadraplex (four wires) is indicative of heavy electrical needs within a structure.

- Large transformers or a bank of transformers is an indicator the electrical needs far exceed a normal electrical service.

- A residential-type gas meter, a large commercial-type gas meter, and a rotary vane gas meter are all significantly different from each other.

Status of a Building

Your ability to evaluate the status of a building can create a foundation for determining appropriate fireground operation risk and priorities as applied to the following four building categories. See chapter 9 for more information on this subject.

- **Occupied:** Defined as buildings that are occupied or have a high probability of being occupied during an incident. In these cases, search operations may be a high priority, either ahead of attack operations or with attack operations.

- **Unoccupied:** Described as a building that is normally occupied but the occupants are likely to not be in the structure. The chance of occupants inside these types of buildings is reduced from occupied buildings, and may place a reduced emphasis on search and a higher emphasis on offensive operations.

- **Vacant:** Likely to still be in an acceptable condition in terms of structural integrity, particularly when a "for sale" or "for lease" sign is displayed on the front of the building. The presence of occupants inside the building is unlikely, search should not be a primary concern, and the emphasis should focus on offensive operations.

- **Abandoned:** Primary indicator that the building has likely outlived its usefulness, and structural integrity may be in disrepair. The risk factor of offensive or defensive operations should be evaluated with an emphasis on fireground safety. If the building is not secure, consider the need for search operations.

Training Resources

The following resources can provide additional information and/or another viewpoint for the subjects discussed in this chapter:

- *Building Construction for the Fire Service*, Francis L. Brannigan, NFPA.

- *Collapse of Burning Buildings*, Vincent Dunn, Fire Engineering Books and Videos.

- *Fireground Size-Up*, Michael A. Terpak, Fire Engineering Books and Videos.

- *Fireground Strategies*, Anthony L. Avillo, Fire Engineering Books and Videos.

- Lightweight Construction: Is Now the Time to Push for Sweeping Industry Changes, Azarang Mirkhan and David C. Comstock Jr. *Fire Engineering*, June 2009.

- Building Construction: Lightweight Steel Framing, Gregory Havel, *Fire Engineering*, January 2008.

- What Floor are We On, Michael Terpak, *Fire Engineering*, March 2006.

- Collapse and Fire Extension in Wood-Frame Construction, Bill Gustin, *Fire Engineering*, April 2006.

- Ancient Buildings Pose Hazards Older than Codes, Gregory Havel, *Fire Engineering*, July 2007.

- Fireground Strategies: The "New Millennium" Multiple Dwelling, Anthony Avillo, *Fire Engineering*, October 2005.

- Fireground Strategies: Fighting Shaft Fires, Anthony Avillo, *Fire Engineering*, December 2007.

- Parapet Collapse: A Serious Danger for Firefighters, Thomas Dunne, *Fire Engineering*, November 2002.

- Operating Safely on Fire Escapes, John Flynn, *Fire Engineering*, March 2008.

- Safe Operations Near Roof Cellular Base Stations, Joseph Viscuso, *Fire Engineering*, March 2008.

- Second-Story Hidden Dangers, Michael J. Lopina, *Fire Engineering*, July 2008.

- Operating at Public Storage Warehouses, James Kirsch, *Fire Engineering*, November 2008.

- The Dangers of Modular Construction, Kevin A. Gallagher, *Fire Engineering*, May 2009.

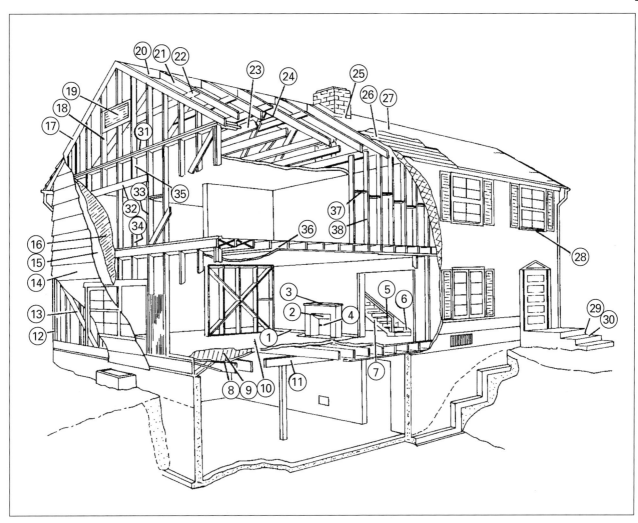

1. Base molding	11. Floor girder	21. Rafter	31. Header
2. Damper	12. Corner stud	22. Collar ties	32. Cripple stud
3. Mantel	13. Diagonal bracing	23. Ceiling joist	33. Trimmer
4. Hearth	14. Siding	24. Knee brace	34. Top plate
5. Handrail	15. Sheathing	25. Metal flashing	35. Cripple stud
6. Stairstringer	16. Vapor barrier	26. Decking	36. Floor joist
7. Stairjack	17. Facia board	27. Roof covering	37. Fireblocking
8. Floor joist	18. Gable stud	28. Sill	38. Stud
9. Rough floor	19. Gable attic vent	29. Riser	
10. Finished floor	20. Ridge beam/ridge pole	30. Step	

Figure 4–70. Basic construction diagram 1

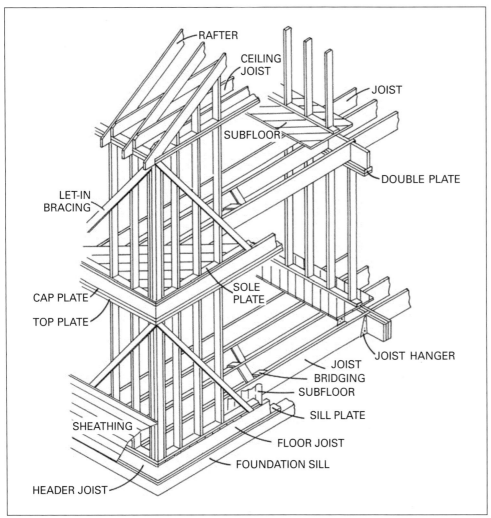

Figure 4–71. Basic construction diagram 2

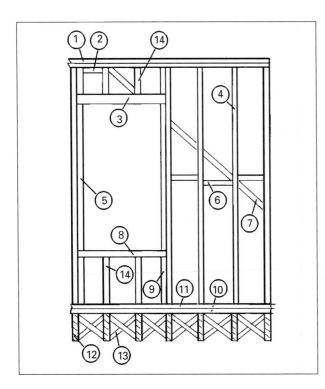

1. Top plates
2. Kicker block
3. Header
4. Stud
5. Trimmer
6. Fire blocking
7. Diagonal bracing
8. Sill header
9. Trimmer
10. Subflooring
11. Sole plate
12. Floor joist
13. Herringbone bridging
14. Cripple studs

Figure 4–72. Basic construction diagram 3

Reading Smoke

Introduction

The mitigation of structural fires is a dynamic process that is normally based on numerous factors that can be summarized as follows:

- Every structural fire presents a different set of conditions and challenges.
- Available fireground time is being eroded as the size of structural members is systematically being reduced.
- The contents of residential and commercial buildings burn significantly different than a few years ago due to the predominance of synthetic materials.
- Every building that is subjected to fire is under demolition until the fire is extinguished.
- Modern personal protective equipment (PPE) is far superior to the PPE of yesterday, yet has the capability to mask the interior conditions of a fireground environment.

Unfortunately, the preceding factors collectively combine to hamper the American fire service with 20,000 injuries, over 100 deaths per year, and an increase in cancer. These figures are more than significant when statistics indicate fireground injuries and deaths are increasing, not decreasing. If the number of structure fires is slowly decreasing (and it is) but the number of fireground deaths and injuries remain about the same each year, then in reality, fireground deaths and injuries are increasing. How could this be when modern firefighters are subjected to superior training programs and high-tech equipment? Additionally, NFPA statistics indicate that 70% of structure fires occur in single-family dwellings (SFDs), and 70% of those fires are confined to one or two rooms. From a simplistic perspective, the answer is the routine SFD fire, as well as other types of structure fires, have dramatically changed.

As delineated in chapter 4, building construction now repetitively uses lightweight materials due to the increasing scarcity of wood and its higher cost. This has directly contributed to rapid collapse of lightweight structural members when they are exposed to fire. However, there is an additional factor that has also contributed to the increasing number of firefighter deaths and injuries—smoke. Because smoke has also dramatically changed over the past several decades, let's begin this chapter by considering the various characteristics of smoke and the importance of why it should be evaluated at every structure fire.

Smoke Characteristics

One of the more popular and relevant national training programs is titled "The Art of Reading Smoke," and for good reason. In the past 40 years, smoke has dramatically changed from two basic viewpoints. Smoke is nothing more than by-products from a fire; therefore, as the materials that burn have changed, the *composition* of smoke has also been altered. As a result, this influences how smoke burns from two basic viewpoints: the degree of heat, and the combustibility of smoke. This is best illustrated by considering the basic elements of smoke. Fundamentally, smoke is a mixture of three basic elements—unburned solid particulates, aerosols, and gases (figure 5–1). Unburned particulates are primarily a combination of the materials that are burning and carbon.

Figure 5–1. Smoke is a combination of particulates, gases, and aerosols.

These particulates are capable of burning if their ignition temperature is sufficient. Aerosols are fine

droplets and include a whole host of hydrocarbons, such as oils, tar, and creosote. Gases are a combination of a multitude of various gases that are released by the materials that are burning and are also capable of burning if their ignition temperature is sufficient. However, fire gases can be much more complicated than particulates because different materials are capable of yielding a wide range of fire gases. Common examples of some of the fire gases that are released during the combustion process are carbon monoxide, sulfur dioxide, hydrogen fluoride, hydrogen chloride, hydrogen cyanide, aldehydes, benzene, acrolein, and sulfur dioxide.

All firefighters should be aware that these fire gases can be detrimental or fatal to any human. However, if the average firefighter was asked to choose the most common and deadly fire gas, carbon monoxide would likely be chosen most often. Although carbon monoxide is well known for its capability to be present in smoke and cause death by asphyxiation if inhaled in sufficient quantities, it is becoming apparent that *hydrogen cyanide* (HCN) may be the most dangerous fire gas encountered by fireground personnel. It is estimated that hydrogen cyanide is over 30 times more toxic than carbon monoxide, and is also flammable. This fact was recently underscored in Rhode Island in 2006 when a firefighter involved at a routine fire in a fast-food restaurant was sent to a hospital and diagnosed with toxic levels of cyanide. Interestingly, over the next 14 hours, two additional fires in the same area were responsible for four more firefighters being hospitalized and also diagnosed with toxic levels of cyanide in addition to one of the firefighters suffering a heart attack. These incidents have been responsible for a new look at this common product of combustion, which may be present in high quantities at structure fires. Hydrogen cyanide can be found in either a liquid or gas form, and is a powerful poison that reduces the capacity of blood to carry oxygen. Bodily organs that are vulnerable to cyanide poisoning are the brain, heart, and central nervous system. The toxicity of hydrogen cyanide is evident by its use in gas chambers and as a chemical warfare agent. Unfortunately, hydrogen cyanide is found in common materials such as wool, paper, wood, and cotton. However, it is also found in elevated levels in synthetic materials such as foam, pesticides, plastics, synthetic fibers, polyurethane, fiberglass insulation, and other similar modern materials. Interestingly, carpets lead the list of culprits.

These materials are common in the typical residential structure fire as well as other types of structure fires, resulting in hydrogen cyanide likely being present in readily detectable amounts in smoke from smoldering fires, free-burning fires, and overhaul operations. Stated from another perspective, hydrogen cyanide is an extremely toxic gas that is released during the combustion process from any material that contains nitrogen, which is commonly found in structural occupancies. Although there are numerous other toxic gases released during the combustion process, hydrogen cyanide is a prime example of why ventilation should be an initial consideration at structure fires, and a primary reason why fireground personnel should not breathe smoke. That statement applies to structure fires, automobile fires, dumpster fires, and other similar incidents. Toxic gases that are inhaled by fireground personnel give credibility to the phrase "when firefighters breathe smoke, they can die."

Smoke that is routinely found on the modern fireground can also be a willing candidate to sudden flashover. Intriguingly, the term *flashover* was relatively unknown prior to the 1970s because the term *backdraft* was well known and understood by most fireground personnel. However, in the 1980s, flashover began to replace backdraft, and today, few firefighters are familiar with backdraft whereas most firefighters are familiar with flashover.

Pre-1970

Prior to the 1970s, conventional building methods and materials were commonly employed. Additionally, the interior contents of residential and commercial buildings primarily consisted of conventional materials such as wood, paper, wool, cotton, and so on. These materials normally burned with a gray-brown smoke (cellulose-based materials) and the smoke would commonly burn around 1,123°F. (This is the temperature at which carbon monoxide self-ignites and is the temperature that is most often tied to the flashover of Class A materials.) Because of the high ignition temperature of smoke from conventional materials, suppression personnel would routinely arrive at a structure fire to find pressurized levels of smoke and an environment that was oxygen deficient. These conditions often presented initial personnel with the potential of a backdraft from the sudden introduction of oxygen when entering a structure. During this time frame, PPE was marginal, but was beginning to improve as a combined result of an aggressive focus to improve the protection of fireground personnel with the advantages of new technology. To summarize this general time frame, fires burned slower than the fires of today, fewer synthetics were present, the term *backdraft* was familiar to most fire suppression personnel, and flashovers occurred less frequently than today.

Post-1970

After the 1970s, building construction and thus smoke from fires in residential and commercial buildings noticeably changed. Conventional materials were steadily replaced by innovative products (defined as synthetic materials) that burn significantly different than conventional materials. Although building codes dramatically improved when compared to the pre-1970 building codes, they do not address the way modern materials burn and the resultant smoke that is released during a fire. To summarize this general time frame, fires now characteristically burn considerably different than the fires of yesterday, and the term *flashover* is familiar to most fire suppression personnel.

Let's review the terms *backdraft* (also known as a smoke explosion) and *flashover* because both of these events are related from the perspective that they are capable of injury or death to fireground personnel, yet are unrelated from the perspective of the conditions necessary for their origins. Popular opinion differentiates between smoke explosion and backdraft, the difference being flaming or not flaming. An explosive percussion surge (non-flaming) is a smoke explosion. A flaming explosion is a backdraft (see figure 5–4).

Backdraft

One definition of the term *backdraft* is "the introduction of oxygen to a confined area that is pressurized with heated flammable gases that are deficient in oxygen that results in an explosive force of significant intensity." To understand the background of this definition, let's review some basic theory. Any fire must have heat, fuel, and oxygen to sustain combustion. In air, oxygen is normally 21% by volume. When heat, fuel, and oxygen are present in a suitable ratio, a free-burning fire will result. If we transfer a free-burning fire to a confined room with no ventilation, the growth of a fire can advance through four phases of progression:

1. Ignition phase

2. Growth phase

3. Fully developed phase

4. Decay phase

In the ignition phase of a fire, the oxygen content is about 21%. This beginning phase of a fire produces some heat, smoke, and fire gases, some of which are flammable. At this stage of a fire, the temperatures in the room are slowly beginning to increase over the ambient temperature of the room. As this process continues, the growth phase of the fire begins where the level of smoke and fire gases increase in addition to the interior temperature, and a process known as thermal layering begins when the column of smoke and fire gases from a fire rise vertically to a ceiling and begin to bank downward. However, as long as there is sufficient oxygen, the fire continues to accelerate. If this process is allowed to continue, the temperature within the fire area increases to the point where combustibles within the fire area ignite. This can also be referred to as a free-burning stage because the fire has sufficient heat, fuel, and oxygen to produce large quantities of heat and fire gases. This is the fully developed phase of a fire. However, if the supply of oxygen is depleted or minimized, this process begins to subside, and the decay phase of a fire will begin. This is also known as the smoldering phase because the oxygen supply has been depleted (13 to 15%), and there are still high temperatures and large amounts of fire gases within the fire area. At this point however, another condition rapidly becomes significant. As smoke and fire gases are heated, they expand. In fact, the volumetric expansion of gases can increase (see figure 5–2), and are capable of creating a significant pressure (up to three times more than the ambient atmospheric pressure of 14.7 psi) inside the room. This internal pressure forces smoke and fire gases from all available openings. At this stage of this process, the room is pressurized, flammable fire gases are high, and oxygen levels are low.

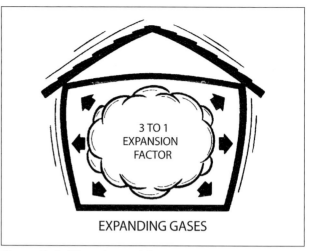

Figure 5–2. When heated fire gases are confined, they create a pressurized environment.

If oxygen is suddenly introduced to this scenario, either by breaking a window or opening a door, the fire triangle (heat, fuel, and oxygen) is completed and an explosion can be expected to occur, often with destructive force to personnel and a structure and/or room.

Trademarks. A potential backdraft is relatively easy to recognize by four distinct trademarks.

1. Unvented fires that are confined by a structure.

2. Smoke that is issuing under pressure from any available openings, or smoke that is visibly being drawn into a pressurized, confined area after openings are made to the involved area.

3. Doors and/or windows that are hot to the touch. Additionally, windows may appear dark (which is a result of the heavy concentration of interior smoke).

4. In some cases where various openings are present, expect heavy, dark, turbulent smoke (refer to figure 5–3) that is a result of the pressurized smoke expanding as it escapes and the inability of the structure to absorb the heat.

Elimination. A backdraft can be eliminated by vertical ventilation over the involved area. This reduces the internal temperature (primarily from the top portion of the room which is the location of the most heat) and vertically exhausts the hot smoke and fire gases to the exterior of the structure without allowing the sudden introduction of oxygen. If horizontal ventilation is the only practical method of ventilation, ventilation personnel should ventilate windows and/or doors from the corner of the structure as this is the safest area for ventilation personnel. The success of this operation is based on the following factors:

- Recognition of a potential backdraft

- Communication between initial entry and ventilation personnel

- Timing between entry and ventilation personnel that results in ventilation preceding entry into the structure; in this case, firefighters preparing to make entry should not be positioned by openings (windows, doors, and so on)

A backdraft is illustrated in the following figures. In figure 5–3, four firefighters have just forced a door for entry into the structure. Notice the heavy volume of pressurized, roiling smoke from the front door and the dark display window to the left side of the firefighters.

Figure 5–4 shows a backdraft has just occurred.

Figure 5–4. A backdraft has just occurred. (LAFD file photo)

Figure 5–3. This door has been forced to gain entry and a heavy volume of roiling pressurized smoke is now visible. (LAFD file photo)

In figure 5–5, the force of the backdraft has blown a firefighter 10 feet backwards (the firefighter on the ground), the company officer is moving to help the downed firefighter (notice the officer's facepiece is smoking), the facepiece of the firefighter to the far right is on fire, and the door opening is involved with fire. Obviously, this company is no longer a viable suppression company.

Figure 5–5. The force of the backdraft has blown a firefighter 10 feet backwards. (LAFD file photo)

Flashover

Before we look at the definition of a flashover, it is important to state that there are abundant definitions without one definition being the accepted standard. As an example, a popular definition of a flashover is "The sudden ignition of exposed combustible surfaces and/or combustible gases in an involved area that results in a sudden and intense rise in temperature." Although this definition is applicable to some fireground flashovers, there is another definition that is not as popular but is also applicable to some fireground flashovers—"Flashover is the ignition of the layer of combustible gases produced by a fire that accumulate close to the ceiling, can quickly spread to the rest of a room, and in some cases, igniting the room contents."

These two definitions share some similarities, but there is a primary difference. The first definition follows the theory that most flashovers are the result of radiant heat from the ceiling that raises the temperature of the room contents to their ignition temperature (radiant feedback), resulting in all contents simultaneously bursting into flame. The second definition is based on the principle that the ceiling area is where smoke (fire gases and particulates) begins to accumulate and then continues to collect along with increasing heat. For this chapter, we will use a definition that incorporates the two preceding definitions—Flashover is the sudden ignition of exposed combustible surfaces and/or combustible gases in an involved area and/or the ignition of the layer of combustible gases produced by a fire that accumulate close to the ceiling and can quickly spread to the rest of a room, both of which result in a sudden and intense rise in temperature. Let's look at this definition by considering a few basic principles.

As a fire burns in a room with the correct ratio of heat, fuel, and oxygen, the heat, fire gases, and other products of combustion rise to the highest point in the room (ceiling area) and then begin to spread horizontally under the ceiling area. As the fire continues to burn, another chain of events simultaneously takes place. As the heat, fire gases, and products of combustion collect at the ceiling area, their temperature increases (sometimes dramatically). Although this rising temperature also increases the temperature of room contents, the highest temperature in the room is at the ceiling area, which can provide the most favorable conditions (heat and fuel) for a flashover (figure 5–6).

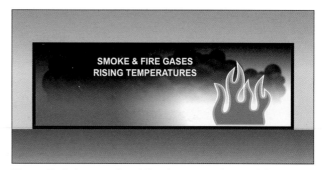

Figure 5–6. In a confined fire, heat, smoke, and fire gases collect at the ceiling area and can provide favorable conditions for a flashover.

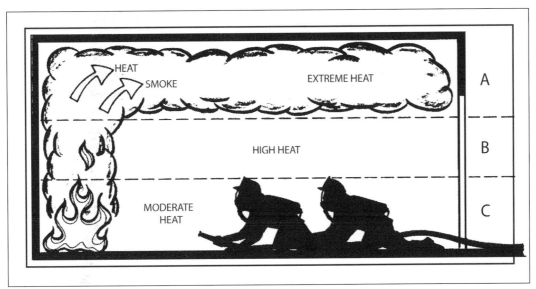

Figure 5–7. In a confined area, increasing heat can create three distinct temperature zones.

Additionally, heat can also bank down in the room and in a short period of time can create three distinct temperature zones, as shown in figure 5–7.

Figure 5–7 shows (A) extreme heat due to accumulated and rising heat from the fire (800°F to 1,000°F); (B) high heat from heated gases within the room and radiated heat (500°F to 600°F); and (C) moderate heat from heated gases and radiated heat, which is the furthest distance from the heat at the ceiling (300°F to 450°F).

These three temperature zones can be easily detected by interior personnel who are making entry into a room that is being heated from a fire. In some incidents, personnel close to the floor will encounter a temperature that can be hot but bearable. If the personnel rise up just several feet, the temperature can quickly become unbearable. This is one reason why firefighters are taught to stay low when entering a heated atmosphere because this is the area of least heat (and the best visibility). As a fire continues to burn and the heat within the room continues to increase, there will be a point where the smoke and fire gases at the upper portion of the room and/or the room contents have reached their ignition temperature. As a result, the upper portion of the room or the contents of the room (furniture, fire gases, smoke, and so on) can flashover and create the following conditions:

- A sudden and intense rise in temperature that can go from 800°F to 1,000°F to 1,800°F or more

- A sudden loss of visibility for interior personnel

- An immediate drop in the oxygen level from 21% to 4%

- An immediate increase in the carbon monoxide level to 4,000 parts per million (ppm)

Depending on the amount of smoke and fire gases within a room, a flashover can be located in the upper portion of the room, or can even result in total room involvement (figure 5–8).

Obviously, any of the preceding conditions can result in injury or death.

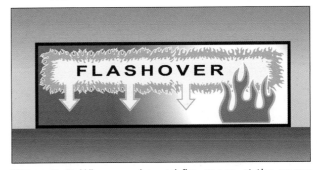

Figure 5–8. When smoke and fire gases at the upper portion of a room reach their ignition temperature, they can flashover and often involve the entire room.

Trademarks. Although the warning signs for a backdraft are relatively easy to recognize, the warning signs for a flashover are often subtle and elusive. Look for the following conditions:

- An environment with inadequate ventilation

- A sudden increase in temperature for no apparent reason

- Personnel being forced to stay low due to increased heat, particularly when extinguishment has not started or been completed

- Visible *tongues* of flame or rollover conditions in the smoke above interior personnel

- Smoke changing to thick turbulent smoke that fills a room

Elimination. If the conditions for a flashover have been identified or seem optimum, two primary

considerations can minimize or eliminate a flashover. But potentially most important, do not let your PPE give you a false sense of security and allow you to overcommit yourself. Stated from another perspective, do not use your PPE as an offensive weapon when it was designed as a defensive weapon.

Any ventilation that is capable of cooling the interior temperature can be effective. As an example, vertical ventilation will cool the upper portion of a room, reducing the temperature that is necessary for a flashover. Although ventilation has always been a top fireground priority, the ability to minimize a potential flashover has increased the importance of ventilation on the modern fireground.

Penciling is a term that describes using a specific type of hose stream to cool the area under a ceiling. By using a nozzle (if available) to direct a short stream—straight stream, not a fog stream—at the ceiling, the water can be turned to steam, thus cooling the upper environment. If this tactic is used,

Figure 5-9. Three firefighters are about to make entry into this structure. (Photo courtesy of Glen Ellman)

Figure 5-10. A short time later, notice the increasing fire under the porch overhang. (Photo courtesy of Glen Ellman)

water that *does* return to personnel is an indicator that the ceiling was not that hot. However, if water does *not* return to the personnel, the ceiling was that hot but two objectives have been accomplished. It is obvious that unless interior conditions change, there is probable cause to anticipate the chance of a flashover. Also, the stream of water likely turned to steam, which is an effective cooling/extinguishing agent and capable of cooling the ceiling area above the personnel (thereby reducing the chance of a flashover). In this case, personnel should consider backing up to a position of safety and communicate conditions to appropriate person(s).

A flashover is graphically illustrated in figures 5–9 through 5–12. In figure 5–9, three firefighters are about to make entry into a single-story SFD.

Notice that fire is visible through the roof on the right side of the photo. Also, there are water droplets around the middle firefighter that indicates the nozzle is in the process of being closed, and the doorway appears relatively clear. Most importantly, notice the small amount of fire that is visible under the porch overhang.

In figure 5–10, there is a small amount of water by the nozzle that indicates the nozzle has just been closed (this is an indicator of the minute amount of time between these figures), and the doorway is still relatively clear. Notice the increasing amount of fire under the porch overhang.

In figure 5–11, as the firefighters are about to make entry, the room is in the process of a flashover. Notice that the greatest amount of fire is at the top of the room and a smaller amount at the bottom of the doorway.

In figure 5–12, the room has totally flashed over with an energy that is graphically visible. Obviously, if the three firefighters would have been just 10 feet inside the structure when the room flashed over, they would have been in serious trouble.

Figure 5–11. As the firefighters are about to make entry, the room is in the process of a flashover. (Photo courtesy of Glen Ellman)

Figure 5–12. The room has totally flashed over and has dramatically changed the fireground environment. (Photo courtesy of Glen Ellman)

If you are caught in a flashover, your visibility will drop to zero, your sense of direction is likely to be compromised, and you have only about three to five seconds to escape. Figure 5–12 is a graphic reminder of these considerations.

Primary distinctions

There are four primary distinctions between backdrafts and flashovers encountered by fireground personnel. In the modern fireground environment, why are backdrafts rare, and flashovers are not? The answer is the smoke! First, remember that smoke from conventional materials can normally have a flash point in the range of 1,123°F whereas the flashpoint of smoke from synthetic materials has a flashpoint of around 800°F to 900°F, or less than the flashpoint of conventional materials. Therefore, when the interior temperature reaches 800°F to 900°F, the conditions for a flashover are optimal. As a result, a room can easily flashover before reaching optimal conditions for a backdraft. Also, a condition known as *ladder fuels* has changed the way smoke burns on the modern fireground. This term primarily refers to the way ignition temperatures of common gases accelerate the flammability of other gases that are present in a compartment fire. As an example, acrolein self-ignites at 450°F, benzene self-ignites at 928°F, hydrogen cyanide self-ignites at 1,000°F, carbon monoxide self-ignites at 1,123°F, and so on. So, let's assume these gases are present in a structure fire, and the temperature is sufficient for acrolein to self-ignite. In turn, the subsequent rise in temperature will cause benzene to self-ignite, which will accelerate the ignition of the bigger gases and cause hydrogen cyanide to self-ignite and so on. This ladder effect is why a room can flashover so quickly today compared to yesterday. The wider range of fire gases that are present on the modern fireground can accelerate the likelihood of a flashover.

Second, a backdraft is a true explosion; a flashover is not. When oxygen is suddenly introduced to an oxygen-deficient environment, an immediate widespread pressure increase that is capable of destroying structural elements can occur. Conversely, when smoke and fire gases reach their ignition temperature, they will readily burn, but at a speed that is slower than an explosion, and will not generate enough pressure to destroy structural elements. However, there will be a sudden and significant rise in temperature that is capable of instantly involving a room/area with fire and will be potentially fatal to anyone in the room/area.

Third, a backdraft is an air-driven event, and a flashover is a temperature-driven event. Fourth, remember that a backdraft is deficient in oxygen, and a flashover is deficient in the proper amount of heat. A backdraft normally occurs during the decay phase of a fire, whereas a flashover normally occurs between the growth and fully developed phase of a fire.

Importance of Reading Smoke

The importance of reading smoke can be viewed from three key perspectives—conditions, personal safety, and assignment.

Conditions

Taking the necessary time to read smoke can be an indicator of what is burning, the amount of material that is burning, extension, the resources necessary to extinguish the fire, the potential of flashover conditions, and other pertinent factors. Remember that until a fire within a structure is extinguished, interior conditions will continue to change, and as a rule, not for the better.

Personal safety

The ability to read smoke (or the environment) can often yield vital information that enhances the ability of fireground personnel to evaluate two key considerations—PPE and SCBA.

Personal protective equipment (PPE). Without a doubt, modern PPE significantly exceeds the capability of the PPE of a few years ago. In addition, some firefighters are now equipped with thermal imaging cameras (TIC) molded into their helmets, and heads-up displays inside of face pieces that give a visual display from the TIC, and vital information such as how much air is left at the current breathing rate, what is the temperature of the environment around the firefighter, and so on. All of this technology can be beneficial, but it can also be detrimental if a firefighter does not have a fundamental understanding of the basics. Therefore, a crucial issue is how interior personnel can stay aware of changing conditions with or without the advantages of their PPE. A modern firefighter must be able to evaluate changing conditions that are separate from the protection of PPE. Before the advent of protective hoods, gloves, and wristlets, firefighters would constantly be aware of environment conditions by the feedback provided by exposed skin on their hands and ears. (This is not to suggest that this method was the most beneficial, but it was fairly accurate.) Unfortunately, the advent of gloves, wristlets, and protective hoods eliminated this ability. Now, most firefighters rely on the perceived capability of their PPE.

However, an important characteristic of PPE is that in spite of its protective capabilities, it will allow some heat to pass through the protective layers to the individual wearing the PPE. Interestingly, if a firefighter stays mentally alert to perceived heat levels within PPE, a significant benefit can often be observed. As a general rule, if the perceived temperature is hot enough to be noticed (you are thinking about it), but a firefighter still feels that it is safe to continue to accomplish the intended task, the temperature around the firefighter is approximately 375°F to 450°F. This is noteworthy from two perspectives. While continuing to advance into the environment, notice if the perceived temperature tends to subside, stay the same, or appears to increase. And if the temperature is approximately 375°F to 450°F, the temperature at the ceiling is likely 800 to 900 degrees, which is approaching the temperature favorable to flashover conditions.

If a firefighter maintains a mental baseline of the conditions that were encountered when entering the structure, then the process of evaluating the environment in combination with PPE can be simplified and provides a measure of safety. Do not put all of your eggs in one basket. The fireground definition of this statement is do not put all of your trust in your PPE, or do not believe what your PPE tells you all of the time.

Self-contained breathing apparatus. In the early 1960s, the advent of self-contained breathing apparatus (SCBA) gave a firefighter the ability to breath fresh air while exposed to interior conditions encountered during suppression operations. Unfortunately, this advantage was limited by the amount of compressed air in the bottle worn by a firefighter. Today, most SCBA consists of a bottle charged to about 4,500 psi and rated to last about 30 minutes under ideal conditions. However, it is well known that under stressful working conditions this rated time is in reality about 15 to 18 minutes.

A dilemma that has plagued the American fire service for decades is the common perception of using the bottle until the low-air-warning device sounds, and in some cases, continuing to use the remaining air to complete a task. Regrettably, this can often result in interior personnel not having enough air to safely exit a structure and/or having to breathe smoke until being able to exit a structure. As a result of this dilemma, *firefighters that run out of air breathe smoke, and firefighters that breathe smoke can die.* Effective January 2007, NFPA 1404, Standard for Fire Service Respiratory Protection Training mandates a new focus on the concept of air management. A5.1.4(2) of 1404 deals specifically with an air management program for a department and individuals that delineates the responsibility of an individual to manage their air consumption during operations in an immediately dangerous to life and health (IDLH) environment. This responsibility is composed of three primary directives:

- Exit from an IDLH environment before consumption of reserve air supply begins.

- Activation of a low-air alarm is a signal that the reserve air is being consumed.

- Activation of a low-air alarm is an immediate action item for the individual and team members.

The 1404 standard requires that fire departments train their members to operate in accordance with the Rule of Air Management, which states:

Know how much air is in your SCBA, and manage that air so that it is possible to leave an IDLH environment before the low air warning alarm activates.

The concept of air management is based on the training and discipline of three basic factors. Potentially most important, invest appropriate time in training scenarios that allow an individual the opportunity to develop a working knowledge of their air consumption rate. Personnel should be able to evaluate the amount of air consumed while performing an assigned task in stressful conditions. This also requires that individuals not only monitor their air consumption rate during fireground operations, but are also aware of how much air is left to exit an IDLH in a safe and timely manner.

Also, personnel should know how much air is in an SCBA at the beginning of a tour of duty and prior to entering an IDLH environment. And finally, personnel should have the discipline to balance the amount of air required to perform an assigned task with the amount of air necessary to exit an IDLH environment.

Because cancer is one of the leading causes of death in the American fire service, the importance of managing a limited air supply (the bottle on your back) in an IDLH environment is long overdue.

Assignment

Although every firefighter assigned to fireground operations should understand the importance of reading smoke, the need to read smoke from a truck company point of view should be a basic skill. Although the reason for this statement should be obvious, there is another less obvious reason, that being lack of protection. In many cases, engine company personnel have a charged hose line during the extinguishment phase of fire suppression operations. This affords some degree of protection if the environment suddenly changes and becomes an immediate threat.

Conversely, truck company personnel (or personnel assigned to truck company operations) can be engaged in interior operations that do not normally require a charged hose line, such as a search. Therefore, it is possible for truck company personnel to be placed in a difficult situation if the interior environment suddenly changes and becomes an immediate threat, and a measure of protection (charged hose line) is not readily available. This is a prime reason why it is even more important for truck company personnel to read the smoke and ensure that forcible exit has been completed.

Reading Smoke

The ability to analyze fire, smoke, and interior conditions can provide valuable information that is necessary to assess the following conditions:

- Operations (exterior and interior)
- Location and extension of fire
- Fireground time that is available
- Your involvement

These four conditions can also assist in evaluating the level of inherent risk to fireground personnel. Obviously, a key to safe risk assessment is the ability to read smoke (and the environment) from outside and inside a structure and is based on the practical application that smoke is the fire talking to you. Additionally, remember that when reading smoke, that determining where you do not see smoke can often be as important as where you do see smoke. To put this statement into practice, consider reading smoke from an exterior and interior perspective. Taking the time to analyze exterior smoke should provide information necessary to determine what operations are needed and in which order. Additionally, evaluating the interior environment can assist in being able to determine how long you should be involved in interior operations. Before we look at the basic elements of reading smoke, let's consider a fundamental starting point that consists of vented and unvented fires.

If a fire is vented either by the progress of the fire or manually by fireground personnel, it should have a lower potential of flashover and a higher probability of safe access (visibility) to the seat of a fire. Although this perspective may not hold true for every fire, let's consider the basic concept by looking at figure 5–13. In this illustration, the fire, heat, and smoke are venting through the roof of the building. (There is a big difference between proper ventilation and only fire showing from one window.)

As this process continues, several events take place simultaneously. As the fire draws heat and smoke up and out of the building, the interior spread of fire will be reduced because the fire is being directed vertically, not horizontally. Along with this event, fresh cooler air from the exterior of the building will be drawn into the building because the fire will create a negative pressure inside the building

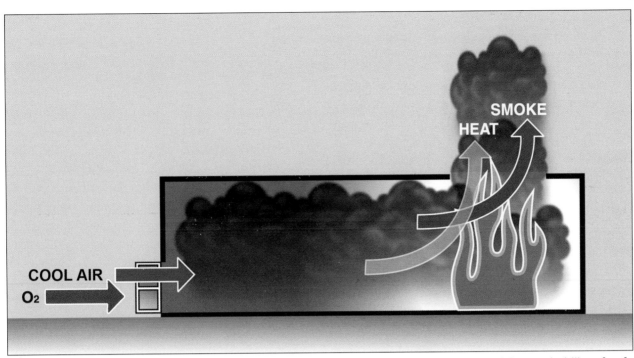

Figure 5–13. Properly vented fires can lower the potential of a flashover and provide a higher probability of safe access to the seat of a fire.

(as compared to the exterior of the building) and draw oxygen from the exterior to fuel the interior combustion process. The introduction of fresh air and oxygen will alter the interior conditions of this illustration by reducing the interior temperature of the environment, thereby decreasing the chance of a flashover. Additionally, it will also increase visibility (because the introduction of fresh air will thin the smoke to some degree) and allow a safer and timelier access to the seat of the fire. Of all of the important factors to consider when evaluating the interior conditions of a structure fire, visibility (or lack of it) should be near the top of the list because visibility can allow timely access *and* egress into and out of a structure, and the ability to find and complete an intended task.

Conversely, unvented fires have a higher probability of flashover and increased access/visibility challenges. Refer again to figure 5–13 and assume this fire is unvented. This fire may not be able to draw fresh cool air and oxygen from the exterior of the building, and a lack of ventilation may result in an increase of heat and smoke within the building, creating two significant hazards that were not present in the vented illustration. As the amount of smoke increases, visibility will decrease. And potentially most important, as the interior temperature

increases, so does the probability of a flashover. As the interior temperature increases to the point that the interior smoke and fire gases are heated to their ignition temperature, the potential of a flashover has been maximized, as has the danger to interior personnel.

With the preceding thoughts on the ability of vented and unvented fires to impact fireground operations, let's continue by evaluating the basic elements of reading smoke from three perspectives: reading the extension of smoke; reading exterior smoke; and reading interior smoke.

Reading the extension of smoke

The ability to read the extension of smoke can be based on many factors, such as where it is, where it is going, and what it is going to impact. If the location or the origin can be determined, then the best route of access to the source of the smoke and the most effective ventilation operations necessary to minimize it can be determined.

If the direction of smoke can be determined, then the pathway the smoke is taking can potentially disclose its spread and ultimate destination. And if the impact of smoke on an environment can be evaluated, then priorities can be established. To overview extension, let's look at three considerations: the six-side concept; the status of the cockloft/attic; and the four primary extension routes and inherent hazards.

Six-side concept. When analyzing any incident to determine the possible extension route of smoke, fire, and other inherent hazards, consider the incident from the perspective of having six sides. As an example, look at figure 5–14, which is a three-story structure with fire visible on the second floor.

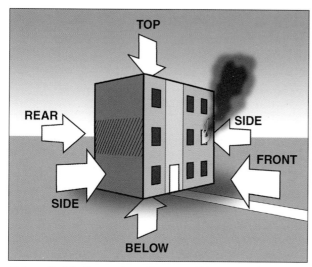

Figure 5–14. To analyze the possible extension route of smoke, fire, and other inherent hazards, consider the incident from the perspective of having six sides.

Applying the six-side concept to determine priorities:

1. Determine the need for the size of the fire on the second floor, extension, search, and ventilation (four sides).

2. Check for extension, search, and ventilation on the third floor (fifth side).

3. Determine if there is any extension to the first floor (sixth side). Although fire and smoke primarily go up, they can also extend downwards.

Status of the cockloft/attic. Determining the location of fire in relation to interior and roof personnel is critical from the perspective of the basic fireground rule, "falling debris always has the right of way." When this rule is applied to fireground personnel, it can determine life or injury from the perspective of

another basic fireground rule —if you can't stand on it, you can't stand underneath it. To summarize these two rules into one simple viewpoint, the location of a fire in relation to fireground personnel should be a key consideration when determining either interior or roof operations. As a simple example, attic vents can be found on most residential and commercial structures because they provide ventilation to an attic area. As a result, attic vents are an excellent indicator that fireground personnel can use to determine the location and/or extension of fire in a structure. Let's assume a fire is located within a structure and is confined within the exterior walls, the floor and ceiling. In this scenario, there is no extension of fire to the attic area, and the potential of structural collapse (for example, the entire building) is minimal.

Conversely, the structure as depicted in figure 5–15 illustrates a radically dissimilar set of conditions. The fire has extended to the attic area and is visible as either smoke and/or fire from the attic vent.

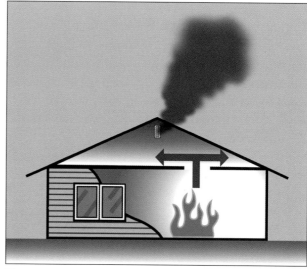

Figure 5–15. Visible smoke from an attic vent(s) is an indicator that fire has extended to an attic area.

This scenario is a graphic example that unlike the previous illustration where fire did not extend into the attic area, there are now three major hazards that fireground personnel can encounter:

- The fire has extended to the structural members that support the roof and ceiling, which in the typical modern dwelling and moderate-sized commercial buildings is normally 2-by-4-inch lightweight trusses.

- Interior personnel now can be subjected to an interior collapse of the overhead structural members.

- Roof ventilation personnel (if used) can also be subjected to a collapse of the attic structural members.

When reading the smoke at a structural incident, look for the presence of attic vents and if present, if the fire has extended to the attic area. This simple consideration can assist in determining the extent of fire and the possibility of a potential collapse of overhead structural members. When the six sides and the attic area have been quickly evaluated, consider the four primary extension routes of fire and smoke.

Four primary extension routes. There are four primary extension routes: basement; single story; multiple story, first floor; and multiple story, top floor.

Basement. Basement fires are potentially one of the most dangerous fires that fireground personnel will encounter. This type of fire continues to claim firefighter lives each year. A basement fire is shown in figure 5–16.

Basement fires have minimal access and egress options because of the below-grade nature of the fire. They often contain a high fire load due to the storage of various materials; and personnel standing on the grade floor above the basement are, in reality, standing on the roof of the fire. If smoke is visible from a basement area, the following hazards should immediately be anticipated:

- The normal extension route for fire and smoke is vertically through those portions of the structure above the basement. Pipe alleys, balloon frame construction, and other similar avenues enhance the upward spread of fire and its by-products.

- Basement fires can be difficult to ventilate, depending on the number of windows and doors that are common to the basement area.

- When attack personnel advance a line into a basement, the open door will enhance the travel of oxygen to the fire. Therefore, expect the fire to become more intense.

Single story. Fires in single-story structures can be relatively straightforward because access and egress of personnel is more simple when compared to other types of buildings. Floor plans can also be straightforward, particularly in dwellings, and extension routes are limited, depending on the size of the structure (figure 5–17).

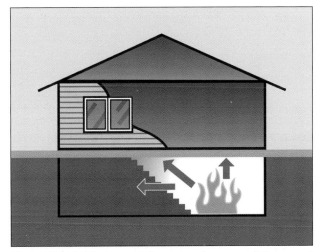

Figure 5–16. Basement fires can be one of the most dangerous fires that fireground personnel will encounter due to its location and multiple routes of extension.

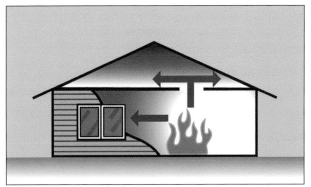

Figure 5–17. Fires in single-story structures can be simplified when compared to other structure fires, but extension into an attic area can complicate the risk of suppression operations.

If smoke is visible from a single-story structure, evaluate the following conditions:

- Read the attic vents to determine the possible extension to the roof structural members. Remember, the only barrier to an interior fire extending to an attic area is the ceiling material. In dwellings, this is normally composed of 1/2-inch drywall, and dropped or hung ceilings in commercials. Both of these materials are not a substantial barrier to fire.

- If fire has extended to an attic area, initial personnel should pull the ceiling (before entering the structure) and extinguish any visible overhead fire. Always keep a fire in front of advancing personnel. Never let fire get behind advancing personnel, and be able to block the entry/exit opening. Remember that fire can enter an attic as a result of exterior lapping from a window into exposed attic vents.

- Ventilation (horizontal, vertical, PPV) can often be rapidly implemented.

- Remember that single-story structures come in a wide variety of sizes, and that there is a significant difference between a 1,700 square-foot dwelling and a 50,000 square-foot concrete tilt-up commercial.

Multiple story, first floor. Fires on the first floor of a multiple-story building are similar to fires in single-story structures; they can be easy to enter and exit, and the space between the first and second floor is minimal and normally of heavier construction (ceiling joists) than attics (figure 5–18):

- In many cases, the flooring between the first and second floors provides a barrier to the rapid extension of fire (compared to ½-inch drywall).

- Smoke can extend to upper floors but is dependent on the presence of vertical voids and passageways from the first floor to upper floors.

- Normally, ventilation is limited to horizontal and PPV.

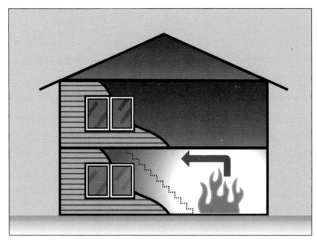

Figure 5-18. Fires on the first floor of multi-story buildings can be similar to fires in single-story buildings.

Multiple story, top floor. When a fire is located above ground, the inherent hazards are dependent upon several key factors (figure 5–19):

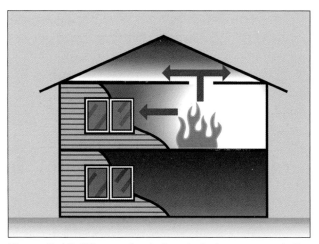

Figure 5-19. When a fire is located above ground, the inherent hazards increase, particularly when the fire extends into the attic area.

- A critical factor in multiple story fires is the distance from the ground to the fire; for example, is the fire on the second floor, or the thirtieth floor? This will have a direct impact on the time it takes personnel to reach the fire. Remember a basic fireground rule for fires above ground—equipment on the apparatus in the street is unusable inside the structure, and may not be easy to retrieve.

- If the fire is on the top floor, then the only barrier between the fire and the attic is the ceiling material. Remember to read the attic vents (if present).

- Horizontal ventilation may be more difficult until personnel reach the fire floor. In some cases, vertical ventilation and/or PPV is a superior method.

Reading exterior smoke

Reading smoke should begin when you stop in front of an incident because *smoke is the fire talking to you*. Before committing yourself and/or resources, take a few seconds and evaluate the five attributes of smoke: color, density, amount, pressure, and dynamics.

Color. Smoke color is indicative of the type of material that is burning if only one material is burning. The majority of smoke being produced at a structure fire is not coming from what is burning but from materials that are being burning and/or heated. As materials dry out, they break down and the smoke color changes. Wood or cellulose materials change to tan, grey, then black as they heat up. Plastics start off smoking white, then emit grey, then black as they heat up. Therefore, smoke color viewed from the outside of a building is really indicative of heat. Whiter smoke is early stage heating and blacker smoke is hotter regardless of the type of material. Also remember that as smoke flows through a building, soot and hydrocarbon particles deposit themselves on surfaces. Therefore, black smoke can actually change to a lighter color before it leaves the building if enough distance or resistance to smoke is present. Additionally, lighter smoke is an indicator of fewer particulates in the smoke and darker smoke is an indicator of more particulates. Remember that particulates will readily burn at the right temperature. The color of smoke can be easily divided into the following four categories:

- **White to light with little or no heat or direction:** In figure 5–20, the fire in this commercial building has been almost extinguished. The white smoke on the left side of the building has minimal direction or pressure and no heat. In this case, this is an indicator of residual smoke from the fire being extinguished in this portion of

the building. The potential of this smoke suddenly flashing and burning is extremely low. Visibility will likely be moderate inside this portion of the building. Additionally, the smoke on the right side of the building is an indicator that all of the fire in this portion of the building has not been extinguished. The moderate color, pressure, and heat associated with this smoke is an indicator of a moderate hazard (or potential) of the smoke suddenly flashing and burning. Depending on the amount of smoke inside the building, visibility will be minimal to moderate.

Figure 5–20. The white smoke on the left side of the building is an indicator of residual smoke and a low potential of flashover. (LAFD file photo)

- **Smoke with some color, pressure, and heat:** In figure 5–21, the smoke from the left side of this building displays some color, pressure, and heat. This is an indicator that the fire in this portion of the building has not been extinguished and visibility will be moderate at best. The potential for a flashover is also moderate. However, the smoke on the right side of the same building is starting to turn black. It displays notable pressure and heat and exhibits the characteristic signs of expanding as it escapes the building. The smoke in this area of the building is a candidate to flashover, and visibility in this portion will be non-existent.

Figure 5–21. The smoke on the left side of this building displays some color, pressure, and heat, and is an indicator the fire in this area has not been totally extinguished. (LAFD file photo)

- **Smoke with significant color, pressure, and heat:** In figure 5–22 there is heavy black smoke issuing with great pressure from the building. This is an obvious indicator that the fire is well advanced in the building. The black color of the smoke indicates that significant heat and/or petrochemical compounds are probably involved. This condition is often referred to as *black fire*. This term indicates that although fire is not visible, there is sufficient heat within the smoke to pyrolize wood (and enhance the removal of gang nail plates or soften glue in lightweight trusses, if present). The smoke in this portion of the building is becoming a prime candidate to flash and burn with significant intensity. Visibility in this building will definitely be non-existent.

- **Fire (or the red stuff):** In some cases, heavy fire will be visible inside a building. Although radiated heat from the fire can be greater than the previous examples, this type of fire will not readily flash and burn as it has completed its goal of burning the interior portion that is visible to suppression personnel. Until this type of fire is extinguished, interior visibility (although moderate) is not a primary issue until the fire has been extinguished.

Figure 5–22. The heavy black smoke with notable pressure and heat is an indicator of a well-advanced fire and a high potential for a flashover. (LAFD file photo)

When reading the color of smoke, be aware that smoke from a fire involving oil based roofing materials such as tar paper and composition shingles will give off a thick black smoke that can easily darken lighter colored smoke and give a false reading.

Density. The density of smoke can often be a graphic indicator of the potential flammability of the smoke and the visibility (or lack of visibility) within a building. As an example, figure 5–23 shows a firefighter standing on a balcony surrounded by a moderate amount of cool, light-colored smoke. Notice that the firefighter is visible through the smoke.

Figure 5-23. The cool, light-colored smoke is an indicator of a low potential of flashover and some visibility inside the structure. (LAFD file photo)

If the firefighter is relatively near the area of the fire (which, in this picture, he is), it should be apparent that this smoke is an indicator of three considerations: the chance of a flashover is relatively low; although limited, there will be some visibility inside the building; and the fire has either been extinguished or is about to be finally extinguished.

Remember, when reading smoke that is some distance from its origin, it can cool and become lighter, thus giving a false reading.

Amount. The amount of smoke is normally not associated with the color of smoke, but it can be a graphic indicator of the quantity of fire, its extension, and its potential flammability. To keep this in its proper context, remember that it takes the combustion process to make smoke, and smoke is capable of burning. Let's apply that statement to quantity of fire, extension, and flammability.

- The relationship of the perceived size of a fire in relation to the size of a building must be considered. As an example, a fire of a given size in a routine building can visibly produce a large quantity of smoke, whereas the same fire in a large commercial building may not exhibit the same quantity of smoke, as the size is able to contain more of the smoke without it being visible. A large quantity of smoke visible from a large building (for example, concrete tilt-up) is a good indicator of a large fire.

- The amount of smoke from a building can be an indicator of the extension of fire. As an example, imagine a fire in a basement of a three-story dwelling with heavy smoke showing from several basement openings only. This is a good indicator that the fire is contained to the basement area. Now imagine that heavy smoke is showing from several basement openings and there is moderate smoke showing from windows on the top floor. Obviously, this is an indicator that the smoke (and fire) has extended from the basement to the top floor. Always compare the area of most smoke to any area of lighter smoke and the distance between these two considerations. Also, always look for areas to where smoke has not extended; this can be as important as where smoke is visible.

- Lighter smoke or smoke that has some transparency indicates a cooler smoke with less combustible material; heavy smoke is an indicator of more combustible products within the smoke. The amount of smoke can also be a graphic indicator of the amount of fire. As an example, a hot, free-burning fire will likely yield minimal smoke, whereas an oxygen-deficient fire can yield a significant amount of smoke.

Let's apply the preceding considerations to the fire shown in figure 5–24. This incident was a fire in a lumberyard office building of approximately 3,000 square feet.

Figure 5–24. After breaking several windows, the large amount of escaping dark, pressurized smoke is an indicator that the fire is now a well-advanced fire. (LAFD file photo)

After positioning attack lines, and while waiting for water, the personnel standing on the stairway have broken out several windows (big mistake). Notice the large amount of escaping smoke from the broken window openings, the remaining unbroken dark window, and pressurized smoke from the eaves. This photo is a graphic example of a large, well-advanced, and hot-burning interior fire. As a point of interest, look at the difference in smoke colorations starting from the top of the photo (initial smoke after windows were broken) and progressing downward to the broken windows (which now display hot, black, roiling smoke as the fire gains momentum).

Pressure. Smoke color is a nice helper, but pressure trumps color any day of the week. The pressure of smoke exiting a structure can be an excellent indicator of the amount of fire within the structure in combination with the amount of confinement. If there is a lack of openings to the exterior of the structure, then expect smoke to display visible pressure as it forces its way to the exterior. The key to reading pressure is threefold:

- Fire makes smoke; thus the greater the pressure, the more fire within the building.

- Always consider the size of the building. A larger building (square footage) can contain more smoke than a smaller building, thereby potentially concealing the true size of a

fire. Conversely, fires in small- to moderate-size buildings can be fairly accurate in indicating the amount of fire within a building.

- The rate of curl or expansion of escaping smoke and fire gases is an indicator of a hot fire. A hot fire will increase the pressurization of the smoke and fire gases, which will be driven out of all available openings, expanding as it escapes.

Let's apply these considerations to figure 5–25. This incident was a 1,700 square foot SFD with an obvious internal fire and no ventilation.

Figure 5–25. After forcing the door to this structure, the large amount of pressurized smoke is an indicator of a large unvented fire. (LAFD file photo)

As initial attack personnel forced the door for entry, they were confronted with a large amount of pressurized, hot, roiling smoke from the doorway. Notice that the personnel have not been able to advance into the structure due to the escaping smoke and heat, and notice how the smoke at the top of the photo is light colored whereas the smoke at the doorway is beginning to display the expanding, roiling characteristics of a fire that is approaching either backdraft or flashover conditions.

Dynamics. The dynamics of reading smoke can be easily evaluated prior to entering a structure and can be graphically demonstrated by figure 5–26. This commercial building displayed a working fire upon arrival of initial companies.

Figure 5-26. After forcing the overhead door to this commercial structure, the smoke began to be forcefully drawn into the structure. As a result, oxygen was fed to the fire. (LAFD file photo)

After forcing an opening in an overhead door, there was high heat, dark smoke down to the floor, and smoke was being forcefully drawn inwards. Approximately 30 seconds after this photo was taken, the interior of the building flashed over. This photo is a reminder that after forcing entry and prior to entering a structure, the smoke (or environment) can often give a clear indication of the fire and/or interior environment. Let's look at the following three examples of the dynamics of smoke that can be based on what is initially observed prior to entering a structure and/or after the observation of initial conditions. These dynamics are static, pressurized, and negative.

Static. As depicted in figure 5–27, assume you are the initial company about to enter this structure, and you observe smoke that is static or *residual*.

If you are about to enter the area of the fire, you can assume that either there is a minimal amount of fire or the fire no longer exists because it takes the combustion process to make smoke. If this is the environment you are about to enter, it obviously presents the least amount of danger of the three dynamics because the environment has stabilized and is not being rapidly degraded by fire.

Pressurized. Assume that you are the initial company and are about to enter the structure depicted in figure 5–28. Upon forcing the door, you observe smoke being forced from the interior to the exterior of the building. At this point, you should read the smoke factors of color, amount, pressure, and density.

These four factors will be an initial indicator of the status of the interior environment and the fire. As an example, if you encounter a slight pressurization

Figure 5-27. Static smoke conditions can be an indicator of minimal fire conditions.

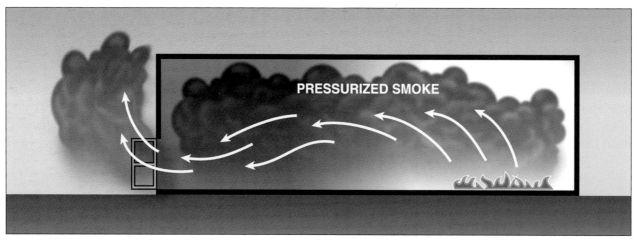

Figure 5–28. Color, amount, density, and pressurized smoke conditions can be an indicator of the status of the interior environment and the fire.

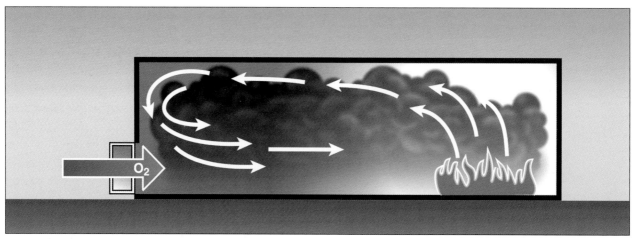

Figure 5–29. If exterior air and interior smoke are visibly being drawn into a structure, it can be anticipated the environment will rapidly degenerate.

of relatively cool smoke of moderate density, the fire is probably not well advanced and the chance of a flashover is low. However, if you encounter a large amount of roiling dark, hot smoke, you can assume you are about to enter a well-advanced fire and an environment that may hold the conditions (or increasing conditions) necessary for a flashover. Pressurization is the next most dangerous dynamic after static smoke.

Negative. Assume you are the initial company that has forced a door to the structure in figure 5–29, and you observe that exterior air along with interior smoke at the door are being drawn into the structure.

At this point, you should evaluate the following conditions:

- As you opened the door, you likely gave the fire what it was looking for prior to your arrival—oxygen.

- The sudden introduction of oxygen *will* accelerate the fire.

- The interior environment can be anticipated to rapidly degenerate, and the conditions necessary for a flashover can be expected to increase.

- In this case, operations designed to improve the interior environment (such as ventilation) should be initiated *prior* to entering the structure.

Obviously, this scenario presents the greatest hazard of the three dynamics we have discussed. Interior visibility will be poor to non-existent.

Reading interior smoke

After the smoke and/or environment is evaluated prior to entering a structure, it then becomes mandatory for personnel engaged in interior operations to constantly monitor the environment. It is highly likely that the interior can change until the fire is extinguished, extension has also been extinguished, and ventilation operations have removed the contaminants. Reading interior smoke should focus on the three basic considerations of the initial company, later-arriving companies, and unvented attic fires.

Initial company. The initial truck company on-scene at a structure fire is confronted with numerous considerations, three of which for this discussion are interior conditions, logistical priorities, and time available to complete the necessary support. Upon arrival, you need to determine the conditions and what priorities (or operations) will be necessary to provide support for suppression operations. It is difficult to prioritize necessary support operations unless the extent of the problem is initially evaluated. With all of the responsibilities that comprise support operations, priorities can be simplified if they are categorized into primary and secondary operations. Primary responsibilities are most often composed of forcible entry/exit, ventilation, search, and ladders. These four operations are necessary to ensure that attack operations are timely and successful. Secondary support responsibilities include utilities, additional ladders, salvage, and so on. Then determine how much time will be required to perform primary support responsibilities and if additional companies or personnel are required for completion. Remember that if additional companies or personnel are required, it will be necessary to evaluate how much time it will take for those resources to become engaged in suppression operations. There is a significant difference between several minutes and ten to fifteen minutes.

Later-arriving company. Later-arriving truck companies are also confronted with numerous considerations, three of which are conditions, priorities, and time. Determine the conditions from the perspective of progress or a lack of progress of the incident, and/or the initial truck company. As an example, if the second-arriving truck company (or personnel assigned to logistical operations) finds the incident is winding down and initial priorities have been or are being completed, then secondary considerations should be evaluated. However, if suppression efforts seem to be unsatisfactory, then the priority shifts to what is most needed. Logistical support can then be evaluated from the perspective of primary or supportive operations. As previously mentioned, primary logistical operations are, in most cases, ventilation, forcible entry/exit, ladders (remember a minimum of two), and search. Make sure these four priorities have been addressed. If they haven't, determine how the additional company can address the most important priority(ies). If primary logistical support has been addressed, then secondary support should be addressed. Secondary support operations consist of additional ladders, utilities, operations necessary to augment primary logistical operations, and so on. Finally, if necessary, evaluate the amount of time required to complete logistical operations. This includes the amount of time necessary for additional resources (in addition to the first and second truck companies) to arrive on-scene to augment or complete fireground operations.

Unvented attic fires. Unvented attic fires are confined in an attic area and are not vented to the exterior of the attic. Although these fires are easily recognized, they present several noteworthy hazards that must be carefully evaluated. As a result of a noteworthy fire load (ceiling joists, rafters, and so on) and a lack of early detection, these fires can be well advanced in a short period of time. As a consequence, the lack of ventilation will result in a rapid and significant buildup of heat, which will be radiated in two directions: up and down. The downward radiation of heat will raise the interior temperature of the portion of a structure that is nearest the ceiling, and can enhance the potential of a flashover without advance warning conditions.

If personnel pull the ceiling to provide an opening for attack operations into the attic fire, the opening can also become an exhaust opening for the expanding steam. This can be a dangerous condition for attack personnel.

In figure 5–30, arriving companies found a vacant single-story SFD with conditions as pictured.

Figure 5–30. The hot, black, pressurized smoke from the attic vent is an indicator of an unvented attic fire that can be extremely dangerous to interior personnel. (LAFD file photo)

Notice the pressurized smoke from the eaves, and the doorway that shows smoke only at the top. Most importantly, however, notice the hot, pressurized, black, roiling smoke from the attic vent. This is a clear indicator that there is an advanced unvented fire in the attic that makes interior conditions extremely dangerous until the attic fire is suppressed. Unvented attic fires can be best suppressed by either vertical

ventilation or by using piercing nozzles. Vertical ventilation will vent the heat, fire gases, and smoke vertically out of the attic, thereby allowing a hose stream to be directed from the underneath side. A piercing nozzle can be deployed through the ceiling (which leaves the ceiling intact and provides a barrier between personnel and the fire) and use the resultant steam to suppress the confined fire with minimal danger to suppression personnel.

Developing Indicators

Personnel should be constantly evaluating the interior of an environment until the fire has been extinguished or the environment has been properly ventilated. First and potentially most importantly, however, before you ever enter a hazardous environment, know how much air is in your SCBA and how long it will last under stressful conditions.

Prior to entering a contaminated environment, evaluate the observable conditions and make a mental note to serve as a baseline comparison within the environment. Remember that under adverse conditions, modern PPE has the capability to shield (or conceal) the environment from a firefighter who is likely focused on fireground activities. Therefore, if a baseline has been established prior to entering a hostile environment, then the process of being able to evaluate the changing conditions (and in some cases, separate from PPE) can be simplified.

Three-strike rule

The three-strike rule is primarily based on the co-principle of firefighter safety and risk assessment. However, a basic presupposition of this rule is that fireground personnel make a mental note of the conditions they are about to encounter as follows:

- Strike one: Smoke with heat. Remember that smoke with heat has a greater inclination to flash and burn than cooler smoke.

- Strike two: Less than acceptable visibility. For this discussion, the practical definition of visibility is that you can find your way out of a problem in a timely manner, if necessary.

- Strike three: The environment is not improving. If the baseline conditions have not improved or are deteriorating several minutes after entry, then the concept of risk analysis should be re-evaluated.

Let's apply the three-strike rule to a simple scenario. Assume that you are about to enter a commercial structure with a pike pole for the purpose of pulling ceilings to find an overhead fire. Just as you enter the structure, you observe the interior conditions consist of moderate smoke, heat, and visibility. The visibility is moderate, so it takes several minutes to locate the area of the overhead fire. While trying to pull the ceiling material, it becomes apparent the level of heat is increasing and visibility is decreasing. The ceiling material is difficult to pull down, so you continue to expend a significant amount of energy and time to accomplish your task because there are now several other firefighters assisting you with the stubborn ceiling. After several more difficult minutes, you now notice visibility has changed from moderate to non-existent (strike one), the level of heat has significantly increased (strike two), and although the low air alarm has not activated, you are able to determine you have about 1,100 psi of air left in your SCBA. In addition to these conditions, you begin to contemplate what path you will take to exit the structure because you cannot see and you made multiple turns to arrive at your current location (strike three).

From this simple scenario, the fictional firefighter should reflect on the amount of time that will be necessary to safely exit the structure. Notice that this scenario does not focus on immediately leaving the structure, but challenges the firefighter to consider the prospect of being able to exit in a safe manner as opposed to exiting with insufficient air, being caught in a flashover, and becoming a national statistic.

Training Resources

The following resources can provide additional information and/or another viewpoint for the subjects discussed in this chapter:

- The DVDs *The Art of Reading Smoke* and *The Art of Reading Smoke Practice Sessions* by Dave Dodson provide an excellent overview of analyzing various examples of smoke and their inherent hazards. They are available from Fire Engineering Books and Videos.

- *In The Mouth of the Dragon, Toxic Fires in the Age of Plastics*, Deborah Wallace, Avery Publishing Group, 1990.

- *Air Management for the Fire Service*, Mike Gagliano, Phil Jose, Casey Phillips, and Steve Bernocco, and available from Fire Engineering Books and Videos

- Ladder Fuels, http://www.cdc.gov/niosh/npg/npg.html

- Air Management: Know Your Air Consumption Rate, Dominick Marino, *Fire Engineering*, October 2006.

- Cyanide Poisoning: How Much of a Threat? Curtis Varone, *Fire Engineering*, March 2006.

- The Breath from Hell, Mike Gagliano, Phil Jose, Casey Phillips, and Steve Bernocco, *Fire Engineering*, September 2006.

- Roundtable, Air Management, *Fire Engineering*, September 2006.

- Flashover: The Sudden, Silent Killer, William E. Clark, *Fire Engineering*, June 1994.

- Toxicology of Smoke Inhalation, Gill Hall, *Fire Engineering*, August 2009.

Forcible Entry

Introduction

When applied to fireground operations, *forcible entry* can be defined as the act of entering by using force against opposition or resistance. This characterization is based on the fact that fireground incidents often require force to gain entry into a structure and would seem to indicate that forcible entry is a one-dimensional fireground operation. In reality, however, forcible entry is two-dimensional. The second dimension of forcible entry is dependent on a significantly different priority—firefighter safety and the ability for the firefighter to exit the structure. Therefore, we need a definition for *forcible exit* that includes numerous openings that are large enough to allow suppression personnel to quickly exit a structure when necessary.

Interestingly, most fireground acronyms that prioritize fireground tasks either fail to list forcible entry or place forcible entry towards the end of the list of priorities. As an example, consider RECEO, which stands for rescue (search), exposures, confinement, extinguishment, and overhaul. Although search and extinguishment are clearly mentioned, forcible entry is not. Firefighter safety is the top fireground priority, so in most cases (when applied to structure fires) forcible entry should be listed as the next most important priority after firefighter safety and before search and extinguishment because before search and extinguishment are started, it is necessary to be inside a structure.

So, if forcible entry is often necessary to allow personnel to enter a structure, then what is the next most important fireground priority after forcible entry? If firefighter safety is still our most important priority, it would then be forcible exit. If it is important for firefighters to get inside a structure, it should be *more* important for firefighters to be able to get out of a structure in a timely manner. It is a fact that a high percentage of fireground deaths and injuries are a result of not being able to quickly exit a structure. Therefore, the emphasis in forcible entry should be quick entry and quick exit. When these two perceptions are applied to modern, lightweight construction and a predominance of synthetic materials inside residential or commercial structures, we should look at modern fireground priorities from a different point of view.

Although forcible entry and forcible exit is simple in theory, applying the techniques necessary to enter a restricted area in a timely manner are often difficult. The fire service is continually challenged by new and innovative security devices specifically designed to restrict entry. Ideally, entry should be as easy as turning a doorknob or opening a gate, but this isn't normally the case. The high level of security that is routinely found in commercial and industrial occupancies is now also commonly found in some residences.

When unique or new security problems are encountered, innovation and imagination must often be combined with good judgment, common sense, and experience to produce a successful outcome. This requires the ability to recognize and select effective and safe forcible entry techniques for a wide variety of situations. The trend for the foreseeable future will be toward stronger and more innovative security mechanisms designed to keep out unwanted persons. Therefore, the knowledge required to handle forcible entry problems will change as building construction and security considerations also continue to change.

Size-Up

A forcible entry size-up begins long before an emergency. A combination of fire prevention, pre-fire planning, and training can increase knowledge and proficiency in regard to tools, equipment, and forcible entry techniques before implementation is necessary. Prior to initiating any forcible entry operation, a size-up is essential to determine which operations are appropriate. At a minimum, a size-up should encompass the reason for entry, the location of the emergency, the construction of the obstacle, the intended method to surmount that obstacle, the resources available, and the appropriate number and size of entry/exit openings.

Reasons for forcible entry

Forcible entry doesn't necessarily mean that access must be provided to a structure that is on fire. Forcible entry operations can be used for numerous reasons. Emergency incidents may include medical service, fireground operations, elevator extrication, and the like. In these incidents, the focus is normally on speed and safety. When speed is a primary concern,

property damage may be secondary. Forcible entry personnel must constantly evaluate the value of speed necessary to abate the emergency versus any damage caused by forcible entry operations.

Non-emergency incidents may necessitate entry to investigate a reported smoke detector or water-flow alarm, lockouts, and so on. In such incidents, the focus is normally on security and conserving property. Because the focus in a non-emergency incident isn't primarily on speed, forcible entry methods should be selected on the basis of efficiency that will result in the least amount of property damage. Before leaving the incident, consider whether the premises should be appropriately secured against unauthorized entry. Regardless of the type of forcible entry operation, success is generally the collective result of judgment, experience, and available equipment.

Location of entry

After the location of the problem has been determined—be it a structural fire, a broken water pipe, or an EMS call—evaluate forcible entry strategies that are suitable to gain access. Techniques necessary to provide access into structures commonly pertain to doors and windows. However, other forcible entry operations may need to be applied to walls, fences, and gates to provide access to the exterior of a structure before forcible entry operations to gain entry into a structure can begin.

Construction awareness

Determine the construction of the device or object that is restricting entry to a specific area. A prior knowledge of construction methods and security devices enable the use of the most appropriate point of entry (for example, the weakest point of the construction or security device) as well as the appropriate methods and tools.

Method

Forcible entry operations generally consist of the following methods:

- Conventional forcible entry using common tools such as an axe, Halligan, crowbar, bolt cutters, or battering ram to force, cut, or breach an object to gain entry

- Power-tool entry using rotary saws, rabbit tools, air bags, and so on to cut or force objects and security devices such as steel doors, metal bars on windows, and case-hardened padlocks
- Through-the-lock forcible entry that focuses on pulling out a locking device (such as a lock cylinder) and tripping the lock with an appropriate tool with minimal damage

Personnel and equipment resources

When determining the reason for a forcible entry operation, the personnel and equipment available to accomplish a given operation must also be considered. Forcible entry techniques are constantly improving due to technical innovations that simplify and expand the capabilities of fire service personnel.

Magnitude

One of the most important considerations in formulating a forcible entry size-up is the size and number of openings and exit points that are needed. If forcible entry is necessary to allow suppression personnel to enter a restricted area, then forcible entry operations should be required to create openings that are large and numerous enough to allow personnel to easily enter the restricted area, and more importantly, also be able to easily leave the restricted area. That means that any structure fire should require a minimum of two ways to easily exit a structure, particularly the routine fires.

Without a doubt, most firefighter injuries and deaths on the fireground would be significantly reduced if they were able to exit a structure in a timely manner. Three prime examples of this fact are the Joe Dupee incident in Los Angeles, the Brett Tarver incident in Phoenix, and the Worcester, Massachusetts incident.

Operations

Timely and safe forcible entry operations are enhanced by the following.

Flexibility

Forcible entry personnel should be able to gain entry in an acceptable period of time. If an operation requires significantly more than five minutes or so, or if acceptable progress is not being made, personnel should be flexible enough to try another method, tool, or move to another site.

Communications

Communications are often essential to the success of forcible entry operations. Emphasis should focus on notifying an incident commander if a difficult challenge has been conquered in a timely manner, or if an attempt has failed, will be delayed, or will consume an excessive amount of time. Additionally, if conditions prohibit creating multiple or sufficient exit openings, the incident commander should be notified so fireground operations can be reevaluated in a timely manner.

Safety equipment

The appropriate safety equipment will be dictated by each incident. Protective equipment with an emphasis on hand and eye protection will be required in most instances. However, when forcing entry to structure fires or contaminated atmospheres, personnel should wear self-contained breathing apparatus (SCBA) with a mask on or readily available as a precaution. An additional concern is the use of tools around flammable atmospheres. In this case, use extreme caution when using power or hand tools that may cause sparks.

Tools

Although numerous tools are available, modern technology continues to produce sophisticated equipment that can simplify and reduce the time and number of personnel required to perform forcible entry. An example is the rabbit tool. A common method to force entry through residential or commercial swinging doors is often accomplished with a Halligan, flathead axe, and two personnel. With the rabbit tool, one person can force residential and commercial swinging doors, particularly inward swinging doors. Although two people are necessary for some forcible entry operations, one person who has access to the following tools can be effective in most instances (figure 6–1):

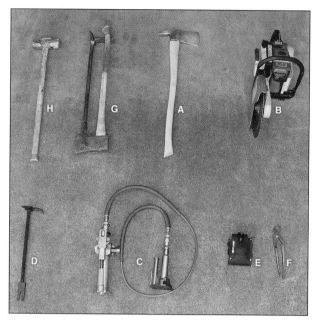

Figure 6–1. Common forcible entry tools

- **Pickhead axe (A).** The pick head axe is one of the most flexible tools in the fire service, and every fire department has them in abundance. This tool can be used for cutting, prying, striking, breaking glass, and so on. Not pictured is a scabbard that allows a firefighter to wear the axe on the hip and allows both hands to be free for other uses. The scabbard is pictured in chapters 15 through 18.

- **Rotary saw (B).** A rotary saw is very effective in cutting most metals such as bolts, metal bars, metal fencing, padlocks, and so on. As detailed later in this chapter, carry the saw with the blade in the outboard position (as depicted in the figure). Prior to using a rotary saw to cut metal, be aware of two noteworthy considerations. The composite blade (aluminum oxide) will wear quickly, and depending on the thickness of metal and duration of completing an intended cut, it may be necessary to replace the blade during the operation. In this case, it is necessary to allow extra time

and preparation with extra blades. Also, composite blades are prone to disintegrate, often sending pieces of the blade through the air at high speeds. For this reason, blades that are currently available are superior when length of cutting time and safety is considered. These blades are known as multi-use blades and are impregnated with industrial diamonds. As a result of the advantages of the multi-use blades (particularly safety), this author can no longer recommend the composite blades. In this chapter, references to rotary saw blades primarily refer to multi-use blades. For a closer look at this problem, see Minimizing Risk In Forcible Entry Operations by Bill Gustin in Training Resources.

- **Rabbit tool (C):** This hydraulic tool is capable of exerting about 10,000 psi of force and can easily open numerous swinging doors.

- **A tool (D):** This tool looks similar to a small Halligan and can be used (on a smaller scale) for prying, forcing some swinging doors, pulling cylindrical locks, and so on. The end opposite the forked end of the tool is shaped like an A and is often more effective than a K tool in pulling cylindrical locks. Also known as an officer's tool.

- **Key tools (E):** These can be used for tripping locks in through-the-lock operations.

- **Channel locks (F):** Besides being a great multipurpose tool, it can also be used for pulling slats on overhead rolling steel doors and removing some cylindrical locks.

- **Irons (G):** Although this combination normally requires two persons, it is pictured because it is considered the standard for forcing inward and outward swinging doors. Composed of a Halligan and flat head axe.

- **Maul (H):** This tool is also pictured because some departments prefer it over the flat head axe due to its superior weight.

- **Wire cutters:** Although this tool is not pictured, it has become popular in many departments due to the proliferation of hung or dropped ceilings. If a dropped ceiling collapses during an incident and a firefighter inadvertently becomes entangled in the wire that was used to suspend the ceiling, disentanglement is extremely difficult unless a set of wire cutters are handy.

It should be noted that some parts of the country present forcible entry problems that can be classified as extremely challenging. As a simple example, notice the number of locks on the interior of the door on the Seinfeld television show, which is an accurate depiction of some forcible entry challenges on the East Coast (conversely, most residential doors and locking devices on the West Coast are not that difficult). As a result of specific forcible entry challenges, specific tools have been developed by some departments to meet these challenges and are not illustrated in this chapter. Additionally, most truck companies have been affected by reduced staffing. This has resulted in a common staffing level of three persons. As a result, having one person able to accomplish most forcible entry operations is an important consideration for truck companies with minimal staffing levels. Therefore, this chapter primarily focuses on common techniques that involve one person and readily available tools.

Techniques

The following sections will present an overview of the various techniques and operations that can be applied to various types of windows, doors, locking mechanisms, fences, gates, and razor wire.

Windows

Today, numerous types of windows may be encountered. Each can present a unique circumstance requiring a specific technique. It is often easier to force a window than a door, and entry through a window may permit a door to be opened from the inside. However, compared to doors, windows offer several disadvantages. They are generally smaller than doors, and may restrict entry to personnel with protective equipment, particularly SCBA. Also, they may be blocked by furniture inside the structure.

Construction. As opposed to doors, windows are simple construction, normally composed of three basic components (figure 6–2):

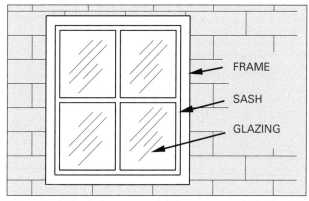

Figure 6–2. A window assembly is composed of a frame, sash, and glazing.

- **Frame:** The structural case or border into which a window is set.
- **Sash:** The metal, wood, or plastic framework that surrounds and supports the glazing.
- **Glazing:** Made from glass and/or thermoplastic compounds. Referred to as glass.

Window glass can be found in various configurations as follows:

- Plate (or annealed) glass is the least expensive type of glass and is used where strength is not required, as in most standard residential windows.
- Heat-strengthened glass is stronger than plate glass but not as strong as tempered glass and is used where greater strength is necessary.
- Tempered glass is heat treated to increase its strength—about five times over that of plate glass. It is commonly used in glass doors and some windows. Tempered glass can often be identified by the word "tempered" in one or both of the lower corners.
- Laminated glass (or safety glass) normally consists of a layer of polyester film sandwiched between two layers, adding

strength to the glass. Common applications are auto windshields and some glass doors.

- Wired glass has a wire grid embedded in it, significantly increasing its strength for security and fire resistance purposes.

- Thermoplastic compounds (Plexiglas, Lexan, and so on) can be significantly stronger than glass of the same thickness. Such compounds are often used to replace glass when security is a concern.

- Air-blast-resistant windows are a retrofit by the General Services Administration (responsible for managing federal office buildings, courthouses, and other governmental buildings) and are designed to mitigate the effects of flying glass in the event of an explosive terrorist attack. There are different methods being utilized to accomplish this degree of protection, but the most popular is a mechanically attached security film as illustrated in figure 6–3.

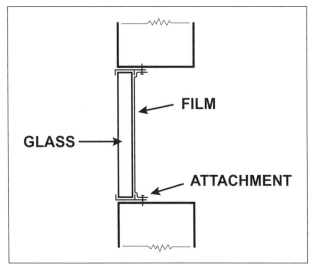

Figure 6–3. Air-blast windows are used to minimize the effects of flying glass inside a building.

Another method used is a blast curtain made of a high-strength synthetic fiber mesh that deploys over a window before the glass is projected into a structure.

- Hurricane-resistant glass is a specific type of glass that has become increasingly popular on the eastern seaboard after hurricane Andrew. This type of glass is typically composed of laminated glass and may have a thicker laminate than many air-blast-resistant window systems. Hurricane-resistant glass is resistant against a 9-foot-long, 2-by-4-inch piece of lumber traveling at 30 mph, and a small object traveling at 90 mph.

- Ballistic-resistant glass is polycarbonate clad with glass and is designed to resist forced entry by firearms of varying degrees.

- Glass-coating films or polyester window films are commonly 4.0 mm thick glass-coating films that are applied to glass windows in new and old buildings before or after the glass has been installed, and are used for energy conservation, noise reduction, and protection from falling glass in earthquakes.

Types of windows. There are four basic types of windows: stationary, sliding, pivoting, and swinging (figure 6–4).

Stationary windows are firmly mounted and cannot be opened. Sliding windows move in a horizontal or vertical direction, such as double-hung or ranch windows. Pivoting windows pivot in the middle of the sash to swing vertically or horizontally. Swinging windows (casement, awning, hopper, projected, and jalousie windows) open to the exterior or interior of a structure.

Considerations. Before forcing any window, evaluate the following:

- **Safety:** Wear appropriate protective equipment.

- **Purpose:** Determine why the window is being opened or forced. Windows opened for ventilation are selected for their capability to remove heat and smoke. Windows opened for access should be selected for size (as large as possible) to enhance the freedom of movement of personnel.

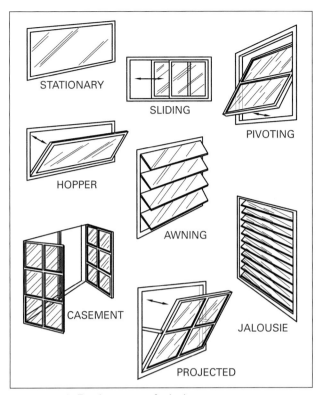

Figure 6–4. Basic types of windows

have not been locked, or that are being held by several layers of paint. Do not break glass unless necessary. If it is essential to break the glass in a window from the exterior, use as little force as possible to limit glass being thrown into the structure because personnel may be inside for other operations.

A commonly recommended method to break glass is to stand on the upwind side of the window, hold the axe handle above your head, strike the glass at the top with the flat side of an axe, and work downward. This method, however, puts personnel in the path of breaking glass. It is also ineffective on tempered glass, windows with polyester films, hurricane windows, thermoplastic compounds, and so on. A more effective and safe way to break window glass is to stand on the upwind side and strike the lower corner with the pick end of a pickhead axe, watching the reaction of the glass (figure 6–5). This keeps personnel away from the window and out of the path of the breaking glass, and indicates what type of glass is being encountered by the way the glass breaks.

- **Construction:** Determine if the window opens vertically or horizontally; inward or outward; if it is plate, heat-treated, tempered, laminated, or a specialty glass; if it has a simple or substantial locking mechanism.

- **Time:** Determine how much time is available (type of emergency), and how much time will it take to open a particular window.

- **Position:** When forcing windows, personnel should stand to the side on the upwind side to increase the odds that any issuing smoke, heat, or fire will move away from personnel. Additionally, a broken window can allow a sudden rush of fresh air to a fire and increase the potential of a smoke explosion. Unless absolutely necessary, do not stand in front of a window while breaking glass to gain entry or provide ventilation.

- **Force:** Before using more invasive tactics, try to open a given window manually. Occasionally a little force is sufficient to open those that may have weak locks, that

Figure 6–5. Windows can be broken effectively by striking the upwind lower corner with the pick-end of a pickhead axe.

Breaking glass. Depending on type, glass will break in different ways:

- Plate or annealed glass will break in large, sharp shards.

- Heat strengthened glass will break in smaller shards than plate or annealed glass.

- Tempered glass will crystallize and collapse, or may be gently pushed inward using the flat portion of the axe head, which will cause the window to collapse.

- Laminated glass, glass with window film, and wired glass will break. However, the glass will adhere to the wire, plastic laminate, or film and remain in the sash. To remove the glass, cut or break it (with an axe or other similar tool) across the top first, and then down the sides. Remove the glass and wire, laminate, or film with the axe.

- Polycarbonate (Lexan) is 250 times stronger than safety glass; it cannot be broken in a traditional manner. If entry is necessary, use a carbide-tipped wood blade on a rotary or chain saw. Make the horizontal cuts first and the vertical cuts second by moving the saw at a moderate speed. If the saw isn't moved at a moderate pace, the blade may bind.

- Hurricane-resistant glass cannot be broken with traditional methods, but can be easily cut with a chain saw with carbide teeth or rotary saw with a multi-use blade. However, beware of small pieces of glass that will be forcibly ejected by the saws, and at the minimum, be sure to wear face and eye protection to minimize the presence of any exposed skin.

- Air-blast-resistant windows can be broken and/or breached with conventional forcible entry tools, although the process will take a little longer than conventional windows. Conventional forcible entry tools are defined as a pickhead axe, Halligan, and so on. Also, remember that after the glass has been broken in this type of window, the security film or mesh covering the window will have to be removed before access and egress are possible. This will take additional time.

- Ballistic-resistant glass can be a very tough customer to break or remove, depending on the specific type. Tests have indicated that carbide chain saws were slow and rotary saws were slightly better than the carbide chain saw. Conventional forcible entry tools were not effective. Tests have also indicated that emergency escape is not an option for occupants or firefighters. This type of glass should immediately raise two noteworthy concerns: because this glass is a viable option for protection of occupants (commercial and high-end residential homes), one should consider what additional money would be spent to guarantee the same level of protection for exterior doors; and if this type of protection is encountered at a structure fire, the incident commander should be made aware that the windows (and likely exterior doors) are not able to be opened in an acceptable period of time. For additional information on these types of windows, see firefighter ingress/escape procedures on the Web at http://final_report_package.pdf.

When any window has been broken, use an axe or other appropriate tool to clear all of the remaining pieces of glass from the sash. Clear out the top pieces first and then the broken pieces from the sides. Never use your hands to remove broken pieces of glass.

Broken windows can result in pieces of glass on the interior and exterior of a structure, and can form a slippery, unsafe surface, particularly on sidewalks and hard floors. This is why it is not recommended to initially break the glass in a glass door (particularly in a commercial building) that is used as the primary entry/exit point as personnel will be walking over shards of glass on concrete which can be extremely slippery. Lastly, remove any screens, curtains, or shades that may interfere with the operation.

Forcing stationary windows. Small- to moderate-sized stationary windows can easily be broken with an axe. When breaking large, display type windows (for example, auto dealerships), remember that large shards falling directly outward can be a danger. Personnel must stand to one side and use an appropriate tool such as a pike pole. To break tempered glass, strike the lower corner with the pick end of a pickhead axe. The glass will crystallize and fall downward.

Forcing sliding windows. Double-hung windows can be opened by inserting a prying tool at the center of the lower sash and prying upward (figure 6–6). This will normally break the lock between the upper and lower sashes. If prying is unsuccessful, break one of the panes, clean out all of the glass in the sash, then reach in and unlock the lock.

Figure 6–7. Ranch windows can be forced by prying the window apart in the middle.

Figure 6–6. Force double-hung windows by inserting a prying tool under the lower sash and prying upward.

For ranch windows, use a prying tool to pry the window apart at the middle because these windows can normally be separated to either disengage or break the lock (figure 6–7). These windows are relatively easy to open, so additional locks (locking bolts, doweling placed between the sash and frame, and so on) may be installed.

In this case, it may be necessary to break and clear all of the glass from the sash, and then to reach inward to disengage the locking mechanism or remove the doweling.

When using a prying tool to force a window, remember that residential windows with aluminum or the new vinyl sashes are normally easy to force due to small and/or lightweight locking devices. However, windows with steel sashes (commercial applications) often require more force due to substantial and/or multiple locking devices.

Forcing pivoting windows. Whereas vertical or horizontal pivoting windows may be opened by prying between the sash and frame, the presence of an interior manual opening mechanism can prevent this and requires you to break a glass pane. It will then be necessary to clear all the broken glass and to reach in to operate the mechanism manually.

Forcing swinging windows. Factory (projected), hinged (hopper), or casement windows can best be opened by breaking the lowest pane of glass, clearing all of the broken glass, reaching inward and upward to unlock the latch, and operating the opening mechanism. If the window doesn't have an opening mechanism, it should be possible to open the window manually after the latch has been deactivated.

Louver (awning) or jalousie windows are often referred to as louver windows, but they are actually different types. Louver windows consist of glass sections about a foot wide. This window is best opened by removing or breaking the bottom pane and reaching inward to turn the handle. If this isn't possible, break or remove all of the glass in the frame.

Jalousie windows, on the other hand, consist of small glass sections about 4 inches wide. Normally they are the most difficult type of window to force open. To do so, remove or break sufficient glass panes at the bottom, reach inward, and operate the mechanism manually. If this isn't possible, break or remove all of the glass in the frame.

Forcing energy efficient windows. Taking a step towards improved insulative capabilities and reducing exterior sound, energy efficient windows are becoming more common, and in some states, are required on all new construction. A common application is a sash of wood, metal, or vinyl that supports two panes of glass separated by an inert gas such as argon. The glazing can be various types including a polycarbonate type material. A popular style of this window in residential applications uses glazing with heat strengthened glass in vinyl sashes that are set into vinyl frames. Commercial applications can use wood or metal frames that support tempered glass or polycarbonate materials.

Forcing these windows is dependent on the material used for the frame and glazing. Glass can normally be broken using conventional methods, and polycarbonate will need to be cut with a power saw. Vinyl sashes can be easily forced by prying (leverage) because the vinyl will bend slightly and locking devices are easily pulled out of the vinyl. Vinyl sashes (depending on the style) can also be removed from vinyl frames by using removal latches. However, wood and metal will require more force using the aforementioned methods. See Forcing Vinyl Windows in the Training Resources at the end of this chapter for more information on vinyl windows.

Barricaded windows. These types of windows have been covered (or boarded over) with various types of materials and are commonly used for security on vacant structures, and for protection in states that are subjected to hurricane type winds. Although plywood and OSB are the most common materials, a wide range of substitute materials can be found, particularly in southern and eastern seaboard states. The obvious concern with this add-on type of construction is all exterior openings have been securely covered, thereby eliminating ingress/egress.

As an example, figure 6–8 shows a piece of plywood that was used to cover the window in this concrete block building. The plywood has been securely attached to the block, and as an extra measure of security, a 2-by-4 has also been attached to the block and the plywood. The plywood and two-by-four will be difficult to remove from the exterior, and virtually impossible from the interior if timely egress is necessary.

Figure 6–8. This window has been barricaded with plywood and supported with a two-by-four.

To remove material covering a window:

1. Determine the type of material (wood, metal, fiberglass, and so on) covering the window as this will indicate the type of tool necessary for removal.

2. Determine the method used to attach the material to the window or building as there are different methods that can accomplish this task. As an example, the material can be simply nailed or screwed to a building, or it can be attached by lag bolts that run through the material to two-by-fours on the interior of the window.

3. Determine the most appropriate method to remove the material as most wooden materials can be removed by prying or cutting with a power saw. Remember, if all windows and doors to a vacant structure are barricaded, the sooner the barricades are removed (doors and windows) the safer the operation for interior personnel will be.

4. Barricades that are nailed or screwed to the structure can be pried off with some effort, but OSB has a tendency to tear and shatter, thereby requiring more time and effort for removal. Power saws (particularly with carbide teeth)

can remove most barricades in the least amount of time and effort expended. However, the depth of cut should only be deep enough to cut the thickness of the barricade material as the window may be in close proximity behind the barricade.

5. When forcefully removing barricade-type materials, particularly plywood and OSB, allow for the material to suddenly separate from the building and travel outwards. This cannot be stressed enough.

One type of barricade deserves special consideration. If four-carriage bolt heads are observed on the exterior of the barricade material, this is a indicator of two two-by-fours on the interior of the structure that are used to strengthen and clamp the barricade material to the structure. This configuration is commonly known as Housing and Urban development (HUD) windows. The two-by-fours can run either vertical or horizontal on the interior. Using a power saw (for this operation a chain saw with carbide teeth is more effective than a rotary saw due to its light weight and lack of the gyroscopic effect), make two cuts in the material as follows: From the upper left corner, make a diagonal cut slightly deeper than the material and down to the lower right corner. Repeat, cutting from the upper right corner and diagonally down to the lower left corner. This will form an X in the material. The key to this method is to allow the saw to cut the material and then cut deeper when the saw indicates it is cutting more depth. This will indicate the saw is cutting the material and the two-by-four behind the material. If this method is done correctly, the saw will have cut the barricade material into four triangles and cut the two-by-four into six pieces. The material and two-by-fours can now be removed.

See Tricks of the Trade, Removing HUD Window Covers in the Training Resources at the end of this chapter for more information on HUD window covers.

When appropriate, clear barricaded-type materials from windows to provide more exit points for interior personnel.

Security bars. As a protection against vandalism, security bars are often used on windows and doors (security bars on doors is discussed in the Doors section later in this chapter). Although these enhance security, they can also present a serious problem because they restrict ingress and egress. During a structure fire, security bars prevent occupants and suppression personnel from using a given window (or door) as a means of escape. Security bars can be mounted outside an exterior wall and over the windows of residential or commercial structures, or in some commercial structures, they can be mounted in the wall over inset windows. Normally, security bars are attached to a masonry structure by lag bolts in lead anchors (creating a compression point), or by having their ends embedded directly into the masonry walls. For structures of wood frame and frame stucco construction, lag bolts are used to attach the horizontal bars to the interior studs (figure 6–9).

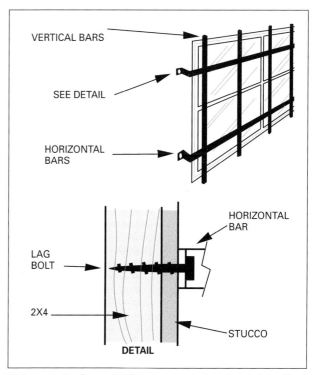

Figure 6–9. Security bars are composed of vertical bars and/or horizontal bars attached to a building by lag bolts.

Frame-stucco and frame-wood structures. Use a rotary saw to cut the horizontal bars near the left or right anchor points; then bend the bar assembly away from the window (figure 6–10). Cut the top anchor point first (as it is the most difficult), and the bottom anchor point last.

Figure 6–10. Most security bars can be removed by cutting the horizontal bars near the lag bolts and bending the assembly away from the structure.

Figure 6–11. In many cases, horizontal members use a slip joint to facilitate installation.

When cutting horizontal members, look closely at the bar to be cut to see if there is a slip joint just inward from the outside portion of the security bar assembly (figure 6–11). In many cases, manufacturers will use solid bar stock from the anchor point to the front portion of the security bar assembly.

A hollow bar is then inserted over the solid portion and continues across the face of the bar assembly until it is inserted over the solid bar stock on the other side of the security bar assembly. This is done so the assembler can adjust the width of the anchor points to match the location of the interior studs of a building. Therefore, when cutting the horizontal bars, cut the hollow portion (if present) of the bar assembly because it is easier than cutting the solid portion of the bar assembly. Obviously, this leaves some of the bar assembly in the window, but it is faster and easier on a composite blade than cutting solid bar stock, particularly if multiple security bars must be cut.

Cutting all of the anchor points allows the entire bar assembly to be removed, time and conditions permitting. However, when this is attempted by one person, the bar assembly will fall outward when the last anchor point is cut. Therefore, personnel must stand to one side while cutting the last anchor point.

It is also possible to cut the bottom horizontal bar near both bottom anchor points and then to bend the bar assembly upward and away from the window. However, this method is not recommended due to the possibility of the bar assembly falling back downward over the window.

Use a sledge to strike the horizontal bars near the left or right anchor points. When the lags have been pulled from their moorings, bend the bar assembly away from the structure. However, it is difficult for personnel to consistently strike the upper horizontal bar (near the anchor point) with a sledge to remove the lag bolt.

Use a pickhead or flathead axe to strike the horizontal bars near the left or right anchor points (figure 6–12). Remember—top first, bottom last.

This method can be effective in driving out the lag bolts, and in most cases requires less time than a rotary saw or sledge. When the lags have been pulled, personnel can bend the bar assembly away from the window.

Figure 6–12. An axe is effective at removing the lag bolts from a structure. When free, bend the assembly away from the window.

In tests with a rotary saw, sledge, and pickhead axe, the axe was consistently able to allow removing security bars in the least amount of time from a frame wood, frame stucco, and masonry building using the aforementioned procedures.

Security bars that are openable from the inside of a structure (identifiable by exterior hinges on the bars and a bracket, or brackets that connect the exterior of the bar assembly to the interior release mechanism) are most easily removed by cutting the brackets that connect the bar assembly to the interior release mechanism. Use a rotary saw; then simply open the bar assembly away from the window.

Masonry construction. Security bars that are mounted outside an exterior wall and over a window are easily removed by cutting the ends of the horizontal bars with a rotary saw and bending the assembly away from the window (as in figure 6–10). When bending the bars away, the remaining lags will often pull out, releasing the entire assembly.

Exterior security bars can also be removed by striking the ends of the left or right horizontal bars near the anchor points with an axe or a sledge. This method is very effective because the lags are screwed into lead anchors. This assembly is held together by compression against the masonry construction. A sharp blow with an axe or sledge near the anchor point should easily break the masonry construction around the anchor and result in the anchor being pulled outward (figure 6–13).

Figure 6–13. Lead anchors are used to secure lag bolts in masonry construction. Because the lag bolt is secured by compression, it can be easily removed.

Place a large crowbar, rabbit tool, or other similar tools that can exert force between the anchor points and masonry wall, and pry outward.

Security bars mounted directly into a masonry wall can be more secure than those mounted outside the wall, and they can offer less access to striking and prying tools. If necessary, bar assemblies that offer minimal access can be removed by cutting their imbedded ends (outside the masonry wall) with a rotary saw and removing them completely.

Doors

Structural forcible entry operations normally focus on doors as a means to gain entry into a building. There are a number of reasons for this. Doors are the normal point of access to a structure, and they open to the normally traveled routes. Also, usually they are the largest openings. This enhances deploying resources through them.

In some commercial structures (for example, most concrete tilt ups), access can be gained only through doors due to a lack of windows. Doors will allow personnel to enter a structure at a lower level (such as closer to the floor) than will a window. This is important if heat and smoke are present.

Door Assembly. The major components of a door assembly are as follows (figure 6–14):

- **Door:** Composed of a single material (for example, wood, metal, glass), or a frame that supports various materials (as in a metal frame surrounding glass).

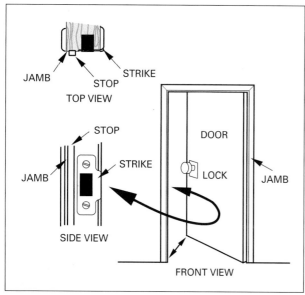

Figure 6–14. A door assembly is comprised of a door, jamb, lock, stop, and strike.

- **Jamb:** The structural case, border, or track into which a door is set. A jamb supports and may contain the stop for a door. Jambs are normally constructed of wood or metal.

- **Strike:** A receptacle that receives a deadbolt or latch from a locking mechanism. Strikes are constructed either as a metal plate that covers the receptacle in a doorjamb or as an opening in a door frame (such as a slot in double aluminum-framed doors).

- **Stop:** A protrusion that stops a door and keeps it from swinging past the jamb.

- **Lock:** Various types of locking devices are used to provide security (figure 6–15).

A key-and-knob lockset can be identified by a keyway in the doorknob. This is one of the most common locks on residential doors. Rim locks are mounted on the interior of a door. These are used as a secondary lock and can be installed by the average person. A tubular dead-bolt lock consists of a surface-mounted cylinder that contains a locking and operating mechanism and dead bolt inside a door that projects into a receptacle in the door jamb. A mortise lock is a lock that is installed into a mortise (cavity) in a door.

Exit locks, also known as panic hardware, are commonly used in public assembly buildings where a door must be easily openable from the inside to provide a timely exit. The entire lock assembly is mounted on the surface of the door and is activated by pushing on a bar, pad, or by turning a handle.

Auxiliary locks are installed to supplement a primary lock, or may be used as a primary security lock (padlock, slide bolt, police locks, and so on). These locks can be located on the inside of the door and not be openable from the outside.

Types of doors. There are four basic types of doors: swinging, sliding, overhead, and rotating. Swinging doors are hinged on one side and swing either to the left or right, inward or outward. Sliding doors move horizontally (either left or right) to open or close an opening. These can be large doors on tracks in commercial warehouses, sliding glass patio doors in residential dwellings, or pocket doors in interior walls of various types of structures. Overhead doors are mounted above an opening and travel vertically or swing upward. These doors are normally used in residential applications (garage doors) and

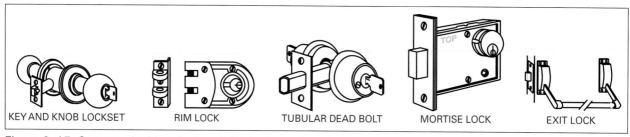

Figure 6–15. Common locking devices fall into five categories.

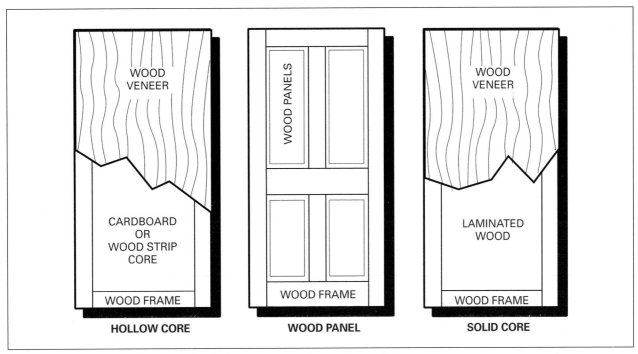

Figure 6–16. Wooden doors may be hollow core, wood panel, or solid core.

commercial occupancies to cover various types of openings in interior and exterior walls. Revolving doors are found in some department stores and airports, and are normally used to conserve energy.

Door construction. Basic door construction involves three basic materials: wood, metal, and/or glass.

Wood. Wooden doors may be hollow core, panel, or solid core (figure 6–16).

A hollow-core door consists of a solid wood frame surrounding a core of wood strips or cardboard, and is covered with a wood veneer. This is a cheap, lightweight door. Wood-panel doors use a solid wood frame inset with panels. The panels can be of wood, glass, or synthetic materials (masonite or pressboard that resembles wood). The panels are the weakest portion of the door. A solid-core door is strong, employing laminated wood covered with a wood veneer.

Metal. Metal doors may be kalemine or hollow (figure 6–17).

Kalemine doors are metal clad with a thin metal covering over a solid wood core. Hollow metal doors are constructed of 14- to 16-gauge steel and use steel ribs between the skins for stability. These doors can be more substantial than wooden doors.

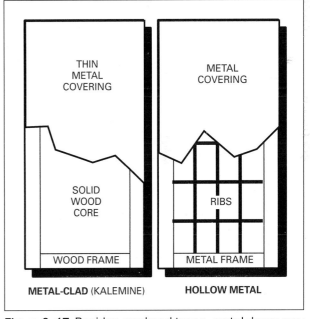

Figure 6–17. Besides overhead types, metal doors may be either metal-clad or hollow.

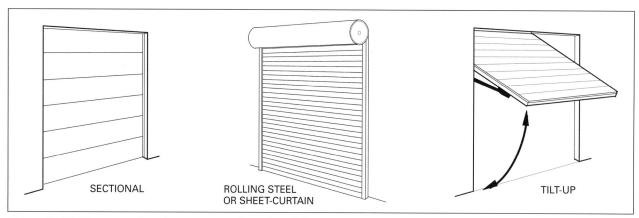

Figure 6–18. There are four basic types of overhead doors.

Overhead doors. There are four types of overhead doors: rolling, sheet curtain, sectional, and tilt-up (figure 6–18).

Rolling steel doors are characterized by interlocking steel slats that are manually or mechanically raised or lowered by winding on or off a storage drum at the top of the door. A sheet-curtain door is a solid-piece door made of thin corrugated metal, and is also manually operated by winding on or off a storage drum at the top of the door. These doors are similar in appearance to rolling steel doors. Sectional doors have sections made of thin metal interconnected by steel hinges, or wood sections made from a frame surrounding solid wood panels. The sections are interconnected by metal hinges. These doors normally travel vertically in tracks mounted in the door jamb and then across a ceiling of a building and are often assisted by counterweight springs above the door. Tilt-up doors are solid-piece doors made of thin metal that is hinged at the center with spring-loaded hinges. These doors operate manually or mechanically, and swing overhead to open.

Glass doors. There are two basic types of glass doors: frameless and aluminum-framed (figure 6–19).

Frameless glass doors use tempered glass of about ¾-inch thickness, with an aluminum cross member (stile) at the top and bottom to support appropriate locking mechanisms. Aluminum-frame doors employ an aluminum frame around heat-treated, tempered, or laminated glass. Locking mechanisms are normally mounted in the vertical frame adjacent to the door handle used to open the door.

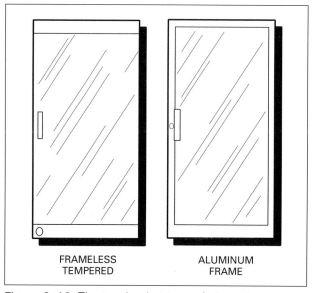

Figure 6–19. The two basic types of glass doors

Door considerations. Before forcing any door, evaluate the following:

- **Safety:** Wear appropriate protective equipment.

- **Present condition:** Try opening the door before forcing it. Don't assume it to be locked. Try before you pry.

- **Purpose:** Determine why the door is being opened. If the door is being forced to provide access for reasons other than fire suppression, place emphasis on selecting the easiest door and causing the least damage. However, if a door *is* being forced to provide access for fire suppression operations, emphasis should be placed on location, damage, size, and time.

Location entails forcing the correct door. Generally, access for fire suppression operations should be created away from a fire so as to advance from the unburned to the burning portion of the structure (if possible). Consider the *damage* caused by the fire versus that which will result from forcible entry. Depending on the size and type of fire, the damage caused by forcible entry will usually be secondary. Create an opening as large as possible (*size*) to enhance the deployment of resources and the timely emergency exit of personnel, if necessary. Evaluate how long (*time*) it will take to successfully force the door.

- **Construction:** Evaluate the type of door and frame construction. Determine if the door is constructed of wood, metal, or glass; determine its weakest portion; determine what tools will be necessary to force entry.

- **Direction of travel:** Determine if a door swings inward, outward, moves vertically, moves horizontally, or rotates. For example, to determine whether a given door swings inward or outward, evaluate the location of the door in the wall and the location of its hinges. A swinging door swings outward (toward exterior personnel) if the door is flush with the exterior of a wall and the hinges are visible (figure 6–20B). A swinging door swings inward (towards interior personnel) if it is recessed into the wall and the hinges are not visible (figure 6–20A).

- **Security:** Determine what kind of lock or device provides security. Remember to compare the relative strength or weakness of a locking device versus door construction.

- **Involvement:** When a door is opened to a heated-contaminated atmosphere, the sudden flow of oxygen can accelerate a fire or cause a smoke explosion. Under these conditions, do not stand in front of the door when it is opened suddenly. If necessary, maintain control of it when it is forced. A strap or line can be attached to doorknobs on inward-opening doors. A charged hose line should be in place for advancement into the fire area *before* the door is forced.

If an interior door is to be opened to a fire area, ask yourself whether heat and smoke will extend to uninvolved portions of the structure. If appropriate, consider how much heat is behind the door. Check for smoke issuing around it and its surface temperature. Before you try a doorknob, remember that metal doorknobs are excellent conductors of heat.

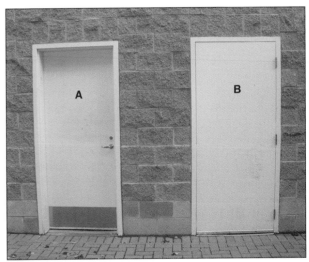

Figure 6–20 A and B. Outward and inward swing doors can be identified by their placement in a wall and the presence or lack of hinges.

Forcing a door. If it is necessary to force a door, identify the type and its direction of travel. Evaluate the strength of the door, its locking mechanism, and the wall. It may be easier to open a plasterboard wall surrounding a metal-clad door than to force the door itself, depending on the locking mechanism encountered. As an example, a firefighter can make a small opening in the plasterboard adjacent to a metal clad door in a metal frame. The opening then allows the firefighter to reach through to the other side of the door and open the door. Patching the plasterboard is relatively simple compared to replacing the door and jamb if force was necessary to force the door. Determine the type of forcible entry and appropriate tools. Remember that conventional forcible entry will focus on the door, and through-the-lock entry will focus on locking mechanisms. Power tools are normally used for difficult situations and/or ease of use by the power they can provide. Normally, personnel should not kick a door to force entry. This practice is dangerous from the perspective of potential injury and the tendency of personnel to fall inward as a consequence of kicking. However, if it is necessary to kick a door open, consider that this procedure is most successful on interior hollow-core wood doors with a single doorknob-type lock, and that a mule kick (personnel face away from the door and kick backward) will keep you from going through if the door opens suddenly.

Through-the-lock entry. Through-the-lock forcible entry methods focus on removing and tripping the lock. Key-and-knob locksets are the most common type of lock found on swinging doors, and are composed of a locking doorknob with a keyway inside the knob. Most of these locks can easily be forced. An axe can often be used to break off the doorknob by striking the top of it (figure 6–21). When the doorknob has been removed, insert a key tool or screwdriver to activate the latch mechanism inside the lock.

Figure 6–21. Use an axe to remove the knob of a key-and-knob lockset. This will allow a tool to be inserted to activate the internal latch mechanism.

The mechanism often consists of a half moon-shaped hole and is easily moved to open the latch. Otherwise, place the forked end of a Halligan over the knob. Place a wooden wedge between a Halligan and door to act as a fulcrum. Push the Halligan

toward the door to remove the doorknob. Insert a key tool or screwdriver into the lock and activate the latch mechanism.

Another technique is to place a small screwdriver at approximately a 45-degree angle to the doorknob, between the junction of the keyway and the doorknob. Deliver a sharp blow against the screwdriver to remove the keyway from the doorknob and enable you to insert a key tool or screwdriver into the lock to activate the mechanism. If necessary, the keyway can easily be inserted back into the doorknob. This method is particularly successful on Kwikset locks, which are the most common key-and-knob locksets in residential and light commercial structures.

Rim locks (or surface-mounted deadbolts) are easily installed on swinging residential doors and provide excellent security. The lock is mounted on the inside of the door, and consists of a spring latch, which may also have a night latch button. This, when activated, will lock the spring latch mechanism, or a vertical deadbolt that interlocks with an appropriate receptacle on the door frame. A keyed cylinder on the front of the door activates the lock with a stem or tailpiece. A rim lock can be recognized by a cylinder that protrudes through the door about ¼ to ⅜ of an inch. This type of lock can be forced using an A tool or sharp K-tool to pull the cylinder. This will bend and pull the metal backing plate and stem out of the locking mechanism, exposing the back of the lock. Insert a key tool or screwdriver into the slot on the rear of the rim lock and turn it to unlock the mechanism. When confronted by a spring-loaded shutter across the slot, use a pointed tool to slide the shutter away from the slot. This will allow an appropriate tool to unlock the mechanism.

If a rim cylinder breaks prematurely while being pulled, use the pick end of a Halligan or similar tool to drive the broken cylinder out of a wooden door, or between the layers of a hollow metal door. Then, drive the rim lock off the back side of the door. When forcing a rim lock, it is important to remember that it might not be locked. Therefore, try the doorknob first. If necessary, force it. If the door is still locked in place, the rim lock is locked and should be forced.

The tubular deadbolt is a popular lock that can provide good security, and is found in four basic configurations: single-cylinder deadbolt; double cylinder deadbolt; high-security deadbolt; and handle set. The single-cylinder deadbolt is keyed from the exterior only, and has a thumb turn handle on the inside. The double-cylinder deadbolt is keyed from both the exterior and interior. The high-security deadbolt has a unique pin configuration, hardened cylinder, and other burglar thwarting designs. This deadbolt is more difficult to remove from the door or cut with a rotary saw. In some cases, it can be differentiated from the cheaper deadbolts by stepped rings on the exterior metal collar, whereas cheaper deadbolts likely have a smooth exterior collar. The workings of the handle set configuration are similar to standard deadbolts, but the deadbolt is separate from an ornamental handle that replaces a standard key in the knob lockset. This set has a separate latch and deadbolt, and provides no more security than a stand-alone deadbolt. See "Door Locks Tougher Than Ever" from *Consumer Reports*, in Training Resources at the end of this chapter for information on ratings of popular door locks.

Deadbolts are mounted within a door and feature a bolt that extends about one inch or more into the door frame. A large keyed cylinder is on the outside of the door, and another large keyed cylinder or thumb knob is on the inside. This lock is different and normally recognized from a rim lock due to its large cylinder, which protrudes about 1½ inches on the outside. The tubular deadbolt cylinder can be too large for a K-tool, whereas a rim lock can be pulled by either an A- or a K-tool.

A tubular deadbolt can be forced by using an A tool to pull the cylinder and expose the latch mechanism. Then, insert a key tool or screwdriver into the mechanism to turn or move the bolt away from the door strike. Otherwise, drive the forked end of a Halligan between the door and cylinder. Pushing the Halligan toward the door will pull the cylinder from the door, exposing the latch mechanism. Insert a key tool or screwdriver into the mechanism to turn or move the bolt away from the strike. If hardened screws have been used to hold the exterior and interior cylinders together, separating the exterior cylinder from the door can take considerable force.

If time is not a factor, a cylinder lock can be drilled; This is effective on residential and commercial deadbolt lock cylinders. Basically, a hardened drill bit is placed above the keyway and near the junction of the keyway and cylinder. After the hole is correctly drilled with the aid of a lubricant, insert a slot blade screwdriver into the keyway and turn to unlock the deadbolt. See "Drilling Lock Cylinders" in the Training Resources at the end of this chapter for more information on this method.

The mortise lock is one of the oldest types of locking mechanisms in use, and is easily recognized by a locking mechanism that has been installed into a cavity in the edge of the door. In residential applications, a mortise lock may consist of a doorknob bolt, dead bolt, or both. When both types of bolts are used, a key must be inserted into the keyed cylinder and turned twice. The first turn activates the dead bolt. The second turn activates the doorknob latch bolt. In commercial applications, a mortise lock normally consists of a pivoting or retracting dead bolt that is activated by turning a key in a keyed cylinder. In both instances, a keyed cylinder is threaded into the body of the lock mechanism and held by a small set screw. A cam on the back of the cylinder activates the lock mechanism when a key is inserted and turned.

Mortise locks can be forced. Prior to removing the lock cylinder, visualize a clock around the cylinder with six o'clock side-by-side with the keyway (figure 6–22). The keyway is normally vertical in residential applications, and can be horizontal or vertical in commercial applications. Using the imaginary clock face as a reference, pull the cylinder with an A- or a K-tool. As explained in the section on metal-frame glass doors as used in commercial occupancies, the cylinder can also be removed by using channel locks or vise grips. Insert a key tool into the cylinder hole with the end positioned over the five or seven o'clock position. If the mechanism is at five o'clock, move it to seven o'clock, or vice versa.

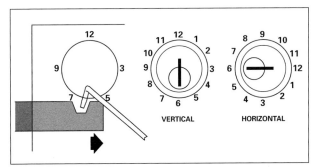

Figure 6-22. After removing the cylinder of a mortise lock, insert a key tool and move the bolt to the five or seven o'clock position to open the lock.

Normally the correct way is determined by noting the position of the detent in the mechanism, or by moving the detent from its present position to the opposing position. If there is a spring lock on the lock mechanism, depress the spring with a key tool and then move the detent mechanism from five to seven o'clock or from seven to five o'clock.

Forcing sliding doors. Residential pocket doors are normally hollow and lightweight. If secured by a locking mechanism, that mechanism should also be lightweight and easy to pry. Sliding patio glass doors as used in residences are normally secured by locks that are only accessible from the interior. If a door is equipped with a single lock (normally in the door handle), the door may be opened by inserting a prying tool between the door frame and jamb, then prying the door away from the jamb. This will break the lock, which consists of pot metal in an aluminum frame. If additional locks have been installed, it may be necessary to break the glass to enter. Remember to look through the glass to see if a dowel has been placed between the bottom portion of the door and opposing jamb. If a dowel (or other similar device) is present, the door cannot be forced open and the glass will have to be broken.

Older commercial occupancies can have large sliding doors that can be quite substantial in warehouse applications. If such a door is secured from the interior, it may be easier to force entry through another door and open this door from the inside. If a door is secured from the exterior, as with a padlock, the security device can be forced using the appropriate method and tool.

Forcing swinging doors. When door hinges are exposed on metal and wooden outward-opening doors, the hinge pins can sometimes (unless designed otherwise) be removed by forcing them out of their hinges. If a door is loose within its frame, use the hinges to pull the door outward until the latch pulls free of the strike. Conversely, if a door is tight within its frame, it may be necessary to insert a prying tool between the frame and hinge to pry the door outward until the latch pulls free of the strike.

If necessary, exposed hinges can also be cut with a rotary saw. To facilitate this, cut the bottom hinge first, the top hinge second, and the middle hinge last, cutting completely through all the hinges from the outside. With this method, you will be standing upright when the last hinge is cut. If a hinge is only partially cut through from the inside, it will jam and prevent the door from being removed. Pry the door outward until the door latch pulls free.

You can also use techniques to force swinging doors with a Halligan and flathead axe. Position the bevel (outside radius) of the forked end of a Halligan toward the door and at the junction of the door and stop, just above or below the lock and about 30- to 45-degrees away from the door. If there are multiple locks, place the forked end between the locks. Drive the forked end between the door and jamb by striking the Halligan with the heel of a flathead axe (figure 6–23). If this isn't easily accomplished, place the bevel (outside radius) toward the doorjamb and drive in the forked end.

Figure 6–23. Inward-opening doors can be forced by driving a Halligan between the door and jamb.

This will spread the door and jamb, creating a gap. Remove the Halligan and again place the bevel side of the forked end toward the door, then drive the forked end between the door and jamb. Repeat this process while increasing the angle of the Halligan away from the door until the forked end is securely between the jamb and door, and past the inside molding. This will place the Halligan approximately 90 degrees to the door. Push the Halligan toward the door, forcing the door open. If opening into a fire environment, be able to control the door.

This can easily be done by attaching a cord or strap to the doorknob (figure 6–24).

Figure 6-24. When forcing inward-opening doors to a fire environment, control the door with a cord or strap attached to the knob.

Figure 6-25. Inward-opening doors can also be forced by using the adz end of a Halligan.

Another method is to place the adz end as close to the lock as possible and between the door and jamb. If necessary drive the adz inward to obtain a purchase. By moving the Halligan outward, the door should be forced open.

One person can use a Halligan. Using a baseball-bat swing, bury the pick into the doorjamb. It may take several tries to place the spike into the jamb correctly. Apply a downward force on the tool if the pick is in a jamb on the right side of the door, and an upward force on the tool if the pick is in a jamb on the left side of the door, as in figure 6–25. Force the door inward to open. Strong or multiple locks will decrease the success of this operation. Additionally, using a baseball-bat swing in a restricted space such as a hallway can be difficult.

Forcing outward-opening doors. Forcing outward opening doors requires different techniques with a Halligan than inward opening doors. Position the adz between the door and jamb, just above or below the lock and with the handle parallel to the door. If there are multiple locks, position the adz between them. Drive the tool between the door and jamb by striking it with a flathead axe (figure 6–26). Do not drive it past the doorstop. Pull the tool away from the door, forcing the door open.

Position the forked end between the door and jamb with the bevel side towards the jamb. Drive the Halligan inward to obtain a purchase. Pull the Halligan toward the jamb to open the door.

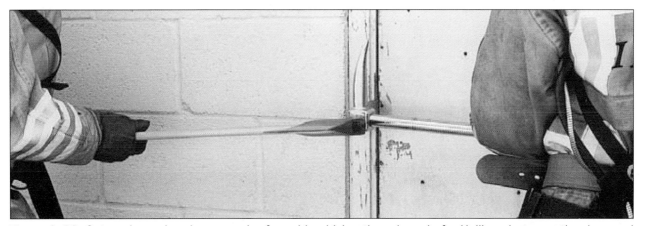

Figure 6-26. Outward-opening doors can be forced by driving the adz end of a Halligan between the door and jamb, then prying outward.

Power tools. Using power tools specifically designed for forcible entry can allow one person to conquer some doors in a minimal amount of time. For example, the rabbit tool can be effective in forcing inward-opening doors. Insert the expanding jaws between a door and jamb, and then expand them until the locking mechanism pulls free of the jamb. On inward-opening doors, additional effort is necessary to force the expanding jaws between the door and jamb because the jaws must first be driven past the stop (figure 6–27).

opened unless the bar is removed. The presence of a security bar can often be recognized by a pair of carriage bolt heads on the exterior of the door.

If there is a deadbolt or key in the knob lock on the exterior of the door, place a blade from a rotary saw between the jamb and door and cut the deadbolt and/or latch. Then, with a rotary saw, place a metal cutting blade at a 45-degree angle between one of the pair of bolt heads. As the blade cuts into the door, it will also cut through the shanks of the carriage bolts (figure 6–29).

Figure 6-27. Rabbit tools can be used by one person to force inward or outward swinging doors.

Figure 6-28. Bars are often used on the inside of outward-swinging doors to provide increased security.

Forcing outward opening doors with a rabbit tool may not be as successful if the jamb and door cannot be expanded enough to pull a deadbolt and/or latch out of the jamb. Normally, the door and jamb can be widened enough to pull a latch free as this only takes about one inch. See Training Resources at the end of this chapter for more information on this subject.

Security bars. Doors equipped with security bars fall into two categories: interior and exterior.

Wood or metal outward-swinging doors may have an interior bar of wood or metal for additional security. The bar is placed horizontally across the interior side and is supported by two brackets on the door (figure 6–28). Because the bar extends past the door jamb on both sides, the door cannot be

Figure 6-29. An interior security bar can be defeated by using a rotary saw to cut the exposed bolt heads from their shanks.

Repeat this process for the other pair of bolt heads. After the four bolt heads have been cut from their respective shanks, drive a Halligan between the door and jamb. As the door is forced open, the four bolt shanks will be pulled out of the door, allowing the security bar assembly to fall on the floor and the door to be fully opened.

Security bars over the exterior of doors are somewhat different than security bars over windows in that they are not solid to the structure but must swing outward (normally) to allow passage through the door opening, and it is usually necessary to remove or open them to allow initial access to the door into a structure. These types of security bars usually consist of two parts: a metal jamb that is secured to the doorway opening, and a metal frame that swings outward and is connected to the jamb by hinges. The frame supports the horizontal and vertical security bars to complete the assembly (figure 6–30).

The lock assembly that is mounted in the frame-bar assembly is usually found in three configurations: a key in the knob type lock (less secure), tubular deadbolt-type lock (more secure), or a combination of both locks as shown in figure 6–30 (the most secure). The deadbolt-type lock is a secure locking device because the deadbolt can have a throw up to two inches, whereas the key in the knob lock may only have a throw of half an inch. Both of these locks are normally mounted in a metal box that is supported by the security bars, a portion of the frame, and may or may not have a metal plate that covers the gap between the frame and jamb, thereby protecting access to the deadbolt and/or latch.

There are several types of security bar assemblies that are commonly used and various ways to open them. First, verify that the lock or locks are engaged. If there is a visible gap between the frame and jamb exposing a portion of the deadbolt and/or latch, a rotary saw blade can be inserted between the frame and jamb to cut the deadbolt and/or latch. If the gap between the door and jamb is tight, use a Halligan to widen the gap to prevent a rotary saw blade from binding. In figure 6–31, a metal plate covers the gap between the door and jamb. If only a key in the knob lock is present, this operation will take less time than cutting through two locks.

In this example, it is still possible to cut downward from the top to the bottom of the plate, and deep enough to simultaneously cut the latch and deadbolt.

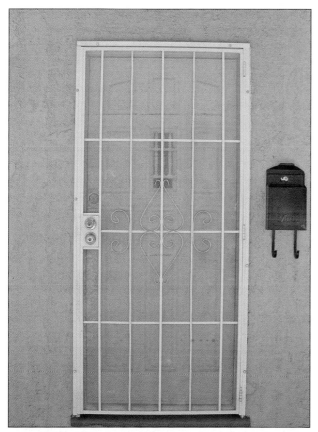

Figure 6–30. Security assemblies over exterior doors are common but must be removed to allow access to the door behind the assembly.

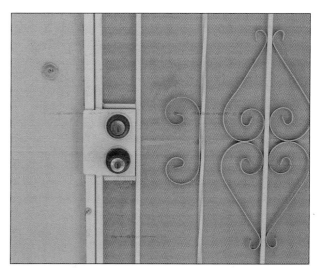

Figure 6–31. In some cases, a metal plate covers the gap between the door and jamb.

Because the hinges on the frame jamb are normally exposed (outward opening door), they can be cut with a rotary saw, separating the moveable frame from the solidly mounted jamb. Cut the bottom hinge first, the top hinge second, and the middle hinge last. The frame should now be able to be removed from the jamb. Be aware that some of these doors use piano-type hinges that are mounted on the back of the frame, travel the full vertical length of the door, and are not visible from the exterior of the door yet still allow the door to open outwards.

Some security doors are equipped with a solid piece of metal that runs the length of the door and covers the entire gap between the door and jamb (figure 6–30) In this case, it is still possible to cut through the protective plate and deadbolt and/or latch as in the prior example. Obviously, it is necessary to be familiar with the types of security doors that can be found in your area.

If it is deemed that cutting the protective plate and deadbolt and/or latch would be too time intensive, the horizontal and/or vertical bars that support the metal box that encloses the locking mechanism(s) can be cut where the bars are attached to the metal box. When these are cut, the metal box and its locking mechanism(s) can often be pulled out and away from the jamb, releasing the lock from the strike.

Forcing aluminum-framed glass doors. Commercial structures often have aluminum-framed glass doors in a single- or double-door configuration. A single mortise lock with a retracting or pivoting bolt (normally made by Adams-Rite, figure 6–32) is used to secure a single door to a frame, or to secure double doors to each other.

Because there is normally a ¼-inch gap between a single door and frame, or between double doors, a rotary saw can be used to cut the retracting/pivoting bolt. If the lock cylinder doesn't have a rotating collar, use vise grips or channel locks to turn the cylinder counterclockwise. You will feel resistance during the initial portion of the turn. This is necessary to break the set screw holding the cylinder. After the set screw is broken, the cylinder may often be turned counterclockwise by hand to remove it (figure 6–33). A key tool, screwdriver, or similar tool can then be used to trip the lock.

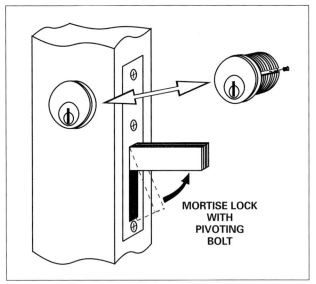

Figure 6–32. Most aluminum-framed glass doors are secured by a mortise lock with a pivoting bolt. The lock cylinder is only secured by a small set screw.

Figure 6–33. When the set screw is broken, the cylinder can be easily removed from the door frame. The lock in the door frame can now be tripped.

If the cylinder has a rotating collar (used to defeat such techniques), drive a flat-blade screwdriver into the cylinder keyway. Use channel locks to turn the screwdriver shank counterclockwise to break the set screw (figure 6–34). The cylinder can now be removed by turning the screwdriver and cylinder counterclockwise.

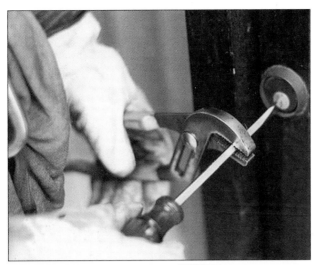

Figure 6–34. If the lock cylinder is protected by a rotating metal collar, drive a flat-blade screwdriver into the keyway. Use channel locks to turn the cylinder counterclockwise for removal.

Figure 6–35. A push bar can be easily removed by striking it with an axe at the junction of the bar and door frame as this is the weakest point of the assembly.

Another method is to use an A-tool or a K-tool can also be used to remove the lock cylinder. After the cylinder has been removed, use an appropriate tool to trip the lock. Depending on the type of lock and the throw of the bolt, it may also be possible to insert a prying tool to pry the bolt out of its aluminum strike. This operation can break the glass, however, and tempered glass may explode outward. Personnel should not stand in front of such a door when using a prying tool. It is common for these doors to be equipped with a horizontal aluminum push bar. If the glass is broken to provide entry, this must be removed so that ingress-egress won't be impeded. Strike an axe near the junction of the bar and door frame to remove it (figure 6–35).

Forcing frameless glass doors. Commercial applications commonly use frameless tempered glass doors in single- or double-door configurations. A single mortise lock (Adams-Rite) is normally used in the bottom aluminum cross member (stile) of each door. To secure a door, a pivoting bolt extends downward into a cavity in the floor. These doors should not be broken due to their high cost of replacement. A rotary saw can often be positioned to cut the bolt between the bottom of the door and the floor (figure 6–36).

This is one reason why a rotary saw that is primarily used for forcible entry operations should be carried with the blade in the outboard position.

Figure 6–36. A rotary saw that is primarily used for forcible entry should be carried with the blade in the outboard position so the blade can be positioned and used in tight spaces.

This allows the blade to be inserted in areas with minimal space considerations. It is also possible to remove the lock cylinder with vise grips, channel locks, or an A- or K-tool, and then to trip the lock manually.

Exit (or panic) hardware is a common security device in various occupancies. Normally a horizontal bar across a door, when pushed, will activate a vertical rod that engages the top door frame and/or floor, or a lock that engages a side door frame. Since exit hardware cannot be accessed from the exterior by common forcible entry tools, an effective tool can easily be made from a car antenna (the metal in coat hangers is too soft). On double doors, normally this tool can easily be slipped in the gap between the doors and used to hook and pull the horizontal bar, thereby opening the door (figure 6–37). If there is a gap between a single door and frame, this tool can also accomplish the same operation.

Figure 6-38. A chain saw can be used to activate a push bar if the operator is familiar with the type of panic hardware present as well as safe operation of a chain saw.

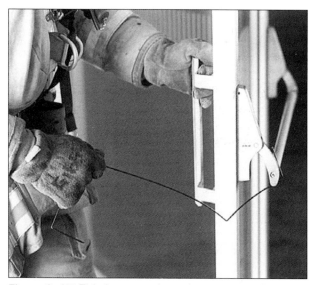

Figure 6-37. This homemade tool can open a door with minimal effort.

This tool is most effective on exit hardware that has a horizontal bar instead of a handle or pad. Another method to open interior panic hardware having a horizontal bar mounted to wood or most metal-clad doors (common on public assembly-type occupancies such as theaters and auditoriums) is illustrated in figure 6–38. About chest high, cut through the door with a carbide chain saw.

When the chain has penetrated the door, cut downward using a *light touch* on the saw until the chain contacts the metal horizontal bar. Release the throttle and lock the chain brake. Rotate the saw forward, pushing the bar downward, unlocking the locking mechanism. Although this method can be effective, it is dependent on a prior knowledge of your district and an appropriate operational knowledge of your equipment.

Forcing cellar doors. Although a cellar door (also known as a Bilco door) covers stairs that allow access to or egress from a basement and is not often considered a true door, it is nonetheless a type of door that opens outward and is normally securely locked. From an intruder's viewpoint, access to a cellar is a potential opportunity to gain easy access to upper floors of a structure. Therefore, cellar doors are normally composed of metal and substantial locking techniques, and can be a formidable opponent against being forced in a timely manner. Because there are countless numbers of these doors across the country, and because they can provide quick access and ventilation to a basement, let's look at forcible entry techniques for these doors. A quick size-up should be composed of the following:

- Determine the door construction and if it is metal, one or two panels.
- Determine if there are any locking mechanisms visible. Look for exterior locks (padlocks, and so on), carriage bolt heads, or a lack of any locking devices. In this case, the locking devices are internal.
- Determine what tools are necessary to force the door(s).

If there is a visible locking device such as a padlock, chain and padlock, or other similar devices, determine the weakest point for cutting with a rotary saw (see the following section on cutting padlocks) and make the necessary cut(s). If carriage bolt heads are visible, this is an indication that a locking device is bolted to the underside of the doors. The bolt shanks can be cut with a metal cutting blade on a rotary saw by placing the blade at a 45-degree angle to the bolt heads (for a pictorial account of this operation, see the previous section on interior security bars on outward opening doors). When the bolt heads are cut off, the bolt shanks should be pulled out of the door because the doors are opened outward.

If there is no evidence of any locking device, this is an indicator of an interior lock assembly that lies underneath both doors and acts to lock the doors together. Determine which door is the overlapping door. The lock assembly is likely about midway up the doors or in the middle portion of the doors. Make a vertical cut, about 12 to 14 inches long, through the overlapping door and deep enough to cut the locking mechanism under the door, and about 3 feet from the bottom with a rotary saw (figure 6–39, dotted line A). This should free the doors and allow them to be opened.

Another method uses a rotary saw to make two parallel cuts from the top of the doors and down to the bottom of the doors (figure 6–39, dotted lines B). (Because this method requires the blade to cut a significant amount of metal, it is best accomplished by a multi-use blade.) Make the first cut top to bottom and on the left or right side of the doors.

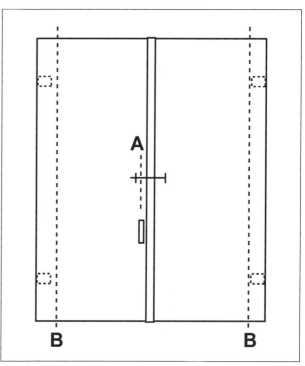

Figure 6–39. Cellar doors (or Bilco doors) can be forced and opened by knowing where to cut the most advantageous part of these types of doors.

Be sure to cut inward from the outside edges about 12 inches to miss the interior hinges, and cut deep enough to cut through the doors. Repeat on the other side of the doors. The doors should fall inward, or can be pushed inward. It is vitally important to be familiar with the type of cellar doors and associated locking devices that are predominant in your area.

Overhead doors

Overhead doors are commonly used to cover various types of openings, such as loading docks and garages in commercial structures. These types of doors can use internal and/or external locking mechanisms that consist of lock cylinders, padlocks, slide bolts, or manual mechanisms. If forcible entry operations are unsuccessful against exterior or interior locking mechanisms, entry must be forced through the overhead door.

A common technique is to make a triangle, teepee, or inverted V cut in the door. Although this method will provide entry, it has multiple drawbacks. It requires at least two large diagonal cuts. Also, it normally results in an opening that is narrow at the top and wide at the bottom. This type of opening doesn't facilitate freedom of movement, particularly the emergency egress of multiple personnel. Also, unless the cut portion of the door is removed, personnel will walk over loose metal in the opening. Finally, when a hoseline is advanced through this type of opening, the door cannot be raised unless either the bottom portion of the door or the hose line is removed.

When forcing overhead doors, it is vitally important to ensure a door cannot inadvertently close on interior personnel. Therefore, always evaluate the type of door and the possibility of the door closing and what measures would be necessary to ensure the door does not inadvertently close. This can be easily accomplished with a pike pole under the door, vise grips on the door tracks, and so on.

An alternative method that can result in a larger opening with less effort and time than a tepee cut is known as the Kemper cut and can be accomplished as follows.

Cutting rolling steel doors. These doors are constructed with steel interlocking slats, and roll around a spring-loaded, manual, or mechanically operated drum at the top of the door. Use a rotary saw to make a cut down the center of the door. Start the cut about head height and continue down to the iron channel, referred to as the bottom bar, at the bottom of the door. Initiating a vertical cut about head height will allow interlocking slats above any damaged slats, and above most locking mechanisms, to be pulled. If any slats have been bent or damaged (which is rare) from being struck by a vehicle, they won't pull free. Make a small angular cut at the bottom of the vertical cut to create a small opening as illustrated in figure 6–40. This small opening is optional but enhances inserting a rotary saw into the door to cut the channel at the bottom if desired.

If it is necessary to cut the bottom bar, place a rotary saw into this opening and cut through the channel. Use channel locks, vise grips, or other similar tools (not your hands) to pull a slat from either the left or right side of the vertical cut. When a slat is pulled free, all of the lower slats (below the one pulled) will fall to the ground and can easily be pushed away from the opening. Pull an opposing slat on the other side of the vertical cut to release all slats on the opposing side, and

remove all of the cut slats from the opening (figure 6–41). When pulling slats in these types of doors, it is important to remember that most rolling steel doors are constructed with a guide on every other slat.

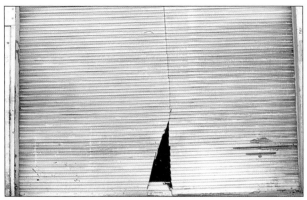

Figure 6–40. For overhead rolling steel doors, make a cut down the middle with a rotary saw. An optional operation is to cut a small opening at the bottom as illustrated.

Figure 6–41. When a cut has been made down the middle of the door, the interlocking slats can be removed to provide a large opening.

Slats without guides are easily pulled free. There-fore, if a slat cannot be pulled free, move up or down one slat to bypass the slat that cannot be pulled (as it has a guide that is in the track alongside the door). However, some doors have been constructed with guides on every slat. Slats with guides can-not be pulled free with the door intact. If a slat or slats cannot be pulled free, use a rotary saw to cut through several slats on both sides of the door next to the door frame (after the center cut). Several slats can now be pulled free as illustrated in figure 6–42. This will allow the lower slats to be removed from the track and door opening.

Figure 6–42. If interlocking slats cannot be pulled free after making a center cut, cut several slats on each side of the door to allow the cut slats to be pulled free, and other slats to be removed from the guides.

When the appropriate slats have been pulled free, the upper portion of the remaining door should raise if it is spring loaded. If the door needs to be manu-ally or mechanically operated, it can now be opened completely. If the door cannot be raised, this method will still open a significantly larger opening than a teepee cut and in less time.

Cutting sheet curtain doors. Sheet curtain doors look similar to rolling steel doors from a distance, but up close there is a noticeable difference. Instead of using steel interlocking slats, they are constructed of thin metal corrugated panels joined together

(welded) to form a solid-piece door. This continu-ous panel will roll around a spring-loaded, manual, or mechanically operated storage drum at the top. This door can be forced by using the same procedure as described in the next section on sectional doors.

Note that both the rolling steel doors and the sheet curtain doors are unlike sectional and tilt-up doors because both doors are relatively flexible and are required to roll around a storage drum at the top of the door; sectional and tilt-up doors do not. Therefore, sectional doors and tilt-up doors can be thicker and may not be capable of being cut com-pletely through with a rotary saw.

Cutting sectional doors. Sectional doors are com-posed of metal or wood sections that are hinged to each other, and normally travel vertically (in tracks mounted in the doorjamb), or vertically then hori-zontally (in tracks mounted in a wall and ceiling). Sectional doors will be addressed in two categories: ordinary and wind resistant. This is due to their construction and potential difficulty of forcing with conventional forcible entry tools and/or equipment.

Ordinary sectional doors can normally be com-pletely cut through with a rotary saw, allowing a firefighter several options to open the door. Remem-ber that a rotary saw with a new 14-inch composite blade is limited to a maximum of a 5½-inch depth of cut. Next to the left or right side of the door, make a vertical cut from about head height and cut down-ward to the channel at the bottom of the door. Make a small angular cut at the bottom of the vertical cut to create a small opening as illustrated in fig-ure 6–43. Place a rotary saw into this opening, and cut through the channel at the bottom of the door (bottom bar). This small opening enhances inserting the rotary saw into the door to cut the bottom bar without cutting the concrete underneath the door.

This operation isn't necessary for wood sectional doors. Complete a horizontal cut the width of the door at approximately shoulder-to-head height, as illustrated on the metal door in figure 6–44. If nec-essary, balance the rotary saw on your shoulder to support its weight and simplify the operation.

Figure 6–43. For sectional or sheet curtain doors, make a vertical cut and a small opening at the bottom of the door and cut the bottom bar as illustrated.

Figure 6–44. To complete the cutting operation, make a horizontal cut about head high across the door.

A chain saw or rotary saw with a carbide-tipped wood blade or multi-use blade should be used on a wooden sectional door. Do not cut near the joint between the sections on metal or wood doors because

there are steel hinges between the sections on the reverse side of the door that will be cut, unnecessarily slowing the operation.

The cut panel can now be swung outward and away from the opening, creating a large opening (figure 6–45).

Figure 6–45. After making a vertical and horizontal cut, the cut panel can now be swung outward and away from the opening, creating a large opening.

If necessary, the saw operator can easily cut any remaining metal rollers that are still securing the door to the track in the doorjamb. The cut portion of the door can be completely removed from the opening. If the door is spring loaded (and it likely is), the upper portion may retract upward, exposing the entire opening. Whether the door needs to be manually or mechanically operated, it should be fully openable.

Be aware that sectional doors are assisted by counterweight springs that may retract when the bottom portion of the door is separated from the top portion. This operation removes some of the weight of the door; however, this has not been a significant concern because the upper portion of the door does not suddenly retract, and most springs lose their tension from use and time.

Wind-resistant sectional doors can be an extremely difficult, time-consuming challenge at best, particularly if a firefighter does not understand how this door construction is significantly different from an ordinary sectional door. Areas of the country that are subjected to high winds (Florida, mid-Atlantic states, and so on) put sizeable strengthening members on the backside of these door that can be of considerable construction and thickness, specifically the horizontal strengthening members that use large steel C channels. The presence of substantial strengthening members can severely complicate cutting through this door because a rotary saw with a 14-inch blade will not cut completely through the exterior skin *and* the horizontal members that are mounted on the vertical members. This type of construction means that a rotary saw must cut through the outer skin of the door, the vertical members, and then the horizontal members. This usually takes a saw that can cut over 8 inches, which a 14-inch rotary saw cannot do. Remember that a 14-inch blade will cut about 5½ inches when new, and that figure can rapidly decrease as a composite blade is used. This is a good example why personnel must be aware of the forcible entry challenges in their area and what measures are necessary to accomplish a given task. The following forcible entry method for this type of door is adapted from the excellent article "Forcing Overhead Sectional Doors" by Captain Bill Gustin, Metro-Dade Fire, *Fire Engineering*, November 2004, and is a must read for departments with this type of overhead door.

Assuming there are no exterior locks on the door (which is normal), make a vertical cut about head height on one side of the door with a rotary saw (figure 6–46A). If the saw feels as though it is cutting more than the outer skin of the door, you are also probably cutting a vertical strengthening member.

In this case, move the saw over 6 inches or so and again make the vertical cut. Make this cut to the bottom portion of the door, and cut through the bottom of the door. Similar to figure 6–43, making a small triangular cut at the bottom (figure 6–46B) will allow the saw to cut the door and not the concrete below the door. If the door needs to be raised several inches to clear a concrete lip that the door closes behind, use a Halligan or other similar tool to raise the door upward before cutting. Now go back and make a second vertical cut, also about head height and parallel about 12 inches from the first cut (figure 6–46C). Now, cut across the top and bottom, joining the first and second vertical cuts (figure 6–46D). This will allow you to remove the outer skin from the door (between cuts A,C, and D), observe the construction behind the outer skin, and be able to insert the rotary saw blade to the appropriate depth to cut the horizontal strengthening members that are visible. Then, complete a horizontal

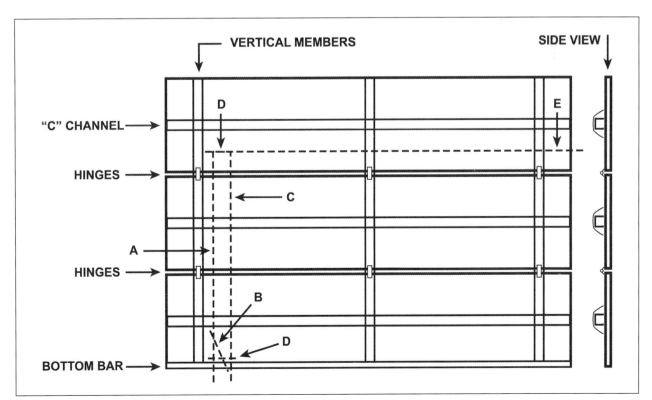

Figure 6–46. To successfully cut and open wind resistant sectional doors, specific cuts must be made.

cut the width of the door (figure 6–46E, from the two initial parallel cuts to the opposing side of the door) at approximately shoulder-to-head height, as illustrated on the metal door in figure 6–44. If necessary, balance the rotary saw on your shoulder to support its weight and simplify the operation (a carbide chain saw or rotary saw with a carbide-tipped wood blade/multi-use blade should be used on a wooden sectional door).

Remember to not cut near the joint between the sections on metal or wood doors because there are steel hinges between the sections on the reverse side of the door that will be cut, unnecessarily slowing the operation. The cut panel can now be swung outward and away from the opening, creating a large opening.

Similar to the previous section on ordinary sectional doors, be aware that sectional doors that are assisted by counterweight springs may retract when the bottom portion of the door is separated from the top portion. This operation removes some of the weight of the door; however, this has not been a significant concern because the door does not suddenly retract, and most springs lose their tension from use and time.

Cutting tilt-up doors. Tilt-up doors are single-piece doors constructed of thin metal and hinged at the middle to swing up or down, similar to a garage door. However, this door can also be extremely difficult to open due to its construction, which, in most cases, is based on the size of the door and the area of installation as follows.

As the size of the door increases, so does the size of the strengthening members on the backside of the door (figure 6–47).

Areas that are subjected to high winds normally result in increased size and number of strengthening members on the backside of this type of door. Most doors of this type cannot be cut completely through with a 14-inch rotary saw blade. Therefore, it is recommended that this door be forced by using the same techniques as described earlier in this chapter for sectional doors. When cutting this door with a rotary saw, it is important to use the largest rotary saw possible and to bury the blade as far as possible to cut the metal braces on the interior side. Additionally, if the springs and hinges on these or other garage-type doors have been exposed to heat

or fire, they may be weakened and incapable of supporting the door. Therefore, it is necessary to support any open tilt-up door, or any other overhead-type door that has been opened, with a pike pole or by using vise grips to lock a hinge so the doors do not unexpectedly close on interior personnel.

Figure 6–47. For tilt-up doors, a large rotary saw that is capable of cutting through the braces on the back side of the door must be used.

Walls

The opening of walls is often referred to as breaching. It is an operation that is generally used as a last resort due to the time and difficulty required to create an adequate opening, particularly in masonry walls. Before breaching a wall, evaluate the following:

- **Type of wall:** Brick, concrete, wood, metal.

- **Stability:** How will the opening affect the stability of the breached wall?

- **Time:** How much time will be required to complete an adequate opening?

- **Available resources:** What is the availability of personnel and equipment necessary to accomplish the operation?

- **Safety:** Prior to and during any cutting operation, remember that electrical wires may be present. Therefore, take appropriate actions to deactivate the electrical service before cutting through any wall, exterior or interior.

Breaching frame-stucco walls. Use a rotary saw with a multi-use blade or carbide-tipped wood blade to complete three cuts large enough to allow easy entry and exit of personnel. Either the carbide-tipped wood blade/multi-use blade or chain saw will cut through the stucco (and a portion of the wooden studs behind it) significantly faster than a masonry blade; moreover, they don't exhibit a tendency to bind as a masonry blade does. If the rotary saw didn't cut completely through the studs (the depth of cut is approximately 3½ inches for a 12-inch blade or 5½ inches for a 14-inch blade), use a chain saw with a carbide-tipped chain to follow in the original cuts to sever them.

Some carbide chain saws are capable of cutting through stucco and wood studs simultaneously and can be more effective than a carbide rotary saw. This is dependent on the type of chain being used. Push the cut portion into the structure (similar to figure 6–48). If this is undesirable or impossible due to storage, shelves, or other obstructions, pull the cut portion outward.

Figure 6–48. To make an opening in a metal-clad building, make three cuts with a rotary saw and push the cut portion into the building.

Breaching frame wood walls. Use a rotary saw with a multi-use blade or chain saw with a carbide-tipped chain to complete three cuts large enough to allow easy entry and exit. Most rotary saws do not cut deep enough (unless a 14-inch blade is used) to cut through the wood siding and studs. Push the cut portion into the structure (similar to figure 6–48). Again, if this is undesirable or impossible due to obstructions, pull out the cut portion.

Breaching metal walls. Sheet-metal walls can be quickly opened with a rotary saw. Complete three cuts large enough to allow easy entry and exit. If the metal is attached to wooden studs, it may be necessary to retrace the original cuts with a chain saw to sever the studs. In this case, the cut portion may be pulled outward or pushed inward (figure 6–48).

If the metal is attached to metal structural members, only cut deep enough to cut the metal siding. Remove the cut portion of metal by pulling outward.

Breaching masonry walls. Masonry walls can be composed of brick, concrete, stone, concrete block, or other similar materials, any of which can present a formidable challenge. Prior to breaching a masonry wall, consider the type of wall, the thickness of the wall, and the presence of concrete and metal rebar.

Masonry walls can be breached most effectively with a battering ram by using the forked end to strike a wall about waist high. Initially, create a small opening that can be slowly enlarged as necessary. Remember that the opening may need to be wedge-shaped (narrow at the top and wide at the bottom) to ensure stability of the wall above. Also, because this can be a lengthy and energy intensive operation, rapid progress may not be possible. Concrete walls and hollow block filled with concrete and metal rebar will require much effort to open a minimal-sized opening. Unreinforced brick masonry walls can be breached quite easily, however.

Miscellaneous obstacles

Padlocks. Padlocks are used in numerous security applications, and consist of a body and shackle (figure 6–49). The body contains a locking mechanism, which may engage one or both sides of the shackle.

A shackle engaged on one leg only is easily opened with a single cut on the shackle, which allows both shackle legs to rotate outward and release. If both legs are engaged (heel-and-toe locking), cuts on both legs are necessary to release the padlock from its hasp. It isn't possible to determine the number of cuts necessary to open a shackle until one cut is made. If the cut shackle legs won't rotate to open, another cut is necessary. For this reason, it is recommended that when cutting a padlock, cut both shackles simultaneously. Lock shackles are made of various materials. Hardened shackles are case-hardened to resist cutting by bolt cutters and other similar tools. Normally they are marked *hardened* or *case-hard*. Shackles without markings are normally softer metals that can be easily cut with bolt cutters. Some training materials advocate that shackles over ¼ inch are case hardened, while shackles under ¼ inch are non case-hardened. This is incorrect because padlocks with hardened shackles under ¼ inch can be purchased at common hardware stores. In any case, both case-hardened and non case-hardened shackles are easily cut with a rotary saw.

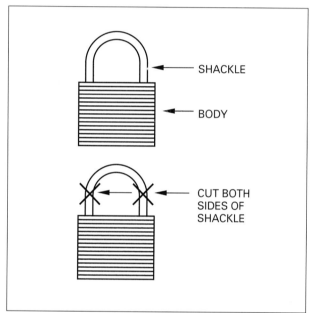

Figure 6–49. Padlocks are composed of a body and shackle. To defeat heel-and-toe locking, cut both sides of the shackle.

Forcing padlocks. Forcing padlocks is simplified by their accessibility—the entire lock is exposed. This makes padlocks vulnerable to a variety of tools and forcible entry methods, among them twisting and prying. The two most common tools used for cutting padlocks are bolt cutters and a rotary saw with a multi-use blade. If a padlock and shackle don't offer a solid surface for cutting, the lock can be secured with vise grips (by another person) while it is being cut. If a firefighter is confronted by a padlock, there are two options available for using the appropriate tool to defeat the padlock:

First, if the incident is non-emergency, then a wider array of tools can be selected as time is not a primary factor. Pipe wrenches, Halligans, flat head axes, dent pullers, bolt cutters, and other similar tools can be effective but usually require that personnel first observe the lock to determine the appropriate tool, and then retrieve and use the tool. Obviously, this takes time that may not be available at an emergency type incident.

Second, if the incident is an emergency and time is a primary factor, then the selection of tools can be minimized. Therefore, in most cases, a rotary saw with a metal multi-use blade is normally the tool of choice due to its wide range of uses unless it is necessary to deal with a unique problem.

If a hasp and mounting plate are not substantial, a prying tool can be used to force the mounting plate away from the mounting surface. The forked end of a Halligan can also be placed over a hasp or shackle and twisted until the hasp or shackle breaks (figure 6–50). When a hasp or mounting bracket is welded to a mounting surface, the weld is the weakest point in the assembly.

If confronted by a hockey puck padlock (or hidden shackle, American 2000, and so on), a rotary saw or cutting torch will be effective if a cut is made toward the top (figure 6–51).

Figure 6-50. Padlocks can often be twisted, breaking the hasp.

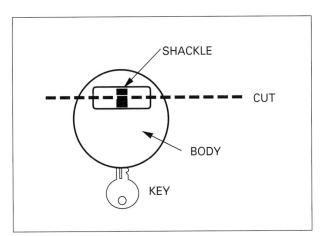

Figure 6-51. The American 2000 or hockey puck lock should be cut toward the top opposite the keyway.

Figure 6-52. When cutting chain-link fencing, cut near a vertical post and maintain tension on the fencing to increase stability.

A large pipe wrench (often, a breaker bar is necessary to provide additional leverage) can also be used to twist these locks off their mounting tabs. However, a piece of strap steel can be formed around the lock (referred to as a *tank lock*) to eliminate the possibility of using a pipe wrench. In this case, a rotary saw or cutting torch must be used to cut the lock apart.

Chain-link fences. When confronted with chain-link fences, personnel must often decide whether it is most advantageous to force entry through the fence or a gate. Emphasis should be placed on creating an adequate-size opening in the least amount of time. To cut a fence with a rotary saw and metal cutting blade, make a vertical cut close to a post. Start the cut about one foot down from the top of the fence to keep the fencing from falling onto the saw operator while the cut is completed.

While the vertical cut is completed, the addition of another person to keep tension on the portion of fence that is being cut (as illustrated in figure 6–52) will keep the saw blade from binding and also enhance the stability of the section being cut.

To complete the vertical cut, finish by cutting the portion at the top of the fence, which was omitted when the cut was initiated. Pull the cut portion of fence back to the next pole. If necessary to create an opening larger than the distance between the two poles, cut the next pole as close to the ground as possible. Pull the cut pole and all of the cut fencing back to the next stationary pole. This will normally create an opening large enough for fire apparatus to pass through (figure 6–53).

Figure 6-53. Cutting the next pole will yield a large opening.

Chain-link gates. Normally, a chain secured with a padlock is used to secure a gate or gates. In this case, use a rotary saw or bolt cutters to cut the chain or lock securing the gate. Single or double gates swing on hinges attached to stationary posts. The hinges are usually clamped to a post and are easily separated, allowing the gate to be opened. To separate a hinge, strike the bottom portion of the top hinge with an axe or similar tool (figure 6–54). This will drive the bottom portion of the hinge downward, releasing the vertical pin and gate.

The gate can now be removed from the bottom hinge and completely opened. If necessary, the gate can be closed and the hinges reattached by moving the hinge upward, re-engaging the vertical pin.

A gate that rides on tracks can also be opened by cutting the chain or lock that secures it. If this is not possible or easily accomplished, the roller guards can be bent outward (with channel locks, as in figure 6–55) at the top and bottom rollers, which will allow the rollers to be lifted off the tracks. The gate can now be swung open. If necessary, the gate can be placed back on its tracks. Gates that are opened and closed by mechanical means (usually a chain) cannot be manually opened unless the mechanical device is deactivated.

Figure 6–55. Gates that ride on tracks can be removed by bending the roller guard outward, allowing the gate to be removed from the track.

Figure 6–54. Chain-link gates can easily be opened by separating the hinges that are normally clamped on the poles.

Normally an electric motor opens and closes chain-link gates with a chain. If a single length of chain is present, it can be cut (with bolt cutters or other tools) or detached (by removing the nut on the threaded shank) where the chain is attached to the fence. However, to ensure the success of this operation, the short section of chain that is under tension from the drive motor to the fence must be released (by cutting, as shown in figure 6–56, or by removing the nut). This will allow the gate to be easily opened. If the long section of chain is cut, the fence won't open due to the drive mechanism, which introduces a gear multiplication factor as the gate is manually forced open. If a continuous loop of chain is used, it will be necessary to cut both the top and bottom loops.

Decorative security fences and gates. Similar to chain-link fences and gates, iron security (or iron decorative) fences and gates are used for a wide variety of security applications. They are found in numerous styles, but are normally recognized by construction that uses iron metal bars—flat, tubular, or square.

Decorative fences. These can often appear imposing due to the style and size of the metal bars. Most decorative fences and gates, however, use hollow bars and supporting members (cheaper), which can easily be cut with a rotary saw. If the bars are

Figure 6–56. If it is necessary to cut the drive chain for a horizontally opened gate, always cut (or disable) the short length of chain that is attached to the fence.

of solid stock, these can be cut with a rotary saw, but the cuts will take longer and a composite blade will wear faster. Either way, cut the top and bottom horizontal members next to two adjacent vertical posts. Four quick cuts will allow the entire section of vertical and horizontal bars between two adjacent posts to be removed easily. This method will open the entire area between two adjacent vertical posts.

Swinging decorative gates. Gates that are manually opened are secured by a lock or combination of chain and lock, which, if not case-hardened, can be cut by bolt cutter, or if case-hardened, a rotary saw. Gates that are mechanically opened (such as found in the entrance to driveways for some dwellings) are secured by an opening/closing arm. The arm can be cut with a rotary saw, separated from the gate by removing the nut/bolt that secures the arm to the gate, or released by the emergency release handle, if so equipped. Remember that some gates are equipped with a "yelp" option for emergency vehicles that will allow your siren to open the gate.

Rolling decorative security gates. These ride on wheels across the ground, and travel horizontally to open or close (figure 6–57). These gates are manually or mechanically operated. A manual gate that rides on wheels can be opened by cutting the chain or lock that secures it. Mechanically operated gates can be opened by cutting the drive chain (as previously described for chain-link gates) or by disabling the drive arm.

Figure 6–57. Decorative gates that ride on wheels can be opened by cutting the chain or lock securing the gate.

Pivoting decorative gates. Center-hinged decorative security gates are identified by spring-loaded hinges attached to either side, which support the gate and assist it in moving up and down (figure 6–58). The gate is operated by an electric motor that runs a chain, or by a gear-driven track that operates an arm attached to the top center portion. Because the gate is mechanically powered, it cannot be opened manually unless the arm attached to the top is released.

This is easily accomplished by removing the nut and bolt (usually a ½-inch nut) that secures the arm to the gate. The pivoting and rolling decorative security gates are common for subterranean parking under office buildings, town homes, condominiums, and so on.

Figure 6-58. Decorative gates that tilt upwards can be recognized by a spring at either side. They can be opened by removing the bolt/nut that secures the arm attached to the top-center portion of the gate.

Note that rolling and pivoting security gates can often be opened by a unique method that is based on the sensor wires imbedded in the pavement behind the gate. As a vehicle approaches the gate from the inside of the structure, it passes over sensor wires in the pavement that activate the gate, allowing the vehicle to pass through the opening. Try tossing a metal tool (for example, a Halligan) under the gate and across the sensor wire area. Usually, the gate will open. However, it is recommended that a small rope be attached to the tool in case this method does not activate the gate. In this case, *slowly* pulling the tool back across the sensor wire area will normally open the gate.

Accordion (folding) security gates. Accordion security gates are commonly used to provide security for doors and windows, particularly display windows, in commercial occupancies. Fortunately, these gates are not of heavy construction, and are easily opened after the accessible security devices or a portion of the gate has been forced or cut. Various types of locks are used to secure these gates. Locks

may vary from simple padlocks to the hockey puck lock. Depending on type, most can be forced using a rotary saw. As bolt cutters are not normally effective against case hardened locks, the rotary saw is a better first choice. Metal brackets on the gates or structure are secured together by a padlock and are easily cut with a rotary saw, bypassing the padlock mechanism. These brackets can also be broken by inserting the forked end of a Halligan over the bracket and twisting until the bracket breaks.

If it isn't possible to cut the brackets or lock, use a rotary saw to cut the gate. Cut downward and next to the wall, as illustrated in figure 6–59. When making these cuts, it is important to cut the diagonal members next to a vertical member.

Figure 6-59. If necessary, accordion gates can be completely opened by cutting the members along one side.

Cutting the diagonal members away from a vertical member can allow the member being cut to vibrate, which can cause a composite blade to bind or break. Because the metal is not substantial, cutting these members is relatively easy, and the entire gate may be opened quickly. It is also possible to cut the bar or bars supporting the security devices, which removes the locking mechanism and supporting bars.

Razor wire. Razor wire was developed to provide additional security to property and structures. Razor wire is constructed of aluminum, aluminum strengthened with high-carbon steel wire, or steel. It is inexpensive and comes in a variety of styles. Easily installed, it may or may not be supported or entwined with barbed wire. Usually it is under varying degrees of tension and is extremely effective in accomplishing its intended purpose. Regardless of the style encountered, razor wire can present a challenge to personnel considering entry, and poses a serious hazard of severe lacerations and incisions.

If necessary to provide quick access over the wire instead of underneath, throw a salvage cover or other heavily constructed material over the razor wire before attempting entry. Canvas offers the best resistance to the sharp barbs. When covered, a ground ladder or aerial device can be placed on top of the wire, or used to flatten it under the cover before allowing personnel to cross it. Hands-on operations with razor wire should be employed only when necessary or as a last resort. Razor wire made from aluminum is easily cut, and can then be separated with great caution by personnel, if absolutely necessary. Remember that razor wire may be under tension and can spring backward when cut. If razor wire is entwined with barbed wire, the barbed wire should also be cut, enabling the removal of the entire assembly. Razor wire reinforced with steel wire (which is visible) must be cut with bolt cutters. However, this type of wire can be under significant tension, which can result in the wire springing back and unraveling when cut. Cutting this wire is *not* recommended.

Training Resources

The following resources can provide additional information and/or another viewpoint for the subjects discussed in this chapter:

- Minimizing Risk in Forcible Entry Operations, Bill Gustin, *Fire Engineering*, April 2005.

- Forcing Vinyl Windows, Richard Andersen, *Fire Engineering*, September 2004.
- Tricks of the Trade, Removing HUD Window Covers, *Fire Engineering*, Daniel M. Troxell, July 2001.
- Temporary Hurricane Shutters and Firefighting Operations, Leigh T. Hollins, *Fire Engineering*, June 2005.
- Door Locks, Tougher than Ever, *Consumer Reports*, October 2005.
- Drilling Lock Cylinders, *Fire Engineering*, Mitch Brooks, September 2004.
- Forcible Entry, Using a Set of Irons, *Fire Engineering*, John Tobin, April 2005.
- Hydraulics vs. Halligans, *Fire Engineering*, Nicholas A. Martin, March 2007.
- Forcing Overhead Sectional Doors, Bill Gustin, *Fire Engineering*, November 2004.
- Burglar Bar Removal: Strategies and Tactics, Jeff Crow, April 2005.
- Forcible Entry Techniques for Private Dwelling Security Bars, *Fire Engineering*, Daniel M. Troxell, December 2000.
- The Other Side of Taxpayers, David DeStefano, *Fire Engineering*, November 2007.
- Tips for Improving Effectiveness in Forcible Entry, Bill Gustin, Part 1, Part 2, and Part 3, *Fire Engineering*, April 2008, June 2008, and August 2008.
- When Seconds Count, Cesar Fabal, *Fire Engineering*, March 2008.
- Back to Basics, Uses for Irons, Damon Tobin, *Fire Engineering*, March 2008.
- Hold the Door, Robert Rowely, *Fire Engineering*, March 2008.
- Whatever Happened to Using a Halligan Tool, Mitch Brooks, *Fire Engineering*, April 2004.
- Trimming a Window, Michael N. Ciampo, *Fire Engineering*, December 2009.

7

Elevator Extrication

Introduction

Transport: To carry or convey from one place to another; the act or a means of transportation; a system of transportation. These definitions describe an action that takes place millions of times every day, and may be accomplished by various methods such as automobiles, aircraft, trains, and ships. When methods of transportation are considered, the common elevator is usually omitted. Interestingly, this mode of transportation is possibly *the* most common means of transporting people from one place to another. Unfortunately, this simple method of transportation does not always operate the way it was intended. Elevators that are transporting people and unexpectedly cease operation always yield predictable results—trapped occupants want out, and the sooner the better.

Although the concept of elevators transporting people is relatively simple, extrication of trapped occupants can be difficult at best. Consider the following possible conditions: an elevator has stopped between floors; the loss of electrical power has resulted in a dark elevator with no ventilation; passengers may be hysterical, panicked, or experiencing a medical problem. These conditions are further complicated by the following facts:

- Elevator technology constantly changes.
- There are more than 60 elevator manufacturers.
- Each manufacturer has produced a variety of elevators.
- It is not uncommon for a single elevator to have features and special equipment offered by several manufacturers, resulting in a wide variety of elevators with unique characteristics.

Elevator Overview

Electric and hydraulic are the two types of elevators commonly in use today. A typical electric elevator (figure 7–1) consists of the car, a hoistway for the car to travel within, cables attached to the car, counterweights, vertical tracks, emergency safety brakes, and electrical equipment (winding drum and electric motor) to raise or lower the elevator.

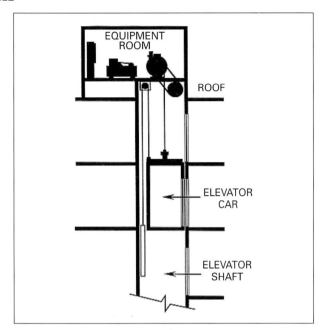

Figure 7–1. A typical electric elevator

A drum-type of electric elevator uses a winding drum that consists of a car attached to one or more cables that pass around a winding drum and to a moving counterweight. Counterweight cables are on one side of the drum and car cables are on the other side with the drum being grooved for cable movement. Both sets of cables run in the same grooves (counterweight cables unwind when car cables wind and vise versa). Operating machinery is located in either a basement or on the roof of a building. Limitations in the length and diameter of the winding drum restricts this type of elevator to 150-foot lifts, and machinery for this application requires more space than other types of elevators. This type of elevator is also very slow. For these reasons, it is no longer produced.

Traction-type electric elevators are used for higher lifts and greater speeds than drum-type electric elevators and use a traction sheave instead of a winding drum. Cables are attached to the car and pass over a traction sheave to counterweights. In this configuration, cables passing over the sheave unwind as fast as they wind, and car speed is dependent on the size of the sheave and electric motor speed. There are two types of traction machinery: geared and gearless. High-speed direct traction (or gearless type) machinery consists of a slow-speed direct-current motor directly coupled to a traction sheave with a brake wheel mounted on the motor

shaft; Geared traction machines use a high-speed motor geared to a traction sheave through worm gears with a brake wheel between the worm gears and motor. Because there is less weight and machinery with traction-type elevators, the machinery room is usually located in a roof penthouse or above an elevator shaft in high-rise buildings.

Similar in general design to the electric elevator, hydraulic elevators (figure 7–2) use hydraulic power instead of cables and sheaves, and are generally limited to six-story buildings.

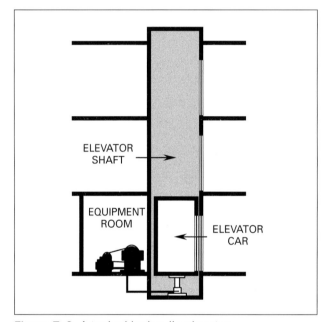

Figure 7–2. A typical hydraulic elevator

However, newer rams use a telescoping configuration that allows an elevator to serve buildings about eight to ten stories tall. In older installations, a car is raised or lowered by a water-pressure-powered ram (although newer installations use oil pressure to power a ram). Some of the less-modern hydraulic elevators may still use cables and counterweights. Controls and equipment (machine room) for a hydraulic elevator are normally found in the basement or lowest level of a building. The machine room contains the two primary components: a controller or relay panel (electronics necessary to run the elevator), and a hydraulic power unit. The hydraulic power unit consists of a reservoir for the oil supply, hydraulic pump, and valves to raise or lower the elevator car.

Doors

Modern elevators utilize two sets of doors for operation and passenger ingress/egress. The door on the car is referred to as the cab door and travels with the car. The door that is seen from the hallway of each floor of the building is referred to as the hoistway door. This door is a part of the building, and there is one on each landing. Doors can be opened or closed by electric motors, or manually for emergency incidents. Safety devices are located at each landing to prevent inadvertent door openings and to prevent a car from moving unless a door is in a closed and locked position. All types of doors are designed with a safety feature that causes a car to stop whenever a door is opened.

There are four basic types of doors used on elevators (figure 7–3):

- **Center-opening doors:** These consist of two power-operated panels that part simultaneously with a brisk, noiseless motion. A variation of this type of door is the two-speed center opening door that consist of four power-operated panels (two on either side of center) that part simultaneously with the doors that are closest to the center moving twice as fast outward as the adjacent door.

- **Two-speed doors:** These consist of two power-operated panels that are geared together. One door moves twice as fast as the other door so that both doors will meet concurrently in the open position.

- **Single-slide doors:** These are power-operated, single-panel doors.

- **Swing-hall doors:** These are manually opened with a single-slide car door, power-operated or collapsible gate.

One other type of door is the vertical bi-parting door. It consists of an upper and lower section that part horizontally in the middle as the upper portion raises upward and the lower portion travels downward. These doors can be found in some freight elevators and are generally not used for passenger elevators unless the residence is in a warehouse.

Elevator doors are normally opened by a power unit that is located on top of a car. When an elevator car is at or within a few inches of floor level (this area is often referred to as the door lock zone), the power unit moves the car door open or closed. A pick-up arm (clutch, vane, bayonet, or cam) contacts rollers on the hoistway door which releases the door latch on the hoistway door, opening the door. The door rollers and pick-up arm may be different on various elevators but they all work on the same principle.

Door restrictors have been installed on elevators since 1980 (although some older elevators have been retrofitted with these restrictors) that prevent the cab doors from being opened more than four inches when an elevator car in not within its landing zone (18 inches below or above the floor landing). The two basic types consist of a mechanical (clutch-type or angle iron) or electromechanical restrictor (electrical solenoid). For more information on these safety devices, see the "Mechanical Elevator Door Restrictor" article in Training Resources.

Brakes

Elevator brakes are similar to automotive brakes and are provided for normal operation. A centrifugal-force governor is provided on most elevators to guard against overspeeding when a car travels in excess of 20% of top speed. If this occurs, the governor will activate a governor safety stop device. Safeties are installed at the bottom of a car and

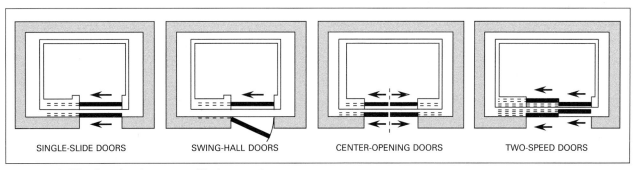

SINGLE-SLIDE DOORS   SWING-HALL DOORS   CENTER-OPENING DOORS   TWO-SPEED DOORS

Figure 7–3. The four basic types of hoistway doors

occasionally on counterweights to provide positive emergency stopping when activated by the governor. If necessary, abrupt stops are possible with this equipment. On elevators capable of high speeds, wedges are forced between two jaws gripping the vertical rail with increasing force to slow a runaway car to a gradual stop.

Controls

Generally, a fire service control is required on elevators constructed after 1973. The control, commonly referred to as the "firefighter's key," should not be a part of a building master key system, but should be keyed for all elevators within a building, and not attached to other building keys. The fire service control should operate as follows:

Phase 1. This is sometimes known as the "firefighter's emergency recall." Inserting the firefighter's key into the lobby key switch and turning to the ON position will place all elevators in Phase 1 mode. This recalls the elevators to the lobby (with or without passengers), non-stop. Upon arrival, the elevator doors will open and the elevator(s) will remain as long as the switch is in the On position. While in Phase 1 mode, the key can be removed and used inside the elevator car to operate the elevator (see Phase 2 mode). Remember that some elevators can be placed into Phase 1 mode by activation of heat or smoke detectors in the building.

Phase 2. Phase 2 can be activated after the implementation of Phase 1 by removing the firefighter's key from the lobby key switch, placing the key into the firefighter's service switch inside the elevator car, and turning to the On position. When in Phase 2 operation, the lobby key cannot override it. The elevator can now be manually operated by selecting an appropriate floor button and depressing the Door Close button until the door is completely closed. If the Door Close button is released before the door is completely closed, the door will automatically re-open.

When the elevator arrives at the selected floor, press the Door Open button to open the doors. The button needs to be continually depressed until the doors are completely open. To hold the elevator at the floor where personnel exited the elevator, turn the key to the Hold position.

Phase 3. The Phase 3 mode is linked with the fire/smoke alarm system (if present). If a smoke detector is activated in the lobby portion of a building, the elevator cars will not return to the lobby but will automatically be recalled to a predesignated floor. To access the elevators, personnel will have to go to the predesignated floor.

Applications

Traction and hydraulic elevators are used as passenger or freight elevators. The following sections provide information about both.

Passenger elevators

Passenger elevators are designed to quickly move passengers to different levels within a multi-story building. Modern passenger elevators are completely automated and are under the control of an electronic computer, which constantly evaluates the needs and demands of the system. Computers constantly make adjustments, move cars, and work to meet current demands. Cars are sent to the area of greatest need and when cars are not needed, they are allowed to rest or sleep.

Freight elevators

Freight elevators are generally less complicated and serve a different purpose than passenger elevators. Larger than passenger elevators, freight elevators can be as large as 12 by 14 feet and have a carrying capacity of up to three tons. Generally, freight elevators are separate from the main lobby of a building and can have a street or alley access. Not being under computer control, freight elevators are more simple to operate and control than passenger elevators, and normally service an entire building from the lowest to highest level. Freight elevators stop at all floors (up and down) unless a special over-ride control is activated, and normally are slower than passenger elevators.

Rescue Procedures

Before discussing any kind of procedures, understand that it is imperative for firefighters to spend appropriate time talking with their local elevator companies and/or persons who respond to an elevator emergency. Responding firefighters need to develop a working knowledge of elevators and options that are available for elevators within a specific area. Additionally, unless firefighters practice extrication procedures and/or firefighter operation with the elevators in their district (weekends are best), then the operation and/or extrication of trapped persons in an elevator can be an unsafe and futile operation.

Prior to initiating elevator rescue procedures, dispatch and on-scene information should be carefully evaluated. When an alarm is received, a knowledge of the type and height of occupancies in the geographical area of dispatch will indicate the probable type of elevator (hydraulic or electric), the applications (industrial/commercial, apartments, multi-story office buildings, high rise, and so on), and the potential of trapped victims (depending on time of day). On-scene information should consist of the following *minimum* considerations:

- Determine if there is an inoperative elevator.

- Determine if the inoperative elevator contains trapped occupants; if so, determine their condition.

- Determine if an elevator repair person has been notified, and if so, what their estimated time of arrival (ETA) is. If a repair person has not been notified, determine who has the repair contract, and notify an appropriate repair person (if necessary). This information can normally be found in the elevator equipment room. Remember that normally an ETA for a repair person is usually a "best guess." If a large area has suffered a power failure (or other similar problem), smaller customers are often a last priority, and the length of ETA can affect the type of extrication if one is necessary.

- Determine the location of the inoperative elevator (between floors or at a landing).

- Determine the type of elevator (hydraulic or electric).

- Determine the location of the elevator machinery room (normally, above an electric elevator and below a hydraulic elevator).

When it has been determined there is an inoperative elevator with trapped occupants, extrication procedures should focus on using the simple solutions first and working towards the more difficult operations. The following considerations should be initially implemented:

- Position personnel equipped with portable radios (set to a specific tactical channel for this incident) at the following locations:

  - Elevator equipment power room. This provides access to the main line disconnect switches and elevator equipment.

  - Floor where elevator is inoperable. Elevator extrication procedures will be coordinated from this location.

- Establish voice contact (either through the elevator doors and/or the emergency telephone within the elevator if present) with the trapped elevator passengers.

- Determine if trapped passengers are injured, in need of medical assistance, or if conditions are stable. Do not extricate passengers using forcible entry methods unless they are in need of *immediate* medical attention, and/or an ETA is unreasonable. Forcible entry methods can be dangerous to passengers, fire service personnel, and cause unnecessary damage to an elevator.

- If conditions are stable, inform the passengers they are safe and will be removed from the inoperative elevator in a short period of time. It is also advantageous to constantly keep trapped passengers informed of operations that are being utilized to remove them or return an inoperative elevator to normal operation.

- If the alarm/emergency bell has been activated, have the trapped passengers silence the alarm because the noise is not conductive to a smooth extrication.

- Have the trapped passengers verify the status of the emergency stop switch in the elevator. Remember that if an elevator is stalled due to a malfunction that can be quickly corrected (overheated relay, loss of power from an activated main line disconnect switch, and so on), it is necessary for the emergency stop switch to be in the normal/run position before power will be returned to an inoperative elevator.

- Instruct the passengers to push on the cab door to ensure it is closed.

- Instruct the passengers to push the Door Open button (if so equipped). It can be beneficial for firefighters to simultaneously push the floor button in the hallway.

- Instruct the passengers to attempt to restart the car by using the car controls.

- If the elevator car is within a few inches of the landing (door lock zone), instruct the passengers to try to manually open the cab door. This may require some effort because the cab door operates the hoistway door through a clutch mechanism. However, cab doors are designed to move with about ten pounds of force. Moving the cab door should release the latch on the hoistway door and allow both doors to be opened.

- If the elevator is equipped with fire service control, use the firefighter's key and attempt to recall the elevator to the lobby.

Personnel in the machinery room should:

- Check the electrical circuits (main line disconnect switch, fuses, and so on) to verify if power is on or off to the inoperative elevator. Occasionally, circuits are tripped due to overheating and can be safely reset. It is not recommended that personnel replace fused breakers and/or fuses because they are not electricians.

- If there is power to the elevator, turn the power off for at least 30 seconds and then turn back on again. This can reactivate an inoperative elevator by allowing relays to reset (at least 30 seconds is necessary to clear any previous programming in the elevator computer). It can be advantageous to repeat this process for a total of three

times. If the car is within a few inches of the landing, the door may be able to be opened.

- An elevator is normally supplied with two separate power sources: 480 volts supplies the motors, and 110 volts supplies the lights, ventilation, communication, and alarm system. It is recommended that the main power/switch be shut off. Evaluate shutting off the 110 volts because this will eliminate the lights and ventilations system in the elevator.

- If there is a Door Open button in the elevator, instruct the passengers to again push this button to see if this opens the doors. If this step fails, deactivate the main line disconnect switch and try to manually open the doors.

- If an elevator is equipped with a recall system, a key can be inserted into the recall key slot; when activated, the elevator should return to the ground floor and open its doors. If the elevator returns to the ground floor but does not open its doors, instruct the passengers to push the Door Open button (if so equipped).

If the aforementioned operations are not successful, the following considerations will overview various operations that can be utilized to remove trapped passengers from an inoperative elevator. Selection and use of a particular operation should be based on the needs and conditions of each incident and the expertise of personnel.

Do not attempt to rescue trapped passengers from an inoperative elevator unless power to the elevator has been disconnected and all lock-out/tag-out procedures have been implemented. If possible, a person should be assigned to the main line power disconnect switch until an extrication is completed.

Hoistway access opening

Depending on the codes and ordinances of a specific area, some elevator hoistway doors are equipped with a hoistway access opening (located at the upper right portion of the right door in figure 7–4).

If so equipped, a lunar key (obtained from a responsible elevator company) can be inserted into this opening, which allows the key to engage a rod or paddle on the backside of the hoistway door. Engaging the rod/paddle with the lunar key will deactivate the locking mechanism and allow the hoistway door and then the cab door to be manually opened. This operation can be used if the elevator car is within or out of the door lock zone.

Figure 7–4. Hoistway access opening in upper right portion in right door

Figure 7–5. Using a small air bag to break the door interlocks

Air lifting bag (center opening doors)

Air bags can be used to force open center opening doors by inserting a Halligan, crow bar, or other similar tool at the top portion of the doors and forcing the two doors apart until a small air bag can be inserted. Instruct the passengers to move to the back of the car, face the rear of the car, and ensure the electrical power to the car has been de-energized. Slowly inflate the air bag (figure 7–5) until the doors are forced open. This operation will break the interlocks at the top of the doors and cause little or no damage to the doors. The broken interlocks are easily repaired.

Note that a rabbit tool can also be used to spread the doors in the same manner as air bags; however, there can be some damage to the doors depending on the use of the tool.

Moving an elevator

As a moving elevator approaches a floor landing, a pick-up arm on the elevator car will engage rollers on the hoistway door(s) (upper right portion of the right door in figure 7–6) and allow the hoistway door(s) to be opened.

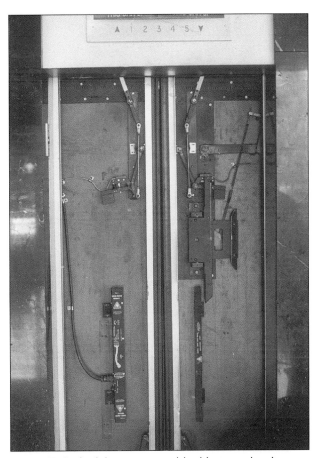

Figure 7–6. A pick-up arm and locking mechanism on the cab door

Depending on the manufacturer and type of elevator, the hoistway doors will open if the elevator is within the door lock zone. This can be a few inches or up to approximately 18 inches above or below the floor level. Normally, the door lock zone is designated as a few inches above the landing (floor level).

The major difference in rescuing trapped passengers from hydraulic or electric elevators is the method used to lower the elevator to floor level or the door lock zone and allow the pick-up arm to engage the rollers on hoistway door(s). The cab and hoistway doors can then be opened and can be accomplished as follows.

Hydraulic elevators. The following two methods for moving an elevator are termed a standard operating procedure (SOP) in some texts and departments, but as operations "that should not be used" by selected texts. Although this author has personally and successfully used these methods for years, the categorical statement they should not be used should be weighed against the training/expertise of personnel, the need to quickly extricate trapped persons in an elevator, and an inordinate ETA for an elevator repair person, which may needlessly put a company out of service until an elevator repair person is on scene:

1. De-energize all electrical power to the inoperative elevator. Direct the passengers to move to the rear of the car and keep the cab (inner doors) closed.

2. Locate the bleeder (lowering) valve. The lowering valves are located on the hydraulic power unit. There are generally three valves on this unit—one to raise the car, one to lower the car, and one that will level the car to a floor landing. Most valves have a manual lowering valve built into the main valve body that consists of a T-handle, knurled nut (in figure 7–7, the knurled nut must be removed to access the manual lowering screw on this EECO DL-3 main valve body), screw (slot head) with a locking nut, allen-headed screw, and older installations have a ¼- or ½-inch globe valve.

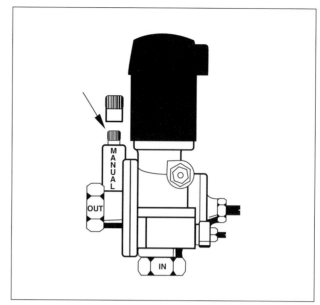

Figure 7–7. A knurled nut or cap must be removed to access the manual lowering valve on this EECO main valve body.

3. Most valves are marked ML, MAN, or MANUAL near the lowering valve. No valve should be fully opened. Slowly open just enough to hear fluid flowing (figure 7–8) and the car will slowly lower.

Figure 7–8. Opening the lowering valve to lower a car to the floor landing

4. When the car is at or within several inches of the floor landing, discontinue bleeding the hydraulic fluid by rotating the valve clockwise until the valve is seated. This will stop the car.

5. Direct the passengers to manually open the cab door. This will release the hoistway door and allow the cab and hoistway doors to be opened.

6. When the trapped occupants are removed, secure the elevator by closing the cab and hoistway doors. Shut off *all* power to the elevator and notify a responsible person that the elevator cannot be placed back in service until repaired by a qualified person.

Remember that some elevators require the electric power to be restored for the doors to be opened and that the hoistway door may not automatically open. In this case, it may be necessary to manually open the hoistway door.

Occasionally the mechanism on the cab door that engages the hoistway door to facilitate its opening fails. This leaves the cab door open and the hoistway door closed. If this condition occurs, a simple solution is to have the occupants of the stalled elevator operate the opening mechanism on the hoistway door.

Electric elevators. When operating in this type of elevator machinery room, do not step on the smoke ventilation grating over the shaft because it may not be designed for your weight. Have two personnel in the equipment room; they should be equipped with two pry tools and a pipe wrench or large channel locks. One person should be on the floor where the passengers will be removed.

1. De-energize the electric power to the inoperative elevator. Direct the passengers to move to the rear of the cab and keep the cab door closed.

2. Locate the friction brake on either side of the shaft (figure 7–9) and determine if the brake moves in or out.

Figure 7–9. Locating the friction brake

3. Place a pry tool between the brake caliper and spring, one on either side of the brake. Working in unison with personnel on each pry tool, slowly and methodically release the tension on the brake. After this is accomplished, rotate the shaft with a pipe wrench or channel locks (figure 7–10). While rotating the shaft, verify with personnel stationed at the inoperative car location that the car is moving in the desired direction (up or down). If not, then rotate the shaft in the opposite direction.

Figure 7–10. With the friction brake disengaged, rotate the drive shaft to raise or lower the car to the floor landing.

Due to counter weighting and the number of passengers in the car, the car *may slowly* begin to move upward when the friction brake is released (this is easily controlled by pressure exerted on the friction brake), and it is generally easier to move the car upward.

4. When the car is at or within several inches of the floor landing, discontinue rotation of the shaft. This will stop the car. Set the friction brake by releasing pressure.

5. Direct passengers to manually open the cab door. This will release the hoistway door lock and allow the cab and hoistway doors to open. Remember that some elevators require the electric power to be restored for the doors to be opened and that the hoistway door may not automatically open. In this case, it may be necessary to manually open the hoistway door.

6. Secure the elevator by closing the cab and hoistway doors. Shut off *all* power to the elevator and notify a responsible person that the elevator cannot be placed back in service until repaired by a qualified person.

Gib blocks

Gib blocks are attached to the bottom of hoistway doors, ride in tracks that are in the floor landing, and are installed about every 18 inches across the underside of a door panel (figure 7–11). They are constructed from various plastic or phenolic materials and provide stability to the hoistway door as it opens and closes. Normal hoistway doors can be expected to utilize two gib blocks. Personnel can gain entry into an inoperative elevator by removing the gib blocks as follows.

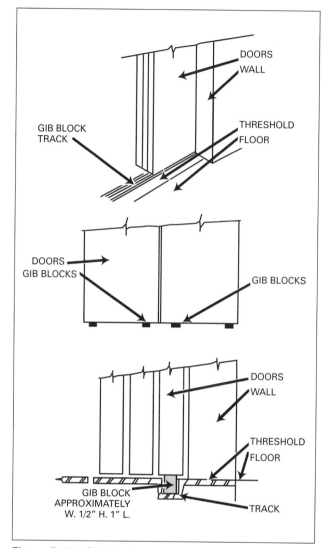

Figure 7–11. Gib blocks

Position personnel in the elevator equipment room, the floor landing where the elevator is inoperative, and the floor landing above the inoperative elevator. If the elevator is between floors or is not even with the floor landing, the elevator may be lowered to line up with the floor landing. If the elevator is lowered to the floor landing and the elevator doors will not open, continue with the following steps; however, note that this extrication method cannot be utilized when the inoperative elevator is located at the top floor of a building or between floors, and cannot be moved:

1. Shut off the power to the elevator.

2. On the floor above the elevator, cut the gib blocks on one of the hoistway doors or the door if only one door is encountered.

This is easily accomplished by using tools such as a plumbers saw, hacksaw, reciprocating saw, air chisel, and so on. If the gap between the bottom of the hoistway door and floor is insufficient, the door can easily be pried upward to provide the necessary clearance.

3. Utilizing the top of the hoistway door as a hinge, the bottom of the door may now be *slowly* pushed into the elevator shaft, allowing personnel to look down at the inoperative hoistway door linkage/locking mechanism, allowing personnel to use a poling tool, pike pole, or other similar tool to release the locking mechanism (figure 7–12). This will release the locking mechanism and allow the inoperative hoistway doors to be easily opened. The cab door(s) may then be opened by the passengers or rescue personnel, freeing the trapped passengers.

Figure 7–12. Use a pike pole or poling tool to release the door-locking mechanism.

When the gib blocks are cut, the hoistway door can dislodge from its upper track if it is pushed too far into the shaft. However, this must be deliberately done to dislodge the door. A good rule of thumb is to not push the door more than 18 inches into the elevator shaft.

Elevator power is not restored until the elevator is restored to proper working condition. Further, ensure that the hoistway door with the cut gib blocks is appropriately posted. To restore the hoistway door, have elevator personnel replace the gib blocks (several dollars each) that have been cut and put the door back on its track.

Adjacent car rescue (common shafts)

When an inoperative elevator shares a common shaft with an operational elevator, it is possible to affect a rescue from an adjacent elevator. Although this is a viable operation, it is dangerous and should only be used as a last resort:

1. Take control of all elevators in the common shaft. De-energize electrical power to the inoperative elevator.

2. Position an operational elevator adjacent to the inoperative elevator, and deactivate the emergency stop button in both elevators.

3. Open the emergency exit in the roof of the operational elevator, and using a short ladder, allow personnel to climb through the emergency exit to the roof of the elevator. The top emergency exit panels are secured by thumb screws and are normally arranged so the cover can only be opened from outside of the elevator. (This means the elevator may need to be accessed from the top). The cover opens outward and the space above the cover is unobstructed by elevator equipment. Frequently, a rope or chain ladder is provided in a metal box on top of the elevator. One end of the ladder is fastened to the elevator structure and can be lowered through the exit opening after the cover is removed.

4. Personnel may now step over to the inoperative elevator and open the emergency exit. Using a short ladder, enter the car and remove/assist the trapped passengers from the inoperative elevator. Carefully assist and direct the passengers to the roof of the operational elevator. Enter the roof emergency exit to the interior of the operational elevator.

Alternatively, you can also perform the following:

1. Take control of all elevators in the common shaft. De-energize electrical power to the inoperative elevator.

2. Position an operational elevator adjacent to the inoperative elevator and deactivate the emergency stop button in both elevators.

3. Open the emergency side exit in the operational elevator. This will allow personnel to open the emergency side exit in the stalled elevator. Side emergency exits are usually at least 16 inches wide and 5 feet high, are located so there is free access to the side exit of an adjacent car, will open inward, and are either hinged or removable. The removable emergency side exit panel is held in place by at least four fasteners, so arranged that they can be operated by hand from both the inside and outside of an elevator. The hinged emergency exit panel is provided with a lock arrangement that can be operated from the inside of an elevator by means of a removable key and from the outside by means of a non-removable handle. (The key is kept on the premises by the person responsible for the maintenance and operation of the elevator.)

4. After the side emergency exits have been opened in the inoperative and operational elevators, a short plank (such as a ladder) should be placed through the exits (between the elevators). Trapped passengers can now be removed from the inoperative elevator to the operational elevator.

Breaching

Occasionally, an elevator car may be stalled in a blind or express hoistway (any enclosed shaft of three or more floors), and as a final resort it is necessary to breach a shaft wall. Elevators of more sophisticated design have access panels from blind floors (floors without hoistway doors). These panels may be as close as every three stories or as far apart as 80 feet. Panels can be helpful, but should not be depended upon because they may be hidden or covered.

After all primary means of extrication have been exhausted or extrication is immediately required due to serious injury or imminent danger to passengers, wall breaching should be considered. Remember, breaching requires personnel, labor, time, property damage, planning and coordination. Shaft panels are on the side of the hoistway where the car doors are located. These panels can be removed and used in several ways.

1. Locate the car by looking into the shaft through a panel. If close enough, the car doors can be forced open and passengers removed through the panel.

2. Panels can permit entry into a shaft and access to the top of a car escape hatch. If there are no shaft panels and it is necessary to breach the hoistway wall, the breach opening(s) should be made on the elevator door side of the shaft and as near the door as possible. The wall opening should be large enough to expose the entire elevator door opening. After shaft entry has been made, car doors can be opened or top of car entry made. If it is necessary to use a ladder from the wall opening to the top or side of an elevator, life lines should be considered for use on rescue personnel.

Other Elevator Emergencies

Not all elevator emergencies involve persons trapped in a stalled or inoperable elevator. Consider the following scenarios.

Fire in the machinery room

When a fire occurs in an elevator machinery room, the probable seat of the fire will be the driving motor. Disconnecting the main line and auxiliary power switches will stop car movement and de-energize the driving motor. Whenever possible, avoid the use of water on elevator machinery. Electrical fire can be extinguished with dry-chemical extinguishers. The removal of trapped passengers from the de-energized elevator should be completed in as routine a manner as possible.

Fire in an elevator shaft

Although there is little to burn in an elevator hoistway, there is the presence of grease and lint. However, the main source of fuel is debris in the pit. The principal hazard of any fire in an elevator hoistway is the smoke created by the fire. Smoke can easily rise to an elevator with passengers and cause a degree of panic. Although some electrical equipment is quite sturdy, some equipment is sensitive to smoke and its by-products, and smoke can extend to other levels in a multi-story building.

All elevators in a hoistway with fire should be stopped and evacuated. If the fire is under a car, the car should be brought to the lowest landing possible. If the fire is in the upper portion of a hoistway, the car(s) should also be stopped and evacuated and the shaft opened as near the fire as possible. Remember, the use of water should be avoided or used sparingly.

Earthquakes

As a result of the earthquake in Sylmar, California, in 1971, over 1,000 elevators were put out of service due to the initial shock and aftershocks. The timing of the earthquake has been credited with saving many lives. Had the tremor occurred at a later hour, the number of people killed or injured would have increased. The primary causes of stalled elevators were power outages, counterweights pulling out of their guide rails, counterweights striking cars, roller guides breaking, damaged cables, damaged machine room equipment, and binding doors. In the event of an earthquake, all elevators should be stopped with the main line disconnect switches and all equipment should be checked for damage prior to attempting to remove passengers. If the building has auxiliary power to the elevators, it should also be shut down. Remember to consider the possibility of aftershocks.

Blackouts

During the 1965 East Coast power outage, thousands of elevators were stalled. The greatest problem to trapped passengers was panic in dark inoperative elevator cars. When there is a sudden loss of power, all power switches should be placed in the off position and passengers extricated with appropriate measures. Some buildings have auxiliary power that can be used to move elevators until normal power is re-applied. If necessary, it may be necessary to force doors, or breach walls if an elevator is stalled in a blind section of a shaft that is not near another elevator.

Escalators

Escalators are a simple and common means of transportation consisting of steps, a driving motor, and gearing that combine to form a continuous belt or track running on a set of gears. An escalator traveling between 90 to 120 feet per minute can carry as many as 5,000 passengers per hour. Each escalator in a building (that has multiple escalators) is an individual installation with separate machinery and controls, including a Stop button that may be located externally and directly adjacent to the top and bottom of each unit (figure 7–13), or below the cover plate at the demarcation line. Stop buttons are provided for emergency operation, and an escalator cannot be restarted without the use of a key-operated switch located in a covered compartment at the bottom of an escalator.

Figure 7–13. External escalator stop button

Rescue procedures

With modern engineering techniques and designs, escalator rescues have almost become a rarity. However, all escalators are not modern and there are occasional emergencies that generally consist of children inserting their fingers and toes between the stop treads and guard plates, types of shoes that are capable of wedging between the treads, and trapped fingers due to persons attempting to pick up dropped articles. Generally, most injuries are not serious. As with most children, fear is the greatest problem to overcome. With adults, it is embarrassment.

If necessary, a moving escalator can be quickly stopped by activating the Stop button. By clearing the escalator of all passengers, the treads can be moved backwards by hand pressure. Some older installations may require the use of a Halligan or other similar tool to push the treads backwards. When pushed backwards, trapped fingers or toes can be easily removed and the victim removed from the escalator or demarcation step line. Because of victim position or the extent of injuries, it may be necessary to remove the cover plate that covers the step treads at the demarcation line. When plate removal is necessary, the plate screws should be completely removed from the plate and the plate lifted straight up rather than backwards or away from the victim. Following removal of the victim, the cover plate should be replaced. The escalator may then be restarted by pushing the start button located next to the stop button. If any part of the escalator has been damaged, it is advisable to place the unit out of service in addition to the key operated switch as previously mentioned. In this case, the escalator becomes an ordinary staircase.

Training Resources

The following resources can provide additional information and/or another viewpoint for the subjects discussed in this chapter:

- The pamphlet "Emergency Removal Of Passengers From Elevators" is available from the Otis Elevator Company by writing on your letterhead to: Otis Elevator Company, General Service Superintendent, 260 11th Ave., New York, N.Y. 10001.

- Don't Get Shafted, A New Fail-Safe Elevator Rescue Method, Alan Cass, The California Fire Service, October 2002.

- Elevators 101, The Use of Elevators at High-Rise Fires, Doug Leihbacher, *Fire Engineering*, January 2003.

- Elevator Rescue, Gary Siegel, *Fire Engineering*, November 1998.

- Using the Elevator at Structure Fires, Roundtable, *Fire Engineering*, May 1999

- Mechanical Elevator Door Restrictors: What Firefighters Need to Know, Bill Gustin, *Fire Engineering*, August 2003.

- Letters to the Editor, Do Not Take Elevators That Go to the Fire Floor, John Mittendorf, *Fire Engineering*, January 2008.

- A Close Call Encounter of the Elevator Kind, Michael A. Dragonetti, *Fire Engineering*, December 2007.

- High-Rise Firefighting Perils: Veterans Perspectives, Jeff Crow, *Fire Engineering*, October 2008.

- Safety Code for Elevators And Escalators, ASME A17.1.

Ground Ladders

Introduction

Although the sight of an apparatus responding to an emergency with a 100-foot aerial ladder, 104-foot platform, or a 120-foot articulating boom with a platform is an impressive sight, each of these aerial devices can have significant operational limitations depending on their ability to access a specific application. As a result, when aerial devices cannot be utilized, the most effective, versatile, and portable tool available to the fire service that can provide above-ground access are ground ladders. Ground ladders have the capability to enhance fireground operations and are also being used for numerous other emergency incidents such as confined space operations, bridging, trench rescues (figure 8–1), and so on.

Figure 8–1. Ground ladders can be used for numerous operations other than laddering structures. (Photo courtesy of Tim Dungan)

In the science of above-ground fire fighting, effective fireground operations are often dependent on the timely and adequate placement of ladders to facilitate a large number of considerations. These can include rescue operations, ventilation operations, initiating above-ground hose lines, access to upper floors and inaccessible locations, access to potential exposures, emergency egress from a structure, and salvage operations.

Many fire officials believe (and statistics confirm) that the loss of life and/or inordinate dollar loss fires are a direct result of inadequate and ineffective ladder-use operations. It is a fact that ground ladders can set the standard for firefighting operations and, at times, ineffective ladder operations can even negate successful fireground operations. This is easily demonstrated by the fact that above-ground structural fires can require different access challenges than fires located at grade level, and must be quickly accessed by suppression personnel before a search can be conducted and/or a fire can be extinguished. Additionally, the modern fireground has also underscored the importance of safety and timely implementation of ground ladders from four perspectives:

- Modern lightweight construction does not give the same amount of time (when exposed to fire) before collapse, so the amount of time to implement ladders that are capable of enhancing fireground operations has been reduced.

- The modern thermal layer is primarily a result of the products of combustion from synthetic materials and is capable of burning two to three times hotter and faster than the thermal layer from conventional materials. This has resulted in the need to quickly place ladders to above-ground fires to allow trapped occupants a means of known or potential egress.

- The burning rate of the aforementioned thermal layer has also resulted in a common fireground problem—flashover—that in some cases has increased the frequency of interior fireground personnel needing to quickly exit the interior of a structure through any available window.

- Many areas of this country now include townhome/apartment/condo complexes that depending on the layout of the complex, can only be accessed by portable ground ladders because access to the interior structures within a complex is limited, or can best be described as very challenging. Therefore, the importance of ladders at structure fires cannot be overemphasized.

Ladder Selection

Ladder selection consists of two interrelated considerations: initial purchase and ladder characteristics.

Initial purchase

Before any ladder is spotted and then placed to an objective, the needs of a particular area and type of ladders necessary for any specific application must be thoroughly analyzed and wisely purchased. Remember, the initial purchase of ground ladders (which normally coincides with the purchase of new apparatus) will often affect fireground operations 15 to 20 years later. Therefore, taking the necessary time to thoroughly scrutinize apparatus constraints and the needs of a district to determine the most applicable portable ladders can result in significant long-term dividends.

Complement. As an initial starting point, NFPA 1904, 6-1.1 recommends that truck companies carry a *minimum* of 115-feet of ground ladders as follows:

- 1 10-foot folding attic ladder
- 2 16-foot roof ladders
- 1 14-foot combination ladder
- 1 24-foot extension ladder
- 1 35-foot extension ladder

This basic complement will provide the ability to:

- Access attics from the interior of structures (10-foot folding ladder and 14-foot combination ladders).

- Enhance ventilation/overhaul operations on peaked roofs (16-foot roof ladder).

Remember, 10-foot folding ladders, 14-foot combination ladders, and 16-foot roof ladders can also be used as a straight ground ladder if necessary.

- Provide access to the roofs of two and three story buildings and a minimum of two ladders to the roofs of two story buildings (24- and 35-foot extension ladders).

However, note the preceding complement is marginal when applied to three-story buildings. Therefore, the needs of an area that are served by a truck company may alter or add to the basic NFPA complement. Consider the following examples:

- Twelve-foot roof ladders are much easier to implement than 16-foot roof ladders on the roofs of most single-family dwellings. Therefore, the presence of numerous single-family dwellings may require the addition of a 12-foot roof ladder.

- Sixteen-foot straight ladders require less personnel and are easier to implement than 24-foot extension ladders to the roofs of most single-story, single-family dwellings. (Remember: There are more single-family dwelling type structures than any other structure.) Again, the presence of numerous single-family dwellings may require the addition of a 16-foot straight ladder in addition to a 24-foot extension ladder.

- Four-story structures that are not accessible by an aerial ladder may require a 40- to 50-foot extension ladder in addition to a 24- and 35-foot extension ladder.

Extension and straight ladders. When comparing extension ladders to straight ladders:

- An extension ladder can be adjusted to a specific length and also be used as a replacement for shorter ladders (for example, a 35-foot ladder can replace 16-, 20-, 24-, and 28-foot ladders).

- Some extension ladders can require additional personnel to compensate for the additional weight and necessity of extension operations as compared to straight ladders of comparable length. For example, a 24-foot metal extension ladder weighs 110 to 142 pounds and requires two personnel for implementation. A 24-foot metal straight ladder weighs 62 to 76 pounds and can be raised by one person. Straight ladders are normally lighter than extension ladders and can require less personnel and time for implementation.

- Extension ladders can increase the danger to personnel due to extending a ladder in the vertical position (overhead obstructions).

- More straight ladders must be purchased to equal the adjustable length of extension ladders. Additional ladders will also require more space on apparatus.

- Rung spacing on extension and straight ladders can vary from 12- to 14-inches. A 12-inch spacing gives more climbing power, and a 14-inch spacing can provide more freedom when using a leg lock (with full PPE) to work from a ladder.

- Minimal apparatus space constraints can place an emphasis on purchasing several extension ladders to provide the capability of additional shorter ladders. However, the addition of selected straight ladders (in addition to extension ladders) can increase flexibility and require less personnel for implementation.

In summary, ground ladder complements should be easy to use, require minimal staffing to place into operation, and reflect the specific needs of an area, present and future.

Construction. Currently, ground ladders are constructed from metal, wood, and fiberglass. Metal ladders:

- Are good conductors of heat, cold, and electricity

- Are easily repairable

- Come in an extensive variety of sizes and configurations

- Can suddenly fail when exposed to heat or flame over 300°F (loss of annealing); however, loss of annealing can begin at 200°F.

- Are cost effective when compared to wood ladders

The most common ladders are constructed from aluminum. The next most common ladders are constructed from wood. Wood ladders:

- Cost more than aluminum or fiberglass
- Can be exposed to heat and flame and retain strength
- Offer good durability
- May require refinishing of damaged finish, depending on frequency of use
- Are not willing conductors of electricity, heat, and cold
- Cannot permanently bend

The least common ladders are constructed from fiberglass. Fiberglass ladders:

- Are not willing conductors of electricity, heat, and cold
- Are moderately priced
- May chip and crack with impact forces
- May suddenly crack and fail when overloaded
- Can burn when exposed to flame

Because metal, wood, and fiberglass ladders are constructed to meet the same standards (NFPA 1931), selection is often a combination of personal preference and tradition.

Ladder characteristics

There are several distinct characteristics that should be considered before selecting a ground ladder: style, cost, weight, maintenance, and safety.

Style. When reviewing the various fire service ladder catalogs, it immediately becomes obvious that there are numerous styles of ladders—pumper, truss and solid beam, tubular rail, attic, folding, combination, Fresno, and so on. Therefore, it is important to research the advantages and disadvantages of each style of ladder and what attributes will best fit on a particular apparatus. Also be aware of how easily a specific ladder is accessible to personnel, and how it can best serve the needs of a particular district. As a general rule, truss ladders have an open side rail between the rungs, can be lighter and more wind resistant, but may be more expensive than solid beam ladders. Solid beam ladders can have a lower profile and reduced overall cost when compared to truss ladders.

When selecting a particular ladder that comes in different configurations (truss versus solid beam, for example) and results in different weights, the lighter-weight ladder may not carry the same weight load as the heavier ladder, and/or the lighter ladder will exhibit more flex when being used.

Cost. Wood ladders are the most expensive, aluminum ladders are the least expensive, and fiberglass ladders can be similar in cost to aluminum ladders.

Weight. The common perception that aluminum ladders are lighter in weight than currently available truss wood and fiberglass ladders should be re-evaluated when reviewing the following chart assembled from 2009 manufacturers catalogs. Weights will vary depending on the style and manufacturer of a specific ladder:

Ladders	Wood	Metal	Fiberglass
16' straight	49 lbs.	50 lbs.	40 lbs.
12' extension	40 lbs.	29–48 lbs.	45–63 lbs.
24' extension	105 lbs.	72–142 lbs.	99–142 lbs.
35' extension	171 lbs.	122–175 lbs.	139–195 lbs.
10' roof	30 lbs.	22–40 lbs.	30 lbs.
16' roof	48 lbs.	39-56 lbs.	42-46 lbs.

Maintenance. The maintenance of any ladder is affected by frequency of use and adherence to NFPA 1932 that recommends all ground ladders be visually inspected at least once each month and after each usage. Additionally, to restore the surface finish and minimize oxidation, an occasional application of a good automotive paste wax should be used on aluminum and fiberglass ladders. The varnish finish on wood ladders should be inspected at least every six months, and redone at least annually, or sooner, if damage to the finish is noted. This preserves the wood and varnish finish of wood ladders. Experience has demonstrated that frequent use of wood ladders may require additional maintenance to preserve damaged varnish finishes as opposed to normal maintenance for aluminum and fiberglass ladders. However, infrequent use of wood, aluminum, and fiberglass ladders will require similar maintenance considerations. Therefore, as the frequency of use increases, so does the necessary maintenance of any ladder, regardless of construction. As an example, one manufacturer recommends maintenance should be performed every six months unless there is a run frequency of more than 100 per month. In this case, it is recommended that maintenance be increased to every three months or sooner. In any case, most ground ladders do not receive adequate maintenance regardless of the type of ladder and frequency of use.

Extension ladders require about a 75-pound pull to extend any section. More effort than this guide line is an indicator that cleaning and waxing of all contact and locking parts may be necessary.

Safety. When selecting a ladder, safety considerations should be carefully evaluated. Therefore, when comparing metal, wood, and fiberglass ladders, remember that metal is a willing conductor of electricity, whereas wood and fiberglass are not. Consider the following facts:

- Never store a ground ladder where it can be exposed to heat (such as engine exhaust).

- Ladders are rated at their maximum load capacity if they are used in the proper angle of ¼ their working length, have been tested for NFPA 1932, and are secured at the top and bottom of the ladder to prevent slippage.

- All fire service ladders are constructed to meet NFPA 1931 standard.

- Fire service ladders are constructed to carry a 750-pound load with a 4:1 safety factor. As a comparison, industrial ladders are designed to support a 300-pound load with a 4:1 safety factor.

- All metal and fiberglass ladders constructed since January 1984 are required to have heat sensor labels. These labels will turn black if the ladder (or the portion of the ladder with the heat sensor) is subjected to a temperature greater than 300°F. If subjected to this heat (and only for a moment), the ladder can lose at least 25% of its load capacity. As of 1994, NFPA now requires four heat sensors per ladder section. If any heat sensor label turns black, the ladder should be removed from service and tested to determine its safety status before returning to service.

According to the Federal Emergency Management Agency (FEMA), a study that was conducted between 1977 and 1988 indicated that nine firefighters were killed due to electrocution involving metal ground ladders (not including the number of non-fatal injuries attributed to these ladders).

According to the United States Consumer Products Safety Commission (CPSC):

- Commission accident data consistently identifies metal ground ladders as a product commonly associated with electrocutions.

- When accurate depth perception is required, the human visual system is limited in its ability in estimating the clearance distance from ladders to power lines as viewed against the sky. When an object (such as a power line) is seen against the sky, the human eye only sees a dark line against the sky. Unless a person is close enough to see some details on an objects surface, it is almost impossible to accurately estimate its distance. Therefore, visual judgment cannot be solely depended on for estimating critical clearance distances of objects in the sky (such as power lines and ground ladders). This is a primary reason why personnel raising ladders must carefully look for any overhead electrical obstructions. Obviously, obstructions such as vegetation are easier to identify and avoid.

- Because people are task oriented, they tend to move an extension ladder while it is still extended to avoid interrupting the task sequence. Therefore, this condition increases the possibility of contacting power lines with an extension ladder. Most electrocutions associated with metal ladders involved extension ladders.

- The average electrical service lines (120/240 volts) to a single-family dwelling are commonly insulated with a material rated at 600 volts. However, age and exposure to the extremes of weather contribute to brittleness and cracking of the insulation material, which degrade dielectric strength and the ability to prevent electrocution. Therefore, power line coverings are not designed, or, in the case of service drop lines, not adequately maintained to be a reliable protection against an abrasive contact by a ladder.

- In the private sector, the telecommunications industry has significantly reduced electrocution hazards by simply requiring exclusive use of either wood or fiberglass ladders (both of which are inherently non-conductive), and the National Electrical Safety Code prohibits the use of metal ladders or metal longitudinal supports along wooden ladders in the vicinity of electrical conductors. Why the, in the interest of safety, are metal ladders still being used in the fire service?

General Application Summary

When selecting a ground ladder, a confirmation of a specific application, objective height, and available personnel for implementation will dictate the length of ladder(s) that should be utilized. Ground ladder applications can be summarized as follows

12- and 14-foot extension ladders

- Can be raised by one person.
- No other ladders are as effective for use inside structures. Therefore, these ladders should be reserved for use in the interior of structures.
- Can be used for access to roofs of some one-story structures.
- Can be used as a step ladder.
- If able to be separated, can be used as two separate ladders, or as salvage drains by placing salvage covers or plastic sheeting on the main or fly section.

20-foot extension ladder

- Can be raised by one person.
- Can be used for reaching the roof of single-story buildings.
- Can be used to reach second story windows.
- Can be occasionally used to reach lower fire escape balconies.
- Effective for catch basin salvage operations and constructing a drain by placing salvage covers or plastic sheeting on the ladder.
- Can be used for bridging operations.
- Can be used on roofs or floors to span suspected weak areas.
- Can be used as an improvised stretcher.
- If able to be separated, can be used as two separate ladders.

24- and 28-foot extension ladders

- Can be raised by two persons.
- Can be used to reach the roofs of two-story buildings.
- Can be used to reach the windows of two and some three-story buildings.
- Used to reach fire escape balconies.
- Can be used to construct catch basins for salvage operations.
- Can be used for bridging operations.
- Can be used to span suspected weak areas.

35-foot extension ladder

- Can be raised by two or three persons.
- Used to reach the roofs of two and some three-story buildings.
- Used to reach two- and some three-story fire escape balconies.
- Can be used to construct catch basins for salvage operations.
- Can be used for bridging operations.
- Can be used to span suspected weak areas.
- Can be used for an auditorium raise.

40-, 45-, and 50-foot extension ladders

- Can be raised by four to six persons.
- Used to reach the roofs of three and some four-story buildings.
- Used to reach third and some fourth floor windows.
- Can be used for an auditorium raise.
- Can be useful in areas that restrict the use of aerial ladder apparatus. Operational considerations should include the time and personnel required for their implementation.

16- and 20-foot straight ladders

- Can be raised by one person. A 16-foot straight ladder is one of the most effective ladders for single-story residential structures (and numerous other single-story structures).
- Can be used for reaching the roofs of single-story buildings.
- Can be used for reaching most second-story windows.
- Occasionally these ladders may reach lower fire escape balconies.
- Effective for catch basin salvage operations and constructing a drain by placing a salvage cover or plastic sheeting on the ladder.
- Can be used for bridging operations.
- Can be used on roofs or floors to span suspected weak areas.

24-foot straight ladder

- Can be raised by one or two persons.
- Used to reach roofs of one-story and some two-story buildings.
- Used to reach second and some third-floor windows.
- Used to reach fire escape balconies.
- Well suited for catch basin salvage operations and constructing drains by

placing salvage covers or plastic sheeting on the ladder.

- Can be used for bridging operations.
- Span suspected weak areas in floors and roofs.

10-, 12-, and 14-foot roof ladders

- Normally requires the initial placement of a ground or aerial ladder before implementation.
- Can be raised by one person.
- Used to provide stability for personnel working on pitched roofs.
- Can be used as a straight ladder although a roof ladder is often more narrow than a comparable length straight ladder.
- Can be used to span suspected weak areas.
- Can be used to ladder the backside of tall parapets/fascias to allow roof personnel to reach the top of a fascia/parapet and access a ground ladder/aerial device.
- Can be used for bridging operations.
- Can be used for catch basin salvage operations and constructing drains by placing salvage covers or plastic sheeting on the ladder.

16- and 20-foot roof ladders

- Normally requires the initial placement of a ground or aerial ladder before implementation.
- Can be raised by two persons.
- Used to provide stability for personnel working on pitched roofs.
- Can be used as a straight ladder.
- Can be used to span weak areas of construction.
- Can be used for bridging operations.
- Can be used for catch basin salvage operations and constructing drains by placing salvage covers or plastic sheeting on the ladder.

Ladder Considerations

There are many SOPs, general guidelines, specific procedures, and specific needs that are dictated by each incident that are used in selecting the proper type and length of ladder, and then placing those ladders into operation. The following is a review of basic ladder considerations.

Purchase considerations

When deciding on purchasing a two- or three-section extension ladder, remember that each additional section increases the weight of a ladder (from overlap between the sections). Also, additional sections will result in a wider and taller ladder, but will create a shorter ladder than an extension ladder with fewer sections.

Identification

When it is necessary to remove a specific length of ladder from the rear of apparatus, appropriate labels on the apparatus below each ladder and/or butt end of each ladder (figure 8–2) simplifies quickly selecting the proper length of ladder.

Figure 8–2. Marking the ends of ladders can simplify the removal of the correct length of ladder.

Type/size

Always use the proper type and length of ladder. If the length or placement is incorrect, either replace the ladder or correct the problem. Do not let an inconvenience become a problem or potential safety consideration.

Lifting and carrying

When lifting ladders from apparatus or the ground, use the muscles of the legs, arms, and shoulders. Coordination of effort among team members is essential for smooth operations. Mark the midpoint (or balance point) of your ladders. This will enhance quickly identifying the balance point of the ladder when it is carried and where to tie a rope if a ladder needs to be hoisted aloft. When carrying a ladder to an objective and it becomes necessary to change direction of travel but it is unnecessary or undesirable to turn a ladder end for end, a simple pivot to reverse direction of ladder personnel can be used.

Multiple ladders and equipment can be carried from an apparatus to the incident as illustrated in figure 8–3. One ladder has been nested inside the other ladder, selected equipment has been placed on top of the nested ladder, and the ladders/equipment are carried by two personnel to the desired location in one trip without making multiple trips to accomplish the same result.

Figure 8–3. Multiple ladders and equipment can be carried to an incident in one trip by two personnel.

Placement

To complement the use of ground ladders, consider the use of aerial ladders on any structure equal to or larger that a one-story structure.

Ladders left on apparatus are not effective on the fireground. Timely and effective ladder placement should be an initial priority for two basic reasons—it lays the groundwork for a timely and coordinated attack on a structure fire, and when a truck company arrives on-scene at a structural incident, this is normally the only time an officer has the entire company together and can effectively control and coordinate ladder placement prior to committing company personnel to other fireground operations. As an example, the first-in truck company in figure 8–4 has utilized its staffing of three to quickly raise two ladders.

Figure 8–4. The first-in truck company has used a staffing of three to initiate two ladders by placing the aerial to the roof and a 35-foot extension ladder to the second floor balcony. (LAFD file photo)

The engineer has raised the aerial to the roof while the officer and firefighter raised a 35-foot extension ladder to the second-floor balcony. The 35-footer can provide initial and secondary access and egress routes to fireground personnel and thus enhance the perspective of fireground safety.

When appropriate, first-in truck officers (or other appropriate personnel) should relay additional ladder requirements to incoming companies (such as the need for additional ladders, proper location, and so on).

When ladders are used to provide above-ground access and egress, a minimum of two ladders should be raised. However, as the number of above-ground personnel are increased over two or three, consider placing additional ladders.

When it is not possible to secure the top of a ladder from which a hose stream must be used, the ladder should be used at a *flatter-than-normal* angle. The base should also be secured.

When ladders have been spotted to an objective and ascended by personnel, never remove the ladder(s) unless personnel (that have used the ladder) are aware or have approved of its removal.

Ladders placed for climbing should be positioned with the base of a ladder resting evenly on both shoes. If the surface the ladder will be spotted to is sloped, the unsupported down-slope shoe can be supported with a wedge (figure 8–5) to level the ladder and provide four-point contact. Wedges should be placed under the unsupported beam and between the ladder and the objective. This reduces the possibility of someone walking by the ladder accidently dislodging the wedge. Wedges can be easily made from treated wood and carried with a ladder.

Figure 8–5. Wedges can quickly provide four-point contact to stabilize an unsupported beam of a ladder.

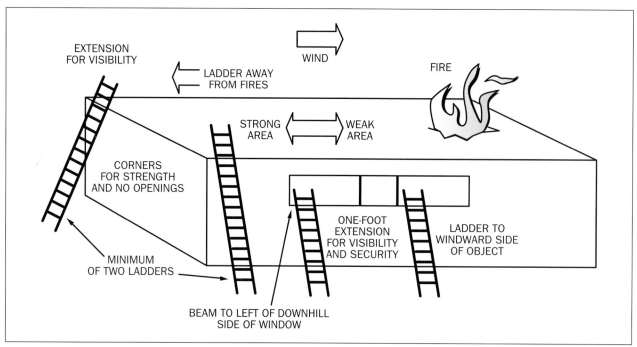

Figure 8–6. Consider placing ladders away from a fire, laddering the corners, using at least two ladders, and extending the ladders for good visibility.

Two wedges should be made and connected with a small cord/rope that can be easily disconnected (remember the minimum two ladder rule?) to facilitate carrying the wedges. The approximate dimensions are 18 inches long by 3½ inches high (at the high end).

Buildings

If possible, a ladder should be placed to the *windward* side of an objective. This will keep personnel on the upwind side of any contaminants. (See fig. 8–6)

As mentioned earlier, there should be a minimum of two ladders to a building. At least two ladders should be placed to a roof, and depending on the location of the fire, at opposite ends or at least two sides away from a fire so there will be more than one way off a roof. On large structures, all four corners of a fire building should be laddered.

Ladder the strong areas of a building or roof. And personnel can start and return to the strongest (unburned) area of a building. Always consider the strong areas of construction and the area that will be traversed. Areas such as pilasters, hips, valleys, and ridges offer excellent areas to place a ladder. Remember that one of the best areas to place a ladder is to a corner of a building. The corners of a

building are considered a prime location for ladders for the following reasons:

- **No horizontal openings.** Windows, doors, and vents are not usually found in corners. Placing a ladder over a horizontal opening is inviting the possibility of a burned ladder and losing a means of egress.

- **Strength.** Structural stability is enhanced where two walls and a roof are tied together.

- **Location.** When roof operations are completed, the location of a ladder can be easily determined by looking at the corners. This is an asset at night, in smoky conditions, or when immediate egress is necessary.

- **Access.** By laddering the corners of a structure, the entrance to a building is generally avoided. Therefore, hose lines that are being deployed through doorways and such are not competing for space with ladder operations.

- **Hips.** Hips converge at corners. Therefore, when laddering the corner of a structure with a hip roof, personnel will step off a ladder and onto a hip.

Ladders should extend above a roof or parapet wall enough distance so they are readily visible (normally, this is not three feet or three rungs). In most cases, the ladder will also be more accessible to personnel (figure 8–7).

Figure 8-7. Ladders should extend above a parapet or roof enough distance so they are readily visible. (Photo courtesy of Tim Dungan)

Windows

When placing a ladder to a window, four options should be evaluated:

- If the tip of a ladder is placed *just below* a window sill, the ladder will not block the window opening for personnel who need to enter or exit through the window. Although this operation enhances egress/ingress through a window opening, remember that personnel inside the building may not be able to identify the window with the ladder. However, this placement is normally preferred for most window operations.

- If a ladder can be placed so the tips of a ladder are *just above* a window sill, the location of the ladder can be seen inside a building and the tips of the ladder will not hinder movement within the window opening. However, if a ladder tip is extended into a window opening much over one-foot, the ladder tip can hinder access and egress as illustrated in figure 8–8.

Figure 8-8. A ladder that is extended into a window much over one-foot can hinder access and egress. (LAFD file photo)

- If a ladder is *not* supported at all four corners (for example, ladder footing is sloped, resulting in a loss of four-point contact), the ladder should be placed on the downhill side of a window sill with the tip of the ladder just over the sill and beam touching the downhill side of the window frame. This will prevent a ladder from moving away from the supporting window frame as it is being climbed.

- If a ladder is placed to the side and above a window sill, ingress/egress through a window opening can be difficult. As personnel transfer their weight from the ladder to the window sill, the ladder can be inadvertently pushed away from the window, creating a dangerous condition for personnel. However, in this position the ladder can be used by personnel (on the ladder) to clean window glass from a window without subjecting a firefighter to falling glass. In no case should a ladder with less than 4-point contact be placed to the side and below a widow sill.

If necessary, ground ladders can be used to ventilate windows. To ensure the safety of this operation, place the tip of the ladder in the top portion of the window and allow the ladder to fall inwards (figure 8–9), breaking the glass from the top downward.

Figure 8–9. If necessary, windows can be ventilated by using a ground ladder to break the glass from top to bottom of a window. (Photo courtesy of Tim Dungan)

Because the ladder will fall into the window, the majority of the glass will be pushed into the structure. Therefore, the location of interior personnel can be a positive or negative consideration.

Footing and securing

Ladders can be footed from the rear or the front to prevent a ladder from inadvertently moving while being used. If a ladder is footed from the rear (between the ladder and objective), the person footing the ladder will not have a view of the objective and conditions and can be struck by falling debris and/or

equipment. Conversely, if the ladder is footed from the front (figure 8–10), the person footing the ladder can watch the building, conditions, and any personnel working on the ladder. Footing a ladder from the front is accomplished by either placing a foot on a beam at the ground or placing one foot on the bottom rung.

Figure 8–10. Footing a ladder from the front can provide noteworthy safety benefits.

When possible, a ladder should be secured (either top and/or bottom) to a stationary object. Unless a ladder does not have four-point contact, securing the base of a ladder is preferable to securing the top of a ladder because the base is more likely to inadvertently move, and securing at ground level is normally an easier operation. Remember there are some conditions that provide a somewhat slippery surface (concrete, fire escape railings, and so on). Conversely, some conditions can assist in securing the base of a ladder (such as lawns and soft dirt).

Parapets and fire escapes

Some structures have tall parapets. If a parapet is of excessive height (over 5 feet) and it is necessary to place a ladder to the parapet to access the roof, a roof ladder should be placed adjacent to the ladder at the top of the parapet to facilitate access and egress from the roof to the top of the fascia (and ladder). If possible, hooks should be opened and placed over the parapet to assist footing for personnel using the ladder.

When laddering fire escapes, the top of a ladder should extend a maximum of one-foot above railings. Ladders should be spotted adjacent to fire escape ladder handrails and strapped to minimize side loads while personnel are climbing the ladder. Avoid laddering to areas that may interfere with the use of fire escape drop ladders. Consider releasing drop ladders for use by occupants and to keep drop ladders from prematurely releasing during an incident.

Roof ladders

Roof ladders should be used on all pitched or sloped roofs where footing is precarious. Remember that pitched-sloped roofs are *always steeper* in reality than when viewed from the ground.

Roof ladders used for roof ventilation operations should be placed on a roof where ventilation operations are anticipated. Therefore, initially position a ground ladder (if possible) near or next to the section of roof to be ventilated (and on the windward side of the anticipated ventilation operation). This will facilitate placement of a roof ladder on the appropriate section of roof (figure 8–11).

This results in a straight path to the ventilation area, personnel will be working on the windward section of roof, and a quick route back to the ground.

Raising/lowering

At any event, the largest ladder that requires the most personnel should be raised first. If possible, ground ladders should be properly spotted, shifted, or moved into the correct position (horizontally) for raising *prior* to being raised (vertically) because ground ladders are most easily and safely maneuvered while on the ground. Also, after a ground ladder is in a vertical position, additional movement increases the chances of losing stability, striking wires and other similar obstacles.

Normally, the operations necessary to lower ladders are the reverse of those used in raising. Be sure the appropriate area is clear before lowering. Check overhead and ground obstructions. If more than one person is responsible for lowering the ladder, the base person is responsible for the safety of the operation. Avoid the tendency to work too fast when lowering a ladder.

Straight ladders are normally lighter and require less time to place into operation than extension ladders. Extension ladders offer a *variable* height factor. Straight ladders with rungs set in the main beams should be raised with the rungs away from a building, and extension ladders should be raised as follows:

- With the fly away from (or on the outside of the main section) an objective for metal and fiberglass extension ladders

- With the fly between the objective and the main section for truss wood extension ladders with rungs set in the main beams

In all cases, consult the appropriate manufacturers recommendation. Some manufacturers can modify a fly out ladder so it can be used in the fly in position if that is preferred.

The two basic methods of raising ground ladders are the flat and beam raises (figure 8–12). Overhead obstructions, limited space, and other similar considerations will dictate the type of raise that should be used.

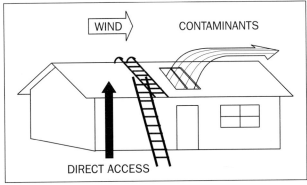

Figure 8–11. Place ground and roof ladders to the windward side of the area of roof to be ventilated.

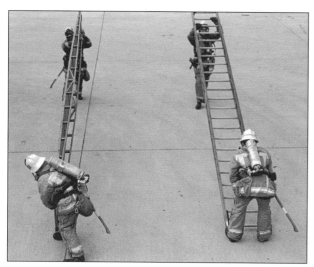

Figure 8–12. The two basic raises for ground ladders are the beam raise (left) and flat raise (right).

Ground ladders that need multiple personnel for implementation require a specific person of a ladder team to be responsible for communications (timing and placement of ladder) among team members. This is usually handled by the base person unless otherwise specified.

Some extension ladders can be separated to make two straight ladders. However, it is not practical on ladders with a halyard. If deciding between different two lengths (20 or 24 feet, 24 or 35 feet, and so on) for laddering an objective, choose the longer ladder for the following reasons:

- The ladder will reach the objective (length).

- The ladder will be heavier and wider. This improves stability and is easier to climb (strength and access).

- Additional extension of the ladder above an objective (visibility).

Climbing

When ascending a ladder, look toward the rungs just above the head. Climb with hands grasping the center of the rungs, body erect, shoulders at an arm length from the ladder, knees in-line with the body, balls of the feet on the rungs. The legs carry the weight and the arms maintain balance and stability. Do not reach up to pull with the arms. Ascend briskly but smoothly, feet and hands working together, taking every rung with the feet and every other rung

with the hands. Descending, look toward the rungs just below the head. Grasp each rung alternating hands. The feet and hands should contact appropriate rungs at the same time to develop a rhythm that is essential to smooth and safe ascending and descending.

When climbing a ladder, at least one hand should grasp the ladder to provide security and stability. Tools can also be carried by straps, slings, or moved with drop cords. As an example, the chain saw in figure 8–13 is attached to the firefighter by a strap allowing the firefighter to grasp the ladder with both hands.

Figure 8–13. Equipment can be carried aloft with a strap, allowing both hands to grasp a ladder while ascending or descending. (Photo courtesy of Tim Dungan)

If necessary, a hand carrying a tool can provide security and balance by positioning the hand and tool (depending on the size of tool) behind a ladder beam with the wrist cocked against a beam.

Instead of climbing a ladder with equipment or tools, drop bags can be effectively utilized in concert with ground ladders to hoist or lower tools and hose lines during *above ground* operations.

Working height

The required working height of a ladder can be estimated as follows (figure 8–14):

- Residential occupancies are approximately 9 feet from floor to floor.

- Commercial occupancies are approximately 10 feet from floor to floor.

Depending on the type of structures in a particular area, the approximate dimensions of 10 feet floor to floor for residential structures and 14 feet floor to floor for commercial structures may be more appropriate. Be familiar with the dimensions that are most applicable to your area.

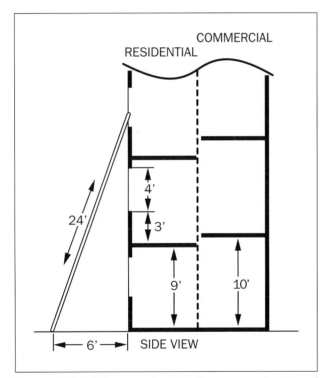

Figure 8–14. When evaluating the proper length of ladder for an objective, consider the distance between the floors as 9 feet for residentials, and 10 feet for commercials.

The average windowsill height above a floor is approximately 3 feet. Allow 4 feet for window height and any extension above an objective must be added to determine the approximate length of ladder. As an example, a 24-foot ladder is the minimum length ladder required to reach the window sill of this second floor habitable occupancy. The base of the ladder should be placed 6 feet (24 divided by 4) from the building for the proper climbing angle.

Climbing angle

Manufacturers recommend a ground ladder be positioned at 75½ degrees to an objective and further advise that this angle will support maximum strength and give the best longevity. However, most firefighters (and training materials) prefer a climbing angle of approximately 70 degrees. This angle allows personnel to climb a ladder in an erect posture while comfortably placing their hands on the rungs of a ladder (figure 8–15).

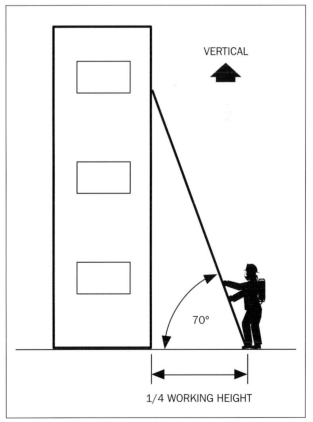

Figure 8–15. To obtain a 70-degree climbing angle, position the base of a ladder one-quarter the height of the ladder from the base of the objective.

To simply achieve 70 or 75½ degrees, the base of a ground ladder is normally placed one-fourth of the working height (vertical distance from grade level to objective) from the base of a building.

Safety considerations

While working from a ladder, personnel should not overextend their horizontal reach. Over extension can suddenly overbalance or twist a ladder and cause a ladder to slide against an objective. If necessary, reposition the ladder, or lock-in to the ladder. This will limit the reach of personnel, particularly when using pike poles. When using power equipment from ladders, particularly rotary saws with their gyroscopic effect, placing the arms as illustrated in figure 8–16 helps to keep balance, limit reach, and enhance safety of the operation. Stability and safety are the key considerations.

Before stepping off a ladder, ensure the stability of the landing area. As an example, although a roof may look normal, fire may have weakened the underside of a roof so it is not capable of supporting additional weight. Stability can be tested by sounding with a rubbish hook, pike pole, hooks, axe, or other similar tool.

Figure 8–16. Securing a ladder beam while using power equipment can improve stability and safety. (Photo courtesy of Tim Dungan)

Training

When summarizing ground ladder evolutions, the most basic safety rule that is commonly violated is the *minimum two ladder rule*. Although most every firefighter is aware of this rule, it is common to find one ladder to a structure with multiple personnel working above ground inside a structure and/or on a roof. Do not let the excitement of an incident obscure basic fireground and training safety considerations! Remember the old adage of "you train the way you fight." Frequently train with all ladders, particularly in challenging locations. This develops proficiency and the ability to pre-plan the appropriate length of ladder for a specific application.

Operational Variations

The following variations can be used (when appropriate and depending on the style/type of ladders available) as an adjunct to standard ladder movement and implementation techniques. These are designed to stimulate initiative in combination with offering additional variations for ground ladder operations that may be affected by unusual or reduced staffing situations. Even if these variations are not used at an incident, they will develop teamwork, improve personnel expertise, and expand operational options.

Repositioning

The base of an extension ladder may be shifted toward a building or to either side after it is lowered into an objective. Do not move the base *away* from an objective unless the sections are strapped together.

When shifting large extension ladders (prior to strapping the fly), one member shall capture the halyard on the front side of the ladder to prevent the dogs from accidentally unlocking, resulting in a premature release of the fly section.

The 16-, 20-, and 24-foot straight ladders can be carried or moved to another location in a vertical position. With some practice, this can be easily and safely accomplished. Lock the top or bottom elbow, and use the other arm to move the ladder horizontally while watching for overhead obstructions. This will keep the ladder balanced overhead and minimize fatigue. It is not recommended to use this procedure on extension ladders due to their additional weight and the potential of an extended fly section becoming inadvertently dislodged from the main section.

Ladders that have been placed to an objective can be adjusted or moved by rolling or sliding a ladder along the objective. When rolling a 28-foot, 35-foot, or other similar extension ladder, it may be necessary to occasionally stop and adjust the ladder because the base will tend to move farther than the top portion of the ladder due to the difference in width of the fly and main section. Sliding a 28- or 35-foot extension ladder will generally require two persons. Move the base slightly in front of the top to help maintain control of the ladder.

Implementation

A 24-foot straight ladder can be raised by one person in the same manner as the 16- and 20-foot straight ladders. Care should be taken to securely anchor the base of the ladder against the objective because the added length and weight of this ladder will tend to lift the base (of the ladder) from its anchoring position as the ladder is raised to vertical. After the ladder has been raised against the objective, the ladder should be adjusted to the proper climbing angle.

If necessary to beam raise a 20- and 24-foot straight ladder with the rungs set in the main beams and *toward* a building, the ladder may be pivoted 180 degrees in the vertical position or may be flipped 180 degrees prior to raising. To flip the ladder, personnel should stop at the intended spot, turn, and face the ladder. The base person will notify the top person that the ladder will be flipped. The base person raises the toe of a boot in preparation to stop the ladder as the top beam rotates down (figure 8–17).

To prevent the top beam from striking the base person in the shin, slightly lower the ladder before allowing the top beam to rotate down. If done correctly, the top shoe of the ladder will rotate down, striking the sole of the base person's boot, and placing the rungs away from the objective (figure 8–18). To complete the operation, ground the shoes of the ladder and continue the raise.

A 20- and 24-foot straight ladder raise (raising the 44) can be used when two ladders are needed and one ladder is bedded inside another ladder (or a 20-foot ladder can be placed inside a 24-foot ladder).

Figure 8–17. Ladders with rungs set in the main beams should be raised with the rungs away from the objective. If necessary, this can be easily achieved by flipping the ladder prior to raising.

Figure 8–18. By flipping the ladder on the ground prior to raising, safety is enhanced, and the rungs will be away from the objective when the ladder is raised to the objective.

As an example, assume a 20-foot straight ladder is bedded inside a 24-foot straight ladder. The base person is positioned at the end of the ladders where the shoes of the 20- and 24-foot ladders are adjacent.

The base person removes the ladders from the apparatus and is positioned between the first and second rungs of both ladders. The top person will be positioned between the fifth and sixth rungs from the top of the 24-foot ladder (figure 8–19), securing the top portion of both ladders. Generally the 44-foot raise will be accomplished in the flat position, which will ensure control of the 20-foot ladder.

Figure 8–19. When practical, simultaneously carry two ladders to an objective.

The ladders are raised to vertical and lowered into the objective. During this operation, it will be necessary for the base person to separate the 20-foot ladder from the 24-foot ladder. After the ladders have been separated, they can be positioned as desired.

A 35-foot extension ladder can be raised with two persons if personnel have practiced and are familiar with this operation. The advantage of this raise is that it can free another person to accomplish an additional operation, which can be a significant consideration within the constraints of minimal staffing, and is conducted as discussed in the following section.

Raising a 35-foot extension ladder. When carrying the ladder to an objective, the base person places their inside arm between the second and third rung from the bottom of the ladder to avoid entanglement with the ladder dogs. The person carrying the top of the ladder places their inside arm between the second and third rung to avoid entanglement with the halyard pulley (figure 8–20).

Spot the ladder and raise to the vertical position as in figure 8–21. Remember to consider the surface on which the ladder is being raised because dirt and grass will tend to foot a ladder better than concrete, which can be a slippery surface.

Figure 8–20. Extension ladders that normally require three personnel can often be carried and raised by two personnel.

Figure 8–21. With practice, the person at the top of the ladder can effectively accomplish the work of two persons as on this 35-foot extension ladder. This can make another person available for other tasks.

The person at the base of the ladder will foot the ladder. After the ladder is pushed to vertical, do not lean the ladder too far toward the objective. If the ladder is not correctly positioned, the ladder will move toward the objective when the person at the base reaches up to lock the dogs. To prevent this condition, the main section of the ladder should be positioned close to vertical. As the fly is extended, it will begin to arc toward the objective. The further the fly is extended, the more it will shift the weight of the ladder toward the objective by the fly arcing toward the objective as it is extended. Care should be exercised using a moving pivot with the two-person 35-foot extension ladder. An unstable surface or lack of proficiency and team work can cause control problems.

Raising a 50-foot extension with four persons. Although there are many variations to this raise, four persons can be effectively utilized—and is often better than six persons.

To initiate a flat raise with the base of ladder *footed into a building*, carry the ladder to an objective placing the base of the ladder near the building. Unpin the tormentor poles and pass overhead to the personnel at the top of the ladder. Readjust the ladder so it is tight into the base of the building. Two persons should be positioned at the ends of the tormentor poles and two persons should be positioned one-foot from the bottom of the tormentor swivels (figure 8–22).

Figure 8–22. Ladders with tormentor poles can be easily raised with four personnel. Use the objective to foot the ladder as illustrated.

Figure 8–23. With the objective footing the ladder, drive the ladder to the vertical position.

Drive the ladder to vertical against the building (figure 8–23). The ladder can be adjusted to provide the proper climbing angle by the two base personnel lifting the base and readjusting to the appropriate spot, allowing the top of the ladder to slide down the face of the building. Pole personnel stay in position in front of the ladder.

When the base of the ladder is spotted and it is necessary to extend the ladder, base personnel shall move one in front and back of the ladder. The person at the pole pulls the ladder out to vertical while the person at the rear pushes the ladder to vertical and the person at the front foots the center of the bottom rung. The ladder can be extended as necessary and returned to the building (figure 8–24). The ladder can also be walked out to the proper position (away from the building) by alternatively pivoting on the ladder shoes.

To accomplish this operation, position the pole personnel as one to the front and one to the side. The base personnel, one to front and one to the back, shall pivot the ladder one turn from the building to allow room for the base personnel in back of the ladder. The side pole person pushes and pulls on the tormentor pole alternately raising one shoe and then the other as the base personnel pivot the ladder to the desired spot. Pole personnel split, one in front and one to the side of the ladder. The front person extends the fly. The rear person locks the dogs, and all personnel lower in to the objective. This method takes a coordinated effort and can be slower than the following flat raise, base of the ladder footed by personnel method.

Figure 8–24. The fly section can now be extended before positioning the ladder against the objective.

Figure 8–25. If the objective cannot be used to foot the ladder, use two personnel to foot the ladder and assist the raise with the tormentor poles.

Figure 8–26. As the ladder is raised to the vertical position, the base members foot the ladder and assist in raising the ladder with the tormentor poles.

To initiate a flat raise with the base *not footed* into a building, carry the ladder to the objective and spot the base of the ladder the appropriate distance from the objective. Unpin the tormentor poles, and leave them on the ground adjacent to the ladder. Two persons foot the base of the ladder shoes with their inside feet. These same base persons secure the tormentor poles with both hands, keeping the poles to the outside of their bodies. The other two persons kneel down at the top of the ladder, face the opposite direction of travel (figure 8–25), grasp the main beams with their inside hand, and raise the ladder, pivoting under the ladder and driving to the vertical position.

The base personnel will foot the shoes and pull on the tormentor poles bringing the ladder to vertical (figure 8–26).

As the ladder reaches vertical, it will be necessary for base personnel to push the tormentor poles to the side to keep them from grounding as they extend past the shoes of the main beams. With this accomplished, one person will become the front pole person and one person will become the side pole person (figure 8–27).

Figure 8–27. The fly section can now be raised prior to lowering the ladder into the objective.

One person will stay at front base and one person will stay at rear base. The front person extends the fly, and the rear person locks the dogs. All personnel lower into the objective.

Step ladder

This operation can be used when operating near high ceilings. As an example, a 24-foot straight ladder with a 20-foot straight ladder that will bed inside a 24-foot ladder will be used. Carry the bedded ladders to the appropriate area. Place the ladders on their beams and remove the 20-foot ladder. Flip the 20-foot ladder and position on the opposite beam (rungs away from the 24-foot ladder if the rungs are

set in the main beams). Nest the fifth rung of the 24-foot ladder in the shoes of the 20-foot straight ladder. Secure the ladders tightly by strapping the fifth rung of the 24-foot ladder to the top rung of the 20-foot ladder with two ladder straps. These straps should be adjacent to the inside of the beams. Strap two 6-foot pike poles inside the beams on the tenth rungs of the two ladders. Two persons foot the bottom shoes of the two ladders, and the ladders are raised (on the beams) to vertical. After the ladders are in the vertical position, adjust the base of both ladders to establish good four-point contact with the ladder shoes to stabilize the ladders. To increase stability, a pike pole may be strapped to the bottom of the 24-foot ladder by raising the 24-foot ladder and placing a pike pole under the shoes. Secure with two ladder straps around the pike pole and the bottom rung of the 24-foot ladder (figure 8–28).

Figure 8–28. Two ladders and pike poles can be used to construct a step ladder. To enhance the stability of the step ladder, strap a pike pole to the foot of one of the ladders.

This operation uses two ladders when both ladders are of the same width and is accomplished similar to the aforementioned step ladder raise. As an example, when a 20-foot ladder will not bed in a 24-foot ladder, a pike pole can be substituted for a rung to strap the ladders together and make the improvised step ladder more stable. Insert a pike pole between the main and truss beams of a 24-foot ladder (if ladder construction allows) between the fourth and fifth rung. Nest the shoes of a 20-foot ladder against the pike pole—one shoe inside the 24-foot ladder and one shoe outside the 24-foot ladder. Using two ladder straps, strap the fourth rung of the 24-foot ladder to the top rung of a 20-foot ladder. Include the pike pole in the ladder straps. As the ladder straps are tightened, the pike pole will bind against the truss blocks giving the pike pole a firm foundation. It may be necessary to hold the top of the 24-foot ladder off the ground during this assembly due to the length of the pike pole.

Raising ladders under overhead obstructions

This raise can be used with straight and extension ladders.

Twenty-foot straight ladder. Two persons place the ladder under the obstruction (rungs up if the rungs are set in the main beams). One person is positioned at the top and bottom of the ladder. The top person faces the bottom person, secures the beams of the ladder, and raises the ladder overhead. The bottom person secures the base of the ladder by a rung and a beam. The top person hops the ladder with the beams in the palms of the hands as the base person slides the shoes of the base along the ground toward the top person (figure 8–29).

As clearance of the obstruction (overhead wires, vegetation, and so on) is obtained, the person at the base will ground the shoes, and the person at the top will push the ladder to the person at the base (figure 8–30).

In extremely tight situations, the last portion of this raise may need to be repeated until the ladder can be safely positioned.

Thirty-five-foot extension ladder. Although two persons can accomplish this raise, three persons will facilitate the raise. This raise is similar to the 20-foot straight raise except two persons hop the

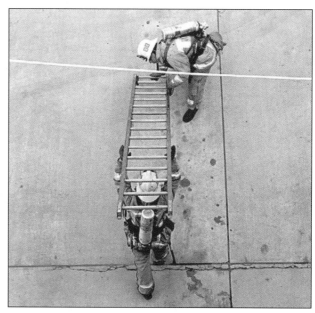

Figure 8–29. As the person at the base moves the ladder forward, the person at the top assists the ladder moving under the overhead objective.

Figure 8–30. With coordination, the ladder will be raised to the vertical position as the ladder is moved under the overhead objective.

ladder—one person on each beam, and one person pushes the base of the ladder. If there is a difference in height of the two persons working the beams, the shorter person should step toward the person pushing the base until the ladder beams are level. This will result in a more controlled ladder raise. When the main section is in position prior to extending the fly, it is advised that one of the rear personnel step to the front of the ladder. This will achieve better ladder stability while the fly is being extended.

Laddering fire escapes

Because there are many ways to ladder fire escapes, personnel must evaluate the needs of the incident in concert with appropriate resources and staffing that is available to accomplish this operation. This section discusses some examples of fire escape ladder operations.

Position an appropriate ladder to the front of a fire escape balcony. Spot the ladder to the hand rail of the fire escape with the top rung one foot over the balcony railing (figure 8–31). The inside beam should be spotted to the hand rail of the fire escape ladder. This will allow personnel to mount or dismount the fire escape ladder.

For multiple fire escape balconies, position a 35-foot extension ladder adjacent to the end of the fire escape balconies. Fully extend the 35-foot extension ladder and lower into the building to allow personnel to access both fire escape balconies. If two 35-foot extension ladders are used in this manner and placed at either end of a fire escape, this will allow four points of ingress and egress to and from the building.

If a ladder is spotted to the side or end of a fire escape balcony, this will keep the ladder free of the curb line on narrow sidewalks. If there is a doorway below the fire escape, this will also avoid the ladder blocking the doorway.

Laddering fire escape balconies with defective fire escape ladders. Spot a 20-foot extension ladder to building adjacent to the end of a fire escape balcony. Fully extend the ladder and lower into the building. Open roof ladder hooks and carry roof ladder up the 20-foot extension ladder. Hang the roof

Figure 8-31. When placing a ladder to a fire escape balcony, spot the ladder at the hand rail of the fire escape and next to the hand rail of the fire escape ladder. This will allow personnel to mount or dismount the fire escape ladder.

Figure 8-32. A ground and roof ladder can be simultaneously used to ladder two fire escape balconies if the fire escape ladder is defective or not present.

ladder on the upper balcony railing by the hooks, and place the inside shoe of the roof ladder between the beams of the 20-foot extension ladder (figure 8–32). If necessary, the 20-foot extension ladder can be respotted so a rung of the extension ladder supports the bottom rung of the roof ladder.

This operation should only be used in life-threatening emergencies because failure of equipment (roof ladder hooks or fire escape hand railings) could cause injuries.

Laddering ships

A unique problem can exist when laddering a floating ship because the ship may not be stationary due to the movement of the water. Most passenger and freight ships will require a minimum of a 50-foot ladder to reach the railings.

The ladder may be spotted and raised in any appropriate manner. However, a problem arises when it is necessary to secure the ladder to a ship (moving object). After the ladder is raised to the ship, the first consideration is to place the tormentor poles in a position to keep them from binding on the dock. This may be done by hanging them over the side of the dock. Next, it is necessary to strap the main section of the ladder to the fly section. Care must be taken to avoid any ship movement from unlocking the dogs before and while the sections are strapped. After personnel have strapped the main to the fly section, a person shall continue to the top of the ladder. When the person is on the deck of the ship, the ladder will be securely strapped to the ship. Raise the base of the ladder and then place the ladder shoes on the platform of an appliance dolly. The beams will rest against the face of the dolly. Securely strap the appliance dolly to the ladder. Emphasis must be placed on securing the ladder to the ship before placing the ladder base on a dolly. If this is not done, the ladder could roll away from the ship.

Laddering aircraft

Laddering aircraft for the purpose of providing access to or from an aircraft will be determined by the height of an aircraft above the ground. If it is necessary to open a fuselage door, select an appropriate ladder that will facilitate access. When the door has been opened, a shorter ladder can then be placed to the bottom of the open doorway to provide the desired egress/ingress to the aircraft.

Bridging

There are several different methods that can be used to accomplish a bridging operation, and each depends on the specific operation and available staffing. The following are several examples of bridging operations.

Straight ladder and drop bag line or equipment line. Place a straight ladder on the ground (rungs down if the rungs are set in the main beams). Take a line and wrap it around both beams between the second and third rungs. If the line is equipped with a clip, snap the clip back on the running line. The running line and clip should be laying on top of the rungs and centered between the truss beams. This configuration is similar to the first half of a roof ladder knot. Two persons foot the base and raise the ladder to vertical. Pivot the ladder so it can be placed across the area to be bridged. Two persons use the line to lower the ladder across the span. As the ladder nears the horizontal, it becomes more difficult to control. The person at the base can assist by placing one hand under the line and pushing up (figure 8–33). This gives the line a greater angle to the top of the ladder and enhances control of the ladder.

Bridging by passing a ladder hand-over-hand at a horizontal angle by three persons. Place a straight ladder (rungs up if the rungs are set in the main beams) adjacent to the area to be spanned. The first person takes a position at the end of the ladder furthermost from the area to be spanned and stands facing the top rung. Bend down and grasp the top rung with the little fingers against the inside of the beams and palms facing downward. The second and third persons position themselves at the other end of the ladder while facing the first member, bend down and grasp the beam. At a signal from the first person, the ladder is raised to shoulder height (figure 8–34). The first person walks the ladder toward the other two persons as they are passing the ladder along at shoulder height hand over hand on the beams.

As the ladder is extended over the span, it becomes more difficult to handle due to the overbalance of the ladder. With some practice, a 24-foot straight ladder can span 21-feet using this method (figure 8–35).

Figure 8–33. Two personnel can use an equipment line or drop bag line to hoist and lower a ladder across a span for a bridging operation.

Fig 8–34. Three personnel can use a ladder to bridge an area to be spanned as illustrated.

Fig 8–35. As the person at the base moves forward, the other personnel enable the ladder to move forward and over the area to be spanned.

This method is simplified by placing the heaviest and/or strongest person at the rung position because this person will have an easier time overcoming the unbalance of the ladder.

Bridging using two ladders as a derrick. This operation can be effectively accomplished by using one ladder as a derrick to lower the bridging ladder, but is the most time consuming and requires the most staffing to accomplish. Place a 24-foot straight ladder on the ground (rungs down if the rungs are set in the main beams). It should be adjacent to the area that will be bridged. Attach two lines to the top of the ladder by wrapping the end of the lines around a beam and rung and clipping the snaps back on the running lines. Raise the 24-foot ladder to vertical while two personnel are footing the base. Place a 20-foot straight next to the base of the 24-foot ladder (with the rungs down if the rungs are set in the main beams). Put the shoes of the 20-foot ladder against the bottom rung of the 24-foot ladder and strap in place using two ladder straps. Take the lines to the top of the 20-foot ladder and attach by using a bite. Slide the bite under the rung, twist once, and pass over a shoe. The two lines are now connecting the top of the 24-foot ladder to the top of the 20-foot ladder. Adjust the tension on each line so they are equal. Have two persons foot the base of the 24-foot straight ladder and two persons tend the lines. Keeping tension on the lines, lower the 24-foot ladder across the span.

Extending an extension ladder. Place an extension ladder on its side. Extend the fly section across the area to be bridged. Lay the ladder flat across the span.

Training Resources

The following resources can provide additional information and/or another viewpoint for the subjects discussed in this chapter:

- Laddering the Fireground, Mathew Rush, *Fire Engineering*, December 2001.
- Ground Ladder Chocks, Mathew Rush, *Fire Engineering*, March 2007.
- Leading with Ladders, Jonathan Smith, *American Fire Journal*, February 2001.
- Placing Ground Ladders, Roundtable, *Fire Engineering*, July 2001.
- *Inspection and Maintenance of Ground Ladders*, NFPA 1932, chapter 4, 1999 Edition.
- Evaluating the Safety of Ground Ladders, Kurt Kenworth, California Fire Journal, Fall 1985.
- New Technique for Ladder Footing, Michael Ciampo, *Fire Engineering*, April 2004.
- Are Your Butts Marked? Michael Ciampo, *Fire Engineering*, July 2005.
- The Arm Lock Maneuver, Michael Ciampo, *Fire Engineering*, March 2006.

Search and Rescue

Introduction

To paraphrase Webster's Dictionary, the term *search* means "to look through in order to find something" and the term rescue means "to free or save from danger." These two operations are usually thought of as one, but they are in fact two different and distinct operations. Although search is normally considered a top priority at many structure fires, it can also potentially be the most dangerous fireground operation that is performed by fireground personnel.

To underscore this perception, the Roundtable section that is featured on a monthly basis in *Fire Engineering* magazine discussed "The Most Dangerous Fireground Activity" in May of 2007. Due to the number of responses from various ranks around the country, it was necessary to extend the discussion to the June 2007 Roundtable. Not surprisingly, the majority of respondents stated that search was the most dangerous fireground activity. To further accentuate this perception, the majority of fireground articles that were written for *Fire Engineering* magazine between 1995 and 2008 were articles dealing with the numerous aspects of search (for example, conducting a search, search priorities, using thermal imaging cameras for a search, accountability, tag/search ropes, RIT teams, and so on) with articles on building construction running in second place.

Searching for victims within a hazardous environment in a structure that is being weakened by fire has long been an integral priority of fireground operations along with the extinguishment of fire. Therefore, it would be relatively easy to assume that the implementation of a search would be as common as initiating an attack line, forcible entry, or raising a ground ladder; however, this is not the case because the successful implementation of a search demands a high level of training and timely execution. Although numerous fireground operations demand training and timely execution, it is the implementation of these factors in unfamiliar surroundings, hazardous conditions, and minimal time constraints that set search operations apart from most other fireground operations. As an example, consider the following:

- Most search operations are conducted in hot, smoky environments with minimal visibility because if these conditions were not present, occupants would likely be able to exit a structure unassisted.

- Search personnel must often conduct a search by *feel* and not by sight.

- The actual working time available from a typical 30-minute SCBA bottle can vary between 14 to 18 minutes under stressful working conditions.

- Searches are often conducted above a fire, without the protection of a charged hose line, and, in some cases, prior to the implementation of an attack line.

- Flashover is a common fireground problem.

- Modern buildings are collapsing faster than the ones constructed several years ago.

- All structures that need to be searched do not fall into the category of a single-story, single-family dwelling with a common floor plan that is about 1,600 square feet in size.

- Trapped occupants within a hazardous environment can quickly succumb to the effects of heat, smoke, and/or fire. As a result, the time for a search is limited because the average time a human can go without a minimum of 15% oxygen is 4 to 6 minutes, and victims exposed to heavy smoke for 15 to 20 minutes have likely perished.

With these thoughts in mind, let's look at four fundamental characteristics from the perspective of the modern fireground, priority, safety, training, and how they have affected search operations.

Characteristics

Modern fireground

Change is commonly defined as "to put or take in place of something else; substitute; to make different, alter." When applied to the modern fire service as compared to the fire service of just 25 years ago, it is readily apparent that changes have significantly altered apparatus, communications, staffing levels, operational capabilities, and numerous other factors. However, when the definition of change is applied to the fireground of today, there are five areas that

have collectively combined to significantly modify or change current operational considerations (time) and firefighter safety. Those areas are building construction, building materials, smoke detectors, protective equipment, and experience.

Building construction. A high percentage of modern buildings use a form of construction that has reduced the costs of materials and the time necessary to erect a structure. That form of construction is lightweight construction. The simple yet effective truss configuration has allowed multiple small members (such as two-by-three- and two-by-four-inch members) to replace larger structural members in spans that can exceed 70 feet. Although the Uniform Building Code addresses strength, it does not address fireground time, which the modern, lightweight truss has significantly reduced. Another type of building construction that can have a dramatic impact on search operations is the use of thermal pane windows, which are commonplace in a significant number of homes (particularly newer homes) across America. These types of windows use two panes of glass that are separated by an inert gas and provide superior insulative advantages when compared to the single-pane glass windows of yesterday. However, these windows can also be a significant detriment to search operations because they can withstand higher temperatures before failing, thereby allowing flashover conditions to readily occur.

Building materials. Building materials utilized on the interior of *yesterday's* residential and commercial structures were commonly composed of conventional materials (such as paper, wool, wood, cotton, and other similar materials). When these materials ignited and burned, they yielded smoke and fire gases that often required temperatures of over 1,000°F for ignition. With the advent of petro-chemical-based materials (synthetics) commonly used in all modern buildings, the fire service is now faced with materials (such as TVs, carpets, furniture, drapes, and so on) that present a different hazard from a noteworthy perspective—when plastic-based materials are heated, they emit gases and/or smoke that also ignite and burn hotter and faster than the gases and smoke emitted from conventional materials. Therefore, the modern fireground environment is more capable of burning (or flashing over in a shorter period of time) than the fireground environment of yesterday, and also results in less time available to a modern firefighter to conduct a search.

Smoke detectors. The advent of smoke detectors has had a noticeable impact on search operations. Occupants with working smoke detectors can be warned of a fire in progress and can safely exit a structure. Not only has this significantly reduced the number of civilian fireground casualties, but it has also reduced the number of fireground searches that are performed on an annual basis. Although this is a remarkable achievement, it also underscores the need to train and be properly prepared when a search is necessary.

Protective equipment. Just a few years ago, typical protective equipment used by attack personnel consisted of boots, pants, coat, gloves, SCBA, and helmet. Although this combination had its advantages and disadvantages, it did allow personnel to use their ears and back of their hands to evaluate the level of heat in a fireground environment. Now, the typical modern firefighter is additionally protected by a hood and wristlets, which has resulted in protective equipment that totally encapsulates personnel. This has allowed personnel to withstand higher temperatures for a longer period of time, and enabled a quick advance into a hazardous atmosphere. However, these advantages can be easily offset by the same protective equipment masking the conditions that interior personnel commonly used in the past to monitor the fireground environment. It is now not uncommon for a firefighter to remain too long in a hazardous environment.

Experience. The fire service experiences fewer fires today when compared to yesterday. Just as sprinklers, smoke detectors, and other similar factors have collectively reduced the number of fires, the experience level of personnel has also been reduced. Therefore, when a modern firefighter arrives at a typical one-room fire in a single-family dwelling (which NFPA indicates is 60 to 70% of our fires), the excitement level can be abnormally high, creating an environment where personnel forget a few basics, thereby reducing fireground safety and increasing the potential of injury or death. At this point, you are thinking that experience can be enhanced or substituted with training. Although this can be accomplished, ask yourself the following question—has the level of training in your department increased or decreased in the area of basic fireground operations?

Let's combine the preceding five areas of consideration and summarize their impact (or change) on the modern fireground:

- A decreased level of experience and confidence can easily foster an increased level of excitement, dull the fine art of the decision-making process, and encourage personnel to forget the basics.

- An increased ability of modern lightweight buildings to offer minimal fireground time, which a firefighter may not be aware of, may result in neglecting to conduct a proper size-up.

- Flashovers have increased on the modern fireground environment.

Priority

Any discussion regarding search operations should include the perspective of the priority of a search when formulating fireground priorities. As delineated in chapter 1, initially determining fireground priorities is potentially the most important consideration for truck company personnel because each incident is unique and always requires evaluating and prioritizing of fireground operations. Despite the fact that each incident determines fireground priorities, there is one priority that is a constant at every incident: firefighter safety. There are no one-way tickets in the fire service because every company that responds to an incident should focus on two basic priorities: returning to quarters with the same number of personnel that responded to the incident (firefighter safety), and not adding an additional problem to an existing problem. So, if firefighter safety is the top fireground priority, what is the second most important fireground priority? Based on operations that are frequently conducted at structural incidents, forcible entry and/or ventilation should be the next most important fireground priorities. These two priorities are the cornerstone of truck company operations because they are primarily designed to improve firefighter safety. Let's look first at forcible entry. Before a search is ever conducted and water is put on a fire, forcible entry is normally required *before* interior operations are initiated. Therefore, if forcible entry is a common requirement before interior operations, then forcible exit should be the next most important fireground priority based on the principle of *if it is important to get into a building, it should be as important if not more important to be able to get out of a building.*

Now take a quick look at ventilation. Based on the facts that the National Fire Protection Association (NFPA) statistics indicate that flashover is one of the top three fireground problems encountered on the modern fireground and fires burn significantly faster and hotter than the fires of yesterday, ventilation should (depending on conditions) be a requirement to improve the interior environment of a structure *before* the implementation of interior personnel. Therefore, in the interest of firefighter safety, and depending on conditions encountered, ventilation should be considered an equal priority with forcible entry and forcible exit.

After firefighter safety has been maximized by ensuring firefighters can enter and exit a structure in a timely manner, and/or the environment has been enhanced so visibility has been improved and the potential of flashover has been minimized, then search, fire attack, and other interior operations are the next priorities.

To summarize creating priorities based on firefighter safety, the following fireground priority list is recommended as a long overdue replacement for rescue, exposures, confinement, extinguishment and overhaul (RECEO):

1. Firefighter safety

2a. Forcible entry

2b. Forcible exit

2c. Ventilation (the order of these may vary depending on the incident)

3. Fire attack and/or search

4. Other necessary priorities

Note that the order of priority for these items is determined by conditions encountered at each incident.

Safety

The subject of fireground safety can be approached from many viewpoints, so let's take a different approach by applying the following perspectives to search operations.

- Whose safety is most important on the fireground, civilians or firefighters? Based on the concept that suppression resources should not take or add additional problems to an incident, the obvious answer is the safety of firefighters is the number one priority, and civilians are number 2.

- The popular acronym of RECEO defines the R as rescue (or search) and implies that search is the most important fireground priority. Therefore, a search should always be balanced against the hazard to search personnel, or risk versus gain.

- Statistics indicate that suppression resources responding to residential structure fires will most likely (70 to 90%) encounter a vacant structure because the occupants are not home or have vacated the structure prior to the arrival of suppression resources. Therefore, before conducting a search, do search personnel *assume* or *know* a search is necessary?

Now combine the preceding three perspectives into a set of operational concerns that can be directly applied *as a starting point* to search operations:

- Search operations must be conducted in accordance with an ongoing risk analysis that should carefully weigh the chances of a firefighter losing a life against the realistic chances of finding and rescuing live occupants (Bill Gustin, *Fire Engineering*, November 1996).

- There is a significant difference between searching for civilians and firefighters. If suppression personnel are suddenly trapped in a structure, there is no doubt that a search is necessary. However, when arriving at a structure fire, is a search necessary because search personnel feel it is the thing to do, or because personnel know there is someone trapped inside the structure? As an example, compare the difference between several helpful neighbors indicating they think there are trapped occupants (which is often wrong) and a husband or wife who states they just escaped from the structure and their two-year-old son is still upstairs in his crib.

- Based on the fact that firefighter safety is the top fireground priority, determine your level of search commitment from one of the following risk categories:

  1. Do you know, assume, or think persons are trapped, and is a search necessary? Remember that most firefighters killed while searching for persons reportedly trapped died while searching for persons who were not there.

  2. Based on the perception, a search will be conducted.

     Determine if a search can be conducted in a safe and timely manner with minimal risk to search personnel?

     Determine if a search can be conducted with moderate risk to search personnel. In this option, how can you reduce the level of risk?

     Determine if a search can be conducted but with a high potential of risk to search personnel. If this operation is attempted, determine what safety measures can be implemented to minimize the risk hazard.

     A search cannot be conducted due to an unacceptable risk to search personnel. In this case, other options should be considered.

  3. If you want to extend yourself or push the envelope in conducting a search:

     Compare the risk to search personnel and to the occupants.

     Evaluate the environment (structural stability, thermal layer, and so on).

     Determine if the search can be conducted in an acceptable amount of time or if it will be a time and/or resource intensive operation.

     Determine if search personnel have any viable alternatives in case of a problem.

     Determine if search personnel can find an acceptable exit in a timely manner.

Determine if the fire, in the process of being extinguished, will be extinguished in an acceptable amount of time, or there will be an undetermined amount of time before the fire is extinguished.

Training

One of the primary reasons for the difficult nature of a search is infrequent use. As an example, when was the last time you actually searched a structure for a trapped occupant? So, although search operations can easily fall into the category of *out of sight, out of mind*, it is mandatory for personnel who may be engaged in search operations to practice search (and rescue) techniques to develop a workable variety of SOPs that can be applied within a framework of teamwork and confidence. That is why the old adage of *practice, practice, practice* is applicable to this discussion. However, remember that practice does not make perfect; perfect practice makes perfect. So train the way you fight. When it is determined that a search operation is necessary, it is time to be able to immediately implement an organized, pre-planned operation. Anything less can likely be ineffective and extremely dangerous.

Size-up

A size-up is defined as *an estimate of the situation through which you determine what to do and how to do it*. Use this definition when considering elements that should be used in conducting a search size-up.

Time of day

Depending on the type of occupancy, the time of day can begin to set the parameters for a search (figure 9–1). As an example, during nighttime hours, particularly 10 p.m. to 6 a.m., residential occupancies can present high-occupancy loads. This applies to the common single-family residential-type occupancy or large commercial residential occupancies such as motels.

Figure 9–1. The time of day can set the parameters for a search operation.

An important consideration that often specifically applies to commercial residential occupancies is that the number of vehicles in a parking lot can be an indicator of the number of persons that can be expected inside the structure. Conversely, occupant loads can be expected to be high during daytime hours in industrial-, retail-, and commercial-type occupancies. Interestingly, occupants are seldom lost in commercial incidents because occupants are ambulatory on day, swing, or graveyard shifts, whereas occupants in residential structures (particularly at night) are non-ambulatory. However, hospitals, institutions, senior rest homes, and other similar occupancies with non-ambulatory patients can present high-occupant loads during all hours of a day. Additionally, remember the time of year can also affect occupant loads. As an example, summer months can result in low-occupant loads in schools, and major holidays such as Easter, Thanksgiving, Christmas, and other similar holidays can minimize the presence of occupants in commercial and industrial occupancies. However, holidays can increase the number of occupants that may be present in residential homes during daylight and nighttime hours.

Type of construction

The type of construction (lightweight or conventional) will often indicate the structural stability and amount of time a structure can withstand the effects of fire and enhance the potential extension of fire (this is covered in chapter 4). Therefore, conventional construction (as compared to lightweight construction) can often provide additional structural

stability and time when exposed to fire and result in additional time to conduct search operations. Conversely, lightweight construction can result in rapid structural collapse, and open webbing in lightweight trusses can enhance the rapid extension of fire, which can also reduce the amount of time to conduct search operations. Let's look at two examples that clearly demonstrate a diversity in the type of construction and potential hazard. Figure 9–2 is an example of Main Street USA in that these single-family dwellings are easily recognizable as typical buildings that were built in the 1930s and 1940s and were constructed with conventional construction.

Figure 9–2. These structures are recognizable as being built in the 1930s and 1940s.

Let's assume there is a working fire on the first floor of one of these structures, and you are directed to conduct a search on the second floor above the fire. Although the presence of balloon frame construction can spread a fire vertically up the walls, the floor joists between the first and second floor are likely two-by-six- or two-by-eight-inch, full-sized lumber; the walls and ceilings in these structures are covered with lath and plaster. Although you will be searching above the fire, this type of construction can give you significantly more time before collapse when compared to the next example.

Conversely, figure 9–3 is an example of a modern townhome apartment building that was constructed with lightweight construction and numerous open or void spaces (some of which are a result of open web trusses) and other alternative materials that have replaced conventional construction.

Figure 9-3. These modern townhome apartment buildings likely use lightweight construction.

Again, assume a fire is on the first floor and you are directed to conduct a search on the second floor above the fire. In contrast to the previous example, the floor joists will likely be lightweight parallel chord trusses or truss (TJI) joists, and the walls and ceilings are covered with ½-inch drywall that does not last as long as lath and plaster, and fires involving TJI floor joists have resulted in fast failure times. Figure 9-4 is an example of fire quickly weakening TJI joists, yet the floor above looked substantial.

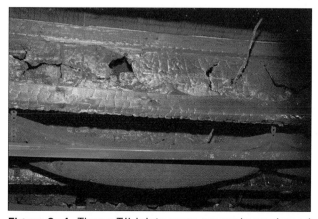

Figure 9-4. These TJI joists were severely weakened by fire and will not support additional weight. (Photo courtesy of William Gustin)

The floor being supported by these weakened TJI joists will not support the weight of firefighters conducting a search. Fires in lightweight construction are a prime reason why search personnel should sound their intended path of travel when searching

above a fire. Additionally, remember that lightweight construction exposed to fire will likely not give you appropriate time to conduct a search.

From these two simple examples, it is easy to see that a fire in the structure in figure 9-2 will burn longer before structural collapse when compared to the structure in figure 9-3. Always consider the potential impact of building construction on structural stability and avenues for fire extension. There is no substitute for developing a familiarity with the buildings in a specific area.

Type of occupancy

In concert with the time of day and type of construction, the type of occupancy also sets the parameters for a search operation. The type of occupancy should be evaluated for the following factors.

Style of occupancy. Try to visualize the basic layout of the interior of an occupancy. Is the structure a single-story, single-family dwelling, a large industrial/commercial occupancy, a multi-story hospital, a high-rise office building, or a commercial/residential building? The complexity (or lack of complexity) of various types of occupancies will require different search considerations. This is why looking at the interior floor plans of the buildings in your district when on fire prevention, EMS calls, and other similar occasions can often provide the opportunity to acquaint firefighters with a working knowledge of the interior floor plan of various types of buildings that are common to a district. There is nothing to replace the familiarity of the interior of a building that you have previously been into. Occupancies can be summarized as follows.

Residential. The definition of a residential occupancy can significantly vary because there are a large number of structures that easily fall under this classification. Some examples of these structures are a common 1,600 sq. ft. single-story dwelling; a 10,000 sq. ft. three-story McMansion; a Holiday Inn Express; a large Victorian-type structure that has been remodeled into a bed and breakfast, and so on. Obviously, the complexity of a search will vary depending on the size of these types of structures. As an example, a common single-story, single-family dwelling can be searched in a timely manner by two personnel due to a moderate size and common floor plans. Conversely, apartment houses and hotels can be large structures that contain multiple floors and numerous rooms, and can require lengthy operations and multiple search teams. Without a doubt, residential-type occupancies present the highest need for search operations as most loss of life from fire incidents occurs in these type of occupancies as they may contain occupants during the day or nighttime hours. However, the occupant load will be greatest between 10 p.m. and 6 a.m. with the occupants generally non-ambulatory (asleep).

Let's look at the diversity of size when applied to a single-family residential occupancy. First, figure 9–5 depicts a common two-story, single-family dwelling of approximately 1,600 sq. ft. A quick size-up of this structure indicates the first floor is the living area composed of a living/dining room, kitchen, and service porch.

Figure 9-5. This structure is an example of a common two-story home of approximately 1,600 sq. ft.

Notice there is an extension on the right side of the house that is likely a bathroom (for example, the small high window on the wall). Just inside the front door would be stairs to the second floor hallway that runs between the two roof dormers and services two bedrooms that are likely separated by a small bathroom. The second floor would be referred to as the sleeping area or floor, and the rooms on the second floor would be relatively small (possibly 8 by 10 feet). A primary search on the first or second floor would be simple because the floor plan is a common plan that should be familiar to personnel responsible for this district.

Figure 9–6 also depicts a two-story residential structure. However, this structure is more complicated than the preceding structure.

Figure 9-6. This structure is an example of a two-story home that is more complicated than the typical residential home.

The house is of an uncommon design or style. Stairs to the second floor are visible on the left side of the structure. This is not a normal configuration unless the house has been converted into a multi-residential occupancy. A quick size-up could indicate a minimum of three occupancies—one on the first floor in the front, another on the second floor and serviced by the stairs on the left side, and another in the back of the structure. However, notice the box to the left side of the front stairs that contains mailboxes for this structure. If you were on-scene, you would quickly notice there are four mailboxes that indicates four occupancies. It is likely this structure has been remodeled from its original design, the floor plan may be a challenge to discern from the exterior, and there can be more occupants within the structure at night than the structure shown in figure 9–5.

Another type of occupancy, shown in figure 9–7, that falls under the classification of a habitational structure is very common—a motel/hotel (or for this discussion a Holiday Inn-type occupancy). This exhibits different noteworthy characteristics than the previous two examples as follows.

Figure 9–7. This residential occupancy exhibits different noteworthy characteristics than a common residential structure.

The structure in figure 9–7 shows there are three stories and at least 50 units (because there are 25 units on the front of the building, there will likely be at least 25 units on the backside of the building). It appears to be a center hallway configuration (delineated in chapter 4). As opposed to a garden apartment configuration, a center hallway can be more difficult to search, depending on the status of the hallways (clear or charged with smoke) which can also affect the ability of trapped occupants to exit the building. These considerations are normally less important for garden apartment type structures.

There are three vehicles in the parking lot on the front side. This immediately raises the question of the presence or absence of a parking lot on the backside and the total number of vehicles if present. Obviously, the parking lot would have more vehicles at night as opposed to daytime hours and is a good indicator of the potential number of persons (plus the office staff) in the building. Vacant hotels can be a prime location to find homeless occupants during any portion of a 24-hour day if the building is not completely secured. Finally, multiple stories in these buildings can enhance the vertical extension of heat and smoke thereby compounding search considerations.

Industrial/commercial. A complete chapter could be written on these types of buildings because the size and complexity can significantly vary from a relatively small commercial building to a large warehouse type structure. One feature all industrial and commercial buildings share are occupant loads, which are normally greatest during the day.

However, some occupancies employ reduced staffing levels during nighttime hours and/or use swing and graveyard shifts. All of the occupants will normally be outside of these buildings prior to the arrival of suppression personnel unless there has been some type of an explosion or other similar problem as the occupants are normally ambulatory. The major hazards that search teams may encounter are large areas with varying floor plans, and the presence of machinery (electrical), tiered storage (collapse), open shafts (stairshafts, loading docks, elevators), hazardous materials, high voltages, mezzanines, and other similar hazards that are dependent on the type of building and occupancy. As an example, an excellent indicator of the interior contents and layout of a commercial occupancy can be provided by the name of the business (if present) on the front of a building (for example, Joe's Pool Supply, Fred's Barber Shop). Again, the number of cars in the appropriate parking lot is an indicator of the number of occupants that need to be accounted for.

Figure 9–8 illustrates a typical *larger* commercial occupancy that appears to be two story but also has a basement for a total of three stories. The name on the upper-right portion of the building (Joseph Winans Furniture Company) indicates the building is likely composed of large display areas filled with furniture (which is a high fire load).

Figure 9–8. A search of this commercial building will require a significant amount of time and personnel.

This configuration will be difficult to search and would take a considerable amount of time and personnel.

Institutions. In these occupancies, emphasis should be placed on the size (particularly the height and number of floors) of the institution, the type of the institution (nursing home, dependent living, hospital, jail, school), and the fact that most institutions (except schools and similar occupancies) will have a high occupant load (staff and patients) 24 hours each day. Additionally, occupants can be non-ambulatory (such as in hospitals) or detainees (jails, for example) and may need to be removed from the structure and facility. This is the primary reason these types of occupancies present the greatest potential for a noteworthy search and even a rescue operation. Sadly, in some cases the staff of these types of buildings will likely be waiting for the arrival of suppression personnel and more than happy to indicate there is a major *people problem* inside the building.

In figure 9–9, notice the sign on the lawn indicates this building is retirement living, which can mean that some occupants are ambulatory and some occupants are non-ambulatory. There can be a large difference among *retirement living*, *dependent living*, and other similar nomenclatures.

Figure 9–9. Retirement and dependent living structures can present notable search considerations.

It is imperative that you are familiar with the different types of these structures in your district and how a search would need to be conducted. Additionally, notice this is a large three-story building. However, it is common for these types of occupancies to be smaller and/or even located in a single-family dwelling that has been converted to this type of occupancy. Know your district!

Public assemblage. The type of public assemblage (restaurants, night clubs, churches, pool halls, and so on) will determine the search potential. Restaurants can have high occupant loads from 6 a.m. to 12 p.m. whereas nightclubs are normally occupied from 6 p.m. to 2 a.m. Generally, conscious ambulatory occupants and sufficient exits can result in reduced search considerations as ambulatory occupants can exit a building before the arrival of suppression personnel. If search operations are necessary, large complicated floor plans in concert with a questionable occupant load can be expected. As an example, the church in figure 9–10 is a common example of a public assemblage occupancy that can be found in numerous districts.

Figure 9–10. Churches are a good example of a public assemblage type building.

Even though most churches are filled with people on Sundays (morning and/or evening), some churches have staff during the week and can also have day schools, and so on. Notice there are numerous vehicles in the right portion of the photo that are parked next to the building.

High rise. High-rise buildings are normally defined as buildings over 75 feet and utilize center core construction with circuit hallways around the core, or center hallways that terminate in vertical stairshafts on either side of a building. Residential high-rise buildings (see figure 9–11) can be apartments, hotels, condominiums, hospitals, and so on. These occupancies are generally characterized by numerous interior compartments of common repetitive floor plans, and a 24-hour occupant load.

Figure 9–11. High-rise buildings can be found in commercial, residential, or a combination of each configuration.

Commercial high-rise buildings are generally characterized by relatively larger open areas, and/or numerous offices, with occupant loads greatest during daytime hours. Regardless of the type of structure, maintenance personnel may be present during nighttime hours. In the event of fire in a high-rise building, danger to building occupants can develop from the following factors:

- The greatest hazard to building occupants is *exposure* to smoke and products of combustion (MGM fire in Las Vegas). Vertical extension of these products is enhanced by poke-through construction, heating, ventilation, air conditioning systems (HVAC systems), elevator shafts, stairwells, and so on.

- Fire in residential occupancies is often limited to the area of origin due to partition walls. However, the presence of work stations in commercial buildings will enhance the extension of fire as demonstrated in the First Interstate Bank fire in Los Angeles.

- Building occupants can panic from a knowledge or belief that a fire is in progress in the building, and occupants may be found during the day, night, and weekends.

Searching a high-rise building can be a time- and personnel-consuming operation. Depending on the type of fire, the entire building may need to be searched to ensure all occupants are accounted for. Therefore, a search should be conducted on a priority basis, beginning in the immediate area of a fire and then advancing to areas directly above a fire (that have been exposed) and concluding with the balance of the building above and below a fire. The appropriate building keys are mandatory to ensure quick access to necessary areas (which includes all elevator cars). Documentation should be utilized to ensure all appropriate areas are searched and to minimize duplication as some high-rise buildings contain a considerable number of occupants. Therefore, removing occupants during a fire may not be practical. If necessary, an alternative to removing occupants is to relocate the minimum number of appropriate occupants to areas of safe refuge as follows:

- If occupants are evacuating a building upon arrival of suppression resources, suppression personnel should attempt to gain control (if necessary) of the evacuation. Initially, the incident commander should determine which stairwell is most suitable for fire attack operations (normally the stairshaft with roof access) and concentrate suppression resources in that stairwell to control the fire. Other stairwells can then be designated as evacuation stairwells. This can often be organized by using a building intercom, or beginning at the second floor, personnel must make their way to each appropriate landing in the evacuation stairwell and attempt to control exiting occupants.

- If necessary, relocate occupants from the fire floor (and at least two floors above and below a fire) to a floor more than two floors below the fire. Depending on the intensity of a fire, it may be necessary to relocate occupants of additional floors above the fire.

Size of occupancy

The size of an occupancy is an excellent indicator of the potential complexity of a structure, its floor plan, and the number of personnel required to conduct a search operation. As the size (single-family dwelling vs. warehouse) and complexity (mini-mall vs. a large multi-story hospital) increases, so does the number of personnel required to conduct search operations. There is a significant difference between searching typical bedrooms and living rooms in the home in figure 9–5, and attempting a large-area search in the commercial building in figure 9–12. The focus on applying a search to various sizes and types of structures is that one type of search operation does not fit or work in all structures.

Figure 9–12. Large commercial buildings can be difficult to search in an acceptable amount of time.

Therefore, ensure that you are comfortable searching a small single-story, single-family dwelling as well as a large industrial building. In summary, size and the appropriate search operation matters.

Floor designation

Buildings with multiple floors can present a significant problem, particularly if a building is on sloped ground that results in different floor levels, depending on which side the building is viewed from. Depending on the layout of a building, it is possible to view a building from the front, quickly determine the number of floors, and then view the same building from the side and/or back and see additional floors that were not visible from the front. This is one reason why it is important to view a building from as many sides as possible. Although this can be easily solved during a 180- or 360-degree size-up, the important considerations are that all fireground personnel are aware of any variance within a particular structure, determine what designation is used for varying floor levels, and ensure that all personnel are aware of the terminology for varying floor levels. If not, then it is possible for search personnel

to miss a complete floor, yet believe that all floors have been searched or accounted for. This is an unacceptable fireground mistake that can also apply to other fireground operations as well as search. A common method to designate multiple floor levels within a structure can be summarized as follows: a basement is less than 50% below grade and will normally have large windows; and a cellar is over 50% below grade and will normally have small rectangular windows (see chapter 4). Remember that basements and cellars that can be entered from the exterior are commonly found under homes on hillsides, and are known as a walkout or a dungeon unit.

Although the preceding method can be used on many buildings, there are some configurations that fall outside the previous classifications. As an example, the home in figure 9–13 is a common structure and would normally be called a split-level home.

Figure 9–13. Although this home is a common example of a split-level configuration, designating each level can be confusing unless a common terminology is used.

Notice the garage is the lowest level, the second floor is above the garage, and the upper grade level (or first floor) is between the garage and second floor level. How would you designate the levels in this home? Based on the previous discussion regarding basements and cellars, the garage would be the first floor, the upper grade floor would be the second floor, and the floor above the garage would be the third floor. However, if a pre-determined method does not apply to a specific structure, it may be advantageous to use the Keep-It-Simple-Stupid (KISS) principle, which would designate the various levels on a size-up as follows: the garage would be the garage level, the second story would be the floor above the garage, and the first floor would be the portion of the home to the right of the garage. Therefore, a radio message to a search team would then be as follows "Truck one, I need a primary search on the floor directly above the garage." As previously mentioned, it is mandatory that all fireground personnel speak and understand the same language, regardless of the method that is used.

Status of a building

As applied to search operations, evaluating the status of a building can begin to create a foundation for determining the risk and need for a search as applied to four basic categories of buildings: occupied, unoccupied, vacant, and abandoned.

Occupied. These buildings are defined as buildings that are occupied or have a high probability of being occupied during an incident. As an example, the townhome complex in figure 9–14 consists of three occupancies, two of which appear to be unoccupied. However, the townhome at the left side of the photo has two cars parked in the driveway.

Figure 9–14. Regardless of the time of day, the presence of cars can be an indicator of occupants within a building.

Even though this photo was taken during the daytime hours, it would be reasonable to assume there is a *high* probability of occupants within this townhome, whereas the presence of occupants in the other two townhomes may be lower. (Remember that some occupants park in their garages.)

Unoccupied. These buildings can best be described as a building that is normally occupied but the occupants are likely to not be in the structure. As an example, the building in figure 9–15 is a restaurant, there is a closed sign in the front door, the building is secure, and there are no cars parked near the building. Additionally, this photo was taken outside normal business hours. The chance of occupants inside this building is remote at best and should not be an immediate concern.

Figure 9–15. A lack of activity and a closed sign on this commercial indicates the presence of occupants is remote.

Vacant. The primary difference between a vacant building and an abandoned building is vacant buildings are likely to still be in an acceptable condition in terms of structural integrity, particularly when a For Sale or For Lease sign is displayed on the front of the building (figure 9–16). In this case, the owner has likely kept the building in a *marketable* condition and there should be minimal surprises. These types of buildings are normally locked to prevent unauthorized entry, and in some cases, may have plywood/OSB materials over the doors and windows to further restrict entry. In either case, the presence of occupants inside this type of building is unlikely and should not be a primary concern.

Figure 9–16. Occupants within secured, vacant buildings should not be a primary concern.

Abandoned. If a building is abandoned, this is a primary indicator the building has likely outlived its usefulness (there is a good chance the building will be demolished), and structural integrity may be in disrepair. This can apply to the roof structural members, floors, stairs (if present), and so on. Additionally, if the windows and doors are covered with plywood/OSB type materials and the building appears to be secure (figure 9–17), then the presence of occupants inside the building should not be a primary concern.

Figure 9–17. It is doubtful there are occupants within abandoned, secured buildings.

Fire location

If possible, the location (and extension) of fire must be determined. This equally applies to horizontal extension in single-story structures and horizontal/vertical extension in multi-story structures. The extension of fire and smoke to additional areas of a structure will indicate areas that should also be searched. When the location of a fire is applied to search operations, a risk estimation can begin to be developed. Let's look at two simple examples.

Assume that a search needs to be conducted on the grade floor of a single-story structure. The advantages of this type of search are the fire is not below search personnel (structural stability), access and egress into and from the structure is simplified as compared to above-ground search operations, and horizontal ventilation can also be simplified as compared to above-ground floors. However, it is important that search personnel determine if fire is overhead (particularly in lightweight construction) because collapsing roof structural members can result in a deadly environment.

Next, assume that a search needs to be conducted on the second floor of a two-story structure and the fire is on the first floor. The advantage of this type of search is that the search team is not on the same floor as the fire. The disadvantages of a search above a fire are access and egress can be more difficult when compared to a search on the grade floor, timely ventilation can be more difficult, smoke above a fire can be more deadly than the actual fire (as was graphically demonstrated at the MGM fire in Las Vegas), and a search crew will be searching above a fire that may be weakening the floor members. This is particularly important when the structure used lightweight construction. (Refer back to figure 9–4.)

Environment

Evaluate the environment of the intended search operation. Determine if it is hazardous, clear, near the seat of the fire, or removed from the location and extension of fire. Always remember that the fireground environment of today is significantly different from the fireground of a few years ago, and the fireground environment of tomorrow will be different than the fireground environment of today. As explained in chapter 5, the fireground environment of today is capable of flashing over in a shorter period of time and with more intensity than the fireground environment of yesterday. This fact can reduce the amount of available time to conduct a search operation and should increase an awareness of interior personnel to constantly monitor and evaluate their fireground environment. Additionally, this is a prime reason why ventilation of any type that can be safely implemented as soon as possible can dramatically improve the interior environment from the following three perspectives: visibility can be improved; carbon monoxide and other noxious gases will be reduced; and internal temperatures will be reduced and minimize the potential of a flashover. Remember that a flashover does not differentiate between search and fire attack personnel.

An additional consideration is to determine the stage of the fire or what is the growth stage prior to conducting a search. From a practical perspective, the four growth stages for this discussion are:

- **Early.** The fire is beginning to grow and has not had a significant amount of time to significantly weaken structural members or produce an environment that is conducive to a flashover.

- **Advanced.** The fire has likely had sufficient time to weaken structural members, produce minimal or no visibility, and present favorable flashover conditions.

- **Fully involved.** The fire is free burning and has had sufficient time to weaken structural members and produce a hot environment that can be conducive to flashover conditions (depending on the amount of ventilation).

- **Smoldering.** The fire does not have sufficient oxygen to freely burn. This stage can be present at the beginning or end of a fire and may be difficult to evaluate its affect on structural members. However, a fire that has smoldered for a sufficient amount of time can easily produce minimal or no visibility, low oxygen levels, high concentrations of noxious gases, and conditions favorable to flashover conditions.

Access/egress

The access of personnel to upper floors of multistory buildings is normally accomplished by interior stairshafts, external ground ladders, and/or aerial devices. These avenues should be evaluated as follows.

Interior stairways can provide direct access to specific floors. However, they can easily be contaminated with heat, smoke, and minimal or no light that can decrease their viability and safety (for example, the Philadelphia high-rise fire). Additionally, if an interior stairway is used to provide access to a floor above a fire, it is possible for fire (until it is extinguished) to extend up the stairway and eliminate it as a means of egress.

External ladders can provide direct access to a specific location (window, balcony, fire escape, roof, and so on) and are generally located in uncontaminated atmospheres (outside a building). However, correct apparatus placement is necessary (and can be time consuming) if an aerial device will be used as opposed to portable ground ladders. If external ladders are used, particular emphasis should be placed on openings that are easily opened and can provide the best access to an area to be searched. As an example, assume a fire is on the first floor of the two-story home in figure 9–18, it is 2 a.m., there are heavy smoke conditions on the first floor with some extension of smoke to the second floor, and it is deemed necessary to conduct a search on the second floor.

Figure 9–18. If fire is on the first floor with some extension to the second floor of this structure, entry can be made through the second floor windows to allow a quick search of the rooms common to the windows.

Instead of entering the first floor and using the stairs to access the second floor, the roof over the front porch can be laddered, providing a quick access to the windows to the second floor and also providing a quick egress route from the second floor. This principle can also be applied to single-story structures such as a common single-family dwelling by entering the window(s) at the rear of a structure (normally the location for bedrooms) to conduct a quick search as opposed to entering the front door to access the bedrooms at the rear of a structure. These operations are often referred to as vent, enter, and search (VES).

Remember to consider openings that can be used as escape routes that have been blocked by security bars and/or multiple locking devices. Of particular interest are vacant structures that have utilized plywood/OSB or other similar materials to secure doors and windows. Any plywood/OSB-type materials that are removed from a door or window should be completely removed to enhance ventilation and provide an easier ingress/egress opening. Additionally, if plywood/OSB-type materials are removed to allow suppression and/or search personnel access to the interior of an occupancy, it should be mandatory to require the immediate removal of other similar materials that may be covering other doors and/or windows that can negate interior personnel utilizing multiple openings for exit openings. Prior to conducting a search, carefully evaluate all openings in a structure (windows, doors, skylights) that can be used for reference points (from inside the structure) and/or entry/exit locations.

Type of rescue

Consider the number, location, and type of occupants to be rescued. The potential number of occupants should be an excellent indicator of the number of personnel necessary to affect a timely rescue. Too, the size of a building is an indicator of the time and amount of resources that will be necessary to conduct a search in an acceptable amount of time.

Occupants who are located on the exterior of a building (fire escape, window, and so on) can often be easily rescued with an aerial device and/or ground ladders. Conversely, the rescue of occupants from the interior of a structure can require a more involved and time consuming operation.

Victims just above a fire are in greater danger than victims below a fire, and victims that are in rooms with visible heat and smoke are a higher priority for a timely rescue than victims that are in uninvolved rooms. As an example, assume you respond to a four-story residential structure with visible people in windows on the third and fourth floors. Additionally, assume there is visible smoke from the third floor windows with the visible occupants and no visible smoke from the windows on the fourth floor with the visible occupants. When suppression personnel arrive on-scene, it is likely that the visible occupants on the third and fourth floors will immediately begin to call for assistance. In this case, who is rescued first? Generally, occupants in a contaminated area take first priority, and occupants in an area that is not contaminated take the second priority (or in some cases can be left in place).

Determine if the search is for civilians or firefighters, and if their location is known or unknown. Although civilians can often be found in common locations (such as bedrooms in residential occupancies at night), they can often choose odd locations (bathtubs, closets, and so on) to escape the presence of heat, smoke, and fire. Conversely, searching for firefighters can be assisted by the knowledge of a previous location, presence of hose lines, personal alarm (PASS) devices, radio communications, and more.

Determine the age of expected occupants and their ambulatory condition. Expect a larger search if a building with non-ambulatory occupants (such as a rest home) is encountered.

Aggressive search

There is some discussion if a search should be conducted in an aggressive manner. Some recent training materials advocate conducting aggressive searches (as an example, see *Fire Engineering*, April 2002, "More Aggressive Searches, More Lives Saved?"). First, let's define the word *aggressive* with Webster's dictionary: boldly hostile; bold and active; enterprising. When this definition is applied to the concept of search, it implies that bold (and aggressive) searching has the potential to save more lives. However, the question of at what cost should also be applied to the concept of aggressive searches. When the issue of the modern fireground environment that features rapid collapse times and flashover is applied to search operations, it quickly becomes obvious that a measured approach instead of a bold approach can be a more realistic and safety oriented step towards maximizing fireground safety by considering the risk versus the benefit concept. However, there are times that an aggressive search is called for, such as when it is *known* that an occupant is trapped inside a burning structure. However, a blanket statement that aggressive searches can save more lives does not appear to consider that the same approach may also needlessly put firefighters at greater risk and additionally may cost some firefighters their lives.

Hose lines

An area of intense debate when applied to search operations is the need for search teams to take a charged hose line when conducting a search. Although taking a charged hose line with a search team is not advantageous in most searches (as it can unnecessarily add to the complexity and time of a search), let's briefly consider a scenario where the presence of a charged hose line can be advantageous. Assume you are the first company to arrive at a working fire in a single-story, single-family dwelling, you are notified there are trapped occupants inside the structure, the most advantageous entry point to begin the search is the front door, and the next company will arrive in three to four minutes. At this point, you are faced with a dilemma: Assuming you will conduct a search, and the next company will initiate an attack on the fire, do you conduct a dry or wet search?

A *wet search* is defined as taking a charged hose line into the structure and placing the line so it would be relatively accessible if necessary instead of taking a charged hose line into all areas that are searched. The advantage of this operation could be a matter of life and death for search personnel based on the fact that the search will be conducted in a hazardous atmosphere that is rapidly degrading without any attack personnel advancing and extinguishing the fire. If the environment were to suddenly exhibit flashover conditions, then the presence of a charged line has the potential to assist personnel in exiting the structure in a timely and safer manner as compared to the next example.

A *dry search* is defined as conducting a search without having the availability of a charged hose line. In this case, the search team would begin to conduct a search before the arrival of the next company who would likely be assigned to attack operations. If the environment suddenly exhibits a flashover condition, there is a partial building collapse, or other similar conditions that are immediately injurious to firefighter safety, the lack of a protection line could be disastrous. Remember that one of the three primary problems on the modern fireground is flashover. To conduct a search in an environment that is going downhill without attack personnel advancing on or suppressing the fire, and without any means of protection is potentially placing search personnel in an extremely dangerous situation. Without a doubt, if you suddenly found yourself in this position, the first thought that would likely go through your mind would be, "I wish I would have brought a protective line!" Granted, this scenario does not apply to all searches, but it does to a selected few searches where a decision must be made to place the occupants safety first, or enhance the safety of search personnel.

Accountability

Although every firefighter should be aware of accountability and how it applies to fireground safety, a popular misunderstanding is that accountability primarily applies to incident commanders, division officers, and passport systems. Although fireground accountability unquestionably applies to the three previous items, it also applies to personnel engaged in search operations from three basic perspectives: search teams should be composed of a minimum of two personnel; search teams often operate in areas remote from suppression personnel and/or may not have the protection of a charged hose line; and exterior command personnel should be aware of personnel conducting a search, their location (or approximate location), and the progress of the search. *Washington Firefighter* published an excellent article in 1999 by retired chief Dennis Compton that can be directly applied to search operations. Selected portion follow.

Regarding company officers, Compton says,

When I was a company officer, the most difficult part of my job at a structure fire was keeping track of my firefighters because, at times, it was like trying to keep kittens in a box. We went through the door and the firefighters would break off from each other if we weren't careful. Company officers are responsible for the members assigned to them. That means the officer who knows who's there, knows exactly where each firefighter is located and knows what each is doing. The officer knows how each is doing, and if they need any help. If the officer knows these things, he or she is the supervisor of that crew.

Could you ever find yourself in a situation as a company officer where your crew is struggling with a task or something? Might you step in and take over what they are doing, and do it yourself? At that point, who is in charge of the crew and their overall safety and accountability? Company officers are responsible for the crew members that are assigned to them at all times—period. Be careful about focusing so much on performing a single task that, as an officer, you lose track of the crew members.

Compton also commented on the buddy system.

The next issue within the concept of fireground accountability is that crew members are responsible for each other. In which module of incident command training did we learn that we are

supposed to stay together in teams at a structure fire? As you know, we didn't learn it in incident command training. We probably learned it in the first three days of basic firefighting school; in fact, maybe the first day.

Compton commented on being responsible for yourself:

The next concept within this fireground accountability model is that we are responsible for ourselves. That's simple; it means that we all know what we're supposed to be doing and we've been trained pretty well to do it. We know what procedures apply in our department; those typically are not kept a secret. We know the safety regulations. We know what we are to do in certain situations and we know we're not supposed to be freelancing. My point is; simply do it. Operate within the agreed-upon structure.

Remember that incident commanders are responsible for branches and sectors, and they're responsible for knowing who's in charge of each, where they are, what they are assigned to do, what other units are in that sector, how they are doing and whether they need any help. Sector officers are responsible for the units assigned to them and for knowing this very same set of things about the crews under their supervision. If they don't, they're simply not in charge of them. Company officers are responsible for crew members, and crew members are responsible for each other; they stay together, they're a team. And finally, we're all responsible for our own individual behavior within the system.

And Compton's thoughts on staying together as a unit are:

*The next component of fireground accountability deals with crew integrity and working in teams of at least two. In order to be **together**, we need to be within voice, sight, or touch of one another at all times. If you aren't, you **aren't together**. Keep this in mind anytime you're working as a team on the fireground.*

Two-In/Two-Out Rule

By now, the fire service is familiar with the Federal Occupational Safety and Health Administration's two-in/two-out rule that requires at least four personnel on the fireground before entering an immediately dangerous to life and health (IDLH) environment. Although there is a debate over the practical application of this law, it is the law of the land, and it is mandatory that fire departments, whether paid or volunteer, are familiar with its application as it can dramatically affect search operations. Specifically, OSHA 29 CFR 1910.134 Respiratory Protection Standard includes the two-in/two-out rule that states the following:

- Members operating inside a structure shall operate under the buddy system and shall at all times remain in direct visual or voice contact with one another.

- Interior structural firefighting operations will not commence until four trained and properly equipped members are present on the scene.

- An interior search can be conducted with less than four trained and properly equipped

members that are present on-scene if there is sufficient cause to believe there are trapped-saveable occupants.

This OSHA standard directly applies to search operations from two basic perspectives as follows: when conducting a search in the interior of an IDLH, a search team shall consist of a minimum of two persons who shall remain in visual or voice contact; and although a minimum of four persons are required to be on-scene before interior operations (such as a search) are initiated, a search can be initiated with less than four exterior personnel if there is *sufficient cause* to believe that a search is necessary. The definition of sufficient cause can be defined as information from a neighbor, cars in a driveway, and other similar examples that would indicate a high probability or a knowledge of someone trapped within a burning structure.

Rapid Intervention Teams

Rapid Intervention Teams (RITs) are not a primary search team, but can be better categorized as a secondary search (and potential rescue) team whose primary focus is to be available during fireground operations to conduct a search to find trapped firefighters, and then affect a rescue to alleviate the problem. When a RIT team is implemented, it is important to have the proper training, tools, and equipment handy and ready for immediate use, and be able to perform the following minimum operations:

- Be thoroughly familiar with basic and advanced search techniques.

- Understand forcible entry and forcible exit. Specific attention should be directed towards determining if a structure needs additional forcible exit openings.

- Determine strategic placement of portable ground ladders; determine if additional ladders are necessary to assist in a timely exit for interior personnel.

- Remove a firefighter from an upper floor window.

- Remove a firefighter from the interior of a structure in a timely manner (may require different types of carries, the use of Skeds, Stokes, and so on). Remember that removing an unconscious firefighter is not a simple *lift and go* operation.

- Be able to hoist or lower a unconscious firefighter.

- Be able to breach walls (may consist of exterior and/or interior walls) and/or be able to enlarge an opening.

It is also important that while a RIT team is staged, an additional size-up of the structure be conducted and the fireground radio channel is closely monitored. A continuous size-up of the incident must be conducted until the incident has been stabilized or until hazardous conditions to fireground personnel have been minimized to an acceptable level.

Equipment

An integral part of any search operation is the utilization of the proper equipment (as required by the incident) such as a turnout coat, pants, boots, a hood, and gloves, which is a basic starting point when entering a hazardous environment.

- **Self contained breathing apparatus (SCBA).** This is mandatory when any type of hazardous atmosphere is encountered. When access to a 30-minute, 45-minute, or 60-minute bottle is available, it may be advantageous to use a 45-minute or 60-minute bottle to provide additional time and flexibility (the size of occupancy is a consideration).

- **PASS device.** Ensure the device is activated before entering a structure.

Tools. The availability of appropriate tools for search operations and ensuring safety can be enhanced by the following considerations.

- **Forcible entry and probing.** An emphasis should be placed on carrying a tool to enable a search team or person to expand

the area being searched (if necessary) for victims and/or forcible entry-exit problems that may be encountered. There are many opinions on the proper tool to carry with most recommendations going to a Halligan, six-foot pike pole, and axe. Remember that a six-foot pike pole can suddenly become much longer in hallways and/or 8-by-10- or 10-by-12-foot bedrooms, and a Halligan can require a significant amount of energy to use for expanding an area of search (sweeping the center of a room, and so on) for extended periods of time. In most cases, a pickhead axe that is worn in a scabbard (see figure 17-19) can enable a firefighter to have both hands free (if necessary) yet can be easily used as a tool to expand a search and/or a forcible entry/exit tool.

However, the most important tool is the one that you are most comfortable carrying, which takes some practice on your part instead of grabbing the most available tool on an apparatus to start a search in a hazardous environment when time is a major consideration.

- **Thermal imaging camera (TIC).** TICs have the capability to enhance a search, but the use of a TIC is totally dependent on the ability of the user to use a TIC to its utmost capability. Obviously this takes practice (remember that practice does not make perfect; perfect practice makes perfect). Several limitations that should be remembered are: a TIC can be very useful while it works (so always carry a spare battery); test your TIC before use; be able to read the image; and TICs do not see through objects such as beds and so on.

- **Tag/search line.** A simple line can be used to ensure contact between two (or more) firefighters while searching, provide multiple lines that can be attached together or branched off a primary line to search multiple areas, or used to attach to the exterior of a structure by search personnel to provide a continuous line from interior search personnel back to the exterior of a structure. The availability of a line can be enhanced with a bag that contains anywhere between 150 to 300 feet of 5/16-inch line with a snap or carabiner at each end (remember that the more line that is carried the larger the bag will be and there is a point that ease of use will be minimized). Therefore, 150 feet is normally sufficient for homes whereas longer lengths may be necessary for districts with commercial structures. For this reason, it may be advantageous to have a 150-foot search line for homes and a longer search line for commercial structures. The bag can be attached with a snap to the rear portion of a breathing apparatus, as in figure 9–19, or carried by a strap that travels over a shoulder.

remembered that tag/search lines are not primarily used to find victims, but are used to provide accountability for search team members, systematically search an area, and/or a method to enhance the ability to quickly exit a structure. To improve this operation, many departments mark the line at specific distances to assist personnel in determining how far they are into a building and which way to follow a line to exit a building. This is easily accomplished by positioning a knot every 50 feet (or any other appropriate distance, such as every 25 feet).

As an example, assume a knot is tied every 50 feet. Starting with the first knot at 50 feet, two knots are tied approximately 10 inches from the single knot as in figure 20A. This process is repeated at every single knot for the entire length of a search line. This knot sequence can also assist disoriented or lost firefighters in determining which way to follow the line to exit a building as follows— one knot/two knots are into a building, and two knots/one knot are *out* of a building. Some departments take this method one step further by attaching a ring at every single knot, which allows for the capability to tether another line to the main line and search an area away from the main line but still retain contact with the main line as in figure 9–20B.

Figure 9–19. A bag with a search line can be easily attached to the rear portion of a breathing apparatus.

When necessary, the bag can be quickly removed from the SCBA bottle and attached to the front of a turnout coat to enhance flexibility and ease of use, or just let the line pay out from the rear or from a bag that is supported by a shoulder. It must be

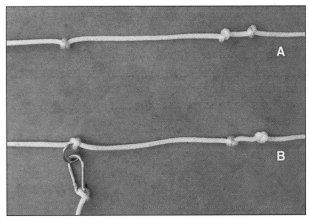

Figure 9–20 A and B. A specific knot configuration can be used to determine which way to follow a search line to exit a structure. Additionally, a ring can be used to attach or tether another line to a main search line.

However, it is recommended that this operation is reserved for search lines that are primarily used for large/commercial areas. Remember to keep operations as simple as possible.

- **Tether.** A tether can consist of either a nylon strap or line (such as 5/16-inch) of any length between 10 and 50 feet, and can be used for multiple purposes. Depending on the environment, number of search personnel, area to be searched and other similar conditions, it can be advantageous to tether multiple personnel during search operations. The combination of hot-smoky environments, lack of visibility, multiple personnel, demanding time constraints, and a lack of familiarity with search operations can provide the necessary ingredients for search personnel becoming separated, lost, or not being able to conduct an effective search operation. A nylon strap or line can be effectively utilized to provide the ability to search larger areas and maintain positive contact between search personnel and/or the main search line during a difficult or hazardous search operation. A tether can also be effectively used to maintain control of an inward opening door (when opening into a hot-contaminated environment), maintain contact with a door when searching behind or around the general area behind a door, and other similar general purpose operations.

  An effective tether can be made from a piece of 12-foot nylon, 1-inch tubular webbing that can also be attached together at the ends (carabiners, Velcro, and so on) to result in a loop of 10 feet to any desired length.

- **Lights.** Each team member should have at least one good flashlight (and two is recommended), and the brighter the better. Many firefighters have one flashlight that is carried by hand and one flashlight that is attached to a helmet. A light allows personnel to see and be seen (by victims or other personnel inside the structure). In atmospheres with limited visibility, a light

(strobe, quartz, and so on) can be placed at the bottom and side of the entrance used by search personnel. This light can aid in identification (and act as a quick reference point) to locate an exit opening from the interior of a building.

- **Knife.** When using straps or lines for contact between personnel or to expand the area being searched, the availability of a knife may be necessary to cut an entangled line or strap.

- **Wire cutters.** Due to the widespread use of dropped or hung ceilings, particularly in commercial buildings, wire cutters can be the only handy tool that can free a firefighter that inadvertently becomes entangled in these wires.

- **Chalk.** Can be used to mark doors to indicate rooms that are being searched, or have been searched.

Strategy

Search operations are composed of two phases that can be accomplished simultaneously or separately as follows.

Primary phase

This phase focuses on speed and minimal time limitations to accomplish a quick search of the general area to be searched. This can be accomplished by an attack team advancing a hose line or by a designated search team. The emphasis is on a rapid advance into the area(s) to be searched in conjunction with a quick traverse around the perimeter and/or quick probing in areas where victims might be found.

Secondary phase

The secondary phase focuses on the thorough and complete examination of area(s) to be searched, which often includes the entire fire structure and may also include the exterior perimeter of a building. The emphasis in this search is the ability to guarantee there are no victims in the fire structure (or at the incident). If possible, do not use primary search personnel to conduct a secondary search. It is important to use a different perspective for this search operation as the primary team will tend to search the same areas again. In all cases, the completion of a primary or secondary search must be reported to the incident commander in a timely manner.

Techniques

Any search operation should be preceded with a basic knowledge of the structure (construction and layout), fire (location and extension), victims (approximate number, age, and location), resources (personnel and equipment), SOPs (training), and remembering that search operations can be a low-frequency, high-risk event, which means that search operations can be an extremely dangerous event. Using the preceding factors as a basic starting point, let's review basic search techniques that should be appropriately considered during a search operation.

A basic search team should consist of at least two personnel; however, the size of a search team can be increased as needed but should not generally exceed two to five personnel per team (depending on the size of a structure and type of search team). A span of control over four (not counting an officer) is usually counterproductive. The key is to use the appropriate number of personnel, depending on the type of search, size of area to be searched, and resources available.

If it is necessary to use multiple search teams, a primary team would be designated #1 and a back-up team would be designated #2 (or other similar designations). If multiple entry points are used, search teams can be designated according to entry points (team 1, 2, 3, and so on; backup teams would then be designated 1a, 1b, 1c, and so on). Each team leader should monitor a rescue channel if used. Other team members should keep their radio volume control turned down to avoid squelch and other unnecessary noises. Keep radio messages to a minimum.

If portable radios are utilized, a search team should be placed on a separate radio channel if possible and should make a quick radio channel check before proceeding with a search operation. This channel should be locked to prevent the channel from being accidentally changed during an incident.

If possible, a backup team with appropriate safety equipment should be positioned near the entrance opening used by the initial search team, particularly on lengthy or large area searches. This team can be separate from a RIT. Remember, the sole responsibility of a RIT is to provide a safety backup for personnel in a hazardous environment. A backup team and/or a RIT team should also monitor the progress of the primary search team, should periodically check on their status, and is responsible for the following:

- Ensure they are not delegated to another responsibility and be ready for immediate entrance into the building. If they are delegated to another responsibility, they must be *immediately* replaced.

- If possible, have radio contact with the search team and periodically verify their status, as in figure 9–21. The larger the area to be searched, the more important this consideration becomes. This can be easily accomplished (and with minimal use of a portable radio) if the backup/RIT team will radio the search team after 5, 10, or 15 minutes (or other pre-arranged signal) have elapsed.

Figure 9–21. A backup team should have radio contact with the interior search team.

As an example, at each five-minute signal, the search team leader acknowledges receipt by double-clicking the portable radio talk button. If the initial search team is not able to verify their status, the backup team/RIT should be prepared to enter the building and assist the search team. Additionally, this method will assist a search team leader in determining the appropriate time left for the search in combination with the size of breathing apparatus being utilized (such as 30, 45, or 60 minutes).

- Position a light (quartz, strobe, and so on) at the bottom and side of the entrance opening to facilitate the identification of the opening (now the exit opening) from the interior of the structure (notice the light in the doorway in figure 9–21). If necessary, a quartz light can also be mounted on the end of an aerial device and used to assist above-ground search operations by placing the light near windows or other openings to allow some degree of interior illumination.

When contemplating the subject of placing search personnel inside a structure, a subject that is not often considered or pre-planned is the ability to immediately remove personnel if the need suddenly arises. Two popular methods that are currently in use are calling a Mayday and three blasts of a horn or horns on apparatus. Personnel that are committed to fireground operations should have a set policy that all personnel are intimately aware of that addresses the following concerns:

- A signal to immediately abandon a building

- Whether or not tools and equipment should be left in place

- An assembly point outside the structure that will allow a timely verification/accountability of appropriate personnel

Remember that as the number of interior personnel increases, so does the time to quickly remove them.

Occupants can be a good source of information for trapped victims and their potential location. Remember to differentiate between humans and pets because pets are often referred to with common names such as Mike, Louie, My Baby, and so on.

When conducting search operations in structures, it is generally agreed that victims should be discovered within about 15 minutes after arrival of suppression companies. Obviously, the larger the structure, the more time it will take to conduct a search with a reduced level of effectiveness (particularly when applied to commercial structures).

Before entering a structure for a search, quickly determine if the structure would be classified as a *tight* structure or an *open* structure because the focus of these terms is your ability to quickly exit a building if necessary. As an example, the church in figure 9–22 (classified as a tight structure) has an entrance in the front and rear only, no side windows, more square footage than the typical residential structure, and will not be easy to exit in a timely manner when the interior visibility is non-existent.

Figure 9–22. This church has minimal entrances and exits for fireground personnel.

Conversely, although the church in figure 9–10 also has more square footage than the typical residential structure, it has numerous windows and multiple doors that can be used for exit points and will be significantly easier to exit in a timely manner as opposed to the structure in figure 9–22.

Victims are most likely to be found close to the point of origin (the area of greatest danger). As victims closest to the seat of a fire are in the most danger, a search should begin as close to the seat of a fire as possible. However, the closer to the point of origin (fire, heat, smoke), the more danger to which search personnel will be subjected.

Statistics indicate that victims above a fire can be in the greatest danger. Therefore, when a multistory building is encountered and fire conditions dictate victims above the fire are in danger, a search should begin above the seat of a fire.

Without specific information and extenuating circumstances, start a search in the living areas of a residential structure during daytime hours. Conversely, start a search in the sleeping areas during nighttime hours.

When initiating a search, attempt to search the high target areas or areas of greatest danger in a priority manner as follows:

- Behind doors
- Bedrooms
- Bathrooms
- Exit pathways

Depending on the type of search, areas out of the *normal* path of travel in a structure may need to be searched. These areas include, but are not limited to, closets, under/in beds (a low bed is likely a bunk bed), bathtubs/showers, corners, and behind doors.

When furniture is encountered, do not move the furniture because this can block your exit path and possibly fall on an undiscovered victim. Instead, leave the furniture in place and search around and behind the furniture.

When personnel are conducting search operations, consider the following advantages of breaking windows if it will not be a detriment to suppression operations:

- Provide more light to the interior of an area being searched.
- Improve visibility.
- Temperatures should be reduced.
- Exterior personnel will know where search personnel are searching.
- Provide a mental uplift for search personnel.

Remember that an opening can become a source of fresh air that can enhance a fire and/or draw a fire to its location. Therefore, try to begin with the most exposed windows and work to the least exposed windows. Additionally, if a primary search is being conducted, the focus is to quickly search an appropriate area or structure, not spend precious time on a stubborn window or horizontal ventilation operations. When windows are not easily opened, consider breaking a window(s), if necessary.

Lights can be turned on to enhance visibility. As you enter a room, slide your hand up the wall without the door hinges until you feel the light switch.

When opening or forcing doors, be aware of the following considerations:

- A warm or hot door is an indicator of the conditions on the other side of the door.
- Open doors slowly. As the door is just opened, evaluate any escaping heat or smoke. Remember that a victim could be on the floor behind an inward opening door. Do not cause additional injury to a potential victim by rapidly and forcibly opening a door.
- If a door is nailed shut, difficult to open, or storage is encountered, that is an indicator the door is not normally used for access, or is a closet.
- Doors that open inward are normally bedroom or bathroom doors. Doors that open outward are an indicator of exit doors, closets, basements, and so on.
- When searching a room, leaving the door open can enhance ventilation and a rapid retreat from the room. Conversely, closing a door while searching a room can keep a fire (outside the room) from worsening conditions within a room and provide a search team additional time to complete a search.

When entering a structure/room with heat and smoke, stay low. This is the coolest portion and provides the best visibility.

The size and type of structure will dictate its floor plan, which in turn will determine the type of search to be used.

Search operations for children and adults are not the same. Children can hide anywhere, and adults will try to escape (avenues of egress). Additionally, when searching for a firefighter, although they can be anywhere, their last approximate location is often known.

When searching commercial buildings, forget the residential mind set for the following reasons:

- Large floor plans require more *time* to be searched.

- A two person search team is probably insufficient.

- The apprehension level of personnel is likely increased.

- Commercials can require considerably more air than residential structures. Big buildings mean big air!

- Fire load will likely be greater than residential structures.

- Specific search patterns may be necessary.

If a search cannot be completed or a specific area cannot be searched, the incident commander must be notified in a timely manner. This will allow another search team to complete the search when appropriate. As an example, attics can be a difficult area to search as they are normally above a fire and are not easy to enter or exit. However, due to the difficulty of being able to easily access an attic, it may be advantageous to search an attic area after the fire has been controlled or mitigated. As a side note, there is a big difference between an attic with a substantial stairway and a fold-down stairway or a small ladder for access-egress.

When entering a centrally located room (such as a bedroom) through a window, remember that the door to the room is normally located across from the window (to provide natural ventilation for the room) and the door in corner bedrooms are normally located in the corner of two walls (remember to look at the common floor plans of homes in your district on EMS calls). Before entering a window, always clear all glass (because the window opening will also be your exit point) and after entering the room, evaluate the benefits of closing the door to the room.

When entering a structure, three factors usually determine the path of search:

- **Information gathered from occupants or firefighting personnel prior to entry:** As an example, occupants who have exited a structure should know who is still in the structure and their approximate location.

- **Attack line:** Assume an attack line is inside a structure and it is suddenly determined the attack team needs to be rescued (flashover, building collapse, and so on). The

most direct path to their probable location is to follow the attack line to the nozzle. To affect a timely rescue, a search team should initially follow and search along the path of the attack line to the nozzle, including the general area around the nozzle. Therefore, it is imperative that personnel practice and become familiar with the concept of following a hose line with only their hands as a reference point.

It is also essential that personnel are able to *feel* a coupling with their gloved hands and be able to quickly verify their direction of travel to the nozzle (interior) or the pump (exterior) along the hose line. This is easily accomplished as follows: assume a nozzle is connected to a *male* coupling; therefore, the hose line behind male couplings lead towards the pump (outside the structure), and the hose line behind female couplings will lead towards the nozzle (into the structure).

With practice, it is easy to distinguish between a male and female coupling by feel alone. A female coupling is twice as long as a male coupling, and the lugs on a male coupling are twice as long as the lugs on a female coupling (figure 9–23).

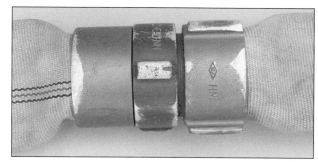

Figure 9-23. The lugs on a male coupling are longer than the lugs on a female coupling. This allows the ability to verify the direction of travel.

- Additionally, before entering a structure for a search operation along a hose line, determine from the proper pump engineer the appropriate hose line to follow into the structure, how many feet of line is into the structure, and what is the type of hose lay (forward or reverse). Remember that a 2½-inch line can be forward or reverse and 1¾-, 1½-, and 1-inch lines are forward only. Every firefighter, whether involved in a search or not, should be familiar with this concept.

- **Area:** It is often necessary to search an unknown area for known or potential victims. Entering the initial entrance opening and turning left or right, following a hose line or search line can determine the path of travel while searching through the area/structure.

Search personnel should keep contact with each other to ensure team effectiveness and integrity. This can be accomplished by visual, voice, or hand contact, or maintaining contact with a tether. Combine the noise of an incident with heat and smoke, a lack of visibility, excitement, time constraints, a separation of 4 or 5 feet between team members, and a difference in *sense of direction* between personnel—the potential for a problem can be significant.

Searching for an activated personal alarm device (PASS) on a missing firefighter can present several different considerations as compared to searching for civilians. The general location of the missing firefighter may be known, and the search team is searching for a sound from a PASS device. Therefore, to assist in finding the PASS device and/or determining the direction of the sound:

- Only the lead team member should determine the direction of the search.

- Open and close doors to isolate the direction of sound.

- Plug one ear to determine the direction of sound.

While searching (particularly in minimal visibility), open shafts, stairwells, loading docks, weak areas, and other similar hazards are dangerous to personnel. Therefore, sound the area in front of the path of travel with a tool or leg. Sounding can be accomplished by personnel moving forward on their hands and knees instead of moving upright. Although this method can be effective, personnel must realize that the head is positioned downward

and the weight of the person can be shifted forward. As an alternative method, personnel can also move forward while simultaneously keeping a leg in front of the body. This method allows personnel to lead with their leg as they move forward, keeping the balance of weight rearward and the head in an elevated position. This method is explained in detail in the February 1996 issue of *Fire Engineering* magazine (see Training Resources).

Search personnel are normally positioned on the floor while searching. Take the opportunity to mentally and physically think about the condition of the floor. Examples are a hot floor is an indicator of fire underneath search personnel, a tile floor in a residential occupancy is an indicator of a bathroom or an entry in an upscale home and aisleways in commercial structures, and carpet is an indicator of living areas in a residential structure and sales areas in commercial structures.

Search personnel should periodically stop, cease breathing for a few seconds and listen for sounds that could assist a search operation, such as cries for help, moans, PAL device, muffled portable radio, and SCBA alarm bell. Combine the noise of an incident with personnel breathing from their SCBA, and the ability to recognize faint sounds can be difficult at best. Additionally, periodically looking out of a window (if possible) can assist with orientation within a structure.

To indicate an area is being searched, has been searched, and to minimize duplication, doors can be marked with chalk (figure 9–24) as follows:

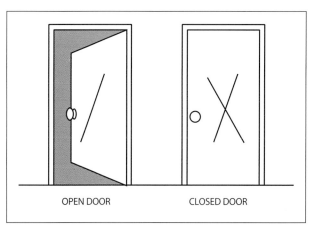

OPEN DOOR   CLOSED DOOR

Figure 9–24. Marking doors with chalk can indicate rooms that have been searched or are in the process of being searched.

- A door with no markings indicates a room has not been searched.

- Upon entering an unsearched room, make a diagonal chalk line across the door. This indicates the room is being searched.

- After searching the room, cross the diagonal line on the closed door to form a large X, put the company identification on the door, and close the door. This indicates the room has been searched and who conducted the search. When heavy or greasy smoke is present, consider marking the bottom portion of a doorway where there is minimal smoke. Railroad or builder's chalk is large, easy to use, and if kept in a plastic container, will stay dry and ready for use.

Should a search team start a search by forcing entry as close to the seat of a fire as possible, start a search by entering through the entrance opening used by attack personnel (which is normally away from the fire), or enter a window above a fire (on multiple story buildings) and search the rooms that are easily accessible from the exterior of a building? Although each of these methods have their advantages and disadvantages, the easiest-to-understand explanation was provided by the late Tom Brennan (see "Random Thoughts," *Fire Engineering*, August 1996) and was termed the To/From-From/To principle:

- On the *fire floor*, go to the fire and search *from* it to your exit (or your entrance point). The advantages of this principle are the search team spends the minimal amount of time in the most dangerous area, the search team will be at their best during the early stage of their search, victims closest to the fire area are in the most danger, the search team will be advancing away from the fire and towards the exit opening, and in most cases, the search team will be advancing and retreating in close proximity to an attack line. The obvious disadvantage is the search team must advance to the fire area and then retreat back towards their exit opening.

- On the floor *above a fire*, search *from* the entrance to the occupancy or area to be searched. In this principle, the search team begins to immediately search either to a secondary exit or back to the primary entrance opening. The advantage of this principle is a search can be conducted in a short amount of time and in some cases by not entering the fire floor to access the upper floor to be searched. This is accomplished by laddering and entering an exterior window to begin the search.

A specific method for a search is commonly known as vent-enter-search (VES). Utilizing this method, a search team would begin a search by forcing entry as close to the seat of a fire as possible, allowing the search team to search close to the fire area in a minimal amount of time. Although this method can be very effective, search personnel can be positioned between the fire and a ventilation opening (normally a window used for the entry portion of this operation). However, search personnel can also be positioned between the vent opening and advancing attack personnel. Obviously, these considerations can place search personnel in an extremely dangerous position.

If it is necessary to breach a room or area through a wall (except masonry walls) due to high heat in hallways, extension of fire, security bars, and so on, the following options can be utilized:

- From the interior of a room, locate the appropriate wall to be breached. Using an axe or power saw, locate a stud. Cut vertically along the adjoining studs and horizontally along the base line. Remove the center stud by striking at the lowest possible point with an axe and pry the cut portion of the wall loose from the top. This will generally provide an opening 32 inches wide in less than two minutes. When walls constructed of drywall and metal studs are encountered, breaching can be easily accomplished by removing the drywall and bending the metal studs apart (or to each side).

- From the exterior of a building, locate the appropriate wall to be breached. If a window is not available, use the aforementioned method. If a window is available, use the aforementioned method under the window.

- Remember, when cutting exterior or interior walls, electrical wires may be present and these methods should only be used as a last resort unless the electrical power to the structure has been eliminated. However, if a quick opening is necessary and electrical power to the structure has not been eliminated, consider the degree of the hazard.

The person in charge of a search operation (search officer) should not participate in actual search methods. This allows the search officer to focus on the search, status of the environment, length of time remaining for the search, crew safety, and the most appropriate exit path. Among other benefits, this can dramatically enhance crew confidence and safety.

Depending on the type of structure, fire, and potential location of victims, consider the use of positive pressure ventilation (PPV) to reduce the presence of heat, smoke, and increase visibility and access to victims. However, when using blowers, consider the potential impact of noise on communications. The use of PPV in search operations is discussed in chapter 16.

Note that a quick attack on the seat of a fire in concert with horizontal and/or vertical ventilation operations can dramatically enhance a search operation and should be a prime consideration.

Search Methods

The following are four examples of methods that can be utilized by search personnel while advancing through a structure. Although some of these methods may never be used at an actual incident, they are submitted as examples of available options that can be used with the understanding that a search would be conducted by a minimum of two persons, a RIT would be used as staffing and conditions dictate, and a light would be placed in the lower portion of the entrance opening. Additionally, assume these searches are conducted in minimal or no visibility which has the potential to dramatically slow and/or cause unexpected difficulties for any search operation. This is why ventilation is so important, particularly for search operations, because it can

noticeably improve visibility that can improve any search operation. These search methods will enhance your search options and dramatically improve personnel expertise and teamwork. The time to preplan search methods and enhance confidence is *before* a search is necessary, not *during* a search in a hazardous atmosphere.

Standard search

The standard search has long been the mainstay of the fire service as most fires occur in common residential structures, most searches are conducted in these types of structures and in rooms that can be called *of common size*, and can usually be capably conducted by two personnel. The advantages of this type of search is all team members are searching, it conforms to the two-in/two-out rule, and the team normally stays together. The disadvantages are all personnel are searching together which limits the areas that are searched in a given time-frame, multiple personnel searching in a typical bedroom can become cumbersome, and the focus is normally on search, not the environment or egress options.

To conduct a standard search, assume a search team of two enters the front door of the single-story residential structure in figure 9–25.

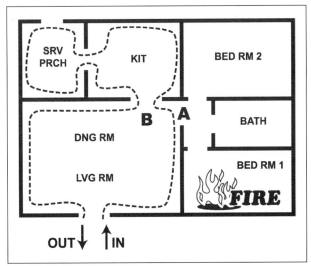

Figure 9–25. A standard search pattern for a single-story, single-family dwelling

At this point, there are two options that are generally used: in the first option, one member would turn right, the other member would turn left, and each would follow the appropriate wall until they meet at the far end of the room, continuing on until all rooms are searched; in the second option, both members would either turn left or right as a team and follow the appropriate wall until all rooms are searched and they ultimately returned to their entry/exit point.

In the first option, the structure would be searched faster, but the searchers need to concentrate on waiting until they meet at the other side of each room before moving on as it would be easy for one of the searchers to be diverted to another room. As an example, one searcher turns left, follows the wall to the doorway to the kitchen (B), and waits for the arrival of the other searcher. Simultaneously, the other searcher would turn right and follow the wall to the doorway to the bedrooms/bathroom. At this point, it would be imperative that this searcher continue on until meeting the other searcher at the doorway to the kitchen and not turn right into the hallway to the bedrooms/bathroom (A). Assuming the searchers meet at the doorway to the kitchen, they would then continue and search the kitchen, service porch, bedroom #2, bathroom, and then exit the structure. The attack team would search bedroom #1 after extinguishing the fire. With this method, it is easy to see that as the size of a structure increases, so does the possibility of the searchers becoming separated and/or becoming inadvertently diverted by multiple doorways. Additionally, this method can violate the principle that search members should stay in voice or sight contact.

In the second option, both search members would turn right and search all rooms (except the room with fire) and then exit the structure as illustrated (figure 9–26).

Although this method would be slower than the first method, the search team would be together during the search. To smooth the progress of this option, consider the following considerations. As the searchers traveled around the perimeter of the rooms, it is beneficial if they maintain an inline position with a wall as a reference (figure 9–27). The first person is responsible for leading the team and maintaining contact with the reference point and the second person maintains contact with the leader and assists in the search.

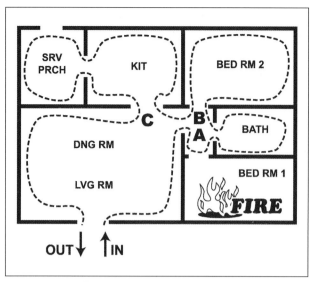

Figure 9–26. A standard search pattern utilizing the right shoulder and inline method

Figure 9–27. Searching in the in-line position can provide accountability.

The inline position will limit the area being searched. However, one firefighter can use a tag line to search away from the leader, or a person can search about 5 to 6 feet away from a search leader/wall (fingertip to fingertip), and more with a tool (additional 3 feet).

Consider when using a tool to expand an area of search, if an object is felt that it will be necessary to crawl over to the object for a positive identification. This is why some training materials advocate feeling with your hands only because it can be a faster operation (if you feel something with your hands, you know what it is).

When a room such as a standard bedroom (or smaller rooms such as bathrooms) needs to be searched, it is more advantageous to have one searcher wait outside the doorway while the other person searches the room. In this case, the person searching the room would enter and turn left or right and make a quick perimeter search which would normally include the center of the room. This should take no more than about 15 to 20 seconds. Upon completing the room, both searchers would continue to the next room. This option would normally be the preferred choice between the first and second options.

If PPV were used in this scenario (and depending on the progress of the fire), position a blower at the front door after the service porch door has been opened as an exhaust opening. With the service porch door open, heat and smoke will be cleared from the living/dining room, kitchen, and service porch. If the search team opens windows (particularly in the bedrooms and bathroom) as they advance around the perimeter of the structure, rooms with open windows will also begin to clear.

Team search

A team search is based on the principle that the search team stays together and uses a search line as a reference point for safety. Although this configuration can be used in various types of occupancies, the focus is that the search team stays together as a team. This method is more advantageous in larger occupancies, particularly commercial occupancies where the floor plan can be challenging along with numerous obstacles such as machinery, desks, aisleways, and so on. The advantages of this type of search is the team stays together, it conforms to the two-in/two-out rule, works well in larger occupancies by allowing searchers to tag to the main line to search additional areas and maintain contact with the main line, and the search line provides a reference to the exit opening. The disadvantages are all members may be searching including the team leader, and the only reference to the exterior is the search line which can possibly become accidently separated (although this has never been personally observed).

To conduct a team search, lets assume a search team of two enters the front door of the commercial structure in figure 9–28, there is a door to the office in the front of the building and a loading dock door in the rear, the fire is in the front portion of

the building, and it is known there is a person who is unaccounted for who was last seen by the area of the fire.

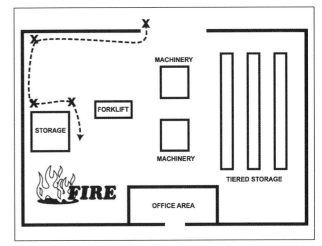

Figure 9–28. A team search operation for a commercial building using a search line as a reference point.

A first consideration is which door to use as an entry/exit point. Remember that when given a choice, larger openings provide more light, better ventilation, and a larger opening for access/egress. Therefore, the loading dock door is the better choice for the search team to enter the building.

As illustrated, the search team would first secure the search line to the exterior of the building/doorway (X), then enter the building and turn right to traverse along the rear wall. If obstacles are encountered, which is likely in a commercial building, two options are generally used: search personnel traverse around the obstacles, allowing the search line to pay out behind them while moving towards the general direction of the target area; and search personnel traverse around the obstacles, but at every change of direction, the search line is secured (Xs). If the search line is not secured as illustrated, it would be easy for the search team to pull the line taut while exiting, which would take the search team along a different route that could hamper their exit. Although it will take additional time to secure the search line at each change of direction, a secure line will lead to the exterior of the building. To best use the search line, the last person on a search team secures the line to the exterior and is responsible to maintain contact with the lead searcher. The lead searcher is responsible for determining the direction of the search. An additional significant consideration is the effort, time, and resources necessary to search commercial buildings that consist of a noteworthy size, floor plan, and/or configuration. In these cases, determine if it would be accurate to admit that the search operation would, in reality, be a recovery operation as opposed to a search operation. As an example of the potential difficulty of this type of search in *larger* buildings, see the article Large-Area Search in *Fire Engineering* magazine by John "Skip" Coleman (listed in Training Resources).

Oriented search

An oriented search incorporates some of the methods of the standard and team search but with one major exception: one member (normally the crew officer) does not search but becomes the orientation point for the other team searchers during the search operation. The advantages of this method are safety of the search team during the search, multiple areas (or rooms) can be searched simultaneously, the search operation is divided into command and search, which allows the search office/leader to focus on the search crew, environment, exit options, and also allows the searchers to focus on the search while being monitored. The disadvantages are the responsibility that is placed on the oriented person and cooperation between all search members. To conduct an oriented search, consider a center hallway residential structure and a large warehouse-type occupancy.

In figure 9–29, assume a center hallway-type structure with a fire in one unit as illustrated, the hallway is charged with smoke, and the search team consists of three personnel. Ventilation of the first floor hallway should be mandatory before, and not later than during a search operation.

Figure 9–29. An oriented search operation for a center hallway-type structure

This will provide immediate relief from the heat and smoke in the hallway and enhance the effectiveness of the search operation. Why spend additional time conducting a search in a hot environment with minimal visibility, when a few minutes devoted to improving the environment can reward a search team with the opportunity to conduct a search in less time and with improved safety? The search team would enter the hallway by the fire and proceed to a spot in the hallway adjacent to the first rooms (1). The officer would direct one of the searchers into the room on the left and the other searcher to the room on the right. As the rooms are simultaneously searched, the officer would stay in the hallway between the rooms until both searchers returned from completing their searches. The team would then advance to the next room which is adjacent from the fire (2) and direct both searchers into the room on the right to search that room. Again, the officer would stay in the hallway adjacent to the room until both searchers returned. If it was not possible to stay in the vicinity of area marked (2) due to the illustrated fire, the search team would move to the next set of rooms. In either case, the team would advance to the next set of rooms (3) and then (4) to repeat the previous oriented search method that was accomplished at position (1) and then in this case, exit at the far end of the building. This type of search would accomplish the following objectives:

- Opposing rooms would be quickly and simultaneously searched in a progressive order, starting from the area closest to the fire and advancing to the other end of the building.

- The officer would lead and direct the search and also maintain a position of safety and control outside of the rooms being searched.

- This type of search would conform to the two-in/two-out requirement.

- The search team would mark appropriate doors of rooms being searched with a diagonal line and cross the line when finished searching the rooms. This indicates rooms to be searched (no marking), being searched (diagonal line), and rooms that have been searched (crossed diagonal line with company designation). Utilizing this method, additional search personnel can be inserted into a search plan, know where to start, and will not duplicate previous searches. Additionally, after the fire is knocked down, the ability to review search operations is enhanced.

To conduct an oriented search in a warehouse-type occupancy, consider a structure similar to figure 9–30 that would be similar to numerous types of occupancies that have either tiered storage or aisleways between display counters such as Walmart, Costco, supermarkets, hardware stores, and so on.

Before we consider a search in these types of occupancies, it is important to realize that the size and/or floor plan will likely consume an inordinate amount of time (particularly if visibility is minimal or non-existent) and negate a meaningful and timely search. Therefore, in these types of structures, it is again important to evaluate the difference between a search and a recovery.

Note that similar to the last search method (team search), the article "Large-Area Search" by John "Skip" Coleman should be thoughtfully considered.

To begin the oriented search in figure 9–30, assume the visibility is minimal at best, there are several occupants who are unaccounted for, and a search team of three will be utilized. It is important to ensure or initiate some type of ventilation to improve the interior environment and visibility. Also, the search team would first develop an initial line (1) from the front to the back of the structure (1) that would be used as a reference point during the search. Some prefer a search line and some advocate stretching a hose line as a reference point. Although stretching a search line would be quicker, it could be pulled in the direction of a secondary line that could create unnecessary problems during a search. Conversely, if a hose line (1¾ or 2½ inch) is stretched

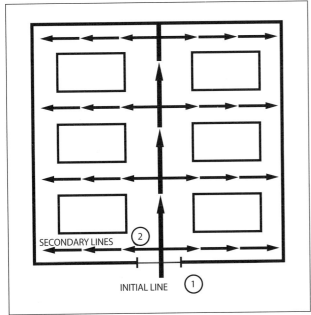

Figure 9–30. An oriented search operation for a large warehouse with aisleways

as a reference point, it would take more time and personnel to develop, but it would be more secure against unwanted movements and maintain its relative position during the search. After the reference line has been developed, the three-person search team would enter the building and search along the initial line to the first aisleway (the officer would stay on the line with a firefighter to either side; if necessary, the firefighters could use a tether to maintain contact with the officer). When the search team reaches the first aisleway, the officer would direct one firefighter to search the left aisleway and the other firefighter to simultaneously search the opposing aisleway while the officer remains on the reference line. Prior to searching the aisleways, each firefighter would attach a tag/search line to the reference line (2) and begin the search. This process allows each firefighter to confidently search to the end portion of the aisleway and then return to the reference line/officer. When both firefighters have returned to the officer, the team moves towards the next aisleway to repeat the same type of search. This method would be repeated until all aisleways have been searched. The search team would then follow the reference line back out of the structure. Although this method would require some setup time, the actual search would be as coordinated and timely as possible under demanding conditions.

Nozzle fan search

Although this procedure is rarely ever used, it is included as an option that can allow a search to be used for two attack firefighters who may be found on a hose line or at the end of a nozzle. This method was successfully used to search for two firefighters on a nozzle that were suddenly caught in a flashover in a restaurant in Los Angeles. This procedure will utilize two personnel to conduct a search utilizing a nozzle for a reference point and methodically searching away from the nozzle if necessary (figure 9–31).

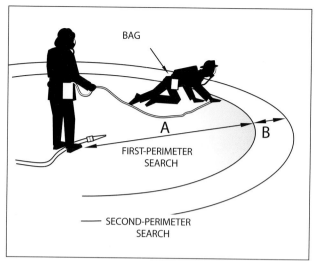

Figure 9–31. A search team can use the end of a hose line as their point of reference, and a search line to systematically search around and away from the hose line.

The search team follows hose line to end of nozzle. If necessary, the team leader can maintain contact with the hose line and tether the other firefighter to the most appropriate side of the line as they advance forward, searching the area around the hose line. The leader stays at nozzle to maintain point of orientation.

The search person gives tether or tag line bag to leader. The person who will search away from the nozzle removes approximately 4 to 6 feet of line and will loop line over itself and attach to a hand with a half hitch.

Keeping tension on line, a member will search the area around the nozzle (A, first perimeter search). If the search is unsuccessful, the person on nozzle will remove another 4 to 6 feet of line, and repeat the operation as necessary (B, second perimeter search). This system allows two persons to *systematically* search an area, maintain contact with each other, and maintain contact with their exit point of reference (hose line).

Rescue Methods

The following rescue methods are examples that can be utilized by personnel to remove victims to a safe location in a timely and effective manner:

Cross arm method

The rescuer crosses his or her arms around the chest of the victim and walks backwards (figure 9–32). This methods works best for civilians as SCBA on a firefighter victim makes this operation difficult.

Mattress/rubbish carriers

Two rescuers can use a mattress carrier, rubbish carrier, sheet, bedspread or other similar materials to carry a victim (figure 9–33). Some departments also use this method to remove unconscious or overweight persons from a structure in EMS incidents.

Figure 9–33. A mattress carrier can simplify the removal of victims.

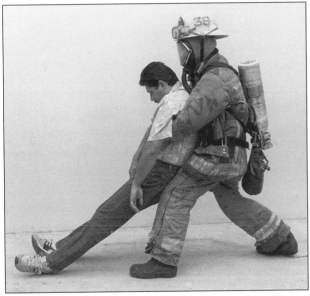

Figure 9–32. The cross arm method can be used to remove civilians from a structure.

Chairs

Chairs can be effectively utilized to carry a victim up or down a stairway (figure 9–34).

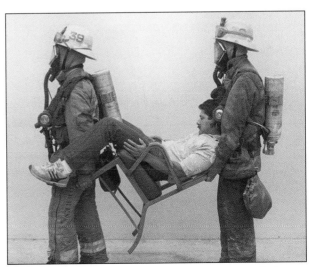

Figure 9–34. Chairs can be used to carry a victim up or down stairs.

Down ladders

Using a nylon strap, join victims wrists together. One rescuer then takes a position at the top of a ladder (figure 9–35). Another rescuer remains in room to assist with placement of victim onto ladder.

Figure 9–35. With practice, a victim can be taken down a ladder.

Rescuer will grasp rungs with both hands and position a knee up high to create a *straddle effect*. Victim is positioned on ladder, face to face with rescuer, strapped wrists over rescuers head and resting on rescuers knee. Rescuer will descend ladder slowly, keeping the victim between the rescuer and the ladder while maintaining constant pressure with the straddle knee. This procedure will lower victim in a secure manner with little strain or risk to the rescuer.

Over the shoulder

To initiate this method, the rescuer stands at the feet of the victim and pulls the victim up to shoulder height in one continuous motion. Although this method can be effectively used by one firefighter, caution must be exercised to ensure the victims face does not strike the rescuers breathing apparatus bottle (figure 9–36).

Figure 9–36. A victim can be placed on a shoulder and removed from a structure.

Firefighter victim

Position the victim on his back, breathing apparatus down. Check the waist strap on the victim for tightness. Grip the victim's shoulder straps near the breathing apparatus frame. Lift the victim and back out of the structure using the strongest muscles (quadriceps) to slide the victim on the lower portion of the breathing apparatus bottle. If the victim is moved by pulling the feet, the position of the victim's head can restrict or close the airway.

Several variations of this method are illustrated in figures 9–37 and 9–38.

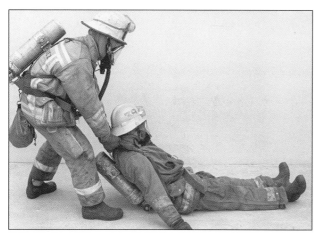

Figure 9–37. A firefighter can be removed from a structure by using the SCBA shoulder straps.

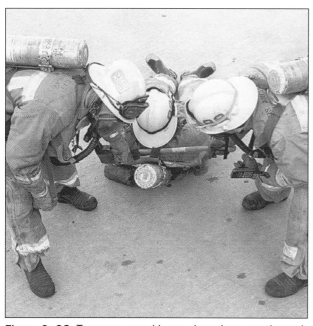

Figure 9–38. Two personnel have placed an axe through the SCBA shoulder straps to assist in removing this firefighter from a structure.

Training Resources

The following resources can provide additional information and/or another viewpoint for the subjects discussed in this chapter:

- Temperature Measurement and Thermal Imaging Cameras, Jonathan F. Bastian, *Fire Engineering*, December 2003.

- Tips for Using Thermal Imaging Cameras, John G. Riker, *Fire Engineering*, May 2002.

- Crab Side Crawl, *Fire Engineering*, February 1996.

- Search Rope Basics, Skip Dorgan, *Fire Engineering*, January 2000.

- The Search Rope, George Howard, *Fire Engineering*, January 1992.

- Search Rope Basics for Large Areas, John Miles and John Tobin, *Fire Engineering*, December 2002.

- Search: The One-Knot/Two-Knot Method, Patrick J. Lavin, *Fire Engineering*, April 2002.

- Fire Ground Accountability, Dennis Compton, *The Washington Firefighter*, Summer 1999.

- How Much Risk is Too Much, Bill Gustin, *Fire Engineering*, November 1996.

- *Incident Management for The Street Smart Fire Officer*, John "Skip" Coleman, Fire Engineering Books and Videos, chapter 11, The Mission Of Search.

- Searching Without a Line—What Would You Do, Mike Lombardo, *Fire Engineering*, May 1998.

- Search, Roundtable, *Fire Engineering*, May 2004.

- Rescue and the First-In Engine, Roundtable, *Fire Engineering*, April 2004.

- The Most Dangerous Fireground Activity, Part One and Two, Roundtable, *Fire Engineering*, May and June 2007.

- The Last of a Dying Breed, Charles R. Angione, *Fire Engineering*, January 2001.

- Two-In, Two-Out-Law of the Land, David H. Cowardin, *American Fire Journal*, February 1999.

- Two-In, Two-Out-Rule, Roundtable, *Fire Engineering*, February 2000.

- Large-Area Search, John "Skip" Coleman, *Fire Engineering*, June 2009.

- The Personal Rope, Tom Sitz, *Fire Engineering*, April 2005.

- Search in Burning Buildings, Steve Shupert, *Fire Engineering*, January 2010.

- Searching for a Conscious Lost Firefighter, Tom Sitz, *Fire Engineering*, July 2009.

- Rapid Preintervention, Robert Goplin, *Fire Engineering*, October 2009.

10 Overhaul

Introduction

Even after personnel have gained control of a fire, extinguishment isn't complete until a post-fire investigation of the total fire area reveals that all of the fire has been extinguished. This fireground operation is known as overhaul. As with other fireground operations, it requires a collective application of personnel, equipment, training, leadership, coordination, and communication to ensure a professional completion.

During overhaul operations, primary emphasis normally is focused on locating and extinguishing any remaining fire; however, overhaul considerations can be simple or extensive, depending on the extent of a fire and structure involved.

This chapter focuses primarily on locating and extinguishing any remaining fire, and preventing rekindles. This chapter also focuses on minimizing property loss.

The major factors involved in determining the type of overhaul are the nature of the occupancy, its stability, the materials present, and the size (extent) of the fire. The main objectives of overhaul operations are to:

- Complete the extinguishment of fire with minimal damage to the building and its contents.

- Determine the most probable cause of the fire.

- Leave insurance adjusters, owners, or occupants a means to determine the loss, and determine the probable cause.

- Leave the building in as safe a condition as possible, and when appropriate, to secure it from unauthorized entry.

- Use the incident as a training opportunity for younger members, applying all of the numerous facets of building construction and related fireground operations.

Strategy

After a fire has been controlled, and depending on the size and/or extent of the fire, the incident commander should survey the building *prior* to overhaul operations for the following reasons:

- To determine the nature and extent of the overhaul. This will allow a review of any unsafe conditions and the ability to develop appropriate safety procedures to be followed during the subsequent overhaul.

- To develop an overhaul strategy with appropriate officers. This includes training subordinate officers in the planning and development of such a strategy.

- To determine the probable fire origin and cause. Not only should the overhaul operation center on complete fire extinguishment, but should also include the cause of the fire, if possible. This consideration may include delaying or modifying an overhaul so evidence is not destroyed during overhaul operations.

- To ensure the availability of appropriate resources for overhaul. This may include the replacement of personnel (who were engaged in the extinguishment phase). Remember that a difficult and/or lengthy extinguishment can tire personnel, and the addition of a lengthy overhaul may cause unnecessary fatigue.

After the building and overhaul area has been surveyed, the incident commander should conduct a strategy meeting with appropriate officers to delegate responsibility for the completion of specific overhaul operations. During this meeting, the following considerations should be evaluated.

Safety

Each fire requires a specific type of overhaul operation. The extent of an overhaul can vary from the simple sorting of materials in the fire area to the continuous wetting down of a total-loss fire until the area has been secured. One factor common to any overhaul is the tendency of fire suppression personnel to relax after the fire has been contained or extinguished. Many overhaul operations, however, are performed under dangerous circumstances. Therefore, the top priority in any overhaul operation is the safety of personnel and the need for them to be constantly aware of their surroundings and its related hazards. Remember that during a fire, the building was under demolition, gravity is constantly working against the building, and water weighs a little over 8 pounds per gallon.

Structural stability

Prior to fire cause determination and overhaul operations, the structural stability of the fire building must be determined. Factors such as the size, extension (horizontal/vertical), and duration of the fire must be evaluated along with the type of construction (heavy timber, conventional, lightweight, unreinforced masonry, and so on) and the type of building. Identify and provide protection for any special hazards, such as holes in floors, open shafts, weakened stairways, and the like. Falls still rank as a leading cause of fireground deaths and injuries. Also, be aware of unsafe walls, rafters, beams, hanging ceilings, and the like (figure 10–1). After looking at this photo, it should be obvious that based on the amount of fire and weakening of structural members, the entire building should be considered a prime candidate for collapse.

However, just because a fire has not been extensive, do not discount the possibility of a collapse.

Figure 10–1. The second floor of this building has collapsed creating unstable walls.

Mezzanines are commonly found in warehouse type occupancies (such as concrete tilt-ups) and can be erected after the occupants occupy the building. Therefore, a mezzanine may or may not be to code, but it will be used for storage that can become wet if sprinklers and/or hose streams are used in the suppression of a working fire. This can result in an overloaded mezzanine that can easily become an elevated collapse hazard.

Be aware of asbestos. Areas where asbestos products are commonly found include sprayed acoustic ceilings, floor tiles, hot water pipe insulation, insulation in boiler rooms, insulated ducting, and old-style blown-in insulation (attics), as well as exterior shakes on older residential structures.

Baled goods can retain a significant quantity of water and may dramatically add to the weight being carried by a structure. Be particularly aware of baled goods above the grade floor in a building (or on a grade floor over a basement) that have become wet. Also, evaluate the type of construction that is supporting heavy materials, particularly if they are wet. There is a significant difference between older and newer construction, and the location of the soaked materials (center of a room vs. the outside edge).

Determine whether the contents of the building are classified as hazardous materials. If so, ensure the proper resources are available with regard to proper handling and disposal procedures.

Check the drainage of runoff water, and determine if it is collecting in an area that may not be able to support the increasing weight.

Be careful to not get caught in the thin wire that holds the metal framework for a suspended ceiling. This type of ceiling is very popular in all types of commercial buildings due to its ease of installation and professional appearance.

Check for the presence of cellulose insulation. As detailed later in this chapter, a fire in an attic that has cellulose insulation can present a formidable challenge.

And finally, unless there is a specific utility hazard, the normal order of priority for handling utilities is:

1. Electrical
2. Gas
3. Water

Consider the importance of a continual evaluation of the structural integrity of a building both before and during overhaul operations. Remember that any inherent structural deficiencies combined with fire will weaken a building and that these, combined with the weight of water, can significantly alter structural integrity. Buildings have collapsed during overhaul operations. If structural integrity is a concern, establish a fire watch and do not allow personnel to enter the suspected area. Depending on conditions, it may be necessary to contact the owner or responsible person about using heavy specialized equipment or even demolishing the building to abate a hazard.

Fire cause determination

It is usually essential that a fire cause investigation be completed before overhaul operations are initiated because the moving of debris can thwart a successful fire cause investigation. Therefore, the preservation of appropriate scene evidence may delay overhaul operations until a fire cause has been determined. When attempting to determine the origin of a fire from perceived indicators, remember that the observations of suppression personnel can provide valuable information. Physical evidence around the suspected point of origin should not be disturbed unless it is essential to alleviate a hazard. If it is necessary to request an arson investigation unit, do so in a timely manner. Consider the possibility of contacting arson investigators on tactical channels or by telephone to obtain assistance during cause determination or to minimize public access to potentially sensitive information.

Assignments

Generally, the success of an effective and safe overhaul depends on the formulation and delegation of an appropriate plan. To minimize duplication and maximize the effectiveness of available resources, and to ensure that the area in question has been appropriately overhauled, the incident commander must determine what overhaul operations are necessary, which areas need to be overhauled, what resources will be necessary, what specific tasks and areas to delegate and to which appropriate officers, and where and how the overhauled material will be placed.

Officers at every level of command must ensure that personnel are closely supervised and that they adhere to appropriate safety procedures. Company integrity must be maintained to enhance accountability and to minimize freelancing. Be sure to use the appropriate number of personnel in a specific area. In small, confined areas, excess personnel will increase the possibility of injury during overhaul.

Ventilation

Unless a fire has been totally extinguished and the contaminants completely ventilated, overhaul operations can be conducted within varying concentrations of smoke, heat, humidity, carbon monoxide, and other noxious gases. During a fire, heated toxic gases rise and collect at the top of a room or building. During overhaul operations, these gases may cool, settle, and combine with additional toxins emanating from smoldering material within the overhaul area, where they will easily be inhaled by personnel.

Additionally, common building materials will retain and radiate heat (particularly concrete) that has been generated during a fire, resulting in elevated temperatures during overhaul. Adequate ventilation during post fire operations can dramatically improve the overhaul environment by minimizing the accumulation of smoke and toxic gases. It can also reduce heat and humidity, improve visibility, and increase the endurance of personnel.

Ventilation can be accomplished individually or by a combination of natural ventilation, horizontal and vertical ventilation using prevailing wind currents, and pressurization by positive pressure. A moderate movement of air is sufficient to cause contaminants within an overhaul area to move in the direction of air flow. This concept is detailed in chapter 16.

Rehabilitation

During fire suppression operations, personnel are often driven to intense levels of physical activity in hot, hazardous environments. When these factors combine with encapsulating protective equipment, plus other factors such as wet protective clothing and hot or cold climates, personnel may become fatigued and need rest before initiating overhaul operations. Heat stress can seriously hamper performance and

limit work time. Because of its insulating qualities and limited permeability, protective clothing will interfere with the evaporative cooling of the human body. Therefore, it is all too easy for a firefighter to retain heat, lose fluids, and become sweat wetted. An incident commander should evaluate each incident for the following:

- Conditions
- Duration
- Exertion level of personnel
- Whether the overhaul is at the beginning or the end of a tour of duty
- Weather conditions

If necessary, select a rehabilitation area to provide personnel with appropriate resting places and liquids, particularly before committing them to a lengthy or difficult overhaul. Rehabilitation measures may involve removing inappropriate clothing, consuming liquids, using fresh personnel, and adhering to the two-bottle rule, which maintains that if a firefighter has used two 30-minute bottles or has worked for 40 minutes, he or she is a prime candidate for a minimum rest of 10 minutes and rehydration before being allowed to continue working.

Utilities

When overhaul operations involve pulling ceilings or opening walls and floors to check for fire extension, electrical service to the structure should be eliminated. This is of particular concern in older wood frame structures that have knob-and-tube wiring, which results in unprotected wiring throughout the structure, and in newer-style apartment buildings that use electric radiant heating, which results in bare wiring in the walls and ceilings.

Gas utilities also should be eliminated to prevent the possibility of an accumulation of gas from becoming a potential explosion hazard. If necessary, the water service should be curtailed to prevent additional water from causing further damage to a structure, thereby creating additional work for overhaul personnel (activated sprinklers are contributors to a growing water problem). This is particularly important when a structure has wood flooring.

Fire watch

During the strategy meeting, consider the importance of designating an officer or other appropriate member to watch the fire area. If a fire watch is necessary, charged hose lines should be in position for immediate use until it has been determined that the potential for a rekindle no longer exists (particularly for fires that are not readily extinguished, such as large-loss fires).

Equipment

Although numerous types of equipment may be employed during overhaul, the availability and use of the proper equipment is mandatory. This section provides an overview of equipment commonly used for these operations.

Self-contained breathing apparatus (SCBA)

The use of SCBA during overhaul can be a paradox of priorities and perceptions. During fire attack operations, the common practice is to use SCBA for protection against hazardous atmospheres. However, during overhaul, the use of SCBA is often not as popular. It is a fact that the same contaminants prevalent during fire attack can also be present during overhaul. Remember that smoke from a structure fire can be composed of numerous toxic gases, some of which are hydrochloric acid, ammonia, hydrogen sulfide, hydrogen cyanide, carbon monoxide, and so on.

It is normal practice that if concentrations of carbon monoxide over 50 parts per million are detected, you should seriously consider using SCBA during overhaul. However, there is mounting evidence that hydrogen cyanide is present in most structure fires and can be a major component in injuries and deaths of firefighters. Regardless of the amount and number of toxic gases that are present, firefighters should take the necessary steps to ensure their personal safety, and the use of SCBA is a major step in that direction. Remember the famous fireground axiom— *it's okay to read smoke, not breathe it.*

Protective clothing

Although the use of full protective clothing is essential in fire attack operations, its use during overhaul should be regulated by the incident commander. Boots, gloves, pants, and helmet are necessary during most overhauls. However, the addition of a turnout coat (if not required) can dramatically reduce the stamina, flexibility, and comfort of personnel due to its additional weight and the clothing's insulating qualities. When appropriate, some departments use lightweight jackets or approved work shirts as alternatives to wet, heavy turnout coats. Consider, however, that T-shirts provide minimal safety protection during overhaul operations and that they are not recognized as safety equipment by most workers' compensation personnel. Additional protective clothing that should be worn (as warranted) is as follows:

- Hood when pulling ceilings with suspected fire above.

- Eye protection such as goggles or face shields for operations such as pulling, chopping, and cutting. However, as face shields are open at the bottom, consider that when looking up, it is possible for foreign material to drop between the shield and face. Therefore, a face shield should not be a first choice for overhaul operations.

- Respirator mask similar to the masks purchasable at paint stores. These masks do not provide protection against the lack of oxygen but will filter most small particulate matter. They can be useful in dusty atmospheres when pulling ceilings, overhauling insulation, and the like.

Hand tools

A number of common tools can be useful in simplifying overhaul operations as follows:

- **Knife:** A common utility knife can be effectively used for opening overstuff materials, cutting burned portions of carpet, and other similar general purpose operations.

- **Drop bag:** Used to lower material (rubbish carriers and such) and tools from above-ground areas and to secure mattresses before removal from a structure. This tool is detailed in chapter 9.

- **Wire cutters:** The presence of hardened wire has become more widespread with its use in suspended ceilings and HVAC ducting. If a firefighter becomes entangled in this wire it can be very difficult to become disentangled. A good set of wire cutters can easily cut this wire in a short period of time. Remember, there is a significant difference between a $7 and $30 set of wire cutters.

- **Channel locks:** A great multi-purpose tool particularly for shutting off utilities.

- **Wedges/door stops:** Wooden wedges and rubber stops can be utilized for numerous applications. Wooden wedges can be used for numerous applications such as holding open doors, keeping the bottom portion of older double hung windows in the *up* position, and plugging fused sprinkler heads (covered in more detail in chapter 11). The wedges are approximately 3 to 4 inches long, 1½ inches high, made from pine (soft woods are flexible, swell when wet, and are more easily inserted between the frame and sash in double hung windows than harder woods), and are easily carried in a turnout coat pocket. A rubber door stop can be stretched between two opposing door knobs. This will keep a door from completely closing and locking, and if necessary, allowing stubborn doors with locks to be closed without locking.

Tools/equipment

There are many tools and pieces of equipment that a firefighter should use for overhaul operations (see figure 10–2):

- **Axe (A):** The pickhead axe is the most versatile tool for cutting, prying, striking, and digging operations. It is an effective general purpose tool.

- **Light (B):** A quartz lamp or spotlight can provide bright illumination. Elevated lights (aerial or platform) will illuminate a large

area and won't shine in the eyes of personnel if directed downward, particularly when overhauling a roof.

- **Mattress carrier (C):** Made of canvas material and approximately 6½ by 9½ feet, it is used to contain and remove mattresses from buildings.

- **Pike pole (D):** This is effective for pulling ceilings and walls and for removing sheathing from roofs; it is also a good general purpose tool.

- **Rubbish carrier (E):** Made of canvas material and 4½ feet square, this is used to remove debris from buildings.

- **Rubbish hook (F):** This hook is effective for opening walls and ceilings; for stripping shingles, shakes, tile sheathing, and plywood material from roofs; and for moving overhaul debris. It is an excellent general purpose tool that can be more effective in overhaul operations than a pike pole.

- **Power saws (G):** Chain and rotary saws are used for opening walls, floors, and roofs.

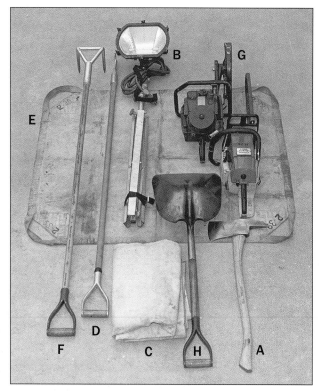

Figure. 10–2. Common overhaul tools

- **Scoop shovel (H):** This is used for carrying debris, moving a variety of loose material, and removing water from carpets and floors.

Four additional tools/equipment that can also simplify overhaul operations should be mentioned as follows:

- **Hook:** Also known as Boston hook, Lowy hook, tin hook, and rake, this is basically an adaptation of a pike pole that has been modified for specific uses such as plaster, older ornamental tin ceilings, and so on.

- **Ladders:** There are several types and lengths of ladders that can provide quick access for personnel into overhead areas (such as attics) for final extinguishment of fire. These ladders are commonly referred to as *inside ladders* because they are primarily used inside of structures.

- **Thermal imaging camera (TIC):** This tool has dramatically changed the ability of fireground personnel to locate fire in hidden/concealed areas that are not readily apparent. Therefore, a thermal imaging camera is virtually mandatory for simplifying overhaul operations, depending on the size and extent of the overhaul. However, the effectiveness of this tool is dependent on two basic considerations: the ability of the operator to make use of the camera to its potential; and not replacing the human factor with the equipment factor. In other words, a thermal imaging camera is only a tool that can *assist* fireground personnel in making informed decisions.

- **Wet-water proportioner:** This is a tool that deserves consideration because of its capability to dramatically simplify numerous overhaul operations. A proportioner automatically meters a wetting agent into an overhaul fire stream. It is a direct-demand, variable-flow proportioner (at a ratio of from 1:6,000 to 1:10,000 parts wet water to water) and is designed to accommodate the varying water flows and pressures of firefighting streams. The addition of a wetting agent lessens the surface tension of water, increasing its ability to penetrate various materials. This agent can be effectively used on baled goods, overstuff materials, cellulose insulation, the underside of wood-shake roofs (wet water does *not* replace overhauling hot spots in roofs composed of composition over shake-shingle materials), and most Class A combustibles, dramatically decreasing the work necessary to overhaul these materials.

A proportioner can be inserted into an overhaul fire stream as follows (figure 10–3):

Figure. 10–3. Wet-water proportioners can increase the effectiveness of overhaul lines.

- A 2½-inch section bypass is connected from a pump discharge gate to the inlet of the proportioner.

- A 2½-inch to a 2½-inch wye is connected to the outlet of the proportioner.

- Two 1½-inch or 1¾-inch lines can be connected to the wye and used for overhaul purposes. This particular method will allow the use of at least two lines as follows: two lines inside a structure for overhaul; one line inside and one line outside of the structure; and two 1½-inch wyes connected together can provide three overhaul lines, if necessary.

A cost-effective alternative to the wet-water proportioner is the wet-water antisiphon sprayer, or lawn fertilizer applicator, which can be purchased at most garden and nursery outlets. This tool can be effective for the application of wet water in situations where a proportioner would be impractical or not cost effective. It is normally attached to a garden hose. Note that you should consider the use of a wetting agent prior to fire cause determination as it can dissolve flammable liquids and hinder detection of their use.

Methods

Although overhaul operations vary at each incident, the focal point is in finding and extinguishing any hidden fire. While accomplishing these objectives, it is essential to consider and constantly evaluate specific potential hazardous building conditions. They can be summarized by weakened ceilings, floors, stairways, roofs, structural members (wood, concrete, and steel), unreinforced masonry brick walls, spalled concrete, and supported heavy objects (air conditioning units on roofs, machinery on floors, soaked baled goods, and the like).

To help find hidden fire, use a systematic approach that considers both the obvious and the potential extension of fire. Begin overhaul operations near the fire origin (first priority) and progress away from it, moving upward (second priority) and downward (third priority), if appropriate. In all cases, as long as there is char or burned material, keep exposing areas until you reveal unburned material. This process is enhanced by considering that hidden fire will often display the following indicators:

- Smoke issuing under pressure as opposed to lazy smoke that appears to be dissipating.

- Materials that appear to be discolored.

- Warm or hot floors, walls, and ceilings. Feeling for warm or hot areas can be enhanced by removing a glove, if appropriate.

- Crackling sounds that would indicate burning material.

- Abnormal odors.

- Darkened streaks and smoke stains near moldings, electrical wall outlets, and so on.

Ladders

The methodology of conducting overhaul operations can be greatly simplified by the use of inside ladders (as these ladders are primarily used inside structures), and pike poles. First, let's look at inside ladders. An inside ladder will allow personnel to access hard-to-reach areas such as attics to check for extension. Ladders or personnel should be equipped with a 6-foot pike pole to facilitate pulling ceilings and opening walls. As an example, a pike pole can be easily mounted on a ladder to be within easy reach. The three most common ladders that are used for inside operations are the combination, the 12-foot, and the 14-foot extension ladders. A general summary of them might be as follows, although the information given can vary, depending on specific manufacturers:

Combination ladder. This ladder can be constructed in a variety of configurations and is also known as a folding ladder, attic ladder, extension ladder, A-frame ladder, telescoping beam ladder, closet ladder, and so on. It is generally about 6 feet tall, which can simplify carrying inside a structure. This ladder offers numerous configurations such as a straight ladder, A-frame ladder, extension ladder, and so on. However, because of its versatility, it may take considerable space to finalize some of its configurations in areas with limited space, such as a hallway.

12-foot extension ladder (telephone ladder). This ladder consists of two 7-foot sections. It is wider at the base, which allows for a more stable platform than a 14-footer. It is also easier to make into a step ladder. Because its base is wider than its top, however, this ladder consumes more space than a 14 footer in closets, hallways, and other tight places. Although not designed for section separation, it can be separated into two 7-foot sections, which may be advantageous in structures with low ceilings or in confined areas.

14-foot ladder. The 14-foot extension ladder consists of two 9-foot sections. It has a narrow tip and base and will make a taller, though less stable, step ladder than a 12-footer. Some 14-foot ladders can be converted into step ladders, which requires sufficient room below the ceiling. Its narrow tip and base facilitate working in tight spaces such as closets and hallways. When the ladder has been separated, one of the 9-foot sections provides a thin, light ladder for use in confined spaces. To increase its stability, consider strapping a pike pole to its base if necessary.

Ultimately, the type of structure, the intended use, and the availability of ladders will determine which ladder should be used. When entering a structure with an inside ladder, normally carry in the ladder top first. This allows easy entry under door headers and will minimize maneuvering once inside.

Pike poles and hooks

Pike poles and/or hooks are often necessary to open ceilings and other areas quickly to check for extension. If an attic fire is not completely extinguished upon arrival, pull the ceiling and extinguish all fire in it near the point of entry into the structure. Never advance into a structure without first extinguishing an attic fire above advancing personnel. Remember that a ceiling is a barrier between an attic fire and interior personnel (figure 10–4). By no means let an extending attic fire get between personnel and the exit opening.

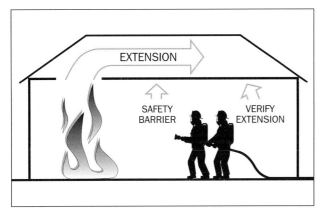

Figure 10-4. While advancing into a structure with fire overhead, never let the fire get between you and your exit.

Overhaul operations

This section summarizes common considerations as applied to the methodology of overhaul operations.

Ceilings and floors. If a ceiling is primarily opened to observe an attic area, the space between two ceiling joists may be sufficient. However, such an opening (usually 16 to 24 inches between joists) won't be sufficient for a firefighter encumbered with SCBA and/or a hoseline that needs to access an attic area. Nor will it serve a quick retreat. Therefore, consider removing one ceiling joist to increase the size of the working area (figure 10–5).

Figure 10-5. Removing a ceiling joist can increase the available working space as well as enhancing safety.

Pulling ceilings in structures with pitched roofs and a room in the attic can be a challenge. An attic room, particularly an add-on, is usually flanked by an attic space whenever the roof slopes to an exterior wall (figure 10–6). A gabled roof will usually have three such attic spaces. A hip roof may have an attic space that extends completely around the exterior. In any case, a small attic space may be found above the ceiling of an attic room. If it is necessary to expose attic spaces around an attic room, pull ceilings (from below) that extend beyond the floor of the attic room above. Normally this will be within several feet of an exterior wall on a side where the roof slopes downward.

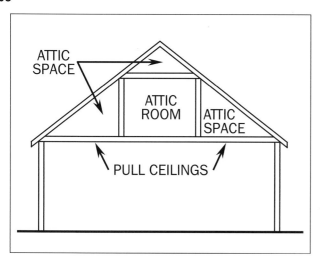

Figure 10–6. Attic spaces may surround an attic room.

Remember that flooring is normally confined to the useful area of an attic room where a person can walk upright. Pulling ceilings in an area with a floor above is difficult and normally will not expose the attic spaces on the sides of an attic room. Do not forget to check the small attic space above the attic room.

Some commercial occupancies have ceilings in the office areas only. Manufacturing or storage areas usually don't have ceilings (depending on the size and type of occupancy) unless the area requires cleanliness or air conditioning. Six-foot pike poles are normally adequate to pull office ceilings.

Suspended ceilings are commonly found in high-rise buildings, modern office buildings, and strip commercials. These can hide heating, ventilation, and air conditioning (HVAC) ducting, as well as electrical and plumbing paraphernalia. Individual panels can be easily removed, and without causing damage, with a 6-foot pike pole. These ceilings are only suspended by thin wires and can easily fail when exposed to fire or sufficient heat.

Public assembly and older strip-type (common attic) commercial buildings are typically constructed with 10- to 16-foot ceilings. These occupancies are often remodeled and may have suspended and/or false ceilings that hide enclosed spaces (and multiple attic spaces) where fire can extend undetected. These occupancies often require the use of a pike pole longer than six feet.

Figure 10–7 illustrates preparing an inside ladder, pike pole, and pickhead axe at a common structure fire. The firefighter assigned to initiate overhaul operations has placed a 12-foot ladder outside the structure next to the entrance. In this position it is out of the way but handy if needed.

Figure 10–7. The inside ladder is out of the way of the entrance opening and readily available. The pike pole and pickhead axe can be used immediately for inside operations.

The firefighter will now advance into the structure and is equipped with an axe and pike pole to begin overhaul operations. Open up ceilings and walls exposed to fire and check for extension. Place particular emphasis on lath and plaster surfaces (as opposed to gypsum board) because wood lath is more susceptible to burning. When pulling ceilings with a pike pole, place the hook end downward and out in front of you. Use short, downward strokes to pull the ceiling away (figure 10–8) rather than down on top of you. If the integrity of a ceiling is suspect, stand in a doorway when pulling ceiling material, as this will afford an area of safety from collapsing material.

Remember, when gypsum board becomes wet, large areas can quickly collapse downward with little warning and effort from a pike pole.

Figure 10-8. Keep the pike pole at arm's length and pull downward with short strokes to direct material away from you and limit exposure to electrical wires.

Walls

When opening lath and plaster and gypsum board walls, minimize damage by first making vertical cuts with an axe (or other appropriate tool) along the inside of two adjoining studs (A) or along the inside of adjacent studs on either side of a center stud (B) before removing the material (figure 10–9). This method will result in a professional appearance and normally requires less effort to repair.

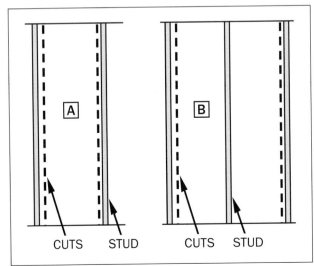

Figure 10-9. Make vertical cuts next to studs to limit damage and to simplify repairs.

Similar to opening ceilings, electrical wires may be present within the walls. Limit the depth of the cut with the appropriate tool to minimize contact with them.

A knowledge of walls is necessary to determine potential areas or pathways of extension of fire. Four types are typically used in building construction: party, partition, division, and exterior (figure 10–10).

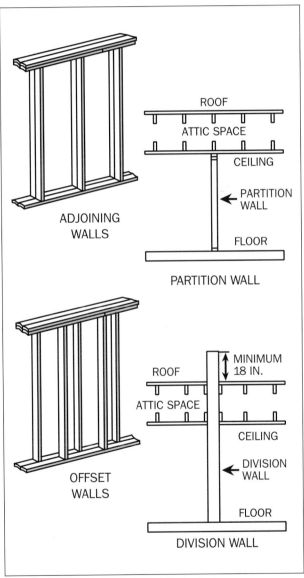

Figure 10-10. Different types of walls can dramatically affect suppression operations and the extension of fire.

Party walls normally separate one occupancy from the next in townhouses, condominiums, and motels, and normally connect the floor and ceiling only. Of these, there are two basic varieties. In the first, the two common walls are joined together and are known as adjoining walls. Adjoining walls will tend to confine a fire between the studs, thus limiting extension. However, in some cases, the two walls can also be separated by an air space of several inches for insulative and sound-deadening purposes. In this case, fire can extend horizontally and vertically between the walls. Adjoining walls may connect the floor and ceiling only, or may project through a roof similarly to a division wall. Offset walls are separated by several inches and the studs are also staggered. In this case, fire can easily travel horizontally or vertically between the offset studs. Offset walls normally connect the floor and ceiling only.

Partition walls are used to divide areas or rooms into smaller areas or to separate one portion of an area from another. Normally they travel between the floor and ceiling only. This type of wall will not restrict the horizontal extension of fire through an attic or other open space above.

Division walls travel through the ceiling to project above the roof at least 18 to 24 inches (figure 10–11). This type of wall is of heavier construction than a party or partition wall and is an excellent barrier to the extension of fire, particularly in an attic.

Figure 10–11. Easily identifiable, a division wall will restrict extension unless breached.

This remains true as long as the wall has not been violated by incomplete repairs, an open fire door, or the like.

Exterior walls comprise the outer shells of structures and are constructed of numerous materials. They also provide the principal support to roof

assemblies. The extension of fire through an exterior wall depends on the type of construction (such as masonry, frame wood, and so on).

Windows. Remove window (1), door (2), base (3), chair (4), crown (5), and wainscoting (6) moldings when they have been exposed to fire (figure 10–12).

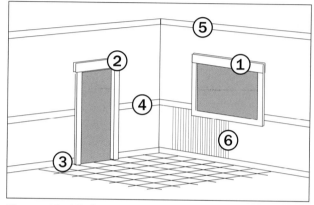

Figure. 10–12. Remove moldings when they have been exposed to fire.

Window and door moldings cover a gap between the casing and wall, which can conceal fire (arrow, figure 10–13).

Figure 10–13. Remove window and door moldings to reveal the gap between the casing and wall.

Focus on older double-hung windows. A cotton sash cord is usually attached to the upper portion of the window, then over a pulley to a hanging lead weight in the cavity next to the window (arrow, figure 10–14). This was formerly used to minimize the effort necessary to raise or lower the window.

Figure 10–14. Double-hung windows have a cavity on either side of the window with a lead weight and sash cord attached to the bottom window to simplify raising and lowering.

Remove the vertical molding to check the cotton sash cord because it will readily burn to the inside of the window casing. Crown, base, chair, and wainscoting moldings can also conceal fire in any open areas behind these materials.

Insulation. If a standard glass-type insulation is present in ceilings or walls, remove it and check both sides, particularly if one side is backed by paper. Of particular interest (and not from a positive viewpoint) is cellulose insulation, which is nothing more than shredded newspaper that has been subjected to a fire-retardant chemical. Although this insulation is supposedly fire retardant, it will burn or smolder from either direct contact with fire, or heat from improper installation. Proper installation requires a clearance of at least 4 inches between the cellulose and a heat source such as recessed light/heat fixtures. If confronted with this type of insulation, remember that a smoldering type of fire can exist under or within the insulation without being visible or issuing visible smoke. If fire is found, do not use a strong stream of water because this tends to disperse the insulation, possibly spreading your problem. Instead, use a gentle spray on the affected

area. A word of caution—if in doubt, take it out! This means, unless you invest the time (either during your investigation or after you have finished your overhaul) to ensure you have eliminated all chance of fire in this type of insulation, there is a strong possibility you will be back. Therefore, unless all of the insulation has been removed (which would be a major undertaking) it may be necessary to re-check the insulated area every few hours—either a fire company or home owner—to ensure it is out. Remember that a thermal imaging camera can be an asset, particularly when the attic area has cooled, but not the final answer as a small, smoldering fire may stay concealed until you leave! If SCBA isn't mandated by SOPs during insulation overhaul operations (particularly fiberglass type insulations), consider using respiratory masks.

Fiberboard is made from shredded sugarcane fiber or wood pulp, and similarly to cellulose insulation, was also used as a cost effective type of insulation. Fiberboard was used in older residential and commercial buildings in ceilings, roofs, and behind siding, will support rapid flame travel, and is capable of smoldering without being detected. Similar to cellulose insulation, remove it or plan on being back for a rekindle. Wet water has proven effective in soaking into fiberboard (due to its fibrous nature), but is not a cure-all for removal of all suspected material.

Basements. When fire has directly involved a basement, check it thoroughly, particularly for areas of vertical extension into portions of the structure above the basement. If fire didn't directly involve the basement, verify the lack of downward extension there as well.

Roofs. On dwellings with wood-shake roofs, several layers of composition shingles over shake shingles, or an existing roof that has been covered with lightweight steel, it may be necessary to strip the exposed portions of the roof to reach an area that is not charred. To remove shake shingles easily from space sheathing, personnel in the attic can pull backward and upward with the pick end of an axe between the sheathing or strike the underside of the shake shingles (between the sheathing) with a square-point shovel. After a roof has been stripped or determined safe, it will be necessary to clear the attic area of any fire debris. Personnel in the attic can use a square-point shovel to remove debris between ceiling joists. When clearing debris between ceiling joists, be careful not to step through the ceiling. Obviously, this operation will be difficult, time consuming, and dirty. Roofs that have been covered by lightweight steel can present a formidable overhaul hazard due to the thin steel and the gap between the steel and the original roof decking. For more information on this product, refer to the end of chapter 18.

When ventilation openings have been cut in roofs or wood floors and decking materials are overhanging a rafter (figure 10–15), thorough overhaul operations will cut overhanging decking material (referred to as a *deadman*) back to a supporting rafter, eliminating the chance that personnel or insurance adjustors will inadvertently step onto the overhang and fall into the attic or space below the floor.

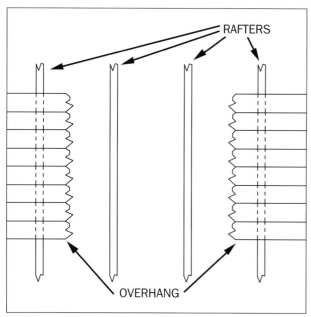

Figure 10–15. Always cut overhanging decking material to improve safety.

When a roof has been opened by ventilation operations or by fire, it may be necessary to provide protection to the interior of a structure from inclement weather. Plastic sheeting can be more effective than salvage covers due to its ease of implementation. Also, you can leave it on the affected premises when you return to quarters. You can easily cover an opening with plastic and secure it to a roof by

stapling it or nailing wood lath over it. To increase the strength of this cover, roll the ends of the plastic (all four sides) in wood lath and nail it in place (figure 10–16 A). Because this configuration provides minimal protection against heavy rain and snow, enhance protection by *raising* the center of the plastic sheeting over the opening with lath or pieces of cut roof decking, as illustrated in Figure 10–16 B and C. Secure the four sides of plastic to a roof with staples by placing wood lath over the plastic and nailing it to the roof, or by wrapping the plastic ends in wood lath and then nailing them.

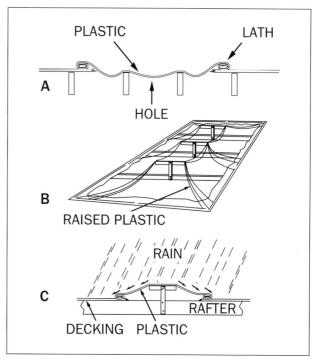

Figure. 10–16. A, B, and C. Plastic sheeting over openings can provide protection against inclement weather.

Horizontal and vertical spaces. Open up horizontal or vertical spaces exposed to fire and check for possible extension. Pipe chases and zero-clearance fireplaces (a fireplace flue with a framed exterior) provide unstopped vertical access for fire extension. These open vertical shafts can also be common to any adjoining areas, such as spaces between floors, attics, and the like. A good example is illustrated in figure 10–17. Look closely at the roof of this four-story condominium complex. Notice there are four separate zero-clearance fireplaces visible from the roof area that can provide a common unfirestopped avenue from the first floor to the attic of this structure. These areas will have to be checked for possible vertical extension of fire.

Figure 10–17. Zero-clearance fireplaces are often un-fire-stopped and a vertical fire channel in a building.

Remember that balloon frame construction can also provide open vertical avenues to an attic.

When a fire has occurred in a kitchen, particularly under or near an overhead ducted exhaust, be sure to check three specific areas for extension of fire: check inside the duct due to the buildup of grease; ensure that the duct goes through the attic and to the roof and does not terminate into the attic; and check the top of the roof where the duct exits because there can be a considerable grease buildup on the roof that will willingly burn if exposed to fire extending up the duct.

Fascias. Fascias are nothing more than a common attic on a building's exterior that can conceal and promote the spread of fire around the exterior of a building. They normally don't have firestopping or sprinklers, and are also common to the adjoining attic. You must open and check this type of construction if extension of fire is suspected. There are three specific areas that must be checked for fire: the interior of the building; the attic area; and the fascia. Remember that if a fire in the interior of a building (with a fascia) extends into the attic area, it can also extend into the fascia area. As illustrated in figure 10–18, you can open fascias in three locations. Opening a fascia in the locations marked A and B will place personnel in the collapse zone.

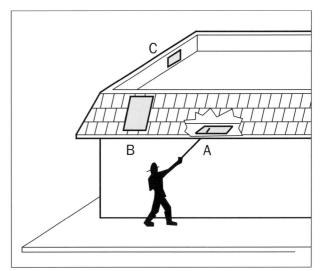

Figure. 10–18. Open fascias to check for fire extension during overhaul operations.

Conversely, if the roof is structurally sound, you can easily and safely check for extension of fire and/or overhaul a fascia by opening its vertical backside portion (C), which is accessible from the roof (figure 10–19). This portion is usually ½-inch plywood or similar material and is covered with roofing material and is easily breached.

If the fascia were to collapse, it would collapse away from personnel on the roof.

Figure. 10–19. In some cases, it can be advantageous to overhaul a fascia from the backside.

Other fire avenues. ABS plastic pipe is now commonly used in plumbing applications. This type of plastic will readily burn if exposed to fire and will act as a vehicle to spread fire throughout any structure that is traversed by it. The combustion process will make copious amounts of smoke and leave pools of melted plastic that are easily identifiable. Open walls and ceilings if you think this type of pipe is present and has been exposed to fire.

Thoroughly inspect areas directly above the main portion of a fire for fire extension. Focus on walls (particularly in balloon construction), the plate at the top and bottom of a wall, and the floor and ceiling areas because these can provide excellent avenues for extension.

Identify and check shafts such as pipe alleys, grease ducts (kitchens), vents, and other similar avenues. If fire has exposed any ducting (air conditioning, heating), expose and check the interior of those ducts as well. This is particularly important when the newer type of plastic-lined HVAC ducting is used because fire can extend inside the ducting and not be visible on the exterior of the ducting.

Post-extinguishment considerations

After extinguishment, firefighters should descend into basement areas to shut off utilities and check for extension of fire, smoke, the potential of overlooked victims, and so on. In buildings with basements, remember that carbon monoxide can settle in this area and become an invisible hazard to firefighters who are not using SCBA.

After a fire has been extinguished, the building should be surveyed for numerous considerations. One of the more important considerations is the type of construction and contents that are being supported by the construction. As an example, assume a working fire in the basement and first floor of a routine dwelling has been extinguished. Prior to overhaul operations, it is observed the floor joists are the typical wooden I-beams, and have been extensively charred from the fire in the basement. It is also observed there are several large pieces of furniture (china cabinet and so on) over this area that are located on the floor above the basement. Based on the degree of char to the aforementioned floor joists, determine if there is a potential collapse that could occur during the ensuing overhaul. Additionally, remember that some rooms in a dwelling are more prone to collapse than others. As an example, bathrooms and kitchens are prime candidates

due to tile flooring, cabinets, sinks, bathtubs, and appliances. Remove baseboards and open walls at the floor line to inspect for hidden fire. Remember that burning debris can fall into the area behind coved ceilings and can travel down to a fire block or the plate at the bottom of the wall.

Remember that overhaul operations can provide an excellent opportunity to train members in building construction and other related tasks. Knowledge of how something is put together is particularly useful for dismantling operations, so don't focus on using brute force.

Tactics

During the post-fire survey, determine the type of overhaul and location for the deposit of fire debris. In some large-loss fires, debris is left in place depending on the type, stability, and content of a given structure. In other types of fires, you may need to systematically remove overhaul debris from the structure to complete extinguishment. However, in most fires, you will often move overhaul debris minimal distances to check for extension, complete extinguishment, and leave the premises in a fire-safe condition. When it is necessary to move fire debris, consider the potential value of that debris to the owner. What may seem insignificant to one person may be priceless to another. Also, any valuables you discover must be accounted for and given to a responsible person, verifiable owner, or incident commander. If possible, contact the building occupant to attest to the location of any valuables before initiating overhaul operations. Consider the following when determining the placement of overhaul debris.

Fire inside a structure

When a fire has occurred inside a structure, there may be times when it is advantageous to leave the fire debris inside the structure. Remember that the old overhaul axiom "when in doubt, throw it out" may not be applicable when the overhaul focus should be on customer service!

When leaving fire debris inside a structure, remember to maintain an appropriate charged line. Use minimal water for final extinguishment. This may mean reducing the size of overhaul lines, particularly when overhauling attics. Always try to minimize additional property loss during overhaul.

Separate burned from unburned material and salvageable material from unsalvageable material (figure 10–20).

Figure 10–21. Keep interior overhaul piles at least 2 feet from the walls.

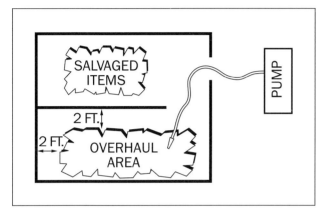

Figure 10–20. Separate overhaul debris from salvageable material.

Remove salvageable items from the fire area, but keep them inside the structure for security. Salvageable items may include furniture, files, books, and paintings. When a chest of drawers or other similar type of furniture has been damaged but its contents have not, use the drawers to move those salvageable items to another location.

Evaluate the potential consequences before placing overhaul piles on combustible floors. Aspects to consider are weight, combustibility of the flooring in case of a rekindle, and additional damage to the floor. To protect it as well as to minimize water damage, place overhaul piles on salvage covers or plastic sheeting. Remember that plastic sheeting will not provide any protection against heat or fire and will quickly fail.

Keep overhaul piles at least two feet from walls (figures 10–20 and 10–21). This provides access around the piles and minimizes extension if reignition occurs.

Separate overstuff from other material. Do not mix overstuff with other fire debris because you often must force water through overstuff material to extinguish deep-seated fire. Ensure that you use appropriate water to complete extinguishment as you add additional burned material to the overhaul pile.

Never pile additional material on top of hot material and then try to soak the entire pile for complete extinguishment. Instead, continue to apply water as you add additional overhaul debris to the pile.

When a fire has occurred inside a structure and it is necessary to remove the fire debris outside, have a charged line in place before starting overhaul piles. Maintain charged lines inside and outside the structure, and assign one member to wet down any hot material brought outside.

Avoid throwing material from heights because the sudden introduction of fresh oxygen can cause flare-ups. If tossing material from a roof is truly necessary, or if material will inadvertently be falling from a roof, do so as a last resort and only after you have taken appropriate action to protect personnel at lower levels.

Take proper precautions when removing overstuff materials to the exterior of a structure. Carrying a smoldering overstuff can introduce a flow of fresh, cool air that can cause it to reignite suddenly and with intensity. This may also be inadvertently enhanced when the implementation of PPV is not coordinated with the removal of overstuff materials.

Separate salvageable items from fire debris and keep them inside the structure for security (figure 10–22).

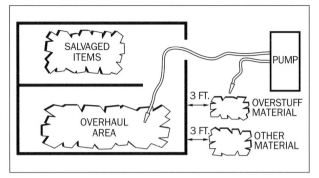

Figure 10-22. Separate salvageable items from fire debris and keep them inside the structure for security.

Figure 10-23. Separate overhaul material from overstuff material.

If necessary, cut open and wet smoldering overstuff material (if possible) before removing it from a structure. Removing a mattress wrapped in a mattress carrier can retard re-ignition and aid in transporting it. In all cases, a supply of water (portable extinguisher) should be available when removing an overstuff. Never use an elevator to remove smoldering overstuff material for obvious reasons.

Separate outside overhaul material by placing overstuff material in one pile (figures 10–22 and 10–23). Open and flood it with water if necessary. Never mix overstuff material with other combustibles. If necessary, consider submerging smoldering overstuff material under water in a trash container.

Place other debris in one pile. Wet it down as you add additional material to it. Never cover hot material with additional material or assume that wetting down a pile of debris will extinguish all of the hot spots within the pile.

Keep overhaul piles at least 3 feet from the structure and other combustibles (figure 10–24). This provides access around the piles and minimizes extension if re-ignition occurs.

Figure 10-24. Keep exterior overhaul material at least three-feet from walls.

Consider the effect of placing overhaul material directly in front of a residential structure. Doing so would not enhance the visual effect of the incident, particularly for returning occupants.

Fire outside a structure

When a fire has occurred outside a structure and the debris will remain outside, maintain an appropriate charged line. Separate burned from unburned material and salvageable from unsalvageable material (figure 10–25). Consider the relative importance of the security of salvageable material if the occupant has been keeping the material outside the structure.

Pay particular attention to automobiles and dumpsters when overhauling them. Modern automobile interiors are finished with plastic materials that produce deadly toxic gases during combustion. Dumpsters have become ideal locations to dump a variety of combustible materials (including hazardous ones), and they can also spawn toxins during overhaul. Before conducting a routine overhaul operation in an automobile or dumpster, consider the probable hazards of smoke from the burned materials and the concept of using SCBA because these incidents can be much more than simple rubbish or auto fires.

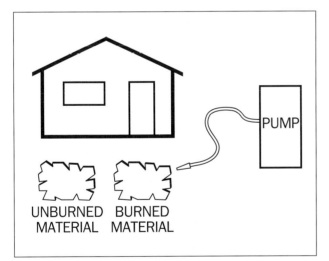

Figure 10–25. Separate exterior burned material from unburned material.

Completion

When you have completed an overhaul operation, remove all but one overhaul line with the other overhaul equipment. This will normally allow hidden hot spots time to accelerate, become visible, and be extinguished with the remaining line. The incident commander should evaluate the structure and then determine the following:

- The overhaul is complete (as planned).

- All fire has been extinguished.

- The appropriate utility companies have been notified. (Fires in restaurants normally require notification of the health department authority.)

- The structure is safe for reentry.

- Before fire personnel leave the premises, the structure does not need to be secured to prevent unauthorized entry.

- The person responsible for the building has been contacted.

- You have offered to walk through the building with the occupant to explain your operations and any appropriate options. Remember, this person is your customer—and a voter!

- The sprinkler system has been restored to operable condition. It must be ready for possible rekindles and because of insurance stipulations.

Take one last look before leaving. The condition in which you leave the premises will reflect the professionalism, effectiveness, and pride of the personnel involved. Consider overhaul to be an opportunity to limit damage, minimize the overall need for repair, and provide the proper environment to allow insurance adjusters the opportunity to determine the loss adjustment. If you have a question about a potential problem, remember that it is sufficient cause to give good reason for your actions to eliminate that suspicion. Bear in mind, too, that the appearance of whatever is left in the wake of an incident is always noticed by the returning owner or occupant.

Training Resources

The following resources can provide additional information and/or another viewpoint for the subjects discussed in this chapter:

- Structure Fire Overhaul, Jeffrey L. Herbert, *Fire Engineering*, March 2008.

- Overhauling for Successful Fire Investigation, Mark Wallace and Dr. John DeHaan, December 2000, *Fire Engineering*.

- Chocking Doors, Todd Connors, *Fire Engineering*, March 2006.

- Seat Belts Carry the Load and Keep Hands Free, Dave Walsh, *Fire Engineering*, May 2007.

- Overhauling When Cellulose Insulation is Present, Jamie, C. Morelock, *Fire Engineering*, March 2006.

- Overhauling Techniques that Save Energy, Michael N. Ciampo, *Fire Engineering*, April 2004.

- How to Avoid Rekindles, Bill Gustin, *Fire Engineering*, November 2000.

- The Punch Technique, Michael N. Ciampo, *Fire Engineering*, November 2008.

11 Salvage

Introduction

Although overhaul, ventilation, and salvage share some of the same objectives in fireground operations, such as minimizing loss and damage from fire and its by-products, each of these operations normally meet their goals in a different manner. The primary focus of overhaul is to ensure that all fire has been extinguished, whereas the primary focus of ventilation is the removal of heat, smoke, and its by-products. In contrast, salvage operations are intended to limit damage incurred from the uncontrolled presence of water from activated sprinklers, leaking water pipes, the effects of fire suppression, and other inherent by-products of fire. As a result, the main focus in salvage operations for this chapter is water control, as well as the protection of a structure and its contents against water damage.

Salvage operations should be an integral part of fire suppression. Attention should be given to it upon arrival of the initial company and continue through extinguishment, overhaul, and the securing of the premises. Although the loss from a broken water pipe is the direct result of water alone, the total loss accruing from a structure fire is the result of two factors: direct loss and indirect loss. Direct loss is damage caused by fire, heat, and the by-products of combustion; indirect loss is damage caused by the operations used to control the fire. Although a portion of fire loss will unavoidably result from gaining access into a structure and extinguishing a fire, it is often the overaggressive, misdirected, or ineffective fire suppression operations that can quickly increase the indirect loss. For example, the indiscriminate use of a heavy stream, even for only a few minutes, will increase loss. The goal in fire suppression operations should focus on suppression in a timely manner in conjunction with minimizing loss. Therefore, the earlier the salvage operations are initiated, the greater the chance of minimizing indirect loss.

However, as the fire service continues to experience budgetary constraints and additional fireground priorities, salvage operations have been relegated to an afterthought instead of a priority. As an example, a routine fire in several rooms in a single-family dwelling normally requires an initial attack line, a backup line, forcible entry, ventilation, and a primary search. To this set of basics, modern fireground priorities now require the implementation of the two-in/two-out constraint and a RIT team. Unfortunately, most fire departments are now staffing their engines and trucks with three personnel on each type of apparatus while still trying to place an emphasis on maintaining a high level of service with a reduced staffing level; therefore, it is easy to see why salvage is often neglected until personnel are available. Yet, the need for effective salvage operations continues to increase. Interestingly, sprinkler systems contribute to this need. Although they reduce fire damage, the water damage from a single sprinkler head can quickly become extensive unless effective tactics for protecting property are employed.

Computers have proliferated widely, and we are increasingly dependent on the valuable information that they store. They are common both in business environments and residences. A moderate amount of smoke and water can quickly cause significant damage to these systems. Proper salvage techniques can help to minimize this.

Public Relations

An important and valuable benefit of salvage is its public relations value. Most citizens have had minimal exposure to the fire service and are unaware of fireground operations other than the ability of firefighters to spray water and cut holes in roofs. The care and skill that can be demonstrated by suppression personnel in the salvaging of property and personal possessions can dramatically enhance a positive public perception of the fire service. This last statement is vitally important from three perspectives.

First, the preceding phrase *the care and skill that can be demonstrated by suppression personnel* indicates this philosophy is a choice. Either suppression personnel want to minimize loss from fireground operations, or they do not. Obviously, this attitude comes from a department administration that supports the concept of minimizing loss at structure fires, and carries this attitude through to each firefighter who is involved in suppression operations.

The second perception is the concept of *loss control*, which centers on the aspect of a full-service system. In other words, minimizing loss begins with suppression operations designed to minimize indirect loss and continues (if necessary) to added-value considerations after the incident. What are added-value considerations? They are the extras, such as putting a tarp over personal belongings before pulling down a ceiling, or picking up trash. It's these thoughtful acts that leave the customer with a favorable impression of your fire department when you return to quarters.

To illustrate the concept of loss control and added value, let's look at two examples. Assume you have extinguished a fire that involved several rooms in a residential occupancy, but it is necessary to pull the ceiling in an unburned bedroom to ensure all fire has been extinguished. In this case, loss control would dictate that the bedroom contents (furniture, bed, and so on) are covered with salvage covers before pulling the ceiling and dropping debris into the room. In this simple example, if the ceiling were pulled first, there would be unnecessary damage to the room contents that could have been easily prevented.

In the second example, assume that you respond to a routine fire in a residential occupancy and quickly extinguish the fire. Post-fire operations then consist of fire cause determination, overhaul operations, and some salvage operations to remaining contents and structure before returning to quarters. This scenario would be considered routine by most fire departments; however, let's assume we introduce the concept of added value to this scenario by including furniture that was wiped dry, furniture that was raised several inches on premade Styrofoam blocks to reduce damage from water on the floor, and dishes/pictures/paintings that were packaged in boxes that displayed your department name on the exterior of the boxes. When you left this scenario to return to quarters, do you think the occupants would remember your extra effort and professionalism, particularly when it was time to vote for a fire department bond measure?

The third perspective is the fact that citizens pay our salaries and operational costs. Previously, it was mentioned that the fire service is currently hampered by budgetary constraints, and their perception of the fire service is often determined by our job performance. Therefore, it is vitally important that fireground operations result in a finished product that leaves a positive lasting impression on the customers that can determine our operational budgets!

A common indicator of a fire department's commitment to professionalism is its approach to minimizing loss and maximizing the service provided to its customers. Remember that the fire service does not have a product to sell, but that the actions of personnel can affect the taxpayers' perception of the services that we provide.

Equipment

Numerous types of equipment can be used for salvage operations. The following is an overview of the more common types.

Protective equipment

Similar to overhaul, the use of appropriate protective equipment should be governed by the prevailing conditions. Self-contained breathing apparatus (SCBA) should be mandatory in contaminated atmospheres. Protective clothing should be regulated by each specific salvage operation and the incident commander or appropriate officer. Boots, gloves, and helmets are necessary during most salvage operations. A turnout coat, however, can dramatically reduce the stamina, flexibility, and comfort of personnel. When appropriate, lightweight jackets, approved work shirts, and other approved clothing can provide the requisite degree of protection and relief from more burdensome apparel.

Channel locks/vise grips

Having an adjustable tool that can perform numerous functions is essential. Channel locks can remove small fittings, pinch off copper tubing, and turn stubborn water shutoff valves. They are also a good general purpose tool for those instances when a specialized implement isn't readily available. Vise grips have the capability to exert a stronger grip than channel locks thereby being able to better loosen stubborn objects.

Ladders

Ladders of varying lengths can be used in conjunction with salvage covers, plastic sheeting, and pike poles to form basins to accumulate or divert water.

Hose

Firefighting hose (for example, 2½-inch) can be effectively used to channel water to a desired location, as illustrated in figure 11–13. It can also temporarily contain and divert water from an activated sprinkler head. To do this, cover a broken or activated head with the female end of a 2½-inch hose and drain it to an appropriate outlet. A bell reducer or traffic cone attached to the 2½-inch hose can be used to cover heads that have protective cages.

Salvage hooks

Salvage hooks are available in S or 2 configurations. The 2 configuration allows the hook to be easily driven into a wall or between moldings and walls. When in place, salvage hooks provide attachment points for salvage covers, plastic sheeting, or salvage cord (which is nothing more than a heavy grade of string).

Miscellaneous equipment

Firefighters need a virtual arsenal of miscellaneous equipment, depending on the situation. Figure 11–1 shows the following necessary tools:

- **Floor runner (A):** Floor runners are made from waterproof cotton duck and are 6 by 18 foot in size. They are used to protect floors and floor coverings, particularly near the entrances used by attack personnel.

- **Pike pole (B):** These come in varying lengths and are effectively used for constructing water drains with salvage covers, opening ceilings for water drainage, and the like.

- **Redwood plugs (C):** Tapered redwood plugs of various sizes can be used to plug broken water lines and some fused sprinkler heads.

- **Salvage cover (D):** Salvage covers are made from rubber-coated cotton duck or lightweight plastic-coated nylon, and are 12 by 18 feet in size. They are used to cover and protect various materials from water damage, channel water, or to form holding basins. Salvage covers with grommets can be hung from nails, salvage hooks, and salvage cord.

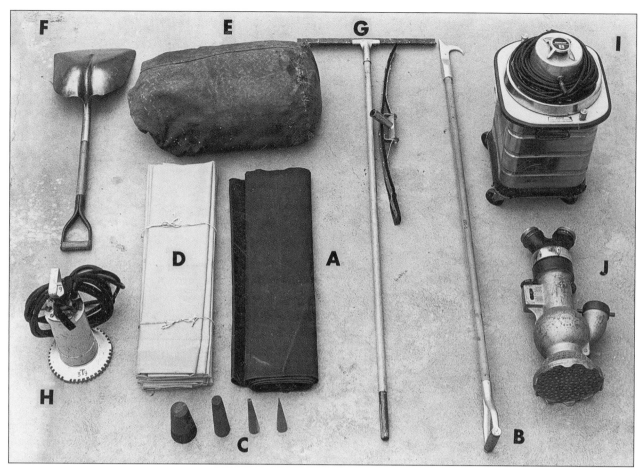

Figure 11–1. Equipment commonly used for salvage operations

- **Sawdust (E):** Sawdust can be used to control, divert, direct, or absorb water. One three-cubic-foot bag is capable of absorbing approximately 30 gallons. Absorbent materials (such as kitty litter) can also be used to quickly absorb water and is more easily cleaned up than sawdust.

- **Scoop shovel (F):** Scoop shovels can be used for spreading and removing sawdust, as well as for removing water from floors and carpets.

- **Squeegee (G):** Squeegees of various sizes and shapes are used either to push or pull water toward an outlet.

- **Submersible pump (H):** Submersible pumps are used for dewatering.

- **Water vacuum (I):** Water vacuums (free standing or back pack) are used to dewater floors, carpets, and other areas where the water isn't deep enough to be picked up by a submersible pump or siphon ejector.

- **Siphon ejector (J):** Siphon ejectors of various sizes can be used for dewatering operations where the water is deep enough to cover the ejector. A pump is necessary to pump water to the ejector.

Additionally, there are four types of equipment/tools that deserve consideration for their ability to dramatically simplify salvage operations:

- **Salvage pan:** When shallow areas of water don't provide enough depth to use a submersible pump or siphon ejector, a salvage pan can be used to provide the appropriate depth. A 32-by-18-by-5-inch pan constructed from sheet metal as illustrated in figure 11–2 will allow water to be squeegeed into the pan via the removable ramps. The water is removed by a submersible pump. When not in use, the ramps are placed in the pan.

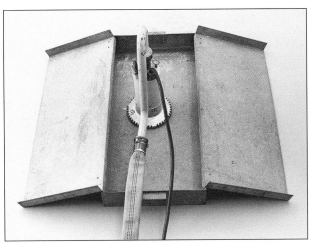

Figure 11-2. A salvage pan allows water to be collected and removed by a submersible pump.

- **Wedges:** Wooden wedges can be used for many purposes, such as holding open doors and windows or plugging fused sprinkler heads as in figure 11–3. The wedges are approximately three to four inches long, 1½-inches high, and are made from pine (soft woods are flexible and swell when wet). Wedges are easily carried in a turnout coat pocket.

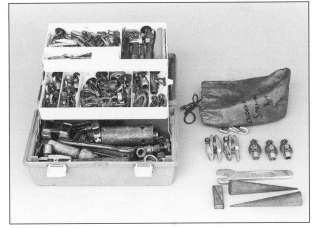

Figure 11-4. A plastic toolbox facilitates carrying common tools used for shutting off and replacing fused sprinkler heads.

- **Sprinkler kit:** Sprinkler systems are increasingly popular in residences as well as in commercial/industrial structures. Consequently, fused sprinkler heads are an increasingly common occurrence. Shutting off and replacing a fused sprinkler head in a timely manner is enhanced by carrying items common to this operation. For example, the plastic toolbox in figure 11–4 facilitates carrying:

  - Spare sprinklers. Remember to replace fused heads with heads of the proper temperature rating.

  - Sprinkler head shutoffs. Wooden wedges, redwood plugs, and other sprinkler shutoff tools are used for closing a fused sprinkler head without shutting down the entire system or when there is a delay in locating or gaining access to the control valve.

  - Wrenches. Due to the various types of sprinkler heads, different types of wrenches may be necessary. These may include flush-type wrenches, pipe wrenches, spud wrenches, and crescent wrenches.

  - Globe valve. A globe valve with a ½-by 6-inch nipple can be used to control the flow from an activated sprinkler head

Figure 11-3. Wooden wedges made from pine are effective in plugging most fused sprinkler heads.

outlet without shutting down the system. While the water is flowing, remove the activated head, screw the ½-by 6-inch nipple into the ½-inch female fitting, turn off the water flow with the globe valve, insert a new sprinkler head into the female end of the globe valve, and turn the globe valve on.

- Easy-out. An easy-out can be used to remove a broken sprinkler head that isn't removable with a wrench or other similar tool.

- Redwood plugs. Redwood plugs can be used to plug a fused sprinkler head and also some broken water pipes, depending on the water pressure.

- The nylon bag in figure 11–4 facilitates carrying *frequently* used sprinkler shutoff tools. This bag can be quickly carried into a structure ahead of the toolbox, if necessary.

• **Plastic:** Plastic sheeting can be effectively used in some instances in place of salvage covers. It is 6 mil thick and 100 feet in length. Wound into a 32-inch-wide roll, it unfolds to a width of 10 feet. Several commercial staple guns and utility knifes are carried in a small bag (figure 11–5) and are used for cutting and attaching (stapling) the sheeting to vertical surfaces when necessary. A wooden rod inserted through the plastic roll with a strap attached end to end facilitates carrying the roll and removing the plastic from the roll.

Salvage kit: Some department that are proactive in their approach to customer service carry pre-made salvage kits that generally consist of (but not limited to) Styrofoam blocks (two to three inches square that can be quickly placed under furniture legs to keep water from being siphoned into them while they are standing in water); several salvage covers, a floor runner, and pre-cut pieces of plastic that can quickly be deployed; and cardboard boxes of various sizes that can be used to hold and protect fragile, expensive, delicate items. For the extra touch, put the name of the appropriate fire department on the boxes.

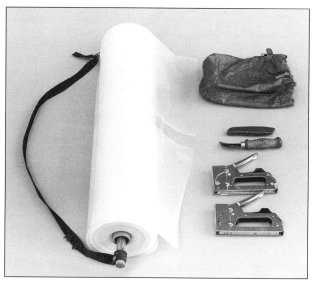

Figure 11–5. Staple guns and knives are used for cutting and attaching plastic covers.

If the preceding salvage kit seems over the top, let's apply a famous business axiom to salvage operations. The business axiom is *if one person is not satisfied with your company/service, they will voice their displeasure to at least ten other people, who will tell ten other people, who will tell ten other people, and so on.* Now, let's apply that axiom to salvage operations as follows: *if one person is happy with your fireground operations, they will voice their pleasure to at least ten other persons, and so on.*

Size-up

Effective salvage operations begin with an accurate size-up and a timely operational plan. The process of determining appropriate priorities from a size-up is simplified with a basic knowledge of several factors:

Pre-fire planning

Pre-fire planning and routine fire prevention inspections offer excellent opportunities to develop a familiarity with building floor plans and contents. Items to be aware of include fragile stock, baled goods, basement drains and sump pumps, floor drains and scuppers, sprinkler system valves, watertight floors, materials stored on the floor in cardboard boxes, stock not on pallets, cabinets and lockers with open tops, water-reactive materials such as chemicals, radioactive isotopes, cyanide, salt baths, and other similar materials.

As most pre-fire inspections are conducted in commercial occupancies, take the opportunity to discuss the following considerations with building management:

- Placing appropriate value on stock on skids, pallets, or the like

- Providing ample space between materials and walls, ceilings, sprinklers, and so on

- Prioritizing the value of the building's contents and determining the location of items that have the most value; locating appropriate water shutoff valves

- Supervising fire protection system; notifying the fire department in a timely manner when confronted with a fire or water problem.

Contaminants

Damage to a structure and its contents is usually caused by water, smoke, and heat. If a water pipe has broken, it will be necessary to locate the appropriate shutoff valve in a timely manner. In a structure fire, water accumulation will depend on the placement and duration of hose streams, while smoke and heat will vary according to a combination of fire suppression and ventilation techniques. The control of smoke and heat is covered in detail in chapters 15 through 18.

Building factors

Building factors that will affect size-up considerations can be categorized as follows.

Construction style. As mentioned in earlier chapters, the two primary construction styles are conventional and lightweight construction. Generally, conventional construction provides greater structural stability when exposed to fire or when supporting additional loads, as in the case of soaked baled goods or accumulated water. On the other hand, lightweight construction or materials will not offer the same structural stability as conventional construction.

Construction features. Construction features affect the degree to which building materials absorb, retain, and are damaged by water.

Floors. Wooden floors are easily damaged by water and don't provide an effective barrier against seepage unless covered by linoleum, tile, or similar materials. Concrete floors can suppress some water seepage; however, expect leakage at cracks (and most concrete will crack, even if the crack looks insignificant), openings for pipe ducting, and the like. As with the type of floors, the number of floors in a building will directly affect the potential flow of water. For example, single-story buildings don't present the same potential for water damage as do multi-story buildings with water on the upper floors.

Walls and ceilings. Walls and ceilings that are composed of plaster and lath-type materials can suppress the flow of water for extended periods of time. However, these materials can easily absorb and accumulate enough water to cause a sudden collapse of significant weight. Gypsum board materials are also absorbent but will become soft, causing a rapid collapse of wet sections and a flood of any trapped water.

Roofs. Although the roof of a structure isn't usually considered to be within the scope of salvage operations, rainwater leaking through a roof can cause additional damage to the interior. Leaks can be the result of faulty roofs, incomplete repairs, and parapet walls with plugged scuppers. Additionally, the excessive accumulation of water on a roof can present a serious hazard of collapse.

Building features. Building features that will affect the potential size and location of salvage operations must be evaluated as follows.

Building size and height. The size of a building is an initial indicator of the potential extent of a salvage operation. In a single-story building, the flow of water is generally limited to a horizontal direction. In multi-story buildings, water can be expected to flow horizontally and vertically. The potential loss to materials that may be affected by a horizontal and/or downward flow must be considered.

Type of occupancy. Consider the type of occupancy and the potential for loss and damage. For example, residentials are characterized by walls, partitions, and hallways that tend to confine and channel the spread of water. Common losses are household and personal items. Commercial/industrial occupancies, varying from simple office buildings to complex manufacturing plants, can be characterized by open areas that enhance the travel and extension of water. Expensive machinery and large quantities of salvageable materials may necessitate extended salvage operations.

Miscellaneous. Other building features that may affect the control and removal of water include floor drains, sump pumps, sprinkler system shutoff and drain, windows and doors, vertical shafts, scuppers, stairways, and so on.

Contents. The contents of a building can be divided into the following basic categories.

Water-reactive materials. If appropriate, specific precautions necessary to ensure the safety of personnel must be emphasized. Identification of these hazards is enhanced by factors such as pre-fire planning, the presence of building personnel (who are familiar with the building contents), the NFPA symbol and the like.

Absorbent materials. These materials are characterized by their capability to absorb and accumulate water, which can result in a significant and hazardous increase in weight. Examples of such materials are baled rags, rolled or baled paper, and baled cotton and cardboard.

Easily damaged materials. Examples of materials that are easily damaged by water are electronic equipment (computers, copiers); dry goods such as flour and paper goods; materials in cardboard boxes on the floor, and fragile stock.

Personal materials. Examples are clothing, mementos, pictures, pets, medications, and so on.

Origin

Determining the origin of the problem consists of four basic considerations:

- Water: the amount of water present and anticipated

- Quantity: the anticipated flow

- Direction: and the materials threatened

- Loss: will begin to indicate the size of a salvage operation and the resources necessary to abate the hazard

Priorities

Salvage operations should begin where the greatest loss is expected. This may or may not be the floor of origin. Additionally, the following factors should also be considered.

Structure fires

For structure fires, salvage operations should generally be initiated with (or as close as possible to) the implementation of initial hose lines.

Multi-story buildings

In multi-story buildings, with water on upper floors that can't be completely contained, simultaneous attention must be given to floors below. Compare the value of contents that have been damaged on the floor of origin to the value of the contents that may be damaged below. Time is occasionally wasted protecting objects that have already been damaged and that will suffer no further consequences if left unprotected. This can tie up personnel and equipment that may otherwise protect valuables in advance of encroaching water.

Residentials

In these types of occupancies, items of the most value normally fall into the following categories:

- Expensive appliances (television consoles, hi-fi equipment, computers, and so on).

- Furniture, particularly antique or collectible-type furnishings.

- Art, with an emphasis on the expensive or collectible.

- Personal pictures; these items may be the most important because family pictures and mementos are irreplaceable in the eyes of the occupant.

- Clothing items may be expensive and hard to replace. Usually they are in out-of-the-way locations such as closets and drawers, which are easily neglected. Do not forget the importance of pets and prescription pills.

Remember that people living in low-income areas value their few material possessions, which are usually uninsured. It isn't unusual to find sums of money in these structures due to a lack of bank accounts. Additionally, these people will usually reoccupy a damaged structure immediately after fire suppression personnel have left.

Commercial/industrial occupancies

The contents in these buildings are normally prioritized in the following order:

Business records. Although a structure can be reconstructed, business records cannot. Of these records, accounts receivable are the most important because they indicate credit extended and are instrumental in keeping a business in operation. Salvage in an office area is of top priority. Computers, desks, and filing cabinets head the list.

Machinery and finished products. Expensive machinery, finished products (which are a result of time and money), and products to be shipped are the next priority of value.

Unfinished stock and raw materials. Unfinished stock and raw materials will normally have less value than the preceding two categories.

Office buildings

In office buildings, the contents of maximum value are normally computers, filing cabinets, and other types of expensive electronic business machines (copiers, photo equipment, desktop publishing equipment). Always evaluate each incident to determine which materials are the most valuable and will have the greatest benefit to a business returning to normal operation.

Operations

The two main objectives in salvage operations are water removal and protection of the structure and its contents from damage.

Water removal

Although the removal of water from a structure can minimize loss from damage, it may also be necessary to minimize in a timely manner any accumulation of water. If water is allowed to accumulate, the weight of the water and water-soaked materials can overload a fire-weakened structure to the point of collapse. For example, consider how water used in suppression operations can add additional weight to a weakened structure:

- A 1½-inch line delivering 100 gallons per minute adds 835 pounds per minute.
- A 2½-inch line delivering 325 gallons per minute adds nearly 3,000 pounds per minute (1½ tons).
- Master streams add between two and four tons per minute.

Although some of this water is turned to steam or will drain from the structure, a significant portion can be absorbed by the building and its contents. Therefore, minimizing damage caused by water can reduce fire loss and increase safety. Removing water from a structure can be accomplished by eliminating the source and by dewatering operations.

Eliminating the source. Water can come from many sources, the most common being water used for suppression. Although it is often necessary to use large quantities, many fires can be extinguished with more judicious applications. In most cases, the measure of an effective engine company is their ability to extinguish a fire with minimal or no runoff. The longer a nozzle is left open or hose couplings inside a structure are allowed to leak, the greater the chances of increasing salvage operations. In addition to fire suppression operations, water can also come from numerous common sources, some of which are overflowing sinks, wash basins, toilets, leaking appliances (for example, washing machines and dishwashers), ruptured water heater, broken or leaking water pipe, and broken, leaking, or activated fire suppression systems. Eliminating the flow from these sources is usually done by locating the appropriate shutoff valve or by plugging the leak. This operation will be covered in chapter 14.

Depending on the incident and source of water, it is often advisable to designate multiple personnel to investigate alternative methods to shut off the flow. For example, in the case of a fused sprinkler head, some personnel can be directed to search for a shutoff while other personnel investigate the feasibility of stopping the flow at the head. This will result in a minimal amount of time and effort being expended, and it will provide an alternative means of control if attempts at the head or shutoff valve are unsuccessful. Always remember two basic salvage rules:

- Time is of the essence.
- The sooner you stop the flow, the less time will be spent in cleanup.

Dewatering operations. Dewatering operations entail removing water from a building. This is normally accomplished by containment and channeling.

Containment. Water can often be contained to prevent it from spreading and to hold it until it can be systematically removed by submersible pumps, siphon ejectors, or drafting. Water can be contained by various means as follows:

- **Sawdust:** Don't overlook this simple but effective method of containing water. Sawdust (or absorbent materials) can be quickly applied from a bag to form a dike, keeping water from extending through a doorway. It can also absorb accumulated water. A three-cubic-foot bag will readily absorb up to 30 gallons. Remember that when sawdust has been saturated, water can begin to leak back onto the floor.

- **Wastebaskets:** A common wastebasket is a readily available means of capturing water. One firefighter can quickly place several wastebaskets under active leaks while other personnel are preparing to relocate contents or place salvage covers. This tactic can be effective in office buildings that have an ample supply of wastebaskets. The timely containment of water will stop additional spread and damage. When possible, resort to the simple and the obvious!

- **Basements:** Because water perpetually seeks the lowest level, areas such as basements, elevator shafts, and loading docks can easily present a number of hazards. Depending on the quantity, a body

of water can exert considerable force against the walls of a structure. Carefully evaluate the structural integrity because water can soften some building materials. Also, a smooth surface may hide the true depth of the water, as well as any irregularities in the floor, such as multiple levels. Unless the depth is known or easily determined, use a pike pole or other probe to determine it.

- **Utilities:** Always consider the potential of electrical and gas utilities in basements and shafts. Elevator shafts may contain machinery (electrical motors), and basements usually contain electrical panels and gas-fired appliances. Remember that water is an excellent conductor of electricity, and that it can also easily extinguish a pilot burner in an appliance, causing a potential gas leak. To avoid such consequences, you may need to eliminate the electrical and gas services to a given structure. Water can collect in a basement, so always consider the area around it for additional leaks. Additionally, if a basement has been flooded, check for the presence of sewage and act accordingly. When assessing the removal of water from below-grade areas, remember that the need to remove it is based on several factors:

  - Is it a fire department responsibility?
  - Is the water creating an *immediate* emergency?
  - Will the damage extend after the fire department leaves, or has the problem been contained?
  - What is the amount of water to be removed?
  - What time and equipment will be necessary to remove the water?
  - Does the fire department need to contain, eliminate, or refer the problem to a responsible party?
  - The potential public relations value.

Remember, using fire department apparatus to draft or pump water to siphon ejectors can result in apparatus being unavailable for other responses. The needs of the incident must be weighed against any public relations concerns.

- **Holding basins:** Holding basins can be constructed from a variety of materials to collect water until it can be removed. They are limited only by available materials, initiative, and resourcefulness. Various methods to construct a holding basin are summarized as follows:

  - A shallow basin can be made from one salvage cover. Roll the sides inward two to three feet. Next, roll the ends inward two to three feet and tuck each corner. This will provide enough depth to operate a submersible pump (figure 11–6). In this configuration, a submersible pump is most effective with a 1½-inch discharge line because it provides the proper back pressure. This method is most effective with rubber-coated cotton duck covers.

Figure 11–6. A salvage cover can be quickly folded to provide a shallow holding basin.

- Additional care must be exercised in tucking the corners of plastic-coated nylon covers, since they are lighter in weight. An effective method for plastic-coated nylon covers is to construct a basin as previously outlined, and then to turn the basin over. The weight of

the water will push downward and outward, effectively sealing the sides and corners

- Three ladders can be strapped together to form a basin. Position them in a triangle and tie them together (by the rungs or beams) with straps, cord, or other similar materials. Two ladders and a pike pole can suffice in lieu of three ladders as in figure 11–7. Next, center and fit a salvage cover over this framework by pressing the cover against the sides of the framework. Tuck any loose outside edges underneath the framework. When constructing ladder basins on upper floors of multi-story buildings, consider the length of ladders and pike poles that will be used, since elevators and stairways may not provide adequate clearance.

Fig 11–7. Ladders, pike poles, and a salvage cover can be combined to form a holding basin.

- Miscellaneous items can also be useful. Collecting water in a timely manner can often require forming a basin out of readily available materials such as chairs, bed frames, night stands, boxes, and other such materials. In some cases, these quick methods will suffice or until more substantial methods are implemented. For example, beds are normally found in habitations, and chairs are prevalent in commercial occupancies. Remove the mattress from a bed frame, and place the frame in the appropriate location.

Otherwise, arrange six chairs back to back. Center and form a cover to fit over and inside the bed as in figure 11–8, or chair framework as in figure 11–9.

Tuck the loose outside edges underneath the framework. If it is necessary to prevent the chairs from separating, tie a rope around them as in figure 11–9. When the basin has been completed, use an appropriate method (such as a submersible pump) to pump out the water.

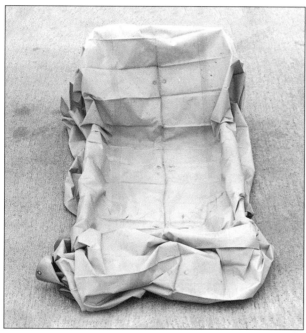

Figure 11–8. A salvage cover over a bed frame results in an effective holding basin.

Figure 11–9. A salvage cover over six chairs results in a quick holding basin. If necessary, use a rope to keep the chairs from spreading outward.

- When there is no practical method to remove seeping water from the floor above, or when it is necessary to protect an entire floor (for example, oak flooring in racquetball courts and aerobic rooms in athletic clubs), it may be necessary to bag the floor. To do this, first determine the size of the room and the number of salvage covers required to cover the floor. Roll the outside edges and press them against the walls to form a six- to eight-inch dike. Seal the edges of intersecting covers by rolling or folding them together as detailed in figures 11–17 thru 11–20. Before bagging a floor, be aware that this operation results in an entire room serving as a catch basin. Ask yourself whether it can support the additional weight. Also, as the size of a room increases, so does the commitment of personnel, equipment, and time needed to complete the operation. Furnishings will significantly increase the difficulty and time necessary for completion. A floor and furnishings can be protected as shown in figure 11–10.

First, group the furnishings and cover them (A). If necessary, seal them with additional covers. Use appropriate covers to bag the floor. Interconnect the covers with a seal (B). Roll the edges of the covers along the walls to provide a dike approximately six to eight inches high (C). If necessary, place an additional cover over the furnishings (D). Cover any appropriate vertical stock on the walls (E).

Channeling. Due to the type of occupancy and incident, it is often advantageous to channel water to a basin inside the structure or to divert it to the exterior. Similar to constructing a basin to contain water, the methods employed are limited only by available materials, initiative, and resourcefulness.

- **Holes in floors:** Occasionally it is necessary to cut holes in floors to remove water. This should be done only to remove a large amount quickly or when there is no other practical option. Before initiating this, evaluate the area below, protect objects as necessary, and construct basins or chutes before cutting the holes. If possible, locate the lowest part of the floor before cutting the hole. In some cases,

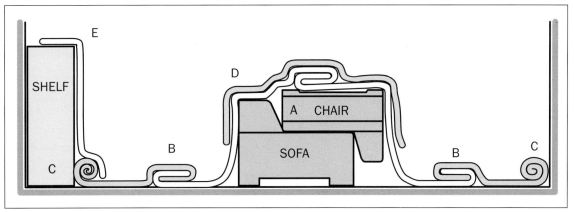

Figure 11-10. Furnishings and a floor can be protected with multiple salvage covers. This can be a time-intensive operation.

it will be necessary to locate the hole near a window of the floor below so that water can easily be diverted through the window and out of the building. The ceiling underneath the hole should be opened first and made larger than the hole in the floor above (figure 11–11). This will keep water from accumulating between the floor and ceiling, potentially collapsing the ceiling. Axes or power saws can quickly cut the desired holes in most floors.

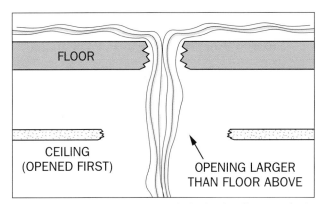

Figure 11-11. When using a hole in the floor to drain water, make a hole in the ceiling first and larger than the hole above.

- **Ceilings:** Accumulated water in a ceiling can be removed by using a pike pole to make a small opening. Cracks will usually converge at the point of most stress in plaster ceilings. Drywall ceilings will usually sag between ceiling joists and

can drop in large sections. Ensure that appropriate salvage measures are completed before opening a ceiling to drain water. Additionally, remember that personnel standing under a wet ceiling should consider the center of the room to be the weakest portion and the edges (junction of the walls and ceiling) and the area underneath a doorway to offer the most strength.

- **Toilets:** If necessary, a toilet can be removed from its junction at the floor, exposing a 4-inch sewer pipe. Before channeling water to the open pipe, form a dike with sawdust or a rolled salvage cover around and behind it to help guide the water to the opening (figure 11–12). If possible, cover the open sewer pipe with some type of screen or strainer to keep it from clogging.

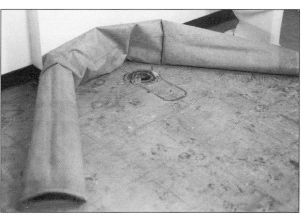

Fig 11-12. A rolled salvage cover behind a drain in a floor will channel water to the opening.

When finished, replace the toilet over the drain to minimize the escape of sewer gas. On multi-story buildings, seal any opening between the sewer pipe and floor with rags, sawdust, or the like. This will minimize leakage past the sewer pipe and floor junction.

- **Drains and pipes:** If available, floor and wall drains can be used to remove water. Their effectiveness is dependent on their size, their distance from the water to be removed, and whether or not they are free of debris (and kept free of debris during the operation). Some multi-story buildings have exposed sewer pipes or roof drains. These can be effective in removing water. Cast iron and ABS plastic can be broken and opened at the floor level. Clear all portions of pipe extending above the floor. Similar to using a toilet drain, it may be necessary to construct a dike behind and around the pipe to help divert water into it. If possible, cover the open pipe with some type of screen or strainer to prevent clogging.

- **Sumps:** Sumps are installed in the low portion of a building. Water can be directed toward this area. If a sump is clogged, remove the grate and clean out the sump. This will often allow the water to flow into the street. If a sump doesn't work, put a siphon ejector or submersible pump in the sump area and direct water to it.

- **Vertical shafts:** Unless necessary, water should *not* be channeled into vertical shafts. There may be machinery at the bottom, and it will later be necessary to pump the water back out. If channeling water into an elevator shaft is unavoidable, consider that hydraulic elevators may have electric-powered hydraulic pumps in or near the bottom of the shaft. Also, on upper floors, ensure that water doesn't run down the inside wall and onto the landing of the floor below. Channeling water into an elevator shaft will normally put an elevator out of service when it might otherwise be used to carry resources to upper floors, and ensure that the elevator

car is *above* the landing where water will be channeled into the shaft.

- **Hose:** Water can be effectively channeled to a desired location by two personnel and one section of 2½- or 3½-inch hose. Stretch a section across the floor and pull it like a net. This will collect and channel accumulated water to the desired location (figure 11–13). In the photo, the hose has been folded back on itself for a length of 25 feet. This method requires far less effort than is expended by using two or three squeegees.

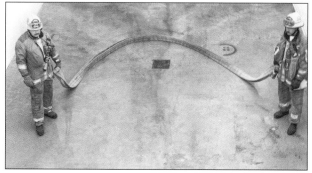

Figure 11–13. A folded section of hose will pull water to a desired location. This operation can replace multiple squeegees.

- **Chutes:** Water draining through ceilings from upper floors can be removed moderate distances through a portal (such as a window) to the exterior by constructing water chutes. When considering this, first consider methods for eliminating the water source. If possible, quickly contain draining water in basins or wastebaskets until the appropriate chutes can be constructed. Constructing a water chute can require a significant period of time to mobilize the necessary personnel, equipment, and space to set up appropriate ladders, pike poles, and other paraphernalia. When constructing water chutes on upper floors of multi-story buildings, consider the length of any ladders and pike poles that will be used. Elevators and stairways may not provide adequate room for access. The head of a water chute should be close to the ceiling and wider than the chute portion to contain splashing water as it funnels downward. The head, needless to say, should be the highest part of the sluice. Also, the water chute should project several feet through the window to keep water from running down the exterior of the building and reentering through cracks and windows. In most cases, basins are easier and require less time to construct than water chutes. Water chutes can be constructed from ladders, pike poles, plastic sheeting, and salvage covers.

- **Ladders and pike poles:** A ladder can be used to support a salvage cover by leaning it against a wall or pike pole supported by an A-frame ladder. Two pike poles can also be used to support a salvage cover by rolling them in the cover. Support this configuration on another pike pole and an A-frame ladder.

As necessary, a chute can be extended with additional covers (figure 11–14). If it is necessary to form the head of a chute (where it is wider at the top and than the bottom), fold a salvage cover to form a triangular shape before rolling in two pike poles. When extending a water chute constructed of pike poles and salvage covers or plastic sheeting through a window (figure 11–15), it is important to consider several points.

Bear in mind the elevation of the chute above the windowsill to ensure proper drainage; the tension on the cover between the pike poles to prevent sagging below the pike poles and a collection of water that could collapse the chute; the distance between the head of the chute and the ceiling to reduce the height of the waterfall splash; the width of the head of the chute, also to minimize splash; and the projection of the chute from the window to keep water from re-entering the building at a lower level.

A chute is not constructed in a short period of time. As a test, two experienced firefighters were given the task of constructing a chute on the second floor of a two-story fire station. The firefighters started from the second floor, went to the apparatus to get the necessary equipment, and returned to the second floor and erected the chute. This process took 18 minutes, which shows that unless water is contained until the chute is erected, it would have

run for almost 20 minutes, and the construction of a chute is not a speedy process.

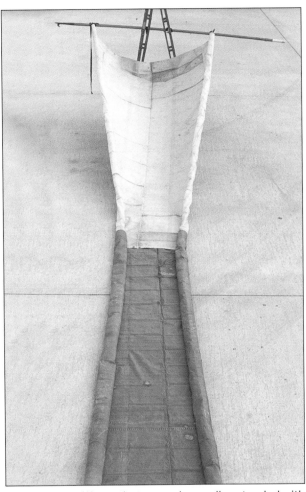

Figure 11–14. Water chutes can be easily extended with additional covers.

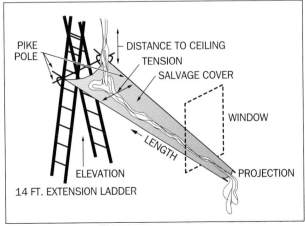

Figure 11–15. Numerous considerations come into play when extending a water chute through a window.

- **Stairways:** Water can be channeled down a stairway after it has been properly covered with salvage covers. These can be rolled and pressed against both sides of the stairway, or they can be hung from the banisters (figure 11–16). Remember to tuck the covers into each step to facilitate walking on them. Plastic is not effective for this operation.

Figure 11–16. Salvage covers can allow water to be channeled down stairways.

- **Hallways:** Water can also be channeled through hallways after sealing them with salvage covers. To ensure a seal, push the rolled edges of the salvage covers against the opposing walls. Salvage covers can also be used in combination with drop bags. To attach the sides of the covers to the walls in a hallway, stretch and tie off each end of a drop bag line about three feet high and down both walls. Hang the salvage cover with 2 hooks on the line.

Splicing salvage covers. When it is necessary to protect a large area or to extend the length of salvage covers, the covers can be spliced together to form a continuous, watertight channel. To splice two covers when water will flow *parallel* to the splice, fold the end of one cover back about two feet and place the other cover over this fold (figure 11–17).

Figure 11–17. If water will flow parallel to a splice, fold the end of one cover back about two-feet and place the other cover over this fold.

The covers can then be spliced by rolling them together as in figure 11–18. If it is necessary to extend the length of a chute, splice the appropriate covers together and roll the sides inward.

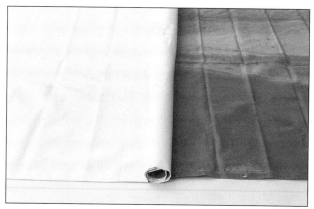

Figure 11–18. The covers can now be rolled together.

To splice two covers when water will flow *against* the splice (from left to right in figure 11–19), fold back the overlaid covers as in figure 11–19.

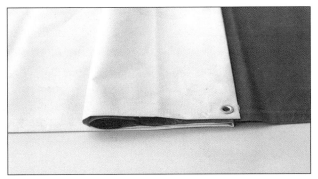

Figure 11–19. If water will flow against the splice, fold back the overlaid covers.

Folding in the same direction, overlay the fold again and press flat (figure 11–20).

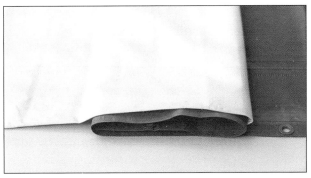

Figure 11–20. In the same direction, overlay the fold again and press flat.

Protection of structure and contents

If it is impractical to channel or divert the flow of water, building contents can also be protected against damage by using salvage covers, plastic sheeting, and floor runners.

Floor runners. Similar to damage caused by water and debris, damage can also be caused to a floor by the ingress/egress of personnel during suppression operations. Dirt, mud, sediments caused by water and ashes, and other contaminants can easily be transferred from boots to carpets and floors,

causing considerable and unnecessary damage. To minimize this, a floor runner or salvage cover should be deployed from the primary entry/exit opening (normally the route of an attack line) to the fire area as soon as possible. This will allow personnel to enter and exit without spreading fire debris past the fire area into uninvolved areas. For example, assume a bedroom on the second floor of a two-story, single-family dwelling is involved with fire, and an attack line has been advanced through the front door. As staffing and priorities permit, a floor runner should be deployed from the front door (figure 11–21) to the second floor hallway, and as close to the fire area as possible during knockdown of the fire. This will result in minimal damage to carpeting on the first floor and stairway of the structure.

Figure 11–21. Floor runners should be deployed as soon as practicable during fire attack operations to maximize their protective value.

Salvage covers and plastic sheeting. Although salvage covers can be used in many instances, plastic sheeting can offer additional advantages when covering building contents. Each has its advantages and disadvantages. The advantages of salvage covers are durability, grommets that enhance hanging where nails or staples cannot be used, increased protection from burning embers, its less slippery surface and greater weight allows the salvage cover to remain in place for water chutes, seals, conduits, basins, and stairways. Also, it enhances rolling the edges. The disadvantages of salvage covers are they need to be removed before leaving an incident, as well as cleaned and refolded before being placed back on the apparatus (in some cases, the material underneath a salvage cover can be exposed to further damage when it is uncovered), large areas need multiple covers, which may need to be spliced, and deploying multiple covers can require additional time and personnel.

The advantages of plastic sheeting include disposability (it can be left at an incident) and size (a roll is normally 100 feet long and 10 feet wide). Also, plastic can easily be cut to length or shape as needed and then nailed or stapled in place. This process is enhanced by its light weight. And in comparable operations, plastic can be deployed with fewer personnel in less time than multiple salvage covers.

The disadvantages of plastic sheeting are its light weight and slipperiness. Sheets of plastic are not well suited to rolling or sealing, they don't provide a safe walking surface for personnel, particularly on stairways, and they offer minimal protection against burning embers. For this reason, evaluate using plastic sheeting to protect horizontal surfaces close to fire suppression operations.

When preparing to cover building contents, the focus should be on materials that could suffer the most loss. Additionally, if the number of salvage covers is limited due to other salvage operations, protect the most valuable contents first. To cover these in a minimal amount of time and with the least resources, it is often advantageous to collect moveable furnishings (or other moveable contents) in compact groups that can easily be covered.

The furnishings should be located away from ceiling openings (for example, light fixtures because water will easily drain from a light fixture). They should also be located away from areas beneath ceilings that will be pulled for fire suppression operations or removal of water. Consider the edges of a room as preferable locations because the center of a ceiling is the weakest point and will normally collapse first. When grouping furnishings at the edge of a room, leave several feet between the wall and furnishings. Be sure to center the group on the heaviest and largest articles, such as sofas, chairs, desks, and filing cabinets. Smaller articles can then be placed on top of the larger articles. Remember to include throw rugs, paintings, the contents of closets and drawers, and clothes. Groups should be sized to be spanned by one cover or one piece of plastic sheeting (figure 11–22).

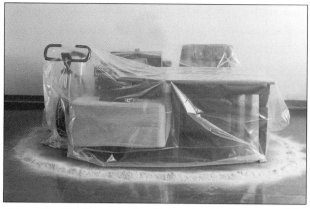

Figure 11–22. Groups should be sized to be covered by one salvage cover or piece of plastic sheeting. Sawdust can be used to form a dam around the perimeter to keep water away from the contents.

After covering a group, tuck the edges under. If necessary, place a sawdust or absorbent material dam around the perimeter to prevent water from seeping underneath the grouping. This is particularly necessary when items such as chairs, sofas, or wood furniture could become damaged when wet.

Along with the use of salvage covers, the problem of water siphoning up furniture legs or fabric touching the floor can also be alleviated by placing an object under the legs of exposed furniture (such as pieces of Styrofoam, cans from the food pantry, and so on). Remember that standing water on horizontal surfaces (for example, tables) can easily create unnecessary damage if left unattended. Wiping up these areas can work miracles toward minimizing loss and enhancing public attitudes.

In single-family dwellings with attic fires, many furnishings can be damaged if the roof collapses or if large ceiling areas are pulled for access by fire streams. Under these conditions, salvage operations can be enhanced by moving furnishings near the walls and covering them as necessary. If possible, the emphasis should be on the deployment of covers before pulling ceilings when fighting attic fires.

Because stationary items such as machinery, shelves, and tiered storage cannot be moved easily, it becomes necessary to use available resources to cover them. Covering shelves or bookcases fastened to walls is predicated on the location of the threat. Water flowing down a wall can seep behind and into the shelving. Therefore, a salvage cover can be draped over the front of shelving or bookcases and forced between the top portion of the shelving and wall, or rolled and pressed against the wall and held in place with nails through the cover grommets. Plastic sheeting is more effective because it is lighter and can be stapled in position or nailed to the wall (studs) through pieces of wood lath. Water cascading onto shelves or bookcases can easily be diverted with a salvage cover or plastic sheeting. A salvage cover or plastic sheeting can be attached to the top of the shelving by nails (salvage cover) or staples (plastic). Other stationary objects such as machinery, storage racks, and the like can be covered as necessary. The protection of vertical objects is normally best accomplished by plastic sheeting.

Deployment

Preparation. Preparation includes having the appropriate salvage resources ready for a timely deployment, selecting the correct resources, and proper deployment.

The methods used to fold salvage covers vary depending on preference, the type of cover to be folded, and the amount of available apparatus space. Cotton duck covers provide maximum protection against fire, but at the expense of a thicker and less flexible cover than vinyl. An accordion fold lays flat and can be easily deployed over objects, or quickly thrown over objects.

Vinyl covers can be easily rolled in a single- or double-roll configuration. A single roll works well for floor runners, and a single or double roll works well for covers. Lay a floor runner flat, fold to the desired width, and begin rolling at one end until finishing at the other end. The runner can be secured by cord or rubber strips made from inner tubes. For covers, use the same method for a single roll. For a double roll, lay the cover flat, fold to the desired width, roll one end halfway and then repeat with the other end. Secure with cord or rubber strips made from inner tubes.

Carrying. If folded, a single-salvage cover can be carried on either shoulder. This frees up both hands for climbing ladders and fire escapes. When carrying two covers, the second is most easily carried on the opposite shoulder from the other cover. This operation also leaves both hands free.

One firefighter can easily carry four covers as illustrated in figure 11–23. Plastic sheeting can easily be carried on either shoulder as illustrated in figure 11–24.

The accordion fold allows more covers to be easily carried by one firefighter than covers that have been rolled.

Figure 11–23. Multiple salvage covers can be simultaneously carried on both shoulders.

Figure 11–24. Plastic sheeting can be easily carried on either shoulder as illustrated.

Implementation. It is recommended that a floor runner be carried with a salvage cover in the apparatus so when personnel take salvage covers into a structure, both the floor runner and salvage cover are together and ready for implementation. As an example, when both are available, the floor runner can be deployed first (protecting the entry way), and then the cover can be deployed for protection of contents. The floor runner can be simply rolled out (if carried in a roll) from the entrance point to the interior of the structure.

One salvage cover can be deployed over small- to moderate-size objects by a single firefighter. Cover the object completely and then tuck in the bottom edges.

When there is no danger of breakage, one salvage cover can be deployed over moderate- to large-size objects by a single firefighter. If the cover has been folded, first center the cover over your forearm and then grasp the bottom fold. With your other hand, palm down, grasp the three or four top folds (figure 11–25).

Figure 11–25. To throw a salvage cover, grasp the bottom fold with one hand, and the top three or four folds with the other hand.

Swing this hand over your shoulder, allowing the folds to fall back on your right hand. While throwing the cover, elevate your arm and retain your grip on the bottom fold (figure 11–26).

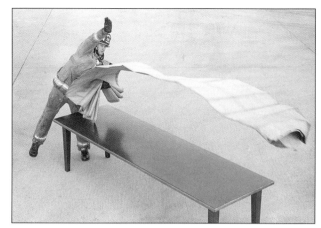

Figure 11–26. Throw the cover over the object.

When it is in place, unfold the cover over the object and tuck in the bottom edges. Obviously, this same cover can be placed on the object to be covered and simply folded out over the object. Rolled covers can also be simply placed on the object to be covered and rolled out over the object to be covered. Rolled covers cannot be thrown over objects that need to be covered.

Two personnel can deploy a single-salvage cover where conditions require careful placement to avoid damaging or displacing valuables. Standing next to the object to be covered, one member holds the cover with both forearms, grasping the bottom fold with both hands palm up. The second person grasps the top fold with both hands down and walks backward until the cover is taut. During this operation, the first person elevates his forearms to hold tension on the cover and to keep it off the floor. Both persons move the cover over the object (figure 11–27) and gently lower it.

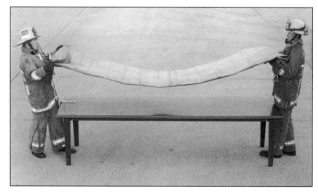

Figure 11–27. For careful placing of a salvage cover, move the cover over the object and gently lower it.

Unfold it to cover the object and tuck in the bottom edges (figure 11–28). This can be accomplished with the accordion fold or rolled covers.

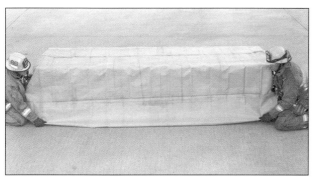

Figure 11–28. Gently unfold the cover to cover the object. Tuck in the bottom edges.

Two personnel can deploy a single salvage cover when it is necessary to shield large objects that afford room to float a cover over the top of the object. As with the previous method, one member holds a cover with both forearms and grasps the bottom folds with both hands. Meanwhile, the second member grasps the top fold with both hands and walks backward until the cover is taut. Both members turn the cover so that the double fold is up and the center of the cover is hanging down. Grasp a fold in each hand and keep the cover elevated (figure 11–29). Both members pull the cover taut.

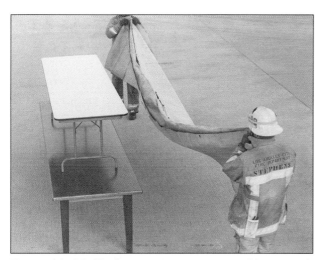

Figure 11–29. To float a cover over an object, each member grasps a fold in each hand and elevates the cover.

With feet braced, inside foot back, elbows high, and palms down, both members work in unison to snap the cover up quickly so that air will fill it like a sail. Let go with your outside hands and float the cover down with your inside hands (figure 11–30).

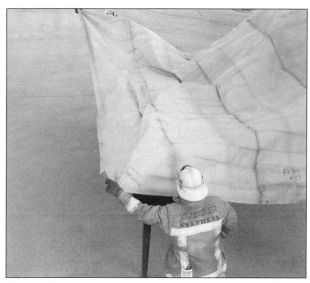

Figure 11–30. In unison, snap the cover up and over the object while retaining control with the inside hands.

Center the cover over the object and tuck in the bottom edges. Practice this method with vinyl covers to see if the accordion fold or roll works the best for your personnel.

Plastic sheeting can easily be deployed by two members. A member stands at one end of the object and holds the handle or strap supporting the roll of plastic while the other member grasps the free end of the plastic and walks backward. When sufficient plastic has been dispensed, the person supporting the roll cuts the plastic with a knife. The roll can be placed on the floor while the plastic is centered and unfolded. Tuck in the bottom edges.

Removal. Salvage covers should be systematically removed from objects to prevent water and debris from causing additional damage. Two firefighters take up positions at the opposite ends of the cover. Fold up both hanging sides. Then, while grasping the ends, raise the cover and walk forward, folding over the raised portion (figure 11–31).

Figure 11–31. To remove a soiled salvage cover, two personnel raise the cover and walk forward, folding the raised portion.

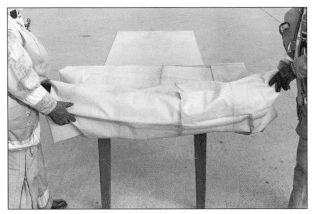

Figure 11–32. When near the end of the cover, fold the opposite end upward.

Figure 11–33. Raise the folds, pivot away from the object while placing the folds over your shoulder, and exit the structure.

When near the end of the cover, grasp the opposite end and fold upward (figure 11–32).

One member raises the folds and pivots away, placing the folds over his shoulder (figure 11–33). The member exits the structure, assisted by the other member if necessary.

Maintenance. As opposed to plastic sheeting, salvage covers do require some maintenance to prolong their longevity. After use, cotton duck and vinyl covers should be cleaned with water and mild soap, rinsed thoroughly, and allowed to dry. Drying is best accomplished by hanging or suspending the cover to prevent mildew, mold, and so on. Before folding and returning the cover to the apparatus, ensure the cover is completely dry and inspect for tears, burns, holes, and other damage. Vinyl covers are more susceptible to damage than cotton duck covers, and cotton duck is more susceptible to dry rot if not correctly maintained.

Training Resources

The following resources can provide additional information and/or another viewpoint for the subjects discussed in this chapter:

- Salvage Size-up, Anthony J. Pascocello, Jr., *Fire Engineering*, August 1996.

- Saving Property: Salvage Operations, Part 1, Anthony J. Pascocello, Jr., *Fire Engineering*, September 1996.

- The Lost Art of Salvage, Bill Sullivan, *Fire Engineering*, April 2001.

- A Salvage Tactic and Tip, Michael N. Ciampo, *Fire Engineering*, November 2005.

- Salvage Basics, Jamie C. Morelock, *Fire Engineering*, January 2007.

12 Electrical Utilities

Introduction

Electricity flows unnoticed into modern homes until a switch is activated, turning on lights and/or household appliances. In commercial applications, numerous functions and processes also depend on the use of electricity. However, before electricity can be successfully utilized for residential and commercial applications, it must be generated, regulated, and safely delivered to appropriate customers.

Electricity originates in generators that are powered by water, coal, oil, or nuclear energy. From generators, high-voltage electricity travels through wires to receiving stations. From these stations it is directed into cities and towns where it is reduced into lower voltages at distribution stations and delivered to customers. Although this basic overview sounds simplistic, it can be complicated from two viewpoints. First, the distribution of electricity is often considered commonplace when in reality, this process frequently consists of electricity starting as one million volts alternating current, traveling thousands of miles and experiencing numerous changes until it enters a common single-family dwelling at 240/120 volts alternating current.

Also, although the number of firefighter injuries and deaths attributed to electrocution is relatively low when compared to other types of fireground hazards, most often firefighters do not place electricity in the proper order of fireground priorities, particularly when electricity displays all of the characteristics of a masked killer—and electrical hazards are present at most structure fires. As an example, most firefighters will commonly place fire attack, search and rescue, ventilation, and forcible entry at the top of their priority list when confronted by a routine structure fire. But too few firefighters pause to consider that a building on fire is a building under demolition, and that demolition can also include the distribution of electricity within the same structure. To consider the importance of electricity from another viewpoint, the frequency of training programs on the magnitude of electrical hazards should be higher.

Definitions

This section is like a mini glossary that defines electric-related terms. It is essential to have a basic understanding of the following terms (figure 12–1).

Alternating current (AC). Current flows first in one direction then the other, reversing itself at 60 cycles per second (North American standard). It is this pulsation that makes it difficult to let go of energized equipment.

Amperage. The quantity of electrons moving through a conductor in one second. If .0050 amps at 20 volts contact a human body, no damage will result. However, 10 amps at 20 volts can cause death. When considering the effects of amperage and voltage on humans, it is the amperage that is primarily responsible for injury and death. The flow of electrons can be compared to the flow of water (gallons per minute, or gpm).

Circuit. The path of electric current, starting from a source and returning back to it. For electricity to be conducted, a circuit must be completed. With electrical power lines, a circuit is completed by going into the ground. Therefore, if a conductor goes from the source of electricity to a ground, electricity will flow along that conductor. If the circuit is broken, for example by a broken wire, then electricity will not flow. However, if a person completes a circuit by touching both ends of a broken wire, the person can become a conductor and electricity may flow through the person's body, causing an injury.

Conductor. A material that contains and freely allows the movement of electrons. Copper and aluminum are typical conductors. A conductor can be compared to a hose that contains the flow of water. Conductors can also be referred to as lines.

Direct current (DC). Direct current flows in one direction only because there are no cycles as in alternating current. Contacting this type of energy results in repulsion and can cause items to be thrown.

Distribution system. Electricity is transmitted through conductors (lines) in overhead and underground systems.

Electron. A negatively charged particle of an atom. The flow of electrons in a conductor constitutes electric current. An electron can be compared to a drop of water.

High voltage. Most utility companies consider any voltage over 600 volts to be high voltage, although voltages over 750 volts are usually identified with the familiar yellow "High Voltage" placards that are normally mounted on power pole crossarms.

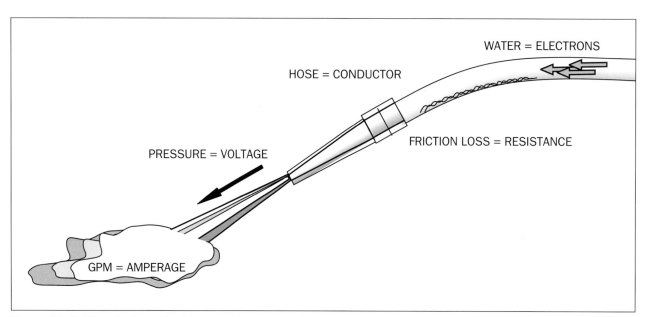

Figure 12–1. Electrical terminology can be compared to the flow of water in a hoseline.

Knob-and-tube wiring. Older services used two open lines into and throughout a structure. This wiring was an ungrounded system that only used a hot and neutral wire, hence, its limit of 120 volts. It was designed to be air cooled, and was insulated by a rubbery type of insulation. Inside a structure, the wires were suspended on ceramic insulators and passed through ceramic tubes into the walls, hence the name knob and tube wiring. This system is no longer legal in new construction.

Quadraplex. Also called quadplex. A four-wire service that consists of a bare support cable and three insulated wires wrapped around the support cable. The three insulated wires are hot, and the bare cable is a neutral. This configuration is a good indicator of a three phase service of 120/240 volts, or 277/480 volts.

Resistance. The quality of an electric circuit, measured in ohms that resists the flow of current. This resistance is due to the conductor's atomic nature, or a load, such as an electric motor. This can be compared to friction loss in a hose.

Single phase. A simple generator will produce a single sine wave. A common example is 120 volts alternating current.

Three phase. Provided by three wires, the voltage of each is 120 degrees out of phase with the others. The benefits of this compared to single phase are a higher current potential that produces more power than single phase, three separate circuits can be run from the service instead of one large circuit, and lower operating costs. Normally used for commercial and industrial occupancies.

Transformer. A device that is used to change electrical energy to a different voltage. By using a voltage within a specific length of wire, the voltage is converted to energy that then increases (step up) or decreases (step down) voltages by specific lengths of wire. See *transformer* in the next section, transmission of electricity.

Triplex. Three-wire service consisting of two insulated wires of 120 volts each and a supporting cable, which is the neutral wire. Supplies both 240 volts and 120 volts AC because each of the insulated wires is 120 volts.

Voltage. The property of electricity that is responsible for moving electrons through a conductor. The 1.5 volts in a battery is not detrimental to the human body, but 120 volts in a common outlet (with the appropriate amperage) can be fatal. Voltage can be compared to the pressure behind water. Voltage is also known as volts. Note that line voltages can vary from 115 volts to 125 volts in a single day. This is why values of 115 volts, 117 volts, 125 volts, and so on are commonplace. This chapter uses the standard designation of 120 volts.

Watt. A measure of consumed power. It is determined by multiplying the amperage of a circuit by its voltage. This is the unit of electricity that is metered and sold by utility companies.

Transmission of Electricity

The transmission and distribution of electricity (volts) is accomplished along a multifaceted system as described in this section (figure 12–2).

Transmission lines (115,000 to 1,000,000 volts)

Electricity can be transmitted to municipalities in lines of 115,000 to 1,000,000 volts. However, if generated locally, it leaves generating plants at 13,800 volts and is increased to 230,000 volts by step-up transformers. It is then distributed throughout a municipality by transmission lines on metal or concrete towers.

Sub-transmission lines (34,500 volts)

Its next stop is a receiving station, which is basically a step-down transformer. The electricity of 115,000 to 1,000,000 volts is reduced to 34,500 volts and sent on. This reduced voltage may be supplied directly to large customers by way of overhead or underground sub-transmission lines; otherwise, it will be further stepped down at a distribution station to 4,800 volts.

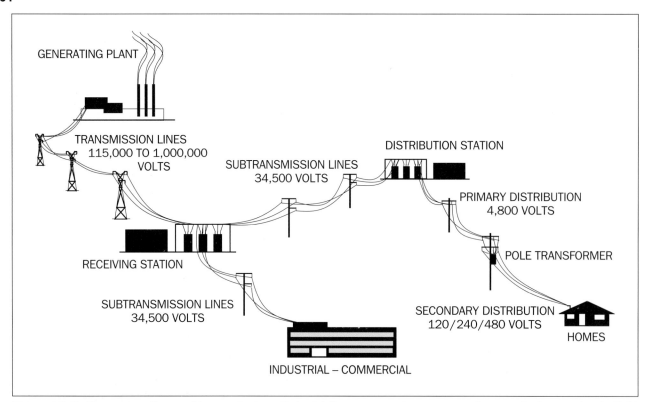

Figure 12–2. The generation and transmission of electricity can experience many changes before a simple light is turned on in a dwelling.

Primary distribution lines (4,800 volts)

The 4,800 volts is then sent through primary distribution lines either underground to a vault (and then to specific consumers) or overhead on power poles, where it may be stepped down yet again and delivered to consumers through secondary distribution lines. Interestingly, 4,800 volts can be more dangerous to personnel than 34,500 volts. In some municipalities, the 4,800-volt system is ungrounded, having relays to clear faults and to isolate trouble in distribution lines. If a problem such as a short is detected, appropriate relays will open, removing power from the problem line. These will automatically reset in varying amounts of time. To unsuspecting personnel, an apparently dead wire can suddenly become live when the relays automatically reset. This is a primary reason why all lines should be considered high voltage and energized unless identified otherwise.

Note that due to a variety of systems, sub-transmission lines can carry between 34,500 and 60,000 volts, and primary distribution lines can carry between 4,800 and 12,000 volts. For simplicity, 34,500 volts for

sub-transmission lines and 4,800 volts for primary distribution lines will be used in this chapter.

Secondary distribution lines (480/240/120 volts)

The 4,800 volts in primary distribution lines can be stepped down to 480, 240, or 120 volts and delivered to consumers through secondary distribution lines.

Transformers

A transformer either reduces or increases voltage. It consists of an iron core, a primary winding, and one or more secondary windings. The connection between the windings is provided by a magnetic field generated by the primary winding. The relationship between the primary and secondary voltages will be the same as the ratio of the turns of the windings. As an example, if the secondary has more turns than the primary, the transformer is operating as a step-up transformer (figure 12–3A). Conversely, if the secondary has fewer turns than the primary, then the unit is operating as a step-down transformer (figure 12–3B).

During operation, a transformer can generate a significant amount of heat, which must be minimized to prevent damage. The cooling methods generally used are air and oil. Oil-cooled transformers commonly use a mineral oil, which can present a flammability hazard and may have PCBs in the cooling oil. Transformers are generally mounted above ground on poles, on the ground on concrete pads (pad-mount transformers), or underground in vaults.

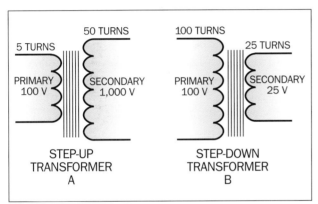

Figure 12–3. Depending on the ratio of the windings, transformers can be used in a step-up or step-down configuration.

Determining the Voltage

Due to the numerous types of conductors and transformers throughout this country, personnel should be familiar with the electrical equipment in their area of responsibility, and must be able to determine the approximate voltages and consequent hazards that may be encountered at an incident. The ability to determine voltages will give you the necessary information to assist in determining a risk assessment of an incident with an electrical hazard. Voltage can be determined by the conductors and transformers as described in the following sections.

Conductors

The voltage of a conductor cannot be consistently and accurately determined by its size (thickness/diameter). The perception that a larger wire necessarily carries a higher current is incorrect. Although

crossarms carrying voltages over 750 volts should be marked with "High Voltage" signs, a high-voltage sign does not differentiate between 750 volts or 34,500 volts (and there is a difference). Normally the highest voltages are on the highest crossarms; however, there are exceptions even to this.

The best method to determine the approximate voltage of a conductor is by the size of its insulators and the distance between parallel conductors. The higher the voltage, the larger the insulators and the greater the distance between parallel conductors. Refer to figure 12–4. Transmission line insulators can be approximately four to 14 feet in length (A); sub-transmission line insulators are approximately 12 to 22 inches in length (B); primary voltage insulators are approximately 4 to 6 inches in length (C), and secondary voltage insulators are approximately 2 to 3 inches in length (D).

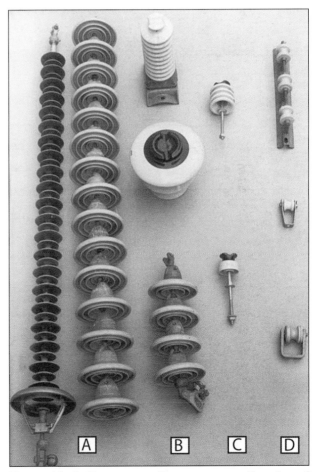

Figure 12–4. For comparative purposes, the size of insulators can be an excellent indicator of the voltage.

Transformers

Although the primary (input) side of a transformer is generally either 34,500 or 4,800 volts, the secondary side (output) can have a multitude of different voltages. Secondary voltages can be identified as follows:

Sub-transmission transformers. In an overhead distribution system, transformers with a 34,500-volt primary side are normally mounted on a platform rack between two or more poles (figure 12–5).

Figure 12–5. Sub transmission transformers are normally mounted on a platform between poles.

The secondary voltages can vary between 4,800, 480, 277, and 240 volts. Unlike primary distribution transformers, the class or ratio of a sub-transmission transformer may not normally be stenciled on it. The size of the insulators can help determine the approximate voltage. The insulators on the primary side (34,500 volts) will be approximately 18 to 24 inches in length. The insulators on the secondary side will be approximately 8 to 12 inches for 4,800 volts, and 3 to 6 inches in length for 480, 277, to 240 volts.

Primary distribution transformers. The input and output insulators on primary distribution transformers (4,800 volts) are similar in size. The class or ratio of a primary transformer may be stenciled on the exterior. As an example, assume a transformer is stenciled with CL 200, U 5670, and 10. The CL 200 denotes a secondary voltage of 120-240 volts, the U 5670 denotes a transportation record number, and the 10 is the size of the transformer (10 kilowatts, or 10,000 watts).

Note that designations may vary in different municipalities. To obtain the maximum benefit from these examples, determine the specific codes (if any) that are used in your area.

Using the factors of determining voltage by insulator size, spacing, and transformers, let's look at several examples.

Figure 12–6 shows (A) 34,500 volts, (B) 4,800 volts, and (C) unmarked transformers reducing 4,800 volts to 240-120 volts or 480, 277, 240 volts.

Figure 12–7 shows (A) 4,800 volts, with (B) the center of the transformer stenciled CL 200 (120-240 volts), U 6576 (transportation number), and 100 (100 kilowatts). The outer transformers are stenciled CL 200 (120-240 volts), U 6678 (transportation number) and 50 (50 kilowatts). This figure shows (C) 120- to 240-volt triplex services to structures.

Figure 12-6. In this example, 34,500 volts can be found at A, 4,800 volts at B, and unmarked transformers reducing 4,800 volts to 480 or 240 volts.

Figure 12-7. In this example, 4,800 volts can be found at A, the stenciling on the exterior of the transformers indicate the output voltages are 120-240 volts, and C is a 120-240 volt service to a structure.

Eliminating Electrical Service

When a working fire is encountered, personnel must consider eliminating the electrical service when there is a possibility that the conductors will be shorted or exposed to fire. Fire extending into walls, ceilings, or floors can burn the insulation off electric wires and damage fixtures. This increases the likelihood that hose streams, tools, and equipment will contact exposed wiring during attack, ventilation, and particularly overhaul operations. Personnel who contact bare wiring risk serious injury or death. The effect of current rushing through a human body in combination with the length of duration of an electrical shock can cause devastating results in four specific areas. These are cardiac arrest, ventricular fibrillation, tissue destruction, and flash burns.

The major considerations toward curtailing electrical service to a structure fire are fire involvement and suspected damage to the structure's electrical system. The safest and most effective method to prevent injury from electric shock is to eliminate all electrical service to a structure. This will also minimize the possibility of damaging transformers and causing a blackout in the surrounding area. Also, it may be necessary in some cases to delay eliminating electricity to larger residential structures such as apartments and hotels until it has been determined that all of the occupants have been evacuated. Residents often rely on hallway lights and exit signs to escape from fire, heat, and smoke. The following sections describe the three principal ways to eliminate electricity from an involved structure.

Depend on the local utility company

A utility company has the equipment, personnel, and expertise to eliminate electrical power safely, allowing suppression personnel to concentrate on other tasks. However, the estimated time of arrival (ETA) of the utility company must be evaluated in relation to fireground operations. If a utility company has given you an ETA of 20 minutes, and suppression personnel are already opening walls and ceilings, either immediate elimination of electrical service may be necessary or curtail overhaul operations until the appropriate utility company has eliminated power to the building.

Shut off the service

If an electrical panel is readily available on the exterior of the building, the service can normally be eliminated by moving the circuit breakers to the off position. To prevent arcing in the panel, shut off the branch circuits (or sub circuits) before deactivating the main circuit breaker. It is important to remember if any breakers have been tripped, as this can be useful information for arson investigators. Some panels are located on the interior, in which case the fire may rule out using the panel to eliminate electrical service. If there is a pad-mount transformer outside the building, it is ill advised to open the cabinet doors and shut off the service by using the switch or by deactivating a fuse. Personnel doing so can be exposed to high voltages. Also, appropriate tools should be used to operate switches or deactivate fuses. Moreover, a penta bolt is used to secure the cabinet doors, requiring a special wrench to gain entry. In Figure 12–8, the main switch is located just under the fuses.

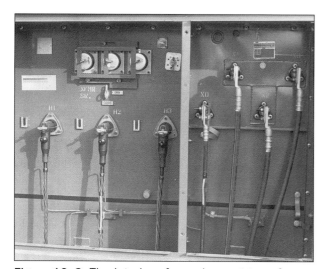

Figure 12–8. The interior of a pad-mount transformer illustrating insulated and uninsulated conductors, fuses, and a main on-off switch.

Meters should not be removed to eliminate the electrical service to a building. Removing a meter may result in the meter flashing or exploding due to a sudden heavy draw of current as the meter clips are pulled out. Also, removing it will expose live line connectors in the panel. Occasionally, too, occupants can run an illegal conductor from the unmetered side of a circuit panel to a remote location, such as a garage. If a meter has been illegally bypassed, removing it won't eliminate the electrical service to portions of the structure that are being served illegally. If the seal securing the ring around a meter has been removed or altered, suspect bootlegging. Always check for additional electrical sources that may enter a building. Finally, new bypass meters only measure the amount of electricity that is used by the customer and won't de-energize the system if removed.

In rare cases, removing a meter may be necessary due to incident priorities. Before electing to do so, take the time to reevaluate those priorities. The speed of the ring (disk) inside a meter indicates the flow of current. A rapidly moving disk means a heavy draw, which will increase the potential of explosion if the meter is suddenly removed. If the removal of a meter is necessary, the appropriate protective clothing/equipment must be worn correctly. This includes turnout clothing, helmet, goggles or shield or face piece, and rubber or leather gauntlet gloves. When removing a meter, do not pull it outward. Instead, pull it upward, toward the source of supply, then out. This will disengage the line connectors at the bottom of the meter before disengaging the top line connectors, which face the source of power. This will lessen the draw of current and minimize the potential of explosion.

Cut the drip loops

If an electrical panel has been damaged, is not accessible, or if the local utility company has an unacceptable ETA (when compared to existing hazards), the electrical lines to a structure may need to be cut. However, only personnel who are trained and properly equipped should accomplish this task. In such an event, the most effective method to eliminate the electrical service is to cut the drip loops if they are readily available, and they are not high-voltage lines. Drip loops are used to prevent rainwater from running down the conductors, entering the structure by way of the weatherhead. Drip loops that have been correctly cut can easily be repaired by the utility company at minimal cost. Cutting them will also lower the chance of fireground personnel contacting live wires on the ground. This operation depends on the type of incident, time constraints, the expertise of personnel, any applicable SOPs, adherence to appropriate safety considerations, and the use of appropriate equipment. The following minimum tools and protective equipment are required when cutting electrical conductors:

- Turnout clothing and helmet, goggles, shield, or face piece

- Rubber or leather gauntlet gloves

- Mechanical axe, approved wire cutters, or hot stick, dry ground and boots

Note that boots commonly worn by suppression personnel (leather and/or rubber) may not afford the degree of protection that they are generally assumed to provide. The conductivity of boots can be increased by the presence of water, carbon black, steel arch and toe protective devices, and metal debris in the soles of the boots.

When cutting wires, do not cut wires that carry over 480 volts, and do not exceed the voltage limits of the tool that is being used. Stand on a dry and nonconductive object. Wood and fiberglass ladders are excellent for above-ground operations. When using ladders, lock in to the ladder while cutting.

If it is necessary to cut insulated wires that are wrapped around a bare wire, spread the insulated wires away from the bare wire (and each other) with the cutting tool before cutting the insulated wires. Stagger the cuts to prevent live ends from touching, and keep the cutting tool from touching any bare wires. Make the cuts at the drip loops, cutting one wire at a time, and starting with the furthest or uppermost wire first. This will prevent crossing over cut live conductors. Cuts should be made toward the source of power and should leave the wires at least six inches long. This will minimize the chance that hot wires will contact each other or the service head, thereby causing a short. This will also allow utility personnel to splice the cut wires without having to run new ones from the pole.

Remember to not cut the bare wire (neutral) that is supported at each end of the conductor. This conductor has a steel core added for strength to support the insulated wires. Cutting this wire can cause the wires to fall to the ground, creating a potential hazard for fireground personnel.

After cutting the service, check for other power lines, since some structures can have multiple electrical sources, particularly on older commercial structures. The safety of personnel is the primary consideration when controlling electrical hazards during an incident. Any electrical hazard is the responsibility of the first person who sees it.

Electrical Considerations

Electrical considerations and the operations necessary to mitigate these potential hazards can be summarized in five categories: dwellings and small commercial occupancies; commercial/industrial occupancies; large apartments and hotels; complex occupancies; and specialized hazards.

Dwellings and small commercial occupancies

Electrical services to these structures (known as service drops) are either overhead or underground (figure 12–9).

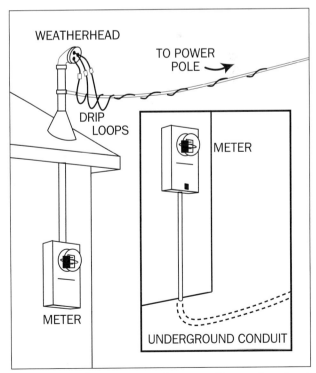

Figure 12–9. Structures can be served by overhead and underground services.

Overhead services. An overhead service will consist of lines running from a pole to the structure in the following configurations:

- Two wire. Older services may only have two open lines coming into the service head. These services supply an ungrounded 120 volts, single phase, and are an excellent indicator of the presence of knob and tube wiring.

- Three wire. Newer service drops will have three lines, one of which is a bare cable and two insulated lines wrapped around the bare cable that supply 120 volts each (for a total of 240 volts, single phase). These services are normally referred to as triplex configurations (figure 12–10) and are most often encountered by fire suppression personnel.

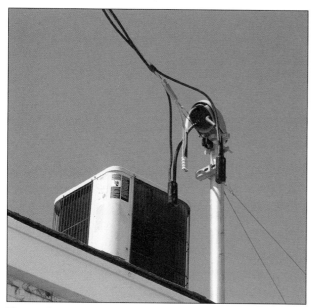

Figure 12–10. A single phase triplex service consists of two insulated wires and a bare cable.

Two- and three-line services generally come from the lower crossarm of a power pole. Typically they are anchored to the structure and configured in a drip loop, and then pass through a service head (or weatherhead) to a meter. The meter is normally outside a structure, although the panel can be inside or outside. Electrical power is then distributed through the structure by branch (sub) circuits. Electrical panels are usually in close proximity (directly beneath) to a service head due to the high cost of conduit and wires between them. Three types of main power disconnects are fused pullout, circuit breaker, and lever. In many older structures, it is necessary to pull a lever to shut off all the power to a structure. Voltages in these structures don't usually exceed 240 volts, single phase.

Underground services. Electrical services can also enter these structures from an underground source. Such wires travel inside underground metal or plastic conduit, terminating in an electrical panel inside or outside of a structure. The meter is normally on the outside because regardless of above-ground or underground services, the appropriate power company can easily read the meter for billing purposes.

Emergency operations. Initially, determine the appropriate method to eliminate the electrical service to a structure. This may include calling the appropriate utility company as soon as possible so their response time to your incident is minimized. If possible, eliminate electrical service at the panel assembly. Remember to shut off the branch circuits before shutting off the main circuit breaker, and look for additional electrical services to the structure. When deactivating branch circuits, remember that arson investigators often want to know what breakers were tripped by the fire. Finally, if necessary and personnel are properly equipped and trained, cut the drip loops. Before cutting overhead services, check the lines at the pole for insulator size and spacing, high-voltage signs, and transformer markings (if appropriate).

Commercial/industrial occupancies

These occupancies can present the hazards of extensive electrical considerations and voltages that range from 120 to 34,500 volts, single or three phase. These voltages are dependent on the electrical needs of a building, and it is common to encounter heavy cable services.

Note that 120-, 240-, 277-, and 480-volt conductors can appear in 1- or 1¼-inch diameters. Remember—don't rely on the size of a conductor to determine its load.

Similar to dwellings and small commercial occupancies, electrical services to these occupancies will also be overhead or underground.

Overhead services. Overhead services can consist of triplex or quadraplex. Quadraplex has three insulated wires wrapped around one bare neutral. It entails a three-phase service that is primarily used for commercial applications. Three-phase provides increased power and reduced costs (for electrical motors and machinery) as compared to single-phase. A three phase service is often recognized by three transformers that are together on a pole and servicing the same structure.

Underground services. Electricity may also enter commercial/industrial structures through underground services and terminate in electrical rooms, fenced enclosures, or pad-mount transformers that can be classified as commercial stations (CS) or industrial stations (IS).

These terms aren't representative of the type of occupancy that they supply. They apply to the transformer, with the primary difference being determined by the incoming voltage. A commercial station receives 4,800 volts, whereas an industrial station receives 34,500 volts. The incoming voltages are then reduced to supply the electrical demands of a given occupancy. Commercial and industrial stations are marked for identification as well as high voltage. The transformer identification number can be preceded by a CS, IS, or some other type of identification code that is specific to a particular utility company (figure 12–11). Pad-mount transformers that do not have a CS or IS preceding the identification number denote a commercial station. Verify the system used by your utility company.

For emergency services, call the appropriate utility company. Determine the most appropriate method to eliminate the electrical service to the structure. When voltages over 480 volts are encountered (either overhead or underground), the power must be eliminated at an electrical panel or by a utility company. If possible, shut down subpanel circuit breakers first, saving the main circuit breaker for last. Remember to check power poles for the following to determine the approximate voltages being encountered:

- High-voltage signs (over 750 volts)
- Insulator size; remember that one side of a crossarm may be primary (4,800 volts) and the other side can be secondary (480-277-240-120 volts)
- Transformer markings, if present

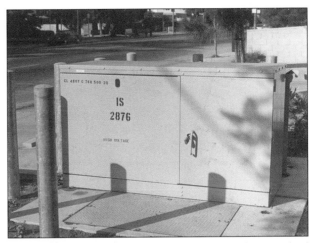

Figure 12–11. Pad-mount transformers are often marked for identification purposes, and to indicate the amount of voltage being fed to the transformer.

In some cases, it may be necessary to eliminate power to certain portions of a building instead of eliminating all of the power. For example, leaving the lights on in a commercial/residential occupancy may be advantageous. Always check for additional power sources to a building because light and power panels may be separate. Do not enter energized electrical rooms or enclosures marked "High Voltage." The appropriate utility company should be called to de-energize these enclosures.

On some older buildings, electrical wires may travel on the exterior of a building. These will run from their anchor point on the building to the inlet to the service head that may be some distance away. The wires will normally run along the inside of a parapet but may be located on the outside of the building or on top of an exterior wall. In this case, be aware that there will be energized wires on the exterior of the building.

Pad-mount transformers can present a fatal hazard if the exterior metal cabinet is moved on the concrete pad, for example, when bumped by a car.

Refer to figure 12–8. Notice that the three incoming conductors on the left side of the cabinet are not insulated. If the metal cabinet is accidentally dislodged and is forced against these conductors, the metal cabinet can be charged to the same voltage as the incoming conductors. This condition cost a policeman his life in San Jose, California. As first responder to a minor traffic accident, he touched the car that was also in contact with a dislodged metal cabinet for a pad-mount transformer.

Large apartments and hotels

These occupancies have electrical services that are similar to commercial and industrial occupancies. The incoming voltages can vary from 120/240 volts to 34,500 volts, and will be fed either overhead or underground. Large apartments and hotels can have both high- and low-voltage services in an electrical room or enclosure, and these enclosures will be found as commercial stations, industrial stations, or possible pad-mount transformers. Additionally, there may be one main electrical panel and meter per building, a subpanel per each floor, a subpanel for each unit, or an electrical meter for each unit as in condominiums, townhouses, and some apartments.

Power elimination to these structures is done the same as large commercial and industrial structures. Depending on the size and type of occupancy, it may be necessary to enlist the services of appropriate building personnel in order to safely eliminate electrical power.

Complex occupancies

Complex occupancies include high-rise buildings, large hospitals, large shopping malls, and the like. Preplanning these occupancies with qualified building engineers or other appropriate personnel is the only practical method to understand the distribution of electrical power within them. The most common electrical service to such buildings is underground. It would be rare to find an overhead service to these occupancies. Voltages in these buildings range from 4,800 volts to 34,500 volts and enter these structures via pad-mount transformers or underground electrical vaults.

These areas may be found outside a structure in a fenced enclosure, inside or outside a building in a designated room, or in an underground vault. These buildings can also have uninterrupted power supplies (UPS) that can supply emergency power to the customer side of a meter in the event of a power loss so that computers, necessary lighting, electronic security measures, and so on, remain operational. UPSs can be generated by two basic sources: numerous batteries connected together to deliver over 500 volts of direct current, and emergency power from generators powered by diesel engines, micro turbines, and fuel cells.

These buildings can have more than one source of power. The UPS sources can be either inside or outside a building. High-rise buildings may have electrical rooms on every floor and will generally be in the same location on each floor. Isolating any electrical problems to a particular floor is the primary consideration in an emergency situation. Do not enter an energized electrical room or enclosure that is marked "High Voltage." Provide your dispatch office with the CS, IS, or other appropriate identification numbers so that a utility company can de-energize the electrical service.

These structures can have other rooms with meters, main panels, and circuit breakers (and are often labeled "Electrical Meter Room") that can be low voltage. These rooms can be entered.

Appropriate information should be obtained from a building engineer when working within these structures. This isn't always possible, however, underscoring the importance of good pre-fire planning and maintaining up-to-date building inventories. The hazards presented by batteries in UPS rooms are the potential for an acid spill, vapors from charging batteries, and bare bus bars (main conductors). The hazards presented by UPS-fueled engines are the obvious presence of the various types of fuels necessary to run the power supply.

In addition to encountering electrical hazards in various types of structures, other, more specialized types of electrical hazards can be encountered. Specialized electrical hazards are normally found outside of buildings. Several examples are transmission towers, poles that support transformers and capacitors, and street lights.

Generating/distribution facilities

These facilities are characterized by steam power plants, receiving stations (figure 12–12), distribution stations, and other such facilities. Personnel should not enter any of these unless accompanied by appropriate utility personnel. If dispatched to a physical rescue within one of these facilities, do not enter while it is energized.

Transmission towers

Transmission towers and lines are found throughout numerous municipalities (figure 12–13). Each tower should have an identification number near ground level that can be given to a dispatch office or utility company for identification purposes in case of an emergency. Remember that transmission lines carry between 115,000 and 1,000,000 volts.

Figure 12-12. Generating-distribution facilities can present extreme hazards to fire service personnel.

Figure 12-13. Transmission towers can carry 115,000 to 1,000,000 volts. They can also produce static electricity generated by the magnetic field from the high voltages.

Wait until it has been declared safe by utility personnel. If necessary, use the apparatus public address system to instruct conscious victims inside hazardous areas because the power within these facilities can present extreme hazards. Do not direct hose streams into these areas unless authorized by utility personnel.

In an emergency situation, consider apparatus placement and overhead lines. If necessary for a fire, secure a water supply and position lines on the uphill and/or windward side of the incident. Protecting exposures may be the primary objective until utility personnel declare the involved area de-energized and safe to enter.

When confronted with an electrical problem, always be skeptical of the reliability of a verbal notification from an appropriate responsible person who is not on-scene. Unless a responsible utility person is on the scene and is willing to go first, consider waiting.

In an emergency situation, instant electrocution can occur upon direct contact or by being in close proximity to transmission lines. If a rescue is necessary on a transmission tower, do not climb or place an aerial device to a tower until approval is obtained from the appropriate utility company. If a conscious victim is on the tower, provide instructions to remain in position until it is safe to affect a rescue.

Transmission towers can present a unique hazard due to a static charge that is present in varying degrees in each tower. The high voltages traveling through transmission lines generate magnetic fields that vary depending on the voltage and atmospheric conditions. A magnetic field can easily generate significant static charges on or near transmission towers. Static charges of 12,000 volts have been measured. Therefore, transmission towers can be deadly to personnel who may be near or on a charged tower.

Poles

Power poles can be constructed of metal, wood, or concrete. In some municipalities, a vertical metal or plastic identification tag will indicate the particular utility company responsible for that pole. Power poles (particularly wood power poles) normally have a ground strap (often encased in wood) on the exterior of the pole for grounding purposes.

Remember that metal poles will conduct electricity. Wood and concrete are considered nonconductors. However, treated wood poles and metal rebar inside concrete poles can conduct electricity. If the grounding strap is broken, such as in a vehicular accident, personnel who touch the auto or pole can serve as the grounding strap with predictable results.

Wires down, pole to pole. Determine the type of wires that are down. They may be transmission lines, sub-transmission lines, primary conductors, or secondary conductors. Remember, the best method to determine approximate voltages is by the size of the insulators and the distance between the insulators. Additionally, due to different types of routing, power can flow either way in a conductor. Therefore, you should always assume downed wires to be live.

In an emergency situation, first secure the area. Keep unauthorized people at least one pole span (100 feet) away from downed wires. Consider all downed wires to be live. Remember that some systems have relays to clear faults and to isolate trouble in distribution lines. If a problem such as a short is detected, appropriate relays will open, removing power from the problem line. These will automatically reset in varying amounts of time depending on whether the relay is loose or tight. To unsuspecting personnel, an apparently dead wire can suddenly become live when the relays automatically reset. This is a primary reason why all lines should be

considered energized unless identified otherwise, and thermal imaging cameras should not be used to detect live wires. Additionally, consider the possibility of hidden hazards associated with downed wires such as energized vehicles, metal patio roofs, chain link fences, telephone lines, and the like.

Personnel who step near or between energized wires on the ground can be electrocuted by a phenomenon known as step potential or ground gradient (figure 12–14). Be aware that although the diagram illustrates two wires on the ground, the same hazard can exist with one wire on the ground. The step potential distance around a wire or between wires depends on various factors, such as the voltage, the type of impurities present, and moisture in the ground. If it is absolutely necessary to walk near a wire on the ground, hop or use numerous short steps.

For this reason, if placing traffic cones around an electrical hazard, always consider the distance between the wire(s) and the cones.

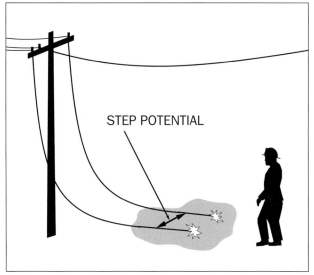

Figure 12–14. Live wires on the ground can create a dangerous condition to personnel known as "step potential."

Wires down, pole to structure. Determine whether the wires are down pole-to-structure or structure-to-pole. Pole-to-structure wires will usually be energized, whereas structure-to-pole wires usually won't. However, check to see if all the wires in the service are down. For example, if one wire is down structure to pole in a three-wire service, power can travel from a pole to the structure in the other wire, through an appliance, and back out of the service to a down wire that is assumed to be an apparently dead wire on the ground, energizing the wire. This is called feedback.

If wires are down, notify all persons in the area of the potential hazard. Be aware of other hazards such as charged chain link fences, metal sheds, telephone lines, and the like. Although it may be advantageous to place traffic cones around the hazard, be aware of the distance between the cones and the downed wires due to step potential. Additionally, always remember that apparently dead wires can suddenly become live and the possibility of step potential.

Pole-mounted transformers. Pole-mounted transformers generally provide a voltage drop of 34,500 volts to 4,800 volts, or 4,800 volts to 480, 277, 240, or 120 volts. Transformers may be mounted alone or in groups of two or more (figure 12–15). It is possible for transformers to explode and burn. In this case, the oil used for cooling inside the transformer is the burning combustible.

transformers containing them may still be located within a given municipality.

In an emergency situation, advise the dispatch office and have the utility company notified. Secure the area. If necessary, apply water in a fog-spray pattern from a safe distance. If polychlorinated biphenyls (PCBs) are present, consider the hazard of runoff from suppression lines.

(See "Hose streams on energized equipment" later in this chapter.).

Pole-top switches. Pole-top switches (figure 12–16) are used to switch 34,500-volt lines for sectionalizing purposes. They are padlocked and are to be opened only by utility company personnel.

Figure **12–16**. Pole top switches are used for sectionalizing purposes and are not to be operated by fire service personnel.

Figure **12–15**. Pole mounted transformers can be mounted alone or in groups of two or more.

In an emergency situation, secure the area. If a switch is burning, advise your dispatch office and ask for utility company assistance.

Although most utility companies have removed pole-mounted transformers containing PCBs, some

Capacitor banks

Capacitors, or banks of them (figure 12–17), are placed in lines at various intervals to boost and to enhance proper phasing of the electricity.

Figure 12–17. Capacitors can store electrical energy, creating a hazard even after they are disconnected from a conductor.

Since capacitors store electrical energy, they can be charged to the full voltage of the line (4,800 volts in this photo), even when they have been disconnected or when power has been curtailed. Capacitors can be oil-filled and may contain PCBs. The same hazards and emergency operations that apply to transformers apply to capacitors.

Street electrical vaults

Underground electrical vaults can be located throughout a municipality. They are accessible through manhole covers and are placed at convenient points in electrical systems to permit the installation and maintenance of cables, transformers, and other electrical equipment. Voltages between 34,500 and 4,800 volts can be encountered in vaults.

A square manhole cover (sewer manhole covers are round) usually indicates an underground electrical transformer vault. If smoke is issuing from holes in a manhole cover, do not attempt to open it. Gas may have formed in the vault as a result of burning insulation or oil, and may be toxic and explosive. The sudden influx of air could cause an explosion. Manhole covers can weigh 300 pounds and have been blown 75 feet into the air and up to distances of 125 feet. Establish a safe perimeter around a manhole opening.

In some locations, there is a second, inner, water-tight cover under the top manhole cover. Make no attempt to extinguish fires involving underground electrical equipment until approved by on-scene utility personnel. Also, never put water into a vault. Use only nonconductive extinguishing agents and only when requested by utility personnel. Do not enter a vault until a qualified on-scene utility representative confirms that it is safe.

Street lights

Street lights may be either low or high voltage, the dividing line being 750 volts. For identification purposes, newer street lights with a photoelectric cell (on top of the light in figure 12–18) are fed by 240 volts. Older street lights without a photoelectric cell can be fed by up to 6,800 volts.

Figure 12–18. Newer street lights can be identified by a photoelectric cell on top of the light housing. They normally operate on 240 volts.

Older street lights use a constant/current transformer, which will try to increase voltages as the perceived load increases. Therefore, severed or cut wires can result in a sudden dramatic increase in voltage. Utility companies consider street lights to be energized at all times.

In an emergency situation, secure the area, and request assistance from the utility company responsible for street lighting. Do not cut exposed conductors due to constant/current transformers that are used to power street light circuits (figure 12–19). Be aware of the possibility that vehicles that have collided with street lighting poles have become charged.

Figure 12-19. Older street lights can be fed by a constant current transformer that can supply up to 6,800 volts.

Traffic lights

Traffic lights don't ordinarily present a significant hazard because they normally operate on 240 volts. However, if mounted on a street light pole, there could also be high-voltage conductors in the pole. In case of emergency, secure the area, and ask dispatch for appropriate utility assistance.

Telephone lines

Telephone lines may be found as a separate system or in conjunction with power lines. If in conjunction with power lines, telephone lines will be located lowest on the pole (figure 12–20). Power for these lines is provided by a telephone company, independent of power lines. The voltages can vary from 45 to over 100 volts, which is used for the ringing circuit of a telephone.

Figure 12-20. Telephone lines are normally found on the lowest crossarm and can have over 100 volts in the lines.

In an emergency situation, advise dispatch and request telephone company assistance. Treat the wires (and any wires contacting telephone wires) as carrying a minimum voltage of over 100 volts, provided that they aren't in contact with wires of higher voltages. Remember that cutting telephone cables can create major damage to telephone communications over a large area.

Neon signs

If there is one area of electrical hazards that could be classified as out sight, out of mind, it is the potential hazards associated with the common neon sign. Although most commercial businesses use this means of advertising, the simple Closed or Open neon sign in the interior of a small business can be a noteworthy electrical hazard. The voltages required for the operation of these signs can range from a relatively low (120 volts) to a transformer that increases the voltage upward of 20,000 volts for the operation of the neon sign.

Neon signs fall into two categories: interior and exterior. Whether the sign is small or large, high voltages are still associated with these signs, and are an additional reason why the electrical utility to a structure should be disconnected during structure fires, particularly prior to overhaul operations.

Separating persons from live wires

Any type of extrication demands cautious, deliberate actions. Personnel who attempt to remove a victim without taking proper precautions may only compound a hazard and can become part of the problem, particularly when an electrical hazard exists.

Try to determine the approximate voltage that is being contacted by the victim. If it can be accomplished in a timely manner, try to de-energize the wires before separating any victim from them. Take appropriate safety precautions if you are able to cut the wire(s). If it is possible to move wire(s) away from a victim, use a fiberglass pike pole or similar non-conductive tool in combination with rubber gauntlet gloves to separate the victim from the energized wires. This scenario is a prime example of the necessity of being able to determine the voltage in a line because it can dramatically affect your risk assessment and decision-making process.

Energized vehicles

Vehicles are normally insulated from the ground by their rubber tires. However, the insulative properties of rubber tires can be diminished by carbon black, metal in radial tires, and foreign debris imbedded in the tires. Vehicles can be energized by live wires by contacting an object that has been energized by a wire, by direct contact with an energized wire, and arcing from nearby wires.

When responding to an incident where a vehicle may possibly be energized, personnel should evaluate the potential of an electrical hazard and the possibility of step potential. When touching an energized object, a person effectively becomes the ground circuit, which can result in serious injury or death. In figure 12–21 for example, a car has struck a metal power pole with considerable force. If you were the first responder to this incident, carefully evaluate if the car is electrically charged. Don't let your own injury—or possible death—be the first indicator that the vehicle was energized.

Figure 12–21. Be aware of vehicles that have collided with a street light or power pole. Before touching the car or occupants, evaluate the possibility of the car being charged with electricity.

Keep aerial devices at least 15 feet from electrical conductors. Some training material recommends 10 feet from conductors; however, when operating aerial devices, unless the operator is smooth during the entire operation, the 10 foot recommendation can suddenly not be enough. Therefore, 15 feet can provide an added measure of safety.

All personnel should be aware of any electrical hazards. Again, never touch a charged or potentially changed vehicle. If personnel are on charged apparatus, they should remain in place and not touch anything. If it becomes necessary to abandon the apparatus, do not step off. Instead, jump clear. When clear, post a member to prevent other personnel from touching the apparatus until the hazard has been eliminated.

Hose streams on energized equipment

Using water on energized equipment presents the danger of the water becoming a conducting medium. The electric current can travel from the source back to the nozzle.

Although the resistance of a water stream is relatively high, dangerous amounts of current can be carried through a solid stream under certain conditions. Fog-spray steams are recommended because the water is broken into droplets surrounded by air, greatly increasing resistance and minimizing conductivity.

Water should only be applied to energized equipment after considering the appropriate risks and when *absolutely* necessary; and always use fog-spray streams. Remember that as the distance between the nozzle and energized equipment increases, so does the resistance of the fog-spray stream, thereby enhancing safety. Also, do not continuously apply water to energized equipment. Use short, quick applications, if possible.

Hybrid vehicles

Due to a combination of rising fuel prices, governmental mandates requiring increased fuel mileage and lower emissions, hybrid vehicles are becoming universally accepted as a viable way to reduce operating costs. Currently, Honda, Toyota, Lexus, and Ford are some of the more popular hybrids available, with other manufacturers developing their own versions. These types of vehicles do not rely on a simple 12-volt battery for electrical power; they use a combination of a small internal combustion engine coupled with a system of 300-volt DC batteries that supply powerful electric motors that operate on 650 volts AC. In the majority of these vehicles, the battery packs are located in the rear with high-voltage cables running under the vehicle to the drive system under the front portion of the car. When conducting an extrication operation, consider the location of the power source and cables that run to the front of the vehicle.

National Color Code

A color code is used nationally to help identify utilities as follows:

Red: Power
Orange: Telecommunications
Green: Sewers and drains
Blue: Water
Yellow: Gas

Training Resources

The following resources can provide additional information and/or another viewpoint for the subjects discussed in this chapter:

- Increasing Transformer Reliability Using Insulating Oil Analysis, Paul Boman, *Fire Engineering*, January 2004.

- Fire Fighters and Electricity, *Pacific Learning Systems*, 1991.

- Electrical Safety on the Fireground, Doug Leihbacher, *Fire Engineering*, June 2006.

- Understanding Electricity and Electrical Dangers, C. Bruce Edwards, *Fire Engineering*, April 1996.

- Electricity 101, Michael Callan, *National Fire & Rescue*, September/October 2004.

- Electrical Hazards Warrant Firefighter Vigilance, Bill Gustin, *Fire Engineering*, October 1999.

- The Dangers of Neon Signs, Ed Spahn, *Fire Engineering*, April 1996.

13 Gas Utilities

Introduction

Natural gas is formed as the result of nature's actions on organic matter over a period of time. It is generally composed of the hydrocarbons methane and ethane. Some of the heavier hydrocarbons commonly found in natural gas are propane and butane, in addition to chemical impurities such as nitrogen, carbon dioxide, hydrogen sulfide, and helium. Natural gas used for the purpose of this chapter generally consists of 97% methane. The other 3% can consist of ethane with trace quantities of carbon dioxide, nitrogen, and sulfur. Helium may also be present at 400 parts per million and is used for the purpose of identifying natural gas. With the presence of helium, seepage from underground storage facilities or breaks in gas lines can be differentiated from seepage in natural underground methane pockets.

Natural gas is used as a common source of heat in residential, commercial, and industrial occupancies due to its clean burning qualities, convenience, abundance, and relatively low cost. Natural gas also supplies approximately one-third of the total energy requirements of the United States. Although the U.S. Department of Transportation considers natural gas as one of the safest types of energy, the fire service occasionally responds to various types of natural gas emergencies.

Properties of Natural Gas

Natural gas is colorless, invisible, has an ignition temperature of 1,165°F, has a flammable range between 5 and 15%, is non-toxic, and is not considered hazardous when inhaled in limited concentrations. However, in quantities large enough to displace oxygen, suffocation can occur. Natural gas is lighter than air, which results in escaping natural gas rising and rapidly dissipating. This is a distinct safety advantage over heavier fuels such as butane and propane, which are heavier than air and, when escaping, will collect in low areas. However, escaping natural gas can be hazardous when trapped by confined spaces such as structures, hollow walls, attics, and so on. Pure natural gas is also completely odorless. For the purpose of leak detection, odorants are added so that as little as 1% of natural gas in the air can be detected. A common odorant is Mercaptan or Thiophane, which are used in concentrations of one pint per million cubic feet of gas. Odorants will not change the chemical properties or hazards of natural gas.

Transmission and Storage

Most natural gas is initially distributed through over 30,000 miles of transmission lines from Texas, Kansas, New Mexico, Oklahoma, Arizona, and the Rocky Mountain area. Transmission lines maintain pressures up to 1,000 psi and use pressure booster compressor stations at specific intervals to maintain appropriate flow and pressures. To compensate for fluctuating demands, gas companies can maintain underground or aboveground storage facilities. Aboveground storage is accomplished in large tanks, and underground storage is accomplished by injecting natural gas into porous rock layers.

Distribution

Natural gas is generally distributed by a combination of transmission lines and distribution mains. Transmission lines bring natural gas into a specific area, vary in diameter from 12 to 36 inches, and contain pressures up to 1,000 psi. Natural gas is then distributed to gas company customers by distribution mains and is carried as medium pressure, which normally does not exceed 60 psi. Older existing distribution mains are made of steel, iron, or copper; however, newer distribution mains are generally polyethylene and are accounting for an increasing percentage of the mains in use. Polyethylene mains are either a pinkish color or bright orange. Distribution mains are normally located underneath city streets, parallel to curbs, under grassy parkways, between the curb and sidewalk, and occasionally in alleys. Generally, fire service personnel do not have the tools or knowledge to shut down transmission lines or distribution mains.

Gas Meters (MSAs) and Equipment

Natural gas is distributed from a distribution main to the customer by a service line (figure 13–1). Service lines run from distribution mains to structures via an outside or inside riser where pressure regulators and meters are located.

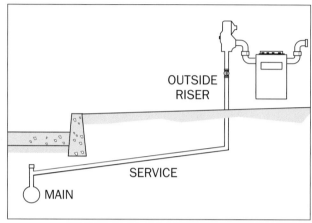

Figure 13–1. Natural gas is distributed from a distribution main to a customer by a service line and riser to a meter and related equipment.

The pressure inside a customer line (lines inside a structure) is low pressure (about 0.33 psi) and is regulated and measured by a regulator and gas meter. Gas companies often refer to gas meters as *meter set assemblies* (MSAs). MSAs can be found in the following three locations:

- Outside sets: MSAs are located outside of structures.

- Underneath sets: MSAs are located underneath structures, stairs, basements, etc.

- Curb meter box: MSAs are located in a vault underneath a sidewalk.

The different types of MSAs are discussed in the following sections.

Residential MSAs

A common gas meter and related equipment is often referred to as a meter set assembly, which consists of piping, shutoff valve, regulator, and meter. This basic MSA is shown in figure 13–2.

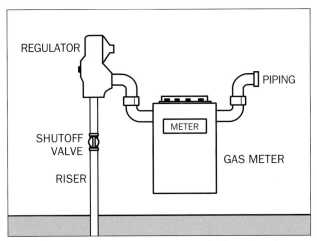

Figure 13–2. A common gas meter and related equipment is often referred to as a meter set assembly and consists of piping, shutoff valve, regulator, and meter.

Figure 13–3. A ring washer on a shutoff valve indicates the service line is a plastic pipe inserted inside an older metal pipe.

The pipe rising vertically from the ground close to the exterior wall of a structure is called a riser. (The riser can also emerge in basements for underneath sets.) Tracing the riser up from the ground, the first fitting is the shutoff valve, called a *stopcock* or *service cock*. This valve is usually of primary concern to personnel responsible for shutting off the gas to a structure. A ring washer on the stopcock valve indicates that the service line is a plastic pipe inserted inside an older metal pipe.

A ring on the stopcock valve (figure 13–3) indicates a branch service (explained later in this chapter), and a band (normally green in color) on the riser near the stopcock valve indicates the service line is plastic.

Next up the riser is a flat circular device called a *regulator*, which reduces medium gas pressure (up to 60 psi) to a relatively low pressure of approximately .33 psi. A few MSAs may not have regulators. Next are pipes and fittings leading to the gas meter that measures the amount of gas used. After the meter, gas flows through pipes and fittings into the structure.

Some MSAs in earthquake-prone areas may be equipped with an emergency vibration-sensitive shutoff device known as an *earthquake valve* (figure 13–4). These valves are marketed and installed by private industry and are normally mounted on the discharge side of the meter of an MSA. Earthquake valves take on various shapes and sizes and can easily be mistaken for any number of gas appliances.

Figure 13–4. Earthquake valves are often used in earthquake prone areas to shut off the flow of gas in an earthquake.

However, personnel should be aware that a swift kick may possibly shut off the flow of gas in the event of a damaged or frozen stopcock. This action should be considered as secondary to shutting off the stopcock and never done in lieu of a normal shutoff procedure.

Industrial MSAs

Industrial MSAs (figure 13–5) are similar to but physically larger than residential MSAs due to the increased volume of gas that is necessary for commercial/industrial applications. Industrial MSAs can be characterized by a number of features.

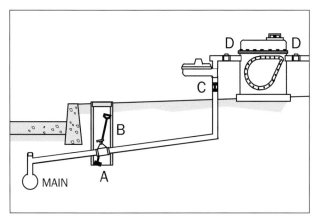

Figure 13-5. Industrial meter set assemblies are similar but larger than residential meter set assemblies.

A curb valve (A) and curb valve extension handle (B) can be utilized to control the flow of gas in the service main. Industrial shutoff valves known as *Nordstrum* valves (D) are used for high-pressure applications, are larger than standard stopcock valves (C), and may require more effort and larger tools to shut off the flow of gas. In figure 13–4, notice the distinct difference between the standard stopcock valve to the left of the earthquake valve, and the Nordstrum valve to the right of the earthquake valve. Some industrial installations have gate valves with wheel handles that may be turned clockwise to close.

Industrial applications that are high pressure can use two regulators in series to regulate the gas pressure. A rotary meter (figure 13–6) can be used instead of a conventional MSA to regulate high pressures and handle quick surges.

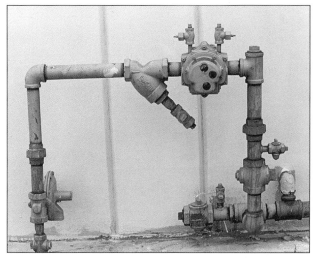

Figure 13-6. A rotary meter can be used instead of a conventional meter set assembly to regulate high pressures and quick surges.

Meters under the sidewalk, called curb meters (figure 13–7), are common in commercial areas. The lid over the curb meter space is made of lightweight concrete or fiberglass set in a metal frame and is easily identified by its gas company markings.

Figure 13-7. Gas meters in commercial areas can be found under the sidewalk, in a curb meter space, or in a vault. The lid over the vault is commonly marked with a gas company logo.

Multiple MSAs

Multiple residential or commercial occupancies are often equipped with multiple MSAs (figure 13–8) that can provide a master stopcock for the building and a separate meter with a stopcock for each unit.

Figure 13–8. Multiple residential or commercial occupancies often use multiple meter set assemblies that provide a separate meter and shutoff for each unit.

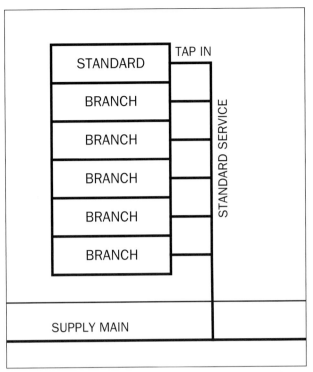

Figure 13–9. Multiple commercial structures under one roof (such as a mini-mall) can be serviced by a branch service. Each occupancy will have a tap-in to the standard service, a meter, and a shutoff valve.

Master stopcocks shut off the entire building and individual meters/stopcocks can control the flow of gas to individual occupancies as necessary. Occasionally, a master meter is utilized to serve multiple occupancies in one structure. In this application, if one occupancy needs to be shut down, the entire building must be shut down.

Branch service

In some commercial areas, multiple structures in a building (such as mini-malls) may be commonly serviced by a single-service pipe, called a *standard service*, from the supply main as illustrated in figure 13–9. The occupancy that is farther from the supply main is designated a standard service, and the other occupancies are branch services. Each occupancy has its own meter, stopcock, and tap-in to the standard service line. A metal or plastic ring around a stopcock indicates that this building is a branch service.

Turning Off Gas

From a fire service perspective, the safest and most direct method to eliminate the unnecessary flow of gas is by shutting off a stopcock type valve. You should note, however, that the local gas company should be notified when gas is shut off to a building or unit within a building. Only the gas company or a qualified plumber should restore service because the system needs to be checked and all pilot lights need to be re-lit. Avoid unnecessary shutdowns. Shutting down a residence causes an inconvenience, and shutting down a commercial or industrial site may cause a major disruption. The following sections discuss how to shut off gas service with different stopcocks and valves.

Stopcocks

Gas companies use several types of stopcocks. Although they differ in size and safety features, they are similar. A basic stopcock consists of the following elements (left to right in figure 13–10):

- **Internal core:** Can be turned to either block or allow gas to flow.
- **Casing:** The outer stationary part.
- **Washer:** Provides proper spacing-pressure between casing and backing nut.
- **Backing nut:** Used to hold the internal core in place.
- **Circlip:** Used on newer stopcocks to keep the backing nut from inadvertently being removed.

Figure 13–10. A basic stopcock consists of a casing, internal core, and backing nut. Newer stopcocks use a circlip to prevent the backing nut from inadvertently being removed.

Residential stopcocks

Shutoff valves can be located by finding the MSA because the stopcock valve and MSA are always adjacent. Initially, look for a "G" that is either chiseled or painted on the curb. A "G" (required by OSHA) marks the location where the service line passes beneath the curb and indicates the side of the building the MSA is located. If a "G" cannot be quickly found, do not waste time looking for one. Next, look along the foundation of the structure. Depending on the particular area of the country, MSAs and stopcock valves are normally located on an outside wall of a structure in a location that is easily accessible to gas company personnel who read meters for billing purposes. If an MSA is not visible along an outside wall, it is usually found in the following locations:

- A crawl space under the building
- Beneath an outside staircase
- In an outdoor closet (particularly new condos)
- In an underground garage
- In a sidewalk or parkway vault at the side of a building
- In a basement with the stopcock valve usually accessible from the outside of the building

To turn off the flow of gas with a stopcock valve, turn the tang crosswise to the pipe (one-quarter of a turn). The tang on some stopcocks can be square. In this case, turn the tang so the line on the tang is crosswise to the pipe. Tangs can be easily turned with channel locks, an adjustable wrench, or other appropriate tools. If a tang appears to be stuck, do not apply too much pressure. If the tang breaks off, it will be difficult to stop the flow of gas. It is recommended that the gas company service a stuck tang; however, stuck tangs can be approached in emergency conditions as follows:

- Use two wrenches—one on the tang, and the other on the backing nut. Holding the tang steady, loosen the backing nut *one turn only*. If the backing nut is loosened too far, the core can fly out, leaking medium-pressure gas. In this case, it may be difficult to replace the core.
- Give the backing nut a tap with a non-metallic hammer, freeing the core.
- Turn the tang one-quarter turn to stop the flow of gas (a half turn will continue to let the gas flow).
- Tighten the backing nut to lock the core in the OFF position.

Nordstrum valve

The Nordstrum valve is a larger version of a residential stopcock. It can be used on high-pressure risers and piping at industrial/commercial sites and comes in various sizes to fit different diameter pipes. The tang on a Nordstrum valve can only be turned one-eighth turn in either direction, ensuring the ON and OFF positions are easy to locate. Depending on the size of the valve, large wrenches may be necessary for this valve. It is easy to see that personnel not familiar with this valve would expect one-quarter turn to shut off the flow of gas, but the valve only turns one eighth of a turn, leaving the impression the valve has not been completely closed.

Gate valves

Some industrial meters are equipped with a gate valve that has a wheel handle. Turning the handle clockwise several revolutions will shut off the gas.

Curb valves

Shutoff valves (referred to as curb valves) can be installed in addition to stopcocks at public assembly occupancies such as schools, hospitals, and churches, and are normally located near the property line by a sidewalk. They are two to three feet underground and under a small metal, concrete, or fiberglass lid with "Gas Company" or "Gas" markings in raised letters. If it is necessary to gain access to this type of valve, lift the lid with a common utility tool (screwdriver) in the hole. If the lid is stuck and additional leverage is necessary, insert the point of a pickhead axe in the hole and apply the proper leverage. This valve is an ordinary gate valve with a faucet-type handle. Turn the handle clockwise to stop the flow of gas. Remember that if this valve is present, the flow of gas can be eliminated by either turning the gas off at the curb valve or meter stopcock

If the flow of gas can be eliminated by either a meter stopcock or curb valve, the meter stopcock should be the first priority because it is possible the curb valve may not service the leak that you are trying to eliminate.

Multiple risers with stopcocks

At some industrial sites with large meters, gas may pass through the meter and into a pipe that goes into the ground. These meters will appear to have a riser on each side. Each riser will be equipped with a stopcock. Turn off both stopcocks.

Branch services

Gas to units within a structure served with a branch service can be eliminated by shutting off the stopcock at the appropriate unit. However, if this is not possible, look for a central valve under a plate near the curb, similar to the curb valve near public assembly occupancies. Close the gate valve clockwise to shut off gas to all related branch services (entire building). Remember that shutting off a single branch MSA will eliminate only the flow of gas to the occupancy connected to the MSA.

Multi-meter installations

In multi-unit residential occupancies (such as apartment buildings) each unit generally has its own gas meter and stopcock valve usually in one or more rows along an outside wall (figure 13–8) or in subterranean parking areas. Each meter should be marked with the corresponding apartment number. The common riser is located at one end of the group of meters and will have a large stopcock. The flow of gas to individual apartments can be shut off by closing the stopcock on the appropriate MSA. If the meters in a multi-meter header are not marked, or the specific apartment is not known, the flow of gas to all meters (and as a result, the entire building) on the common header can be shut off by using the large stopcock on the common riser.

Look again at figure 13–8. There is another set of piping and valves, and a meter to the right of the multi-meter installation. By looking closely, notice there is a rotary vane meter instead of a typical MSA. This indicates there is a higher use of gas through this assembly than the multi-meter assembly, and probably feeds the water boiler/heater in the building. Remember that when gas has been shut off, only a gas company representative should restore the flow of gas.

Gas Leak Emergencies

Gas leak emergencies can fall into a wide variety of incidents, from a simple turning off a valve to an appliance to a backhoe that has ruptured a distribution main, and the resultant leak has ignited and is exposing multiple structures. In all cases, it is imperative that incident personnel quickly determine the extent of the leak (if any), remember that a natural gas leak can present an unexpected and sudden lethal hazard, and should evaluate three primary objectives:

- Safety of civilians and fireground personnel.

- Protection of property.

- Prompt and efficient control of the incident.

Prior to responding to a natural gas leak emergency, invest time with your local gas company to familiarize yourself with their product and their advice about your operations at a natural gas leak. This meeting can pay huge dividends at this type of incident.

In gas leak emergencies, gas company representatives should be notified as quickly as possible because they have the expertise and equipment to handle leaks. The first gas company person to arrive on the scene may be a construction supervisor or a field service representative. They are responsible for assessing the situation and calling for appropriate gas company personnel and equipment. In a major emergency, the second gas company person to arrive will usually be a district management representative who will remain at your command post to communicate with fire department personnel, gas company personnel, and the public.

The initial positioning of apparatus should be carefully evaluated. The initial company should stop short of the incident, windward (upwind) side, and if near a structure, position the apparatus on the windward corner of the building. The corner of a structure is not a normal collapse zone. Avoid parking apparatus over manholes, storm drains, and so on. Apparatus can provide a source of ignition in these collection areas, and if a diesel engine ingests natural gas, it can runaway and destroy itself. Secure a source of water from the windward side, and have necessary protective lines in place.

A two-person recon team should be equipped with full protective equipment, a self-contained breathing apparatus (SCBA), intrinsically safe portable radios, PASS devices in the ON position, direct-reading atmospheric monitor (DRAM), explosion-proof flashlight, and any appropriate tools that can be used to shut off typical natural gas valves. Also, a two-person backup team with full protective equipment, SCBA, and charged hose line should be positioned out of the potential explosion/collapse zone.

Regarding the use of a combustible gas indicator (or similar types of equipment), it is imperative the equipment is calibrated properly and that personnel know how to properly use the equipment.

Minimize potential ignition sources. Disable or do not use doorbells, light switches and thermostats, and appliances. Avoid walking on carpets in dry weather. And never open a valve that has been closed.

Natural gas emergencies can be divided into the four following categories:

- Inside leaks

- Outside leaks

- Leaks resulting in fires

- Certified natural gas (CNG) vehicle incidents

Each leak has its own dangers and concerns, but the greatest danger results from gas leaks inside structures due to the potential for an explosion within a building and possible life hazards. Outside leaks are the next most dangerous. Gas may enter buildings or utility manholes with the resulting potential for an explosion. Gas that has ignited on the exterior of structures is probably the least of our problems. Although it may ignite nearby exposures, the potential for an explosion is minimal.

Tactics at inside leaks (no fire)

Safety is the first priority upon arrival on the scene of a reported leaking natural gas incident. If met by a heavy odor of gas, expect the worst. Expose as few people as possible, meaning prompt evacuation where necessary. A faint whiff of gas or a leak that has come and gone for the past few days may allow more leeway to perform a leak examination. Consider a leak that has come and gone for a few days as a possible exterior leak. The reason behind this consideration is twofold: An exterior leak will follow the path of least resistance and can travel some distance; and dirt will filter the odorant from the leaking gas, hence, the come and go smell. In both cases, call the gas company.

For faint odors, always check the condition of pilot lights as a first action. If a leak is suspected at a specific location, pour soapy water over the suspected area to confirm its presence (watch for bubbles). The two dials to the left of the words CUBIC FEET in figure 13–11 can be used to detect a flow created by small leaks because these dials measure only small quantities of gas. (Make sure all pilot lights are turned off to use this method.)

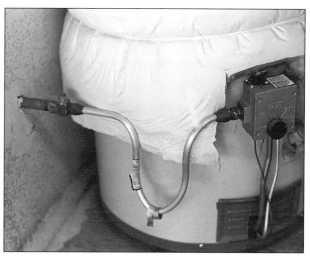

Figure 13–12. This appliance valve is located next to the water heater.

Figure 13–11. On this meter, the two upper dials can be used to detect the flow created by small gas leaks if the pilot lights have been turned off.

When a leak is found, always try to isolate the area as close to the leak as possible. As an example, by turning the one-quarter appliance valve near the water heater in figure 13–12, a leak from the water heater can be stopped, leaving the rest of the premises unaffected.

If that is not possible, move back along the supply piping to the next point of control, generally another one-quarter valve just past the meter called the meter wing cock. A common gas service is often present in multiple-tenant occupancies such as apartment houses or shopping centers. Each tenant has a meter and meter stopcock. Additionally, where the service enters the building there will be a master valve called the service entrance cock that stops the flow to all tenants in the building on the supply line. This valve should be used with discretion—a minor leak at an appliance does not warrant shutting off 20 or 30 units. Conversely, if there is a major leak and there is difficulty determining which meter controls the appropriate apartment, the service entrance cock provides the fastest means of control.

Safety considerations

All that is required to ignite an explosive gas/air mixture is a small spark or an open flame. Something as small as the spark that occurs inside a light switch when it is thrown on or off, or the static spark created after walking across a carpet and then touching a metallic object can ignite a gas/air mixture. Locating the source of a leak, stopping the flow, searching, and venting require a great deal of coordination. Consider where the venting should be done; when and how venting should occur; and if the building should be entered to shut off the gas, or if it should be shut off at the curb valve or stopcock.

Ventilation should be based on the principle that gas is lighter than air, so venting at the upper area of a structure is important. Remember to vent the blind spaces at the top (for example, attics), and consider the use of positive pressure ventilation (make sure you use electric blowers with intrinsically safe motors). When venting, don't cause unnecessary damage, but be thorough. Generally, opening windows may be sufficient.

Note that some gas companies no longer advise customers who report a gas leak to open their doors and windows. The advice is now limited to three points:

- Do not light matches and cigarettes.

- Do not turn lights or anything electrical on or off.

- Stay away from the area or building.

Opening doors and windows in a room where escaping natural gas is present may change the gas-to-air mixture to an ignitable level. This possibility increases the potential hazard to occupants and may not be the correct solution. Evaluate the situation and the opening of doors and windows as necessary for ventilation.

Personnel searching the building for occupants, the source of the leak, and the location of an appropriate shutoff can usually vent at the same time. Do not overcommit personnel. Send in only enough to do the job. Other personnel should stand by in a safe location, such as on the opposite side of apparatus from the building. Generally, begin ventilation operations as soon as possible, but consider how large a leak is present, the sources of potential ignition and their ability to be controlled, and the status of the gas/air mixture. Determine if it is below, within, or above the flammable range. Of all the decisions to be made, where to stop a leak (inside or outside a structure) has the most variables. The practicality of either means must be determined, and a curb shutoff is not always present. The decision to send personnel inside should be made after considering the risks. Unless there is a known life hazard present, expect the worst and treat the situation as the potential time bomb that it is.

Outside leaks (no fire)

Outside leaks can be just as dangerous as inside leaks because gas can take the path of least resistance as it tries to escape to the atmosphere. Quite often that path is along the gas service pipe or other underground lines into buildings or manholes. This migration is insidious because as the gas travels through the ground, it tends to be deodorized as the soil filters out the odorant. The best immediate procedure for responding to an outside leak should include several considerations.

Park apparatus upwind of the leak and away from manhole covers, storm drains, or structures that might contain trapped gas. If the leak is adjacent to structure(s), close doors and windows to prevent gas from entering. Protect exposures close enough to be damaged if gas ignites.

Call the gas company. Only the gas company can shut down a high-pressure transmission or distribution line. Allow the gas company representative the opportunity to assess the situation and call for equipment. In major emergencies, the gas company will send a management representative as spokesperson for the company.

The role of fire service personnel at the site is to maintain public safety. This usually means eliminating potential sources of ignition until the leak is stopped, evacuate when appropriate, and maintain site security. It can take several hours for the gas company to shut off gas in a large, high pressure pipeline.

The only sure way to safely check suspected areas is to use a combustible gas indicator (or other similar equipment) to detect the presence of natural gas and its lower explosive limit (LEL). Many outside leaks are caused by contractors excavating in the area. If an excavating machine has ruptured a line, remember that the line may have been pulled before it ruptured, possibly causing another leak in the same line that is away from the initial rupture.

If a meter or the surrounding pipes have been damaged, check the riser for a bright yellow pipe with a green sleeve, or a yellow plastic ring (indicating the customer has plastic pipe underground). Be aware that the plastic pipe may be broken underground.

Be aware that on rare occasions, pipelines may contain hazardous contaminants such as polychlorinated biphenyls (PCBs). If any liquid flows from a gas pipe, do not let it touch your body. Treat the liquid as a hazardous material, and notify the gas company for clean up responsibilities.

Whenever a gas travels through piping, it can create a static electric charge on that pipe. On steel pipe, that current is drawn off and safely dissipated to ground by the conductive pipe itself. Plastic pipe is an insulator and it is probable that the pipe will have a static charge. A person grabbing the pipe to apply a plug will likely discharge this current, creating a spark. If necessary, fog down the plastic pipe with a wetting agent such as wet water to prevent ignition from static electricity. Water alone tends to bead up and run off of PVC pipe, thus reducing the effectiveness of water eliminating the possibility of a static charge. A wetting agent such as wet water will add an adhesive quality that helps ensure a static charge cannot form. Use water judiciously, and avoid flooding as mud hampers working in the area. Some plastic pipes are equipped with a wire that is used to help work crews locate the pipe from above-ground. This wire provides no electrical charge. Additionally, it should not be relied upon to provide an adequate ground to eliminate the potential for a static spark.

Tactics at leaks resulting in fires

Considerations at a natural gas fire should include the following:

- Protect exposures.

- Generally, let the fire burn until the supply of gas is shut off.

- If necessary, control evacuation and maintain site security.

- Notify the gas company.

- Small fires may be extinguished with dry chemical or CO_2, if necessary, to approach a shutoff valve. For larger fires, fog streams can be used to approach valves. Use care when placing hose streams where excavations have ruptured a gas line. Try to keep unnecessary water out of a pit. Utility crews may have to work in that area to stop the leak and the water could compound their problem.

Tactics at compressed natural gas (CNG) incidents

The use of CNG in vehicles has steadily grown over the past several years due to the escalating rise in gasoline and diesel fuels prices, and the fact that CNG is a cleaner burning fuel in internal combustion engines. It is now common to see city buses, vehicles used in daily utility company operations, government vehicles, and delivery vehicles using CNG as their primary fuel. A storage cylinder (or cylinders) mounted under or in the trunk of a vehicle hold the CNG between 3,000 to 3,6000 psi. Some vehicles that use CNG and gasoline are referred to as dual-fueled vehicles. Dual-fueled vehicles and vehicles that only use CNG are identified by a dark blue, diamond-shaped identification label with the letters "CNG" in white. CNG vehicles do not have to display the Department of Transportation (DOT) hazardous materials label; these are only required for vehicles carrying CNG tanks as cargo. Safety features consist of CNG vehicle tanks approved to pass government safety and crash tests, cylinder overpressure disks, cylinder shutoff valves, and an emergency one-quarter-turn fuel line shutoff valve that is labeled and located in close proximity to the driver's seat. Also be aware that companies using CNG as a vehicle fuel will periodically receive shipments of CNG to restore their supply capabilities.

In the event of a vehicular accident involving CNG with no fire present, it is best to shut off the cylinder valves and the master fuel line valve. In the event of a fire, the tank(s) will likely be exposed to heat or fire. CNG tanks are not filled with a liquid, but rather with a gas, so the entire tank is a hazard. Although the tanks are equipped with an overpressure disk that is designed to fail and reduce increasing pressures within the exposed tank(s), there is still the possibility of a tank rupture if a disk malfunctions. In any event, try to keep water on the tank(s) until the incident is controlled. If the disk(s) do rupture, the tank will empty in a short period of time with the gas escaping in an upward direction.

Training Resources

The following resources can provide additional information and/or another viewpoint for the subjects discussed in this chapter:

- Managing Natural Gas Emergencies, Brian L. Parsley and Gregory F. Schwab, *Fire Engineering*, July 2000.

- Natural Gas Hazards, Frank Montagna, *Fire Engineering*, November 2004.

- Natural Gas Emergencies, Gregory P. Perricone, *Fire Engineering*, May 2005.

- A Tale of Two Gas Leaks, Frank Montagna, *Fire Engineering*, February 2003.

14 Water Utilities

Introduction

Similar to air, water is one of the most abundant substances on earth. It is also one of the most important elements necessary to sustain life. Water is composed of two atoms of hydrogen and one atom of oxygen, hence the chemical name dihydrogen monoxide and the chemical formula H_2O. Although water covers about two-thirds of the earth's surface, it is still necessary to deliver water to residential, commercial, and industrial customers to sustain the common necessities of life. Unfortunately, the presence of calamities such as earthquakes, tornadoes, fires, building collapse, and common problems such as broken water pipes and activated sprinklers often result in the need to quickly control the flow of water to minimize unnecessary damage to life and/or property. It is the *uncontrolled* flow of water that often requires the action of fire service personnel to abate this type of hazard in a timely and professional manner.

Transmission

Water is a liquid that descends from the clouds as rain and forms streams, lakes, and seas. Because water is the lifeblood of every community, it is necessary to supply water for domestic usage as well as fire protection considerations. This is accomplished through a combination of supply systems and distribution systems.

Supply systems

Before water can be delivered for domestic usage, it must first be collected and then delivered to distribution systems. This is accomplished by supply systems that consist of gravity, pumps, and a combination of the two.

Gravity. Gravity systems use a source of water (such as lakes and reservoirs) that are higher than a municipality. The pressure of the water that is delivered into a distribution system within a municipality is determined by the elevation of the water source above a municipality. As an example, a height of 100 feet will create a pressure of 43 psi (0.433 psi per foot of elevation × 100 feet = 43 psi).

Pump. When a source of water is below or at the same elevation as a municipality, or cannot naturally provide sufficient pressure through adequate elevation, pumps are used to supply water at the proper pressure into a distribution system within a municipality.

Combination. Water is initially supplied into a distribution system by a combination of pumps and/or gravity. Reservoirs and/or elevated tanks are located at strategic locations within a system to provide a source of reserve water if necessary. As water demands vary with the time of day and changing weather conditions, a sufficient flow of water at adequate pressure under varying conditions requires the storage of water. Municipalities can use large, open reservoirs to small steel or concrete storage tanks of 10,000 to 10 million gallons.

Distribution systems

Water is delivered throughout municipalities by a distribution system of pipes, valves, check valves, and water meters, that are normally cross connected at specific intervals to supply water to areas that can be affected by repairs, breakage, and other similar problems. Pipes within a distribution system can be classified as trunk lines, mains, and service and customer lines.

Trunk lines carry water from a primary source (reservoirs, pumps, and so on) to mains within a distribution system. Trunk lines can vary from 36 inches to 10 feet, and are constructed from iron or concrete. Mains normally run beneath a street and parallel to a curb, can vary from 2 to 72 inches in size, and are constructed from iron pipe or plastic. Mains can be cross connected to form loops that allow for shutting down a specific main for repairs within a distribution system. Figure 14–1 shows water delivered to individual structures by a combination of service and customer lines. A service line runs between a main and a water meter, can vary from ½ inch to several inches, and is constructed from iron, copper, or galvanized pipe. A customer line runs between a water meter and a structure, can also vary from ½ inch to 12 inches, and is constructed from iron, copper, or galvanized or plastic pipe.

Water utility companies are normally responsible for the maintenance and repair of a water distribution system and service lines. This includes the water metering device and the shutoff valve on the

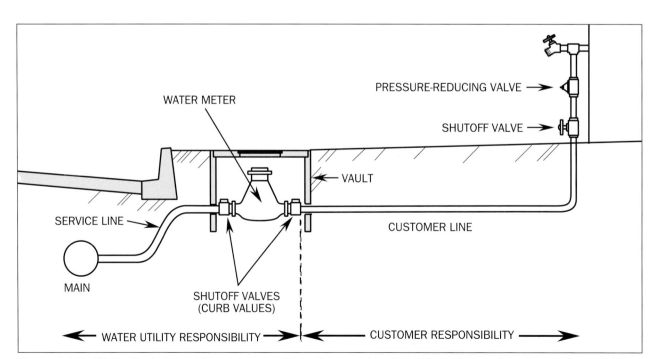

Figure 14-1. Water is delivered to customers through a combination of service and customer lines, valves, and meter.

supply side of a water metering device. Building owners are normally responsible for the maintenance and repair of plumbing from the shutoff valve (on the output side of a water metering device) and all plumbing inside a structure.

Service

The type and size of water service to a structure is determined by a combination of applicable plumbing code requirements and water consumption needs. This combination results in the following types of water service configurations.

- Domestic water only
- Domestic water and water for fire protection through a single service
- Domestic water and water for fire protection through separate services

Water service systems are in part composed of meter boxes, vaults, or manhole covers, and shutoff valves.

The meter box, vault, or manhole covers provide access to water meters and/or shutoff valves. A meter box, vault, or manhole often has a centerpiece or lid with a hole that can be utilized to remove the cover (figure 14–2). The centerpiece or lid normally identifies the type of utility with the word(s) WATER, or CITY WATER METER, and are often marked with blue paint.

Shutoff valves (service valves, curb valves, and so on) provide a means for controlling the flow of water through service and customer lines which may include shutting off the flow of water to repair a service or customer line, water meter, or abate an unwanted water flow problem. The two types of shutoff valves that are commonly utilized are indicating and non-indicating valves.

Non-indicating valves do not indicate an open or closed position and are characterized by ball and cone valves (figure 14–3) and gate valves. Ball and cone valves have a rectangular valve-nut (or tang) attachment on the top of the valve. When the valve is fully open and the valve-nut/tang is turned perpendicular to the valve, a ball or cone rotates 90 degrees to the closed position.

Figure 14-2. A meter box provides access to water meters and shutoff valves.

Figure 14-3. Ball and cone valves have a rectangular valve-nut/tang on top. The valve nut is turned 90 degrees to stop the flow.

Gate valves are non-rising stem valves that have a square valve-nut attachment at the top of the valve stem (figure 14–4). When the valve is fully open and normally turned clockwise (some valves are turned counterclockwise), a gate lowers to the closed position.

Figure 14-4. Gate valves are non-rising-stem valves. Turn the stem clockwise to stop the flow.

Indicating-type valves indicate the position of a valve (open or closed) and are characterized by outside stem and yoke (OS and Y) valves (rising stem-type valves) and post-indicator valves (figure 14–5). Post-indicator valves are connected to an underground gate valve.

Figure 14-5. Post-indicator valves are indicating-type valves. They are connected to an underground gate valve.

Water meters and detector check valves

Water meters are used to calculate water usage, and check valves are utilized to allow the flow of water in one direction only. The type of meter and/or check valve is dependent on the type of water services.

The water meters shown in figure 14–6 are used to calculate water usage for a domestic water service. Water meters are normally directly adjacent to shutoff valves and vary in size from ½ inch to 10 inches (inlet/outlet size).

Figure 14-6. Water meters of various sizes are used to measure domestic water use.

A compound meter is used to calculate water usage for domestic water services that can use low and high water flows (figure 14–7). During low water flows, a compound meter operates as a standard water meter. However, if high flows are necessary, a check valve opens and allows the passage of large flows that are metered separately.

Figure 14-7. Compound meters are used for domestic water services that can use low and high water flows.

Compound meters are 4 inches or larger (inlet/outlet size) and normally directly adjacent to shut-off valves.

The detector check valve with by-pass meter is used to supply water for a closed fire-protection system (figure 14–8). When necessary, a check valve opens and allows water to flow straight through unmetered.

may be located in the same area or vault, or separated in an adjacent area or vaults.

Figure 14–9. A combination service provides water through a meter for domestic use and water through a detector check valve for fire protection.

Figure 14–8. Detector check valves flow water in one direction only, and are commonly used in fire protection systems to prevent water from flowing back into the distribution system. Most are also equipped with a small bypass meter to verify unauthorized use and the presence of leaks.

Pressure-reducing valves

Pressure-reducing valves are normally used in domestic services to reduce high water pressures to an acceptable working pressure inside a structure (such as dwellings commonly using 40 to 80 psi). Pressure-reducing valves are not normally utilized in fire protection services.

In the closed position, a check valve prevents water from flowing back into a water distribution system (for example, from fire department apparatus pumping into standpipe inlets and potentially contaminating the system). Most detector check valves use a small water meter in parallel with the check valves to detect leaks and the unauthorized use of water (see the bottom of figure 14–8). Detector check valves can vary in inlet/outlet size from 2 to 10 inches.

The factory mutual (FM) meter is a high-flow meter that provides water for fire protection purposes. It can vary in size from 4 to 10 inches. A domestic water service may be in the same vault as a FM meter.

Combination water service

A domestic water meter and detector check valve are used for providing separate water services (figure 14–9). A water meter and detector check valve

Location

A typical domestic water service normally consists of piping, a water meter, shutoff valves, and a pressure-reducing valve. A typical fire protection service normally consists of piping, detector check valves, and shutoff valves. The order of placement, however, for water system components between a main and structure can vary depending on the municipality. Additionally, it is important to remember the location of water system components is also dependent on climate. In cold climates, water meters, pressure-reducing valves, detector check valves, and shutoff valves may be located below the frost line or in the basement of a structure. In warm climates, water meters can be placed just below grade level, and pressure-reducing valves, shutoff valves, and detector check valves can be located above or below grade level.

The following sections provide an overview of common structural water service systems.

Single-family dwelling

Domestic water services will normally consist of a ½- to 2-inch single water meter with a shutoff valve on one or both sides. A customer line runs underground from the meter to a riser that enters a structure on the exterior or in a basement. A pressure-reducing valve can be located near the meter or structure. Water service for a domestic sprinkler system is dependent on the size of the sprinkler system. Small systems are serviced by the domestic system, and larger systems may consist of up to a 2-inch detector check valve that is located in a separate vault.

Apartments/motels

Domestic water service normally consists of 1- to 3-inch customer lines and a single 1- to 3-inch water meter with a shutoff valve on one or both sides of it. Similar to single-family dwellings, and depending on water requirements, a customer line runs underground from a water meter to a riser that enters a structure on the exterior or interior. A pressure-reducing valve can be located near the water meter or structure. Water service for fire protection will vary between 2 to 4 inches. The water services for domestic water and fire protection are in separate below-ground vaults or located above ground. A common alternative to a single 1- to 3-inch water meter is a 3-inch equivalent water service (figure 14–10). This system consists of two 2-inch water meters in adjacent vaults that are connected to a single 3-inch customer line that enters a structure on the exterior or basement. In this configuration, the pressure-reducing valve is located near the structure.

Hotels

Domestic water and fire protection services are provided by a combination water service that provides domestic water with a 4- to 6-inch water meter, and water for fire protection with a 4- to 8-inch detector check valve. Shutoff valves are normally adjacent

to the water meter and detector check valves, and a post-indicator valve is normally present between the detector check valves and structure. The water meter and detector check valves can be in the same vault, separate vaults, or above ground.

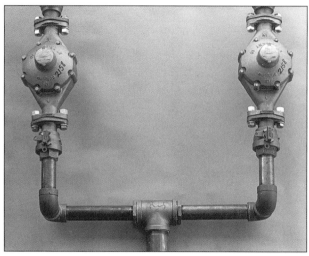

Figure 14–10. An equivalent service uses two water meters connected to a single three-inch customer line.

Industrial/commercial

Water services to these types of structures are not primarily based on the size of a structure, but depend on the water requirements of a structure. As an example, a large warehouse with a small office area and large storage area can have a large (4 to 10 inch) water service for fire protection, and a small (1 to 2 inch) water service for a domestic water system. Conversely, a moderate-size commercial occupancy can have a 2- to 4-inch service for fire protection and a large (4 to 8 inch) domestic water service for specific requirements. Common applications are as follows:

- Domestic water service with a 1- to 2-inch water meter and 2- to 4-inch detector check valves for fire protection in separate vaults. Shutoff valves are normally adjacent to a water meter and detector check valves, and a post indicator valve should be present between the detector check valves and structure.

- Combination water service with a 4- to 8-inch water meter for domestic water service, and 4- to 10-inch detector check valves for fire protection in the same vault. This configuration is used for large water volumes. Shutoff valves are normally adjacent to a water meter, and detector check valves and a post-indicator valve should be present between the detector check valves and structure.

- A commercial occupancy with multiple occupancies (such as mini-malls) can be serviced with water for a domestic water service by multiple, single 1- to 2-inch water meters in separate vaults, which can supply each individual occupancy (referred to as a battery, figure 14–11), or a single water meter for an entire building. Shutoff valves are normally adjacent to water meters in separate vaults or a single meter that serves an entire building.

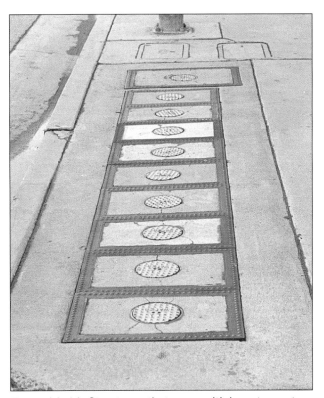

Figure 14–11. Structures that use multiple water meters normally have a battery of vaults in front of the structure.

Water for fire protection can be provided by detector check valves in a separate vault, or a separate or common vault with a water meter that serves an entire building. Shutoff valves are normally adjacent to detector check valves, and a post-indicator valve may be present between the detector check valves and structure.

High-rise

Domestic water and fire protection service is provided by a combination water service that consists of a 4- to 10-inch water meter for domestic water and 4- to 12-inch detector check valves for fire protection in the same vault. Above-ground risers may be present for domestic and fire protection services. Shutoff valves are normally adjacent to water meters and detector check valves in this type of configuration, and a post-indicator valve is normally present between the detector check valves and structure.

Structural Fire Protection Systems

Common structural fire protection systems can be overviewed as discussed in the following sections.

Sprinkler system

There are several different types of sprinkler systems.

Automatic wet pipe system. The pipes in this system are filled with water at all times. When a head is activated, water immediately flows from activated heads. Water is supplied to this system by a gravity tank, pressure tank, automatic fire pump, or a water service system.

Automatic dry pipe system. These pipes are filled with compressed air. When a head is activated, air is immediately released from activated heads. This causes a drop in air pressure, which allows water to enter the system and flow from activated heads. Water is supplied to this system by a gravity tank, pressure tank, automatic fire pump, or a water service system.

Non-automatic system. The pipes are maintained dry. To charge this system with water, it is necessary for fire department apparatus to pump into fire department Siamese connections to supply water to activated heads.

Deluge system. Sprinklers may be open or closed, and all sprinklers are supplied with water by a heat-actuated valve. Water is supplied to this system by a gravity tank, pressure tank, automatic fire pump, or a water service system.

Shutoff valves

Depending on the type of sprinkler system and source of water, shutoff valves are normally located in different areas. Post-indicator valves are located between a main and sprinkler system within a structure. OS and Y valves are located near a sprinkler alarm system. This includes wall-indicating valves (figure 14–12).

Figure 14–12. Wall-indicating valves are mounted on an exterior wall and indicate the closed or open position of the valve.

Shutoff valves are near the outlet for a gravity tank, pressure tank, fire pump, or detector check valves.

Standpipe systems

Standpipe systems are primarily used to provide water for fire protection requirements inside of some buildings and can be used by building occupants and/or fire suppression personnel, depending on the type of standpipe system, and how water is supplied to it.

Standpipe systems include dry, wet, and combination.

Dry standpipes are not directly connected to a water supply and consist of a Siamese connection, piping, and 2½-inch outlets on each floor (stairway and/or fire escape balcony) and roof. Water must be supplied to this system by fire department apparatus. Be aware that the inlets to these systems can be blocked with debris that has been inserted by pranksters.

Wet standpipes are directly connected to a water supply, are primarily designed for occupant use, and commonly consist of 100 feet of 1½-inch unlined single jacket hose, a ½-inch straight tip, and a shutoff. Water is supplied by a water service system, gravity or pressure tank, or a combination standpipe system.

Combination standpipes are directly connected to a water supply and are equipped with 1½-inch outlets with hose and nozzle (primarily intended for occupant use) and 2½-inch outlets for use by fire department personnel. Water is supplied by a water service system, gravity or pressure tank, or a combination of these systems.

Combined sprinkler-standpipe system.

A sprinkler system is connected to a combination standpipe system. Depending on the type of standpipe system and source of water, shutoff valves are normally found in specific locations. Post-indicator valves are usually found between a main and sprinkler system within a structure. OS and Y valves are normally located near a sprinkler alarm system. Shutoff valves are usually located near the outlet for a gravity tank, pressure tank, fire pump, or detector check valves.

Turning Off Water

The ability to eliminate the uncontrolled flow of water in a timely manner can be consistently achieved only by a combination of pre-fire planning and training that can provide the knowledge to quickly locate and identify a water source that needs to be eliminated. To increase the effectiveness of this operation in concert with the type of leak encountered, evaluate the following considerations.

For leaks outside a structure, use multiple personnel to locate the appropriate shutoff location because one person may not find the correct location in the least amount of time. Also when encountering a leak outside a structure, determine if the leak is from the domestic or fire protection service. Domestic water services have a water meter for calculating water usage and a pressure-reducing valve for reducing the water pressure. Fire protection services do not. Remember that *large* leaks may need to be handled by utility company personnel.

For leaks inside a structure, use multiple personnel to simultaneously investigate the interior and exterior of a structure to locate the most appropriate shutoff location. This approach enhances the ability of personnel to locate the appropriate shutoff location in a timely manner. Also, when encountering a leak inside a structure, determine if the leak is from the domestic or fire protection service. If a leak is from a domestic service, determine if the leak consists of hot or cold water. Hot water leaks can normally be quickly turned off at a water heater (or boiler), and cold water leaks may need to be turned off at the source or outside a structure. Finally, determine if a leak can cause or contribute to structural instability (potential collapse) due to the weight of water-soaked materials (building materials and contents).

Occasionally a shutoff valve may be encountered with a sign such as "life support machine in area" or "shut gate system divide." Do *not* turn these valves off unless appropriate approval is obtained.

When closing ball, cone, or gate valves that are 3 inches or larger, it may be necessary to *exercise* a valve to completely eliminate the flow of water because valves can become encrusted with rust and scale. Therefore, it is often necessary to open and close a valve several times to loosen any encrustation and allow a valve to fully seat.

Occasionally a ball or cone valve may be difficult to close due to encrustation. Forcing a stubborn ball valve can break the rectangular valve-nut/tang attachment on the top of a valve, eliminating the ability to close this type of valve. If a stubborn cone valve is encountered, slightly loosen the nut on the bottom of the valve (if accessible) and strike the threaded shaft with a hammer. This will drive the shaft into the valve and free the valve. The valve may now be shut off. Ball and cone valves can be identified by the lack or presence of a nut on the bottom of a valve. Ball valves do not have a nut on the bottom of a valve, and cone valves do.

All shutoff valves should be slowly closed and opened to avoid water hammer and potential damage. Ball, cone, and gate valves that are 2 inches or larger will require specific tools to provide the necessary leverage to close these valves.

Water services outside of structures

Shutting off water leaks that are located outside of structures in mains and service lines are normally handled by utility company personnel. Leaks in customer lines (between a water meter detector check valve and structure) can normally be handled by fire service personnel.

Domestic water services. For ¾- to 1-inch service, a shutoff valve (ball or cone valve) is normally located adjacent to a water meter and may be located above ground or underground. If the valve is underground and not in a meter box, it may be necessary to dig down outside a meter box to access the valve. Turning these valves 90 degrees will eliminate the flow of water. If a valve has an arrow on the valve-nut/tang, the valve-nut/tang must be turned 180 degrees so the arrow is pointing in the opposite direction. In cold climates, a round metal lid is normally located between the street and structure (normally by a parkway, sidewalk, and so on). Removing this lid will allow access to the shutoff valve that is located below the frostline.

For a 1- to 3-inch equivalent service, a shutoff valve (ball or cone valve) is normally located on both sides of a water meter on equivalent services. Remove both vault covers to access the meters/valves. It is necessary to shut off a valve in both vaults to completely eliminate the flow of water past this service configuration.

For a 4-inch and larger service, gate and OS and Y valves are located on both sides of a water meter. When closing these valves, remember it may be necessary to exercise a valve to completely eliminate the flow of water. Large above-ground or underground installations may consist of back flow valves that are inline with the water meter and multiple OS and Y or gate valves. Due to the size of these water services and the possibility of damage to building heating, cooling, and other similar systems, personnel should not close these valves unless absolutely necessary. Normally, the appropriate water utility company should be requested to handle these systems.

Fire protection services. For 2-inch services, a shut-off valve (ball or cone valve) is normally located adjacent to a detector check valve. Turning this valve 90 degrees will eliminate the flow of water past this service.

For 4-inch and larger services, indicating valves (post-indicator valve) can be used to eliminate the flow of water past these valves. Post-indicator valves are located outside of structures, are connected to an underground gate valve, and have a handle or wrench to operate the valve. To operate a post-indicator valve:

1. Remove any padlocks or chains securing the control valve.

2. Turn valve in clockwise direction to close the valve. The valve will visually indicate the position of the valve by an arrow or display which states OPEN or CLOSED. Shutoff valves (gate and OS and Y valves) that are normally located directly adjacent to detector check valves on 4-inch and larger fire services (figure 14–13) are not normally used to eliminate the flow of water past these valves.

3. If necessary, a gate valve should be turned clockwise to stop the flow of water past this service. OS and Y valves (figure 14–14) are turned clockwise (so the stem is not visible) to stop the flow of water.

Figure 14–13. If necessary, these gate valves can be turned clockwise to stop the flow of water to a fire protection system.

Figure 14–14. OS and Y valves are turned clockwise (so the stem is not visible) to stop the flow of water.

Fire hydrants. Water to a sheared fire hydrant can normally be eliminated by locating the gate valve cover between the hydrant and center of a street. Removing the gate cover will provide access to the gate valve that controls the water from the main to a hydrant. Use an appropriate tool to close the gate valve. Remember that these valves may have right- or left-hand threads, and valve stems can be easily sheared.

Water services inside of structures

Determine if an interior leak can be most effectively eliminated inside or outside a structure. Shutting off water leaks that are located *inside* of structures can be accomplished as follows.

Domestic water service—exterior operations. For exterior operations, depending on the type of leak, close accessible shutoff valves near a water meter (or near the globe valve that can normally be found on the vertical riser that enters a structure in warm climates as in figure 14–15). This will quickly eliminate the water service to the interior of a structure until an interior leak is located and repaired.

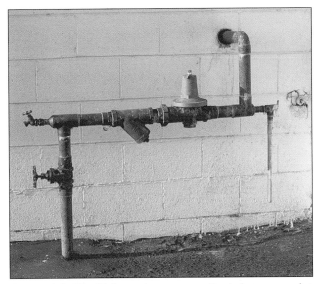

Figure 14–15. Globe valves on vertical risers can be easily accessible and can be used to stop the flow of water into a structure.

Domestic water service—interior operations. There are several common sources for interior cold-water leaks.

Broken pipes. Broken pipes in walls normally require shutting off the flow of water to a structure using exterior shutoff valves near a water meter, etc.

Toilets. Toilets are found in two configurations: residential and commercial. Leaks from residential toilets can be easily eliminated by shutting off the angle valve at the wall near the toilet. In these applications, plastic tubing has replaced metal tubing and is more susceptible to splitting due to hardening.

Leaks from commercial toilets can be eliminated by removing a cap over the relief valve and using a slot-head screwdriver to turn the relief valve screw inwards or outwards. This should let the valve seat and stop the flow of water to the toilet.

Urinals. Leaks from commercial urinals can be eliminated by removing a cap over the relief valve and using a slot-head screwdriver to turn the relief valve screw inwards or outwards.

Sinks and residential appliances. Leaks from sinks and other types of residential appliances can normally be eliminated by shutting off the angle valve(s) at the wall near the sink or appliance.

Manufacturing process. Leaks from a manufacturing process that utilizes water will often have readily available shutoff valves near the process. If these types of valves are not readily found, using exterior shutoff valves may be necessary to eliminate the flow of water.

Hot water leaks. Hot water leaks can normally be eliminated by shutting off the appropriate valve at a water heater or boiler. However, remember to shut off the gas or electricity to the heater to prevent a possible rupture due to expanding water in the heater.

Fire protection service

For exterior operations, depending on the type of fire protection system and location of leak, closing the following valves will eliminate the flow of water past a valve (and will normally stop a leak between a valve and structure) and to a fire protection system inside a structure.

Post-indicator valve. Use the attached handle to turn the valve until the word CLOSED appears in the window on the post.

Gate valves. Remember, it may be necessary to exercise these valves to completely eliminate the flow of water.

OS and Y valves. Similar to gate valves, it may be necessary to exercise these valves (figure 14–16). Closing an OS and Y valve that is located near a sprinkler alarm mechanism will eliminate the flow of water to a sprinkler system.

Figure 14–16. Common on sprinkler systems, OS and Y valves are turned clockwise to close.

NOTE

A national color code has been adopted that significantly enhances the identification of utilities as follows:

    Red: Power
    Orange: Telecommunications
    Green: Sewers and Drains
    Blue: Water
    Yellow: Gas

However, when a valve has been closed, open the drain valve to drain residual water in the system as this amount of water can be noteworthy.

Interior operations include broken pipes and fused sprinkler heads.

Broken pipes. Use exterior shutoff valves (post indicator valves, OS and Y, and so on).

Sprinklers. Use appropriate sprinkler shutoff tools and/or replace fused sprinkler heads.

15 Ventilation Principles

Introduction

The word *ventilation* has several definitions, but can be easily defined when specifically applied to structural firefighting operations as procedures necessary to affect the planned and systematic re-direction and removal of fire, smoke, heat, and fire gases from a structure. However, this definition may be too simplistic when we consider that safe, timely, and effective ventilation operations are often performed in hazardous atmospheres and under dangerous conditions. In addition to the definition of ventilation, ventilation can be divided into two categories: theoretical and experiential. Theory provides a foundation on which to build an understanding of ventilation, and experience provides the practical opportunity to develop necessary expertise. In this chapter, we briefly look at the theory of ventilation; in subsequent chapters, we consider the practical applications of ventilation. Let's begin by looking at the theory and need for ventilation.

Ventilation Theory

Ventilation does not put out fires. It never has and probably never will. However, effective ventilation is often integral to the attack, control, and extinguishment of structure fires. Do all structure fires need ventilation? Obviously not. However, ventilation should often be of primary concern, particularly when fireground personnel must be subjected to heat, smoke, a multitude of hazardous fire gases, and flashover. To understand how fireground conditions can determine the importance of ventilation, let's consider several simple examples. An unconfined fire will draw cold air into its bottom as hot gases, smoke, and heated air rise vertically (figure 15–1).

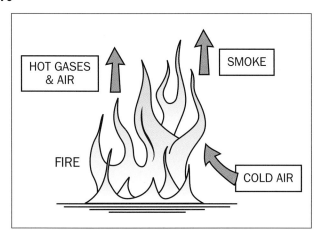

Figure 15–1. As an unconfined fire burns, cool air is drawn to its bottom as heated gases and smoke rise vertically.

Because the fire in our example is unconfined, it is able to draw as much cold air as necessary to sustain combustion, and the products of combustion are free to escape into the atmosphere and not hinder the combustion process. Combustion is most efficient when heat, fuel, and oxygen are present in the proper ratio, which can be easily altered by a lack of heat, fuel, and oxygen. Or the products of combustion are not able to escape and interfere with the amount of available oxygen. As a result of

the proper ratio of heat, fuel, and oxygen, our free-burning fire will emit a minimal amount of smoke and fire gases compared to a fire that cannot draw as much air that is necessary for a free-burning fire.

To illustrate what happens when our free-burning fire is transferred into a confined environment that will not allow a proper supply of oxygen to be available to the fire, and also not let the products of combustion escape into the atmosphere, let's transfer our free-burning fire into a simple, old-fashioned wood-burning stove. With the ash box door and damper open (figure 15–2), the free-burning fire will produce heat, some smoke, and fire gases that are dependent on the materials being burned.

The heated smoke and fire gases are lighter than the cooler air at the base of the fire, so they will rise vertically and exhaust out of the open damper. This allows cold air to be drawn into the bottom of the fire with hot gases and smoke rising vertically. Now, close the damper and ash box door. The rising column of hot smoke and fire gases will reach the highest level in the stove and begin to bank downward (mushrooming), forcing the cooler air to the base of the fire in the stove. As this process continues, a system of air circulation develops. Depending on the material that is burning, the temperature inside the stove can easily exceed 1,000°F. Additionally,

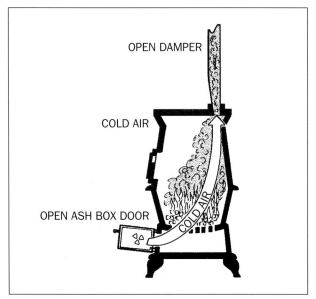

Figure 15–2. A fire in a common wood-burning stove also draws in cool air from the open ash box door as heat and smoke are exhausted through the open damper.

Figure 15–3. With all of the openings closed, the fire will pressurize the interior of the stove.

the level of carbon monoxide increases at a rapid rate while the free oxygen is depleted. Because the stove is airtight with all exterior openings closed, the fire will begin to smolder due to a further deprivation of oxygen. Soon, the stove will completely fill with hot smoke and fire gases, with temperatures possibly exceeding 1,300°F. At this stage, the stove is pressurized (figure 15–3), flammable carbon monoxide is abundant, temperatures are high, and oxygen levels are low.

In this instance, restoring the stove to a proper operating condition would require opening the damper first to exhaust the heated, flammable contents to the atmosphere (figure 15–4).

However, if the ash box door were opened first (figure 15–5), fresh oxygen would be suddenly introduced into the flammable, oxygen-deprived environment with a likely predictable result—enhanced conditions for a backdraft, or the fire would intensify and enhance conditions for a flashover of the contents within the stove.

Now let's apply the preceding example to a fire within a room in a single-family dwelling. As a fire burns in the room, the heat and fire gases rise to the highest available point (ceiling), then begin to bank down (similar to the stove example). If suppression personnel attempt to enter this environment in the early stages when the room is not completely filled with smoke and fire gases, and even if visibility is acceptable (figure 15–6), they will enter an environment that will vary from warm to hot, and encounter a thermal layer at the ceiling that will continue to approach its ignition temperature unless the fire is quickly extinguished.

Obviously, if the fire is not extinguished before the thermal layer reaches its ignition temperature, it can suddenly flash and create an untenable environment for any trapped occupants and suppression personnel.

However, let's assume that this process has continued to fill the room with heat, smoke, and fire gases and fireground personnel have just arrived

Figure 15-5. Opening the ash box door first can cause the oxygen deficient contents to explode through the door.

Figure 15-4. Opening the damper first will allow the heated, pressurized contents to exhaust vertically.

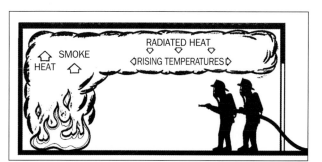

Figure 15-6. If suppression personnel approach a fire that isn't sufficiently ventilated, they can be subjected to rising temperatures and radiated heat.

on-scene. As these fireground personnel prepare to enter the structure, this process has continued to increase the heat and amount of smoke and fire gases within the room (figure 15–7).

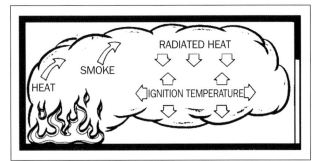

Figure 15-7. As a fire burns inside a structure, heat and smoke rise to the ceiling. As the temperatures increase, heat is radiated throughout the fire area.

Because this is a fire within a structure, and there are furniture and numerous other flammable items within the room, the increase in temperature within the room has also increased the temperature of the contents within the room (which are likely synthetic). As fireground personnel enter the structure, they are potentially entering an environment that is about to suddenly flashover unless the fire can be quickly extinguished and/or ventilation improves the interior environment before entry of fireground personnel.

Finally, let's apply the confined fire example to a multi-story building that is subjected to a fire on the first floor of the building. As the fire initially burns, heated gases and smoke also rise to the highest point within the structure, then begin to bank down to fill other areas (figure 15–8). If this process continues, the structure (although much larger than the preceding example) can also become filled with heated smoke and fire gases.

There will be poor visibility, and the buildup of heat can be intolerable, depending on the status of the fire.

In the three previous examples, suppression personnel were faced with a choice: Enter the structure to suppress the fire without any type of ventilation, and risk encountering a hot, flammable environment that is inherently dangerous; or use ventilation prior to entry to improve visibility and reduce the

dangerous concentrations of heat, smoke, fire gases, and minimizing the potential of flashover.

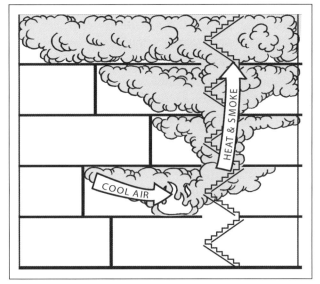

Figure 15-8. When heat and smoke have risen to the highest point, they will mushroom downward and contaminate other common areas within the building.

The Need for Ventilation

The simple examples in the previous section point to a common problem. Whether a fire and/or structure is small or large, fire gases and other natural by-products of combustion can be expected when encountering a fire within a structure. Common combustible materials contain carbon. When burned with an ample supply of oxygen, they produce carbon dioxide, which has little toxic effect on the human body (although carbon dioxide displaces oxygen-rich air and leads to asphyxiation). However, when the supply of oxygen is reduced below 21%, the production of other fire gases is accelerated. These gases include formaldehyde, carbon monoxide, acrolein, sulfur dioxide, ammonia, hydrogen cyanide, hydrogen chloride, chlorine, oxides of nitrogen, phosgene, isocyanates, and so on. Fire fatalities from the inhalation of noxious gases exceed deaths from all other fireground causes combined. Investigations of fire fatalities from exposure to toxic atmospheres have confirmed that carbon monoxide is a primary culprit, although recent studies are confirming that hydrogen cyanide may also be at least as deadly as carbon monoxide, and potentially more deadly. Other studies suggest that other fire gases are potentially just as lethal. (Chapter 5 has more information on the dangers of smoke.)

The inhalation of fire gases is a primary danger to building occupants, who obviously seldom have the benefit of protective equipment. Suppression personnel *are* equipped with protective equipment and SCBA, yet firefighters still die each year from breathing toxic gases. Firefighter deaths per year are slowly increasing among both active and retired firefighters. A portion of a three-part study on the relationship between cancer and firefighting by the Institute for Cancer and Blood Research for a major fire department indicates that the death rate from cancer among firefighters doubled from 17% to 34% between 1950 and 1980. Increased exposure to gases spawned by petrochemical fires could be the cause. Additionally, the incidence of brain cancer as a cause of death was 129% higher among firefighters than the general population, and a 30-year firefighter faces a one-in-three chance of dying from cancer compared to one in five among the general population. During the 1930s and 1940s, firefighters were primarily exposed to gases from wood and paper fires, but as petrochemical plastics and other synthetic materials have proliferated, so has the danger to fireground personnel. The flammability of components has increased, and toxicity has increased accordingly. With all of this in mind, ventilation should be considered as a primary tactic to reduce the deadly hazard to building occupants and fireground personnel. Let's restate that last sentence from a different perspective: Depending on conditions, ventilation operations should be considered on an equal priority, if not higher, than fire attack and search operations by the initial company.

Another danger posed by toxic gases to suppression personnel is the thermal layer. A thermal layer can be defined as a combination of smoke, fire gases, and heat that is capable of burning, often with significant intensity and a sudden rise in temperature. Let's consider three scenarios in which the thermal layer plays a critical part.

Scenario One: The year is 1960. The first-arriving engine company to a structure fire finds a fire in a two-story hotel. A quick size-up reveals that a working fire in one of the apartments on the first floor has charged the apartment and first-floor hallway with heat and smoke, resulting in minimal visibility and increasing heat. Two firefighters quickly develop an attack line into the hallway. Because it is 1960, the firefighters don't have the protection of SCBA and hoods. As they slowly and cautiously advance down the hallway toward the seat of the fire, they stay as low as possible and monitor the levels of heat with their ears. Reaching the open door of the involved apartment, they stop to one side and direct water toward the upper portion of the apartment. Then they quickly close the door. After a short while, they reopen it and observe that the fire has virtually been extinguished. Final extinguishment soon follows.

Scenario Two: It is 20 years later. With the same conditions as in the first scenario, the two firefighters develop an attack line into the hallway. It is 1980, so they don't have the protection of hoods; however, their SCBA enables them to advance toward the seat of the fire while they monitor the levels of heat with their ears. Approaching the open door of the involved apartment, the advancing firefighters notice that conditions in the hallway are rapidly deteriorating and that tongues of flame are showing at the ceiling. Reaching the open doorway, they direct a stream of water with a rotating movement into the apartment. Although the fire is knocked down, the disturbed thermal layer and expanding steam quickly envelop the firefighters, causing extreme discomfort and some steam burns.

Scenario Three: With the same conditions as in the previous scenarios, the two firefighters quickly develop an attack line into the hallway. It is 2010, so they both have state-of-the-art protective clothing, including SCBA and hoods. As they cautiously advance the attack line, their progress is enhanced by their protective equipment. In their desire to reach the seat of the fire quickly, their protective equipment masks the worsening conditions around them. As the firefighters approach the open door of the involved apartment, the apartment and hallway suddenly flash over, enveloping them. Depending on your age and experience, do any of the preceding scenarios sound familiar? Interestingly, these scenarios share a common thread. Let's consider where we have been with the fireground of yesterday and where we are potentially going.

Up until the mid-1960s, suppression personnel generally attacked structure fires within two basic parameters. First, firefighters of that era dealt mainly with conventional construction, which offered increased fireground time as compared to lightweight construction. Second, conventional materials were used in the interiors of structures. When these materials burned, they yielded smoke and fire gases that were slow to ignite and burn as compared to those of today. However, these conditions were primarily responsible for a condition commonly referred to as *backdraft* that was primarily a result of the combustion of conventional materials and oxygen-deficient atmospheres that were abruptly given oxygen when fireground personnel suddenly entered a structure with the aforementioned conditions. During this same time period, and in the absence of SCBA and hoods, suppression personnel were taught to stay low, use the oxygen around the water that was being discharged from the *spray* nozzle, and monitor the environment with their ears and the backs of their hands. Additionally, they would position the firefighter on the nozzle to the side of the entrance when applying water to the seat of the fire, and, for some confined fires, use the principle of indirect firefighting. This meant directing a water pattern toward the upper portion of a room, then closing the door and allowing the expanding steam to help aid extinguishment. As tactics changed over the years, suppression personnel began to advance more rapidly to the seat of the fire, and indirect firefighting was slowly replaced by improved attack tools and speed. (After all, why did you join the fire service? Was it to become an expert in salvage and overhaul operations, or to put first water on a fire?)

Throughout the 1970s and early 1980s, building construction increasingly employed lightweight materials, and petrochemical synthetics proliferated widely in residential and commercial structures. Unfortunately, when these materials burn, they yield heat, smoke, and fire gases that are significantly different from the by-products of conventional materials. Remember that smoke and fire gases are flammable substances that, when heated to the correct temperature, will readily ignite. Typical fire gases and smoke from synthetic materials can ignite at between 800°F and 900°F. By comparison, wood products produce 7,000 BTUs, polyurethane products produce 12,000 BTUs, and polystyrene products produce 18,000 BTUs. Personnel in this era began to encounter environments that were capable of burning two to three times hotter and faster than environments prior to the 1970s. In addition to the changing environment, members were also using protective equipment, which allows advancing attack personnel to withstand higher temperatures for a longer period of time. With it, they can more quickly advance to the seat of the fire. Given this ability, the old method of indirect firefighting, positioning attack personnel away from openings, became little more than a memory. Not surprisingly, a new fireground hazard known as *flashover* started to increase, and firefighter deaths and injuries increased while the frequency of backdrafts decreased.

Today, attack personnel are exposed to two significant hazards that personnel of the pre-1960 era never encountered—synthetic materials and flashovers. Synthetic materials are now common in carpets, furniture, wall coverings, electronic appliances, and all else. When these products burn, they produce smoke and fire gases that are more susceptible to flashover conditions. Hollywood popularized the term backdraft with its movie of the same name, and most fire service personnel can easily recognize the potential indicators of a backdraft. The common indicators of a flashover are less well known. Still, if a class of students from the fire service with various backgrounds and experience are asked whether they have ever seen a backdraft, 2% to 5% will normally raise their hands. When asked about flashover, approximately 70% will raise their hands. Why, then, is flashover a common modern problem, and backdrafts are not, particularly when new structures are tighter than they have ever been, which should enhance backdraft? The answer becomes clear when you consider the changing content of modern buildings and the increased use of synthetic materials. The lower ignition temperatures of plastics and their resultant smoke and fire gases can easily flash over before a typical fire that is confined to one room (70% of our structure fires) can develop the pressurized, oxygen-deficient environment necessary to support a backdraft.

Each year, the United States averages 100 deaths in the fire service. Of that number, approximately 15 normally occur as a result of brush-type incidents. A noteworthy percentage of the remaining 85 deaths occur from flashover, and that percentage is slowly rising. Studies conducted in Sweden and England indicate that fire gases and smoke are more dangerous than the actual fire itself. If fire gases and smoke encountered by the fireground personnel of today have the potential to flashover more easily than the environments of yesterday, is it possible that smoke and fire gases encountered *before* reaching the seat of a fire might be a higher priority to the safety of fireground personnel than the fire itself? Also, at a time when fire environments are enhancing the conditions for flashover, the modern firefighter is totally encapsulated in protective equipment. This has allowed suppression personnel to advance to the seat of a fire quickly, while minimizing their ability to evaluate their environment. If a typical modern firefighter is asked what is used as an indicator to evaluate the fireground environment, a blank stare is a common response. A common tendency is for firefighters to use their protective equipment as an offensive tool rather than a defensive tool. When protective equipment is used as an offensive tool, it allows attack personnel to advance too quickly into environments that can readily flashover.

Three additional concerns related to the fireground environment should be carefully evaluated. Although the initial company arriving at a structure fire will generally give first consideration to attack and search operations, ventilation should be added to that list of initial considerations in the interest of firefighter safety. In some cases, depending on conditions that are encountered, ventilation should be a first priority. Also, if the modern thermal layer and potential of flashover have become greater hazards, what form of ventilation might best reduce the danger? This question is best answered by briefly considering horizontal, vertical, and pressurized ventilation. Although horizontal ventilation is the most popular form due to its safety and ease of implementation, it doesn't take advantage of the natural vertical path of heated gases and smoke. It also has minimal effect on the hottest portion of a room (the ceiling), where most flashovers originate. Therefore, horizontal ventilation clears the region around the opening but has limited potential to prevent flashover. Vertical ventilation takes advantage of the natural path of travel of heated contaminants. It ventilates that portion of a heated environment that needs it the most. However, it also requires more time, staff, and equipment to accomplish.

Positive-pressure ventilation can enhance horizontal and vertical ventilation operations, but it must be used judiciously when fireground operations favor using it. In summary, choose your method of ventilation carefully, and don't rely merely on opening windows and doors to clear a contaminated area as you may be tragically disappointed.

Another fireground concern has to do with the human factor. Ask yourself whether the fireground environment has decreased your endurance and ability to make sound decisions. When the human body is fully encapsulated by protective equipment and is subjected to a hot working environment, the body will try to ventilate itself by sweating. Because there is minimal airflow over the skin, the body will continue to sweat to provide some degree of relief from rising body temperatures. Unfortunately, this results in *sweat-wetting*, which reduces the fluid level in the body, decreasing strength, stamina, and the ability to make sound decisions. This is why numerous departments incorporate some form of ventilation and rehabilitation operations into their post-fire operations.

Benefits of Ventilation

With the preceding review of the need for ventilation on the modern fireground with a strong emphasis on safety, let's consider seven basic tools for how ventilation can assist firefighting operations.

People tool

When heat, smoke and fire gases are directed from the interior of a structure and away from occupants and/or their means of egress, search and/or a rescue potential is enhanced by providing an improved atmosphere for occupants and personnel engaged in these operations. Improving the atmosphere will also enhance the ability of suppression personnel to quickly enter a structure to affect suppression operations.

Exposure protection tool

Reducing the quantity of heat, smoke, fire gases, and fire will relieve dangerous concentrations of heat created by convection and radiation, thereby reducing the exposure of fire and its by-products to other portions of a structure.

Confinement tool

Reduced concentrations of heat, smoke, and fire gases will assist suppression personnel who are confined inside the affected structure by improving visibility and reducing interior temperatures. This will assist in the development of hoselines, accurate placement of interior personnel, locating the seat of a fire, prompt extinguishment, simplified search operations, and so on. Remember, inhalation of products of combustion is still considered a primary cause of injury to firefighters and death to civilians.

Visibility tool

When an atmosphere that is charged with heat, smoke, and fire gases is reduced, visibility will improve and allow suppression personnel to conduct interior operations in a hazardous environment in a timelier manner.

Backdraft and flashover tool

Hazardous environments can be dangerous and/or fatal to occupants and fireground personnel who routinely enter this type of atmosphere to conduct searches and extinguish fires. Ventilation operations can minimize the factors that are responsible for backdrafts and flashovers and significantly reduce or delay their occurrence.

Overhaul and salvage tool

A combination of reducing contaminants and prompt extinguishment will result in a reduction of fire damage. Less than one-half of the nation's fire loss is direct fire loss; water, smoke, and heat do much of the rest.

Safety tool

Of all the reasons for ventilation, firefighter safety is the most essential reason for ventilation, particularly when the basic element of smoke and its inherent hazards are examined.

Types of Ventilation

Let's conclude this chapter by considering how a common ventilation operation can assist firefighting operations. In figure 15–9, for example, a fire is burning in one room of a structure. The fire and its by-products have risen to the ceiling.

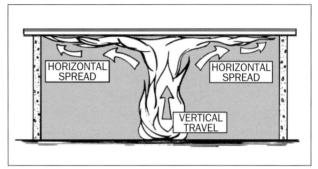

Figure 15–9. Horizontal spread is enhanced when vertically traveling fire meets a horizontal surface, such as a ceiling.

The fire is beginning to spread horizontally, and its by-products will soon mushroom downward, creating a dangerous environment for advancing personnel. In figure 15–10, however, an opening has been created over the fire, allowing the smoke, fire gases, heat, and thermal layer to clear vertically.

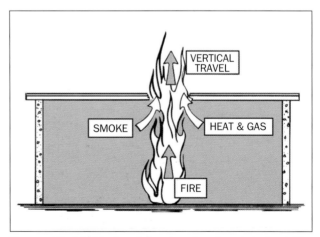

Figure 15–10. Horizontal extension is minimized if a ventilation opening is created over a fire. Additionally, heat, smoke, and fire gases will be reduced in the contaminated area.

This operation dynamically benefits firefighting operations by directing heat, smoke, and fire gases from the interior of a structure and away from occupants or their means of egress. The improved atmosphere also enhances search operations and allows personnel to enter a structure more quickly to initiate suppression operations with a reduced chance of a flashover. A combination of reducing contaminants and prompt extinguishment will result in a reduction of fire damage.

If the modern thermal layer and potential of a flashover has increased in frequency and degree of hazard, and the content of smoke from synthetic compounds can present a significant hazard that firefighters must frequently encounter, what form of ventilation has the potential to most effectively reduce the danger of these considerations? This question is best answered by briefly considering the following five different types of ventilation (natural, hydraulic, horizontal, pressurized, and vertical) and using the strengths of each type to best achieve the desired result depending on each type of incident.

Natural

Natural ventilation is best described as ventilation that is created by fire as it burns through portions of a structure or uses natural openings such as windows, doors, and so on. When present, this type of ventilation is not initiated by fireground personnel, but must be evaluated to determine its effect on an environment/fire.

Hydraulic

From a simplistic perspective, hydraulic ventilation is normally accomplished by a firefighter standing in a contaminated environment and using a spray/fog nozzle to direct contaminants from the interior to the exterior through available openings such as a window. However, although this method can be beneficial, it requires a firefighter to stand in the contaminants while they are being exhausted to the exterior. Obviously, horizontal, pressurized, and vertical ventilation do not normally require this constraint.

Horizontal

Although horizontal ventilation is the most popular form of ventilation due to its ease, potential effectiveness, and speed of implementation, it does not take advantage of the natural path of travel (vertical) of heated gases and smoke. It can also have a minimal effect on the hottest portion of a room (ceiling), where most flashovers originate. Therefore, horizontal ventilation will ventilate that portion around the opening (open door or window) but could have a minimal effect on a flashover.

Pressurized

Pressurized ventilation (for example, positive pressure ventilation) can enhance horizontal and vertical ventilation operations, but must be used when fireground operations favor its implementation.

Vertical

Vertical ventilation will take advantage of the natural travel path of heated contaminants and ventilate that portion of a heated environment that needs it the most (ceiling area). However, vertical ventilation takes additional time, staffing, equipment requirements necessary for completion, and can place ventilation personnel in a dangerous location (over a fire).

The method of ventilation must be chosen carefully. Do not rely on just opening windows and/or doors to ventilate a contaminated area to improve the interior environment and minimize an existing thermal layer to reduce the potential of flashover. Remember that the type of ventilation is critical to the overall effect of the intended operation. The following chapters expand on the different types and methods of fireground ventilation operations.

Training Resources

The following resources can provide additional information and/or another viewpoint for the subjects discussed in this chapter:

- Ventilation: Still Needed, Donald J. Loeb, *Fire Chief Magazine*, November 1996.

- *Fire Engineering's Handbook for Firefighter I and II*, chapter 15, Ventilation, PennWell, 2009.

- For additional applicable articles dealing with smoke, see Training Resources in chapter 5, Reading Smoke.

- Primary Ventilation: A Review, Tom Donnelly, *Fire Engineering*, June 2007.

16 Pressurized Ventilation

Introduction

From the acceptance perspective, this chapter has been the most fascinating to review and update since the initial printing of *Truck Company Operations* in 1998. During the early 1970s, positive-pressure ventilation (PPV) was first used by a few departments in California. As a result of the positive benefits, PPV was introduced to the fire service on a national level in the early 1980s to tepid acceptance. It was generally believed that the introduction of pressurized air to a fire would surely accelerate a fire, and additionally enhance the extension of fire within a structure. However, as more departments either used or conducted experiments with PPV, it became obvious that PPV was a significant improvement over the commonly used form of pressurized ventilation—negative pressure. As the use of PPV increased during the late 1980s and into the early 1990s, numerous credible scientific tests were conducted by a wide range of institutions (North Carolina Department of Insurance, the United States Coast Guard, University of Central Florida-College of Engineering, China Lake Weapons Center, Northrup Aviation, to name a few) and numerous departments such as Austin Fire Department, Salt Lake City Fire Department, Portland Fire Bureau, Ventura County Fire Department, Forth Worth Fire Department, and others with positive results and comments. Not surprisingly, the majority of fire departments in the United States used PPV during the latter portions of the 1990s, along with an increasing number of European fire departments, specifically England and Germany. Recently, an interesting scientific perspective has been introduced by the National Institute of Standards and Technology (NIST), U.S. Department of Commerce, in conjunction with the Toledo Ohio Fire Department, Chicago Fire Department, and the Fire Department of New York, who have taken engineering principles and applied them to fire service PPV tactics with positive results. It is not the intent of this chapter to summarize all of the tests conducted by NIST (April 2008) and other previous organizations, but to simply overview PPV from a practical perspective and list additional information in Training Resources that can be reviewed, if necessary. Needless to say, PPV is now an accepted form of ventilation that can be used during numerous fireground operations. Let's briefly review the basics of both negative and positive pressure ventilation and then take a look at some common and advanced applications of PPV. It should be noted that for this chapter, the term *blowers* is used to denote fans and ejectors.

Pressurization Ventilation Methods

For this chapter, there are three different forms of ventilation: natural, negative-pressure, and positive-pressure ventilation.

Natural ventilation

As a fire burns within a structure, the interior fills with hot products of combustion that rise and fill all of the available spaces. A simple method to ventilate the products of combustion is to use natural convective currents, easily done by opening doors, windows, and other such openings. Although this method may perform satisfactorily, it can depend, among other factors, on the proximity of the opening to the contaminants to be removed, the ability of the contaminants to travel unobstructed to the opening, the number and size of the openings, wind, humidity, and the temperature differential between the interior and exterior of the building. Although natural ventilation is viable within certain limits, portable blowers can dramatically overcome these shortcomings, forcing contaminants through preselected openings.

Negative-pressure ventilation

Assume that the room in figure 16–1 will be ventilated. In this example, the room is filled with various products of combustion. The warmer gases have risen toward the ceiling, and the cooler gases have settled toward the floor. The door and window are closed.

Figure 16–2 shows the same area being ventilated by means of negative pressure. The door has been opened and a blower placed inside the building to exhaust the contaminants, coaxing them to the exterior by creating suction within the room. By opening the window, the by-products will be replaced with fresh incoming air.

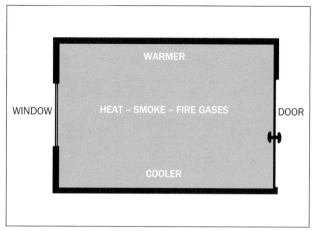

Figure 16–1. When contaminants fill a room, the lighter, hotter gases rise to the top, and the cooler, heavier gases settle to the bottom.

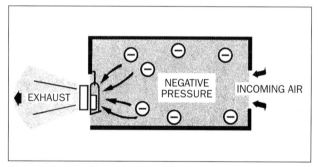

Figure 16–2. Negative pressure involves placing a blower inside a room to exhaust the contaminants to the exterior.

Although this method can prove satisfactory, it has several disadvantages, not the least of which is that personnel must be exposed to hazardous contaminants to position the blower. Contaminants are drawn through the blower, creating a need for additional equipment cleanup and maintenance. Smoke from a wood fire always contains some unburned gases and a fog of unburned tarlike liquids. When these materials come in contact with blowers, they form creosote, which is often difficult to remove from blowers. Also, blowers placed in doorways or hallways can block pathways in and out of the building. To position blowers effectively, you must often suspend them in doors or windows using straps, ladders, or other accessories. Blowers placed inside buildings can add to the overall noise and confusion, hindering communications. Finally, interior blowers aren't efficient at removing contaminants from the tops of rooms. Air will follow the path of least resistance, which is normally a more or less straight line from the fresh air inlet to the blower. This limits the flow of air in the regions near the ceiling.

Positive-pressure ventilation

Figure 16–3 shows the same room under positive-pressure ventilation. This method forces clean, fresh, pressurized air into the room, creating a positive pressure within. As in a balloon, the pressure will be equal at the top, bottom, and corners of the room. When the window is opened, the contaminants from all parts of the pressurized room will exhaust to the exterior.

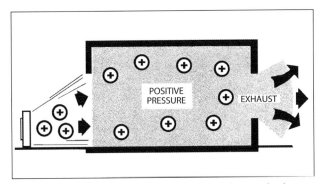

Figure 16–3. Positive pressure involves placing a blower outside a room to exhaust the contaminants to the exterior.

Compared to negative-pressure ventilation, positive-pressure ventilation has a number of advantages. Personnel aren't normally exposed to hazardous interior contaminants while positioning exterior blowers. The contaminants aren't drawn through blowers, thereby reducing the need for cleanup and maintenance. Also, doorways, windows, and halls don't need to be blocked by blowers, and exterior blowers aren't dependent on additional equipment or accessories for setup and operation. Because of this, fewer personnel are required to deploy them. Exterior blowers have minimal impact on interior noise, yet compared to negative-pressure ventilation, PPV is at least twice as efficient at removing contaminants from a building because it can clear the air in all corners of a given room.

Implementing Positive Pressure

As an initial summary of implementing PPV, let's look at the "Ten Commandments of PPV":

1. Have a goal and know how to accomplish it.

2. Have charged lines in place.

3. Determine the location of the fire and the potential airflow route.

4. *Make the exhaust opening first.*

5. Seal the door with a cone of air (if possible).

6. Do not block the entrance opening.

7. Use the correct size and number of blower(s) for each application.

8. Ensure the proper exhaust opening size (if possible).

9. When appropriate, use the principle of sequential ventilation.

10. Consider carbon monoxide when using gasoline-powered blowers.

Effectively implementing PPV depends on controlling the flow of pressurized air between the entrance and exhaust opening (figure 16–4). Of course, proper training of and communication among fireground personnel are also crucial. The following sections detail the specific implementation of PPV.

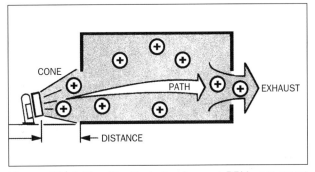

Figure 16–4. To effectively implement PPV, you must control the entrance and exhaust openings as well as the airflow route.

Entrance opening

The correct type of blowers should be positioned so that a cone of pressurized air completely covers the entrance opening. This can be accomplished with single or multiple blowers.

Single blower. Place a single blower so that the cone of pressurized air just covers the entrance opening as in figure 16–5. Optimum placement depends on the size of the entrance opening and the size of the blower. Smaller blowers need to be placed farther away from entrance openings.

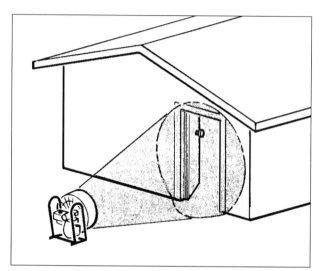

Figure 16-5. It is best to position a single blower so that a cone of pressurized air covers the entrance opening.

Larger blowers can be placed closer because they produce a larger cone of pressurized air. Blowers of the same size but of different makes can produce cones of different sizes, so it is important to know the operational capabilities of your equipment. Regardless of the size of a blower, most are designed so that they can be tilted back about 20 to 30 degrees, enhancing their ability to achieve a proper seal with a cone of pressurized air. Occasionally you may encounter conditions that won't allow you to place the blower at a correct distance. For example, consider a raised porch that has a width of 4 feet—too narrow for a blower that requires 6 or 7 feet. In such cases, position the blower as far away as conditions will allow. If a blower is being used for smoke removal only, the smoke can recirculate from

the unsealed portions of the entrance to the blower, then back into the contaminated area, lengthening the time for ventilation. *Although it is important to seal the entrance opening with a cone of pressurized air, doing so isn't absolutely necessary.* If the blower is being used for fire attack and fire is near the opening, the flames themselves can recirculate, creating a hazard to personnel and the blower. Interestingly, the efficiency of PPV in such cases isn't compromised; only the time factor and potential safety considerations of the operation.

Multiple blowers. Multiple blowers can dramatically increase the volume of the airflow, reducing the time to complete a given operation. For standard entrance openings (such as door openings of 3 by 6½ feet), maximum effectiveness can be achieved by placing two blowers in line with each other. In figure 16–6, Blower A has been positioned about 2 feet from the entrance. This ensures that all of the pressurized air from the blower will enter the building, while still allowing sufficient room for ingress and egress.

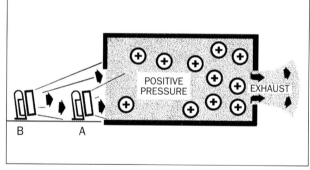

Figure 16-6. Multiple blowers can be placed in a series configuration to increase efficiency.

Blower B, positioned behind Blower A, covers the entrance with pressurized air, forces pressurized air into the building, and increases the capacity of Blower A by approximately 10%. The proper location for Blower B is determined by the distance necessary to cover the entrance with pressurized air. If two blowers of *unequal* size are used in series, place the larger one about 2 feet back from the entrance and the smaller one behind the larger one. This way, the larger blower will provide most of the pressurized air for ventilation while the smaller one covers the entrance. Some blowers are designed to stack vertically.

Place stacked blowers approximately 2 to 3 feet back from an entrance for maximum efficiency and ease of ingress/egress (figure 16–7). Use the larger cone of pressurized air to seal the entrance. For standard entrances, multiple blowers in parallel (side by side) are less effective than multiple blowers in a series or stacked configurations.

Figure 16-7. Some blowers are designed to be stacked.

Figure 16-8. Large entrances may require placing blowers in a parallel configuration to seal the width of the opening.

For large entrances, however, multiple blowers in parallel are better able to cover an opening with combined cones of pressurized air (figure 16–8). The size of the opening will dictate the number of blowers necessary.

Remember that some openings, such as vertical loading dock doors, can be reduced in size by closing them partway. Depending on the number of blowers available, large areas may be effectively ventilated by using a combination of parallel (proper coverage of the opening) and series (increased volume) blowers. Enclosed areas that lack adequate channels for an exhaust can be effectively ventilated by using multiple blowers as depicted and delineated in figure 16–14.

Airflow route

It is imperative that the flow and path of pressurized air between an entrance and exhaust opening be controlled to achieve maximum ventilation. If pressurized air is directed from an entrance to an appropriate exhaust opening without being diverted to other openings, contaminants will be removed in a minimal amount of time. Simultaneously opening unwanted windows and doors won't facilitate successful PPV operations.

Exhaust opening

Exhaust openings can be selected to provide horizontal or vertical channels. *Always create the exhaust opening first; then pressurize the structure or area to be ventilated.* Otherwise, where would the pressurized airflow go until the exhaust opening has been created? Generally, PPV is most efficient when the exhaust opening is between three quarter and one-and-three-quarter times the size of the entrance opening. This depends on the number of blowers, the cubic feet per minute (cfm) rating of the blowers, and the size of the area to be ventilated. For example, apply this general guideline to a single-story, single-family dwelling of about 1,800 square feet. A single 18-inch blower powered by a two- to three-horsepower engine would be most efficient when used with exhaust openings that are three-fourths to about the same size as the entrance opening. By comparison, a single

18- to 24-inch blower powered by a five horsepower engine or greater would require an exhaust opening between one and one-and-a-half times the size of the entrance. Multiple 18- or 24-inch blowers powered by approximately five horsepower engines, in series or stacked, would require exhaust openings of up to one-and-three-quarter times the size of the entrance. In summary, do not spend an inordinate amount of time worrying about the proper exhaust opening size, as you may be limited to the available openings (windows, doors, etc.).

Carbon monoxide

If a gasoline-powered blower is being used and an exhaust odor is noticeable inside the area to be ventilated, it indicates that the exhaust opening isn't large enough. An insufficient exhaust opening allows pressurized air and blower exhaust to accumulate inside the area to be ventilated. The exhaust odor should disappear by increasing the size of the exhaust opening. This also reduces the amount of carbon monoxide in the area being ventilated. Because carbon monoxide in the exhaust of gasoline-powered blowers can be present, remember that water, hydraulic, and electric blowers produce pressurized air without CO as a by-product. Consider using such blowers at incidents in rest homes, hospitals, and anywhere else that respiratory problems among the occupants are a concern.

As discussed in chapter 17, some manufacturers produce flexible tube extensions for gasoline-powered blowers that divert exhaust gases away from the area being ventilated.

Weather

Temperature, humidity, snow, and rain don't have any appreciable affect on PPV. Although cold, damp weather may limit the ability of smoke to rise, it won't limit the ability of blowers to move contaminants horizontally or, in most cases, vertically. Wind can have an adverse affect on PPV, but its affect is a matter of direction and velocity. As in any ventilation operation, maximum efficiency can be obtained by using the prevailing currents to an advantage. It is best to pressurize a structure on the windward side and exhaust the contaminants to the leeward side. If it isn't possible to use the prevailing wind,

PPV has proved effective against winds of up to about 25 miles per hour. Above that, efficiency will be accordingly reduced.

Training and communication

The key to effective PPV is a matter of controlling the entrance and exhaust opening, as well as the path of the interior airflow. These factors can only be achieved in their proper relationship if all of the involved personnel have been properly trained and are aware of the goal. Remember, the fundamental concept of PPV is by creating a controlled flow of air through a given area, the laws of physics dictate that contaminants will follow. If you can't control the flow of air, you might not control the results, some of which may not be anticipated.

Ventilating Structures

To achieve maximum efficiency with PPV, don't open the structure or ventilate all of the rooms simultaneously, as is common practice in some structure fires. Multiple openings can diminish the flow within each room, increasing the time necessary for ventilation. Depending on the size of the area, it may be possible to ventilate several rooms at once with multiple blowers. The most effective method for most structures, however, is by ventilating the rooms sequentially. This means targeting specific areas and ventilating them in a sequence that maximizes the flow of pressurized air while minimizing the time required. For example, assume the dwelling in figure 16–9 to be charged with smoke. All of the exterior doors and windows are closed, and there is no appreciable wind.

A blower has been positioned to cover the front door with pressurized air. An effective sequential PPV plan would be as follows:

1. To clear the kitchen, living room, and dining room, open Exterior Door 1 in the kitchen and ensure that both of the interior doors to the kitchen are open. When these rooms have cleared, close the exterior kitchen door.

2. To clear Bedroom 1, ensure that the bedroom door is open; then open the Bedroom Window 2. When cleared, close the bedroom window or door.

3. To clear Bathroom 1, ensure that the bedroom and bathroom doors are open; then open the Bathroom Window 3. When cleared, close the bathroom window or bedroom door. An alternative method is to ventilate the bedroom and bathroom simultaneously. To accomplish this, ensure that the bedroom and bathroom doors are open, Bedroom Window 2 is closed, then open the Bathroom Window 3. When both rooms have cleared, close the bathroom window or bedroom door.

4. To clear Bedroom 2, ensure that the bedroom door is open; then open Window 4. When cleared, close the window or door.

5. To clear Bedroom 3 or Bathroom 2, repeat the steps for Bedroom 2 using the appropriate windows.

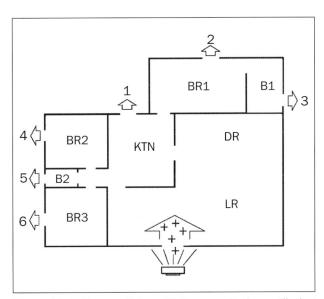

Figure 16–9. Sequential ventilation results in ventilating specific areas in a strategic order.

Basements

If a basement has multiple openings that can be used for entrance and exhaust openings, position a blower to cover the entrance opening with pressurized air, and use the appropriate channel for the exhaust. If a basement lacks suitable exhaust openings, several methods can help provide them. If a frame-wood structure has shiplap siding, the two bottom shiplap boards can easily be removed, exposing the opening between the ends of the floor joists and between the floor and plate. It will be necessary to remove the basement ceiling to complete the opening. If that method isn't practical, create an appropriate exhaust opening on the floor above the basement, and then position a blower to pressurize the floor above the basement. Next, place an additional blower in the interior doorway to the basement. This will pressurize the basement and force contaminants out of the upper portion of the doorway. The by-products will then be carried out of the structure by the flow of pressurized air across the floor above the basement that was previously pressurized (see figure 16–14).

Dwellings

Raised porches are commonly encountered in residences. If the size of a porch doesn't allow a blower to be placed properly, move the blower as far back as the porch will allow. If the door isn't properly sealed, contaminants can exhaust from the unsealed portions and be carried back into the structure. Although this condition can lengthen the operation, the overall effectiveness of the PPV operation isn't significantly compromised. Additionally, remov-

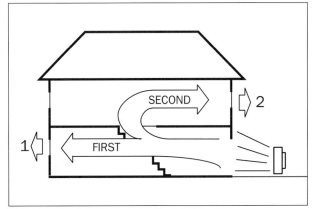

Figure 16–10. In multi-story structures, start at the lowest level and ventilate toward the top of the structure.

ing screens will improve the viability of window exhaust openings by at least 50%.

Single-story dwellings. Use sequential ventilation and any prevailing wind to ventilate horizontally. These structures offer excellent opportunities to experiment with the number and size of blowers to determine their effectiveness in areas of varying sizes.

Multiple-story dwellings. If a multiple-story dwelling needs to be ventilated as in figure 16–10, always start at the lowest level and ventilate toward the top. To ventilate the first floor, ensure that all of the exterior windows on the upper floor are closed, or that a stairwell door to the upper floor is closed.

Position a blower at an appropriate entrance opening; then sequentially ventilate the contaminated areas on the first floor. This will provide maximum pressurized air for ventilation on the first floor and no flow of air on the second floor due to a lack of an exhaust opening. To ventilate the second floor, leave the blower in the same position and ensure that all of the exterior windows and doors have been closed on the first floor. If a stairwell door has been closed, open the door and sequentially ventilate the contaminated areas on the second floor.

Multiple-occupancy structures

Apartment buildings, condominiums, and other multiple-occupancy dwellings can have single or multiple floors, as well as enclosed central hallways, which provide access to numerous rooms within the building. The hallways may be of considerable length, and there may be fire doors at various intervals along the way. If fire doors are present, they may be opened to allow pressurized air to travel along a chosen route, or they may be closed to isolate specific areas. This may serve to keep contaminants from spreading to uncontaminated areas or to divert pressurized air to a specific area. Positive pressure can be effective for distances of over 1,000 feet if the airflow between the entrance and exhaust openings is controlled, resulting in a directed flow and minimal loss. Also, using blowers of five horsepower in series or stacked will dramatically enhance effectiveness.

If the flow of pressurized air is controlled, physics dictates that heat, smoke, and fire gases must

follow. Multi-story residentials can also be effectively ventilated with positive pressure. Pressurize the appropriate hallways; this can be accomplished by directing pressurized air from the appropriate entrance opening to the selected hallway from a pressurized stairwell. Next, ventilate a pressurized hallway by directing pressurized air from the stairwell to the contaminated hallway. Finally, ventilate contaminated rooms or other areas that are common to a pressurized hallway.

Consider the structure in figure 16–11 to be a recently constructed three-story apartment complex with a center hallway on each of the three floors, an enclosed stairwell that is common to each floor, and fire doors separating each floor from the stair shaft.

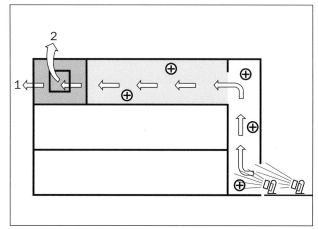

Figure 16–11. Positive pressure is very effective on multi-story center-hallway structures, including apartment buildings.

A fire has occurred in an apartment at the end of the third-floor hallway, resulting in the apartment and third-floor hallway being charged with smoke. Position a blower outside the building to cover the entrance to the appropriate stair shaft with pressurized air. Ensure that the fire doors to the first and second floors are closed; then open the third-floor hallway door to the pressurized stair shaft. With an open exhaust opening at the end of the third-floor hallway, the hallway will be cleared of contaminants. After the hallway has been cleared, close the exhaust opening at the end of the third-floor hallway and open the door to the contaminated apartment. Sequentially ventilate the apartment of all contaminants. If an exhaust opening can't be opened at the end of the hallway, open the door to the contaminated apartment and ventilate the hallway through that apartment. This is most effectively done by opening a patio door, a large window, or by using a room with a great number of windows as the initial exhaust opening. When the hallway and that portion of the apartment have been cleared, sequentially ventilate the other portions of the apartment.

Depending on the floor plan of a given building and the location of the fire, opposing blowers can be used to remove contaminants in the hallway and an apartment (figure 16–12). Blowers of equal or unequal ratings can be used effectively for this operation.

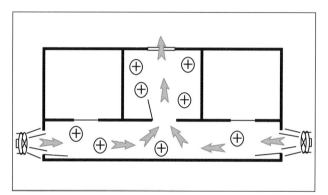

Figure 16–12. Opposing blowers can be used to force contaminants from a hallway and an apartment.

Multiple occupancies that may mandate creative techniques are the older, two-story, center-hallway buildings. These buildings often don't have fire doors,

resulting in vertical paths between the first- and second-floor hallways.

Depending on the location of smoke in the hallways, normal positive-pressure operations may cause smoke to circulate between the first- and second-floor hallways before exiting the building. To alleviate this condition, use two blowers as illustrated in figure 16–13 to enhance the pressurization and equal flow of air on each floor. The removal of smoke can be accomplished as follows.

First floor only. Open Door A to clear the first-floor hallway. Place Blower 1 outside the structure to seal the entrance door with a cone of pressurized air. If necessary, close Door A and open the entrance door and appropriate window of any contaminated apartment on the first floor to clear that apartment.

Second floor only. Open Door or Window B to clear the second-floor hallway. Place Blower 1 outside the structure to seal the entrance door with a cone of pressurized air.

Assuming that Door C enters to a contaminated apartment on the second floor, close Door or Window B and open Entrance Door C as well as any appropriate window of the contaminated apartment to clear the apartment. During the entire operation for the second floor, Door A remains closed.

First and second floor. Open Door A to clear the first-floor hallway. Place Blower 1 outside the structure to seal the entrance opening with a cone of pressurized air. Place Blower 2 as depicted to simultaneously enhance the pressurization of the first and second floor. Several options then become available,

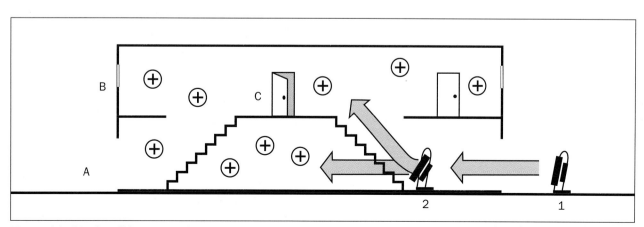

Figure 16–13. Conditions may dictate using multiple blowers to enhance pressurization and airflow in buildings with multiple floors that aren't separated by fire doors.

depending on the size of the building and the location of the stair shafts. The first option is to close Door A and open Door or Window B to clear the first and second-floor hallway. The second option would be to open Door A and close Door B to clear the first- and second-floor hallways simultaneously. The third option would be to close Door A and Door or Window B; then sequentially open the windows and doors of any contaminated apartments on either floor.

Commercial buildings

Commercial buildings are diverse in their size, height, and type of occupancy. However, consider the following factors when contemplating PPV operations.

Sealed areas. Typical modern office buildings present a unique challenge. Unless windows are broken, the offices in these structures normally have only one opening for an entrance and exhaust. To ventilate these areas, use a blower to provide a flow of air past the contaminated opening. This blower can be located outside or inside to provide pressurized air to the interior of the building through a hallway and past the area to be ventilated. Place an additional blower *in* the doorway of the area to be ventilated, as shown in figure 16–14. The blower will create a positive pressure within the area to be ventilated and force the contaminants out of the upper portion of the entrance. The pressurized air flowing past the entrance will force the contaminants to the exterior of the building. If the flow of air past the blower positioned in the entrance opening is insufficient, the exhausting contaminants from the entrance may recirculate.

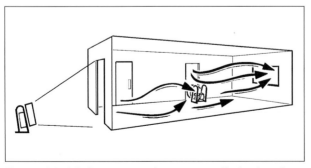

Figure 16–14. Structures that present limited options for entrance and exhaust openings can be ventilated by multiple blowers as shown.

In this case, use another blower to increase the flow of pressurized air past the contaminated opening.

Area. Depending on the type of building, determine the effectiveness of removing contaminants horizontally or vertically, as well as the appropriate path. Large structures or large areas within structures require increased air volume to remove large quantities of contaminants. Consider using larger blowers for these applications. Multiple blowers in series or parallel blowers can provide additional airflow, enhancing the removal of contaminants. Use multiple blowers to pressurize large entrance openings. Some large doors can be partially closed to facilitate pressurization. Remember that office areas should be considered prime locations for ventilation if no life hazards are present. If the records of a business are saved (accounts receivable), so is the financial cornerstone of the business. Some commercial occupancies have large, open areas that are normally difficult to ventilate. These occupancies require a combination of blowers capable of providing adequate volumes of pressurized air, not to mention an understanding of PPV principles. When possible, large areas of a building should be divided into smaller areas by closing partition doors and then sequentially ventilating each contaminated area. Large structures that are composed of smaller areas such as stockrooms, workstations, offices, and the like should be ventilated using sequential techniques in a planned, coordinated operation. When large ventilators, broken skylights, or other openings in a roof will negatively affect PPV operations, consider using them as vertical channels. Close doors and windows below these exhaust openings to ensure that the flow of pressurized air is maximized and directed to the appropriate openings.

Multiple floors. Buildings with multiple floors can be ventilated by sequentially ventilating each floor. Position blowers on the exterior of the building and pressurize the opening to a stair shaft that is common to the contaminated floors. As always, start on the bottom floor and work upward (figure 16–15). Use the pressurized stair shaft to channel pressurized air to each floor as necessary. Sequentially ventilate each floor by opening a door to the pressurized stair shaft and an appropriate window as an exhaust opening.

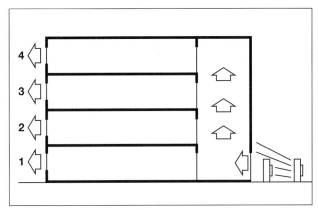

Figure 16–15. In multiple-story structures, start at the lowest level and ventilate toward the top of the structure.

High-rise buildings

Of the various types of structures encountered by the fire service, high-rise buildings (defined as buildings as over 75 feet in height) can be the most challenging, in part because most suppression operations in them occur above the ground. Therefore, logistics and the time required to implement fireground operations become primary considerations. The migration of smoke and toxic by-products throughout a high-rise can often present a greater hazard to life than the spread of fire itself. During fires, the migration of air throughout a high-rise frequently results in the spread of smoke and toxic gases to areas far removed from the actual fire. This can render vertical and horizontal escape routes impassable and hamper fire suppression efforts. Since smoke normally travels upward, the floors above become prime exposures for stratification and mushrooming.

Inherent factors such as poke-through construction, HVAC systems, stair-shaft pressurization systems, vertical stair shafts, and curtain construction can enhance the spread of fire and smoke throughout these types of buildings. If a fire is above the reach of aerial devices, the only methods of vertical access are the elevators and stair shafts. Stair shafts can become contaminated with smoke, heat, and fire gases, and elevators can be dangerous to personnel if they inadvertently travel to the fire floor, so safe access to upper floors can be challenging.

Accordingly, high-rises present unique and varied ventilation problems. Never attempt a haphazard approach to ventilation without specific plans for the effective use of personnel and equipment. Additionally, when ventilation operations are necessary, it is essential to have a clear understanding of the parameters imposed by high-rise buildings.

The following characteristics are common to high-rises and can affect the travel of contaminants within, assisting or hampering ventilation operations:

Construction. As a result of various construction factors, high-rise buildings will leak between floors and from floors into stair shafts. Smoke can travel between floors through unsealed openings, spaces between the floors, and poke-through pipe alleys. Additionally, the pressure generated by a fire (up to three times that of atmospheric pressure) can force smoke through these openings, contaminating areas on other levels.

Doors. Passageways created by open doors allow smoke to travel to additional areas within a building. It is essential to monitor the status of doors that may affect ventilation airflow. Give stair shaft doors special consideration. Doors to stair shaft enclosures that are left open will allow smoke and heat to be drawn in, creating exposure problems for personnel in the shaft and on higher levels.

Windows. Openings created by windows that have been removed will allow smoke to travel horizontally to the exterior. Panels or windows in the exterior walls that aren't openable may need to be broken to create ventilation routes to the exterior. It is essential to communicate with personnel on the exterior of a high-rise before breaking windows; however, if communication with exterior personnel is either impossible or ineffective, and if ventilation is mandatory for safety, break the windows.

Sealed buildings. Most high-rise buildings can be classified as sealed because their external glass panels aren't openable. The internal environments of these buildings are controlled by HVAC systems and are therefore capable of retaining smoke, heat, and fire gases until they are manually ventilated.

HVAC systems. Heating, ventilation, and air-conditioning systems control the internal environments of high-rise buildings. Many HVAC systems incorporate remote-controlled dampers at the supply and return ducts on each floor. These dampers can be closed to prevent heat loss when the building is unoccupied, and they can be opened to supply conditioned air to floors within the building as necessary. Additionally, HVAC systems can form natural channels for the distribution of smoke. HVAC systems equipped with remote-controlled dampers can also serve as smoke-control systems. Unfortunately, many variations are found in smoke-control systems, which may be manually or automatically activated by smoke detectors. Some systems use a combination of both. Unless personnel are totally familiar with a given HVAC system, they should shut down the system until a knowledgeable person can be contacted.

Stair shafts. Stair shafts provide natural channels for smoke and fire gases within high-rises. For ventilation purposes, stair shafts can be categorized in two ways. In the first category are those that access the interior of a building at the bottom and are open to the interior floors open to the roof (figure 16–16A).

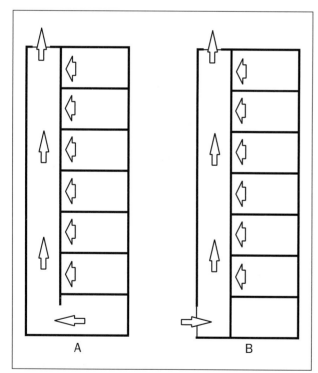

Figure 16–16. For ventilation purposes, stair shafts are either reached from the interior (A) or exterior (B) of a building.

Tests have indicated that, regardless of the atmospheric temperature or humidity, the natural airflow within these shafts is virtually static in buildings up to about 25 stories.

Buildings of this category over 25 stories tall naturally flow minimal amounts of air in an upward direction. Therefore, ventilation efforts directed at removing contaminants in these shafts will likely need to be augmented by means other than natural airflow.

In the second category are those stair shafts that:

- access the exterior of the building at the bottom;
- open to the floors of the building;
- open to the roof (figure 16–16B).

Tests have indicated that, regardless of the atmospheric temperature or humidity, any natural movement of air in these shafts is usually upward and it can be significant. Drafts of 3 to 6 miles per hour can be expected by opening the upper and lower stair shaft doors. Natural vertical air currents that flow upward in stair shafts can be created simply by opening the doors at the ground and roof levels. The resulting air currents can be used effectively to remove contaminants within. Additionally, a natural vertical flow in a stair shaft can be created or augmented by either building shaft pressurization fans and/or PPV. When building pressurization fans are activated (manually or automatically), an upward flow of air will be created by opening a door at the top. The strength of this flow depends on the number and strength of the fans. These fans can be capable of providing flows adequate for ventilation. Positive pressure created by fire department blowers can effectively create or augment the natural flow of air in a stair shaft. By pressurizing the ground-level entrance to the shaft and opening the roof door, contaminants can be vertically discharged. Additionally, positive pressure can be used in conjunction with stair shaft pressurization fans.

Elevators. Elevators can be at the core of the building, or they may be randomly located. Like stairways, elevator shafts provide excellent channels for the by-products of combustion.

Mushrooming and stratification. The formation of stratified smoke and fire gases on floors below the top floor in multi-story buildings is enhanced in high-rises by their height and open vertical passageways. Stratification occurs as heat and smoke travel vertically within a building. These products of combustion rise through any available vertical openings until their temperature is equal to that of the surrounding air. The stratification of smoke and gases generally serves as a lid for other products of combustion, which will tend to bank down below the stratified smoke and spread horizontally to other portions of the building (figure 16–17). This process is common in vertical stair shafts.

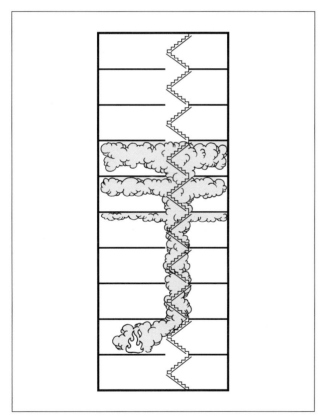

Figure 16–17. The stratification of smoke creates an atmospheric lid, forcing other products of combustion to bank downward and spread to other portions of the building.

Mushrooming is a condition caused by smoke and fire gases that cannot vertically escape a building. The smoke and gases will rise to the highest level possible and then begin to bank down to fill all of the available spaces. Depending on the height of a particular high-rise and the location of the fire, stratification and/or mushrooming smoke and fire gases can contaminate numerous floors and quickly complicate ventilation operations.

Vertical ventilation. Opening the bottom and top doors of a stair shaft can quickly develop a natural upward flow of air. Blowers can augment this flow, regardless of the height of the building (figure 16–18). Such currents can be used to keep contaminants from accumulating in a stair shaft (if the roof opening is kept closed) or to remove accumulated contaminants and keep the shaft clear by opening the roof door.

This operation is currently performed by the Los Angeles Fire Department, which requires that first-alarm truck companies be responsible for pressurization of vertical stair shafts in high-rises with fire. Emphasis is placed on stair shafts that may be used by attack personnel.

Horizontal ventilation. To cross ventilate a contaminated floor, use blowers to pressurize a stair shaft, thereby directing air across the contaminated floor and out an appropriate exhaust opening (figure 16–19). Remember to use windows on the leeward side of the building, if possible.

If necessary, contaminants can also be directed to an opposing stair shaft that opens to the roof (figure 16–20). With the roof door open, the contaminants will be vertically exhausted from the opposing *unpressurized* stair shaft.

Such cross-ventilation operations are effective up to approximately 25 floors in a high-rise. If it is necessary to cross ventilate levels above the 25th floor, you may need to position an additional blower in the pressurized shaft to supply additional air to the floor to be ventilated (figure 16–21) because a certain volume of air will be lost through cracks and other openings in the stairshaft. Such operations may require coordination of personnel both on the roof and on the ground.

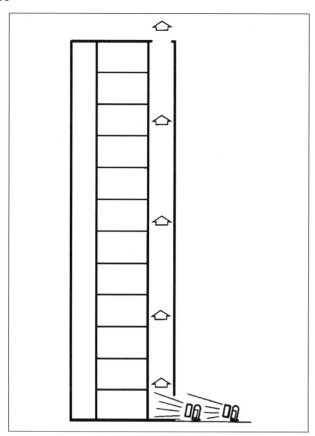

Figure 16-18. The natural upward flow of air in a stair shaft can be augmented by portable blowers.

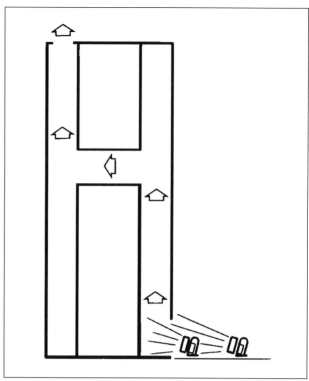

Figure 16-20. If necessary, contaminants can be horizontally directed to an opposing stair shaft that opens to the roof.

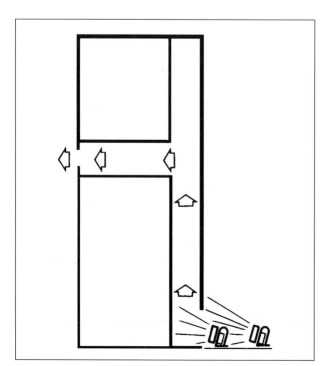

Figure 16-19. Cross ventilation can be accomplished by pressurizing a stair shaft and directing the airflow across the contaminated floor and out an appropriate opening.

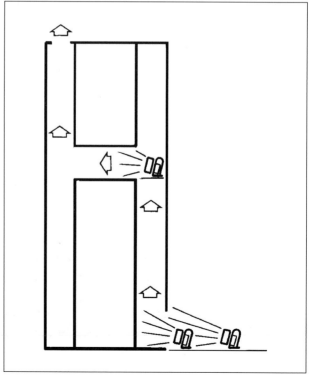

Figure 16-21. Because of leakage in tall stair shafts, it may be necessary to use additional blowers to ventilate above the 25th floor.

PPV in Other Operations

Overhaul

Overhaul is conducted to ensure that a fire is completely out, determine its probable cause, and leave the owner or insurance adjuster some means of determining the loss. Such operations are often conducted in conditions that may be tenable yet hazardous to personnel. Remember that any fire gases that have collected at the top of a room as a result of a fire may have cooled and settled within the overhaul area, where it can easily be inhaled by personnel. Additionally, common building materials retain and radiate heat that has been generated by fire, resulting in elevated temperatures. If some type of ventilation isn't implemented before and during overhaul, toxic elements will accumulate. Positive pressure can provide a flow of fresh, cool air through the area in question (figure 16–22). This will reduce humidity, smoke, radiated heat, and toxic gases. It will also improve visibility.

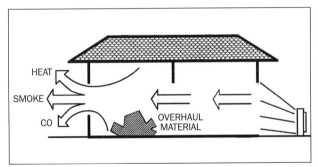

Figure 16–22. Positive pressure provides a flow of fresh, cool air that can dramatically improve the conditions encountered by overhaul personnel.

Depending on the type of fire, concentrations of carbon monoxide can easily rise to 500 parts per million (ppm) to 1,200 ppm during overhaul. Measurements during actual operations have indicated that positive pressure can reduce CO concentrations of 1,000 ppm to between approximately 130 ppm and 220 ppm. A single gasoline-powered blower, however, will introduce approximately 140 to 160 ppm of CO into the structure. Therefore, if high carbon monoxide levels are present, positive pressure will lower them. If the CO levels are low, PPV might actually raise them. The introduction of CO can be controlled by exhaust tube extensions on gasoline powered blowers or by using electric, water, or hydraulic blowers.

The effectiveness of PPV during overhaul operations is determined by the overhaul area and the blowers being used. In general, a moderate flow of air is sufficient to clear toxic by-products out of the overhaul area. This is normally achieved by running blowers at partial throttle. A single blower is adequate for an average single-family dwelling of 1,800 square feet. As the size and power capabilities of a blower are increased, so is its effectiveness in larger areas. Practical experience indicates that PPV doesn't tend to accelerate or spread fire that may be in the overhaul area. However, if fire does begin to flare, reduce the RPM of the blower or shut it off. Otherwise, increase the distance between the overhaul area and the blower before extinguishing the fire. One other note: never use PPV in place of SCBA. Personnel should always use SCBA when confronted by hazardous atmospheres.

Search

It's a fact that positive pressure can potentially enhance a search operation while increasing firefighter safety. This was graphically demonstrated during tests conducted by the North Carolina Bureau of Insurance and the United States Coast Guard. However, there is also concern for trapped occupants who may be located between a fire and the exhaust opening. In this case, the introduction of positive pressure might result in death. Most structure fire victims, however, die of smoke inhalation, not burns. Victims who may be located between the fire and the exhaust opening will most likely have died before PPV is implemented, and victims who are near the seat of a fire will probably have died from heat or contact with the fire. Therefore, although positive pressure can change the direction of fire and heat toward an exhaust opening, it doesn't normally contribute to fatalities. Most victims die of smoke inhalation, so those who are still alive might even be saved by the influx of fresh air that PPV will bring. Additionally, positive pressure also has the potential to enhance a search operation, since it fosters better visibility and reduces the threat of flashover.

Fire attack and PPA

Positive pressure (or positive-pressure attack) can be used as an initial suppression technique prior to introducing hoselines because it will reduce by-products, internal temperatures, and the potential for flashover. It will also improve visibility and reduce the time required to place attack lines. Also, when water is placed on the fire, the PPV currents will carry the steam away from personnel on the attack line. Personnel can often enter an involved area and not have to crawl on the floor when they advance the lines. Overall, PPV enhances firefighter safety. Like other fireground techniques, PPV must be used with care. It requires both technical and practical knowledge. Remember that potential backdraft conditions must not be present and blowers capable of at least 7,000 to 10,000 cubic feet per minute (cfm) provide the most effective results. A single blower of five horsepower is the minimum that should be used on a single-story, single-family dwelling of approximately 1,800 square feet (if smaller less powerful smoke ejectors or blowers are utilized, performance and results will be reduced accordingly). Larger structures or a need for increased efficiency will require multiple blowers either in series or parallel. Pressurized air from the blowers must seal the appropriate entrance opening (if fire is close to the opening), and the exhaust opening, whether horizontal or vertical, should be as close to the seat of the fire as possible. Attack lines should be advanced expediently after implementing PPV.

In this operation, the cone of air must seal the door as fire can be driven back through any unsealed areas around the door, depending on the location of the fire in relation to the entrance opening.

The following scenario offers an overview of a typical PPV-PPA operation in the single-family dwelling in figure 16–23. Assume there is a fire in bedroom #2, and the structure is charged with smoke and heat:

1. After arrival of an initial company, an appropriate person circles the structure to determine the location of the fire and a suitable exhaust opening. If the location of the fire cannot be determined, a combination of the type of incident, experience of personnel, and common sense will dictate the implementation of positive pressure with an attack line.

In this scenario, the proper location of the exhaust opening is both windows in bedroom #2. Remember to remove any screens from windows due to their ability to reduce ventilation efficiency. Exhaust openings can be horizontal or vertical avenues; however, horizontal avenues should be given first priority due to their ease of implementation and reduced exposure to the involved structure.

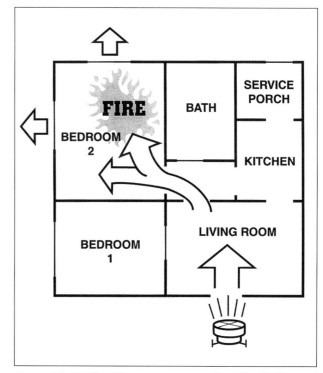

Figure 16–23. Positive pressure can be effectively used as an attack weapon on confined fires.

2. When the location of the fire and exhaust opening has been determined, a blower and attack line are placed at the point of entry of attack personnel (front door). When possible, an attack line and blower should be positioned by an entrance opening that is away from the seat of the fire. This will allow the attack line to be advanced from the uninvolved to the involved portion of the structure.

3. When the exhaust opening, attack line, and blower(s) are ready, the blower(s)

should be started and positioned to cover the front door with pressurized air, after the exhaust opening has been created first.

4. With pressurized air traveling through the entrance and exhaust opening, heat and contaminants between the exhaust and entrance opening will immediately begin to be expelled to the exterior of the structure.

5. After allowing several seconds for the structure to begin to clear, advance the attack line into the structure. Attack personnel should spend as little time as possible in the entrance opening as this will block or restrict the pressurized air being forced into the structure. The reduced heat, smoke, and improved visibility should enable an improved attack on the seat of the fire. As the fire is knocked down, the resultant steam from extinguishment will be carried away from attack personnel and expelled with the heat and smoke.

Although positive pressure can be effectively utilized with an attack line in many applications, it cannot be applied to every confined fire. This technique will only achieve satisfactory results through a combination of proper training and judicious implementation.

Dumpster fires

Each day, the fire service collectively responds to hundreds of dumpster fires. Interestingly, the extinguishment process is the same today as it was 30 years ago, normally consisting of several firefighters dressed in protective equipment (without SCBA) using an attack line to extinguish and overhaul the fire. Although most personnel wear SCBA on the fireground, few wear it on dumpster calls because these incidents are perceived as being only rubbish fires. Unfortunately, the contents of the modern dumpster have drastically changed from 30 years ago. Plastic, fiberglass, and illegally dumped hazardous materials can produce numerous toxic gases that are extremely dangerous when inhaled. For example, one of the most popular chemicals used

as a stabilizer for plastics is chlorine. When this chemical is inhaled, it will willingly combine with moisture in the human respiratory tract, forming hydrochloric acid that can destroy mucous membranes and cause internal bleeding, possibly death.

Positive pressure can dramatically improve the suppression and overhaul phases of dumpster fires by providing a directional flow of pressurized air, and allowing suppression personnel to operate upwind of the fire, away from the deadly products of combustion. Place a blower approximately 8 to 10 feet away from the dumpster before initiating suppression operations. The pressurized airflow will force the products of combustion away from personnel. Using two blowers in parallel increases efficiency. Position the blowers according to the prevailing wind, and ensure that personnel wear SCBA despite any habitual practices to the contrary.

Auto fires

The common auto fire is also generally considered a routine incident that is easily handled on a daily basis. The interiors of modern automobiles, however, have also changed greatly over the last 30 years. Plastics and other synthetic materials can produce copious amounts of deadly toxic gases when exposed to fire. Here, too, many fire service personnel neglect to wear SCBA during attack and overhaul operations. Positive pressure can dramatically improve the circumstances of these incidents. Open the doors on either side of the vehicle and place a blower 8 to 10 feet away on the windward side (figure 16–24).

The pressurized airflow will direct the products of combustion and heat away from personnel. One blower is sufficient for most automobiles. As with dumpster fires, place the blower according to the wind and ensure that personnel wear SCBA.

Figure 16–24. Using PPV from the windward side at vehicle fires allows personnel to operate away from the products of combustion. (Illustration courtesy of Tempest Technologies)

Salvage

Defensive tactics at structure fires normally mean heavy streams and predictable loss. The loss of a building, however, shouldn't necessarily spell the loss of a business. Mere buildings can be replaced. Perhaps the greatest threat that fire poses to a business is that the accounts receivable records might be lost. When initiating defensive operations, focus attention specifically on operations to preserve the office area. This may entail salvage operations designed to protect file cabinets, computers, and other such items. Positive pressure can assist salvage operations by improving the office environment, thereby allowing personnel to initiate appropriate operations. Pressurize the office area with standard PPV techniques. You can use the pressurized air either to remove heat and contaminants or to thwart the contaminants from encroaching on the office. Either way, you will buy salvage personnel sufficient time to initiate the appropriate operations.

Ships and aircraft

Positive pressure is also effective for marine and aircraft fires. In the marine incident depicted in figure 16–25, the contaminants from a fire have been isolated by closing bulkhead doors A and B. Appropriate vertical passageways then channel the flow of pressurized air through and away from the contaminated area.

Similarly, contaminants can easily be removed from aircraft by using fore and aft doorways and standard PPV techniques. A PPV operation performed on a Boeing 737 cleared a zero-visibility smoke condition in the cabin in 45 seconds. If an aircraft is connected to a terminal by a jetway, ensure that the airflow is directed toward the rear of the aircraft so as not to force contaminants through the jetway and into the terminal.

Miscellaneous

Positive pressure has also proved effective in other applications, such as removing flammable atmospheres and minimizing contaminated atmospheres in hazmat incidents. Being able to place blowers in fresh air, away from flammable atmospheres, enhances effectiveness and safety. For example, PPV has become an effective tool in leaks of anhydrous ammonia, which are common in cold-storage facilities. Multiple blowers are used to isolate a leak and minimize extension of the ammonia. A recent innovation, plastic tubing, can be used to channel pressurized air to specific locations that may be difficult to reach with portable blowers. Tubing also allows you to place a blower outside of a problem area while still directing pressurized air to a specific location. Of particular interest is PPV's benefits to rehab because it has been demonstrated that PPV is able to reduce the body temperatures of personnel (using full protective clothing) engaged in fire attack operations. Tests by the United States Coast Guard have indicated that positive pressure can reduce the core temperature of personnel by as much as 5 degrees during suppression operations —significant from the standpoint of endurance and safety. During fireground rehabilitation operations, PPV can lower the ambient temperature around firefighters by approximately 10 degrees and by over 15 degrees when a light water mist is used with the pressurized air.

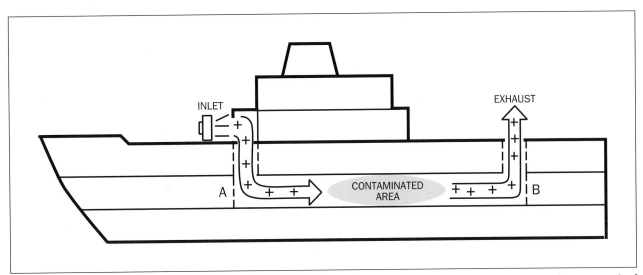

Figure 16–25. Positive pressure has proven effective in marine incidents, which often require the removal of contaminants below deck.

When You May Not Want to Use Positive Pressure

Remembering that PPV isn't universally viable, personnel should evaluate each incident to determine whether PPV is suitable in a given circumstance.

Balloon construction

Unlike platform construction, balloon construction can allow fire to travel unrestricted through a wall into an attic. Positive pressure can enhance this if an exhaust opening isn't opened prior to the pressurization. If an exhaust opening is created prior to PPV, however, most of the pressurized air will flow toward the opening and have little impact on the spread of fire in open walls. Therefore, when you encounter fire extension in balloon construction, the key is to create an exhaust opening before initiating PPV.

Attic fires

Implementing PPV in structures with attic fires should be evaluated in terms of the presence and size of attic vents. If fire has extended into an attic with vents that are sufficiently large to allow a significant flow of pressurized air that may enhance the directional travel of fire, don't use PPV unless attack operations can be implemented in a short period of time.

However, the absence of attic vents will normally allow PPV to pressurize the structure and attic without spreading an existing attic fire. Remember that any roof ventilation openings can also be considered attic vents.

Location of fire

When the location of a fire is unknown, carefully analyze the consequences before using PPV. This is particularly important for attack operations. If the structure is small and relatively simple, PPV is likely to be more successful. As the size and complexity of the structure increase, however, so does the chance that PPV operations may not be as anticipated. Remember, PPV will primarily affect heat, smoke, and fire that is in the path of pressurized air traveling toward the exhaust opening.

Vent-entry-search

Vent-entry search (VES) can be an extremely dangerous operation, which is why it is only used by some fire departments in the United States. Using positive pressure in conjunction with VES (or an outside ventilation man—OVM) operations can increase the danger even more unless the PPV is carefully controlled. Assume that a VES operation has been initiated in figure 16–26. When the window near the seat of the fire is opened, and personnel enter to conduct the search, the application of PPV can make the search area untenable for VES personnel.

If VES operations are to be conducted, and if they are to be near or between the fire and the ventilation opening, they should conduct their search and exit the structure before PPV is initiated. This heightens the need for coordination and communication between PPV and VES personnel.

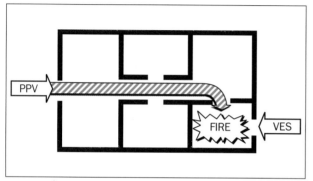

Figure 16–26. Positive pressure can increase the dangers of vent-enter, search unless both operations are carefully controlled and coordinated.

Search

In general, PPV can be a benefit for search operations but should be critically evaluated for its impact on occupant and firefighter safety. The incident commander must carefully weigh such factors as the fireground conditions, parameters of safety, and the amount of time necessary to implement PPV.

Overhaul

If the exhaust of a gasoline-powered blower is detected inside a structure, it indicates that the exhaust opening for the PPV operation is too small. Increase the size of the PPV exhaust opening to minimize the concentrations of CO and other by-products of the blower. Remember to evaluate the use of gasoline powered blowers in facilities that have occupants with respiratory concerns.

Trench rescue

The flow of pressurized air can dry the earth, thereby increasing the chances of collapse.

Questions and Answers about PPV

Do I need special blowers for PPV?

The implementation of PPV is not dependent on a specific type of blower. The basic difference between negative pressure and positive pressure is the location or placement of blowers (inside or outside a structure).

Is there a difference between a smoke ejector, fan, or blower?

A fan, smoke ejector, or blower is nothing more than a motor or engine and a propeller. Therefore, the difference is not in the terminology, but the way they are used. If they are placed inside a structure, they will eject contaminants to the exterior of a structure. Conversely, if they are placed outside a structure, they will force air inside a structure and driving contaminants to the exterior.

Will PPV increase the size and/or intensity of a fire?

Although a blacksmith commonly uses bellows to increase the size and intensity of a fire, PPV does not. Note that bellows are always directed at the seat of a fire, whereas PPV results in cool pressurized air surrounding the entire fire. This will reduce the heat of the fire which needs heat, fuel, and oxygen in the proper ratio to burn.

Will PPV redirect or push fire throughout a structure?

The only significant air velocity is near the entrance and exhaust opening. Airflow within other areas of a building are very low; therefore, the affect on fire is minimal. This is easily demonstrated when standing inside a large warehouse when the building ventilation system is in operation. Although there will be a noticeable movement of air at the entrance and/or exit openings, the movement of air within the building is undetectable. However, if fire is in close proximity to a blower, it can change or redirect the direction of a fire.

Does PPV force or enhance the travel of fire into concealed spaces in a building?

If hidden voids do not have an exhaust opening (an opening that would allow air to flow through the void), there will be no air flow through the void and no impact from PPV on fire within the void. If the void does have an opening that will allow air to flow through the void, the fire can be enhanced within the void. However, if suppression personnel have created appropriate exhaust openings, virtually all of the pressurized air from positive pressure will flow to and out specific exhaust openings and have a minimal effect of hidden voids. However, remember that it is important to open up any hidden void to check for the extension of fire, and the sooner the better.

Does carbon monoxide from gasoline powered blowers create a dangerous environment to fireground personnel?

Although the inhalation of any carbon monoxide should be avoided as carbon monoxide is accumulative, gasoline powered blowers can inject approximately 140 parts per million into a structure, which is a small amount. However, carbon monoxide can be eliminated from gasoline powered blowers by utilizing exhaust tube extensions that direct the exhaust away from the air being drawn into the blower and forced into a structure.

Can PPV be used on every fire?

Positive pressure is the same as any other tool in the fire service arsenal. It should only be used when appropriate and at the discretion of appropriate personnel/officers. If positive pressure is used on a structure fire and the building burns down, it was not the fault of positive pressure. It was the fault of the personnel who used positive pressure incorrectly.

What is the most effective location for a PPV blower, and where is the best location for the exhaust opening?

If possible, always use the direction of a prevailing wind to your advantage; then create an exhaust opening as close to the seat of a fire or contaminated area as possible. Next, create an entrance opening away from the fire or contaminated area if possible and then place the blower the appropriate distance back from the entrance opening to seal the door with a cone of pressurized air. The most effective distance is determined by testing with your specific blower. Recommended average distances of 6, 8, 10, or 12 feet may not be appropriate for your equipment. In fact, 10 to 12 feet is normally too far back from an entrance opening.

Training Resources

The following resources can provide additional information and/or another viewpoint for the subjects discussed in this chapter:

- *Positive Pressure Attack for Ventilation and Firefighting*, Kriss Garcia, Reinhard Kauffmann, and Ray Schelble, Fire Engineering Books and Videos, 2006.

- Roundtable, PPV, *Fire Engineering*, December 1999.

- Thirty-Story High-Rise Tests in Toledo, Ohio, www.fire.nist.gov/bfrlpubs/NIST IR 7412.pdf, 2007.

- Technology Roundup: PPV in High-Rise Buildings, Mary Jane Dittmar, Dan Madrzykowski, Steve Kerber, *Fire Engineering*, May 2008. See NISTIR 7412, and NISTIR 7468, and www.fire.gov

- Positive Pressure Ventilation for High-Rise Buildings, G.D. Lougheed, P.J. McBride, and D.W. Carpenter, Institute for Research in Construction, National Research Council Canada, Research Report 102, August 2002.

- Fog Streams and PPV, Bill Gustin, *Fire Engineering*, November 1997.

- Positive Pressure Ventilation in a Test Setting, Larry, D. Hughes, *Fire Engineering*, December 1989.

- Using Electric Fans for PPV, Richard E. Andersen, *Fire Engineering*, October 2003.

- Pressurized Fire Attack Precautions: The "Big Threee," Kriss Garcia and Reinhardt Kaufmann, www.FireEngineeringUniversity.com.

- Wind-Driven Fire Research: Hazards and Tactics, Daniel Madrzykowski and Stephen Kerber, *Fire Engineering*, March 2010.

Ventilation Rules and Equipment

Introduction

Ventilation operations should be governed by a set of standard operating procedures (SOPs), the values of which will determine the course of a ventilation operation. Personnel must know how to accomplish various types of ventilation operations before the need arises because proper ventilation depends on training, knowledge, and certain safety fundamentals. Other than actual fireground conditions, there is no substitute for training to help you determine how to accomplish these types of operations. Discuss and practice ventilation operations on a frequent basis. If you do so, then basic operations will be automatic at a fire, giving personnel a greater opportunity to focus on the specifics. The fireground will provide the opportunity to hone operational skills.

Size-Up

Before we consider the physical aspects of a size-up, let's reconsider why we do one. To use an analogy, when serving the ball in tennis, you are in control. You control the speed of the ball, its placement, and its spin. Prior to your involvement at an incident, a size-up puts you in control. When sizing up for ventilation operations, ask yourself:

- What is the type of building and the time of day?
- Is there a fire?
- What is its extension?
- Are the occupants endangered?
- Is ventilation needed?
- Is horizontal or vertical ventilation feasible?

To a large degree, when personnel are committed inside or on a building, the building can be in control. So take a few seconds and determine what about a building and fire is important. Although various fireground factors can usually be determined before personnel start to work, a size-up is a continuous process.

Initial size-up

An initial ventilation size-up lays the foundation for abatement strategy, and it can easily be accomplished before stepping off the apparatus. Evaluate the present and future factors that will set appropriate ventilation parameters. For example, some of the variables the roof ventilation team should consider are the building's type and age, the location and extension of fire, the type of roof, the equipment available to you, and other such factors.

Building type and age. Determine the type of building. Suppose it is a multi-story hotel. Depending on conditions, personnel may consider venting the stairs and hallways to aid evacuation. Consider the age of the building, and if it appears to be an older style of construction. If so, determine if it has conventional or unreinforced masonry construction. Being familiar with the basic types of construction will help you determine the approximate time available for ventilation operations and any specific noteworthy hazards.

Location and extension of fire. First determine if there is a fire. If so, determine its location and extension. If there is fire on the first floor of a four-story commercial, going to the roof to create ventilation openings may not be vital. However, if the fire is on the top floor of a four-story apartment house, then venting over it and checking for extension in the attic is of primary concern.

Type of roof. Hopefully you are familiar with the different types and styles of roofs and construction in your area so you can determine the type of construction on which personnel will be working. If roof operations are necessary, determine the strong and weak areas of the roof, where to walk and not to walk, how to use the roof construction to an advantage when cutting ventilation openings, and the approximate time available for ventilation operations.

Ladders. Determine the type of building and roof to help you ascertain the proper type and length of ladders, if they must be used. Decide if an aerial device can be used, if ground ladders will be necessary and if straight ladders or extension ladders will be viable. When considering the type and pitch of a roof, consider the need for a roof ladder.

Hazards. Immediately verify what hazards, if any, stand in the way of ventilation operations. For example, consider the problems and solutions necessary if electrical wires are blocking ladder placement to a roof. Consider the name on the front of a given building, since that will often tell you what you can expect inside.

Continuing size-up

All personnel must remain aware of their surroundings. This means taking time to monitor and verify the changing conditions and, if appropriate, relaying this information to other appropriate personnel.

Area size-up

An integral part of size-up is relaying information from a particular area to other personnel. The roof of a building is an ideal location to observe the building's layout, its physical features, and the status of the fire. Such information can be vital to fire attack personnel, who may be unable to see due to smoke, as well as to the incident commander, who must continually evaluate the fireground.

Plan Ventilation Operations

If ventilation operations are necessary, begin by identifying three necessary areas that consist of the direction that smoke, heat, and fire gases must travel to exit the structure; a method to move the smoke, heat and fire gases; and the type of ventilation opening(s) that you will use. This entails varying degrees of priority depending on the conditions that you encounter. Let's expand on the direction of travel, the method of movement, and the type of opening(s).

Direction of travel

The initial step in planning ventilation operations is to determine the direction of travel, which is either horizontal or vertical.

Horizontal ventilation. For reasons of ease and safety, horizontal ventilation is the most popular ventilation method. It is often accomplished by simply opening doors and windows. Using horizontal ventilation to remove heat, smoke, and fire gases can be effective; however, its effectiveness depends on the size of the opening and its proximity to the contaminants. Before using horizontal avenues, consider several factors. Determine whether the direction of wind (if present) could carry contaminants to uninvolved areas of the fire building or exposures. If the building has multiple stories, and windows will be opened on the lower floors, determine if the direction of smoke will contaminate the upper floors. Also, ascertain the speed or force of the wind. When windows are opened on the windward side, determine if the wind will accelerate the fire and enhance extension. Finally, open the windows on the leeward side of the building first. Windows on the windward side can then be opened to allow the wind to force contaminants from the structure (figure 17–1).

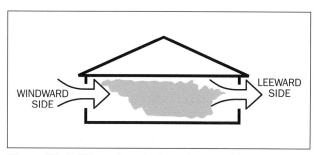

Figure 17–1. When the wind can be used for horizontal ventilation, open the leeward windows first and then the windward windows.

Opening windows close to the seat of the fire is a top priority. This will provide some ventilation to the fire area and allow the expanding gases from suppression operations to escape. Remember that opening windows in a fire area can increase the supply of oxygen and cause the fire to accelerate. Therefore, properly timing ventilation in conjunction with suppression is necessary for safe operations. After windows in the fire area have been opened, open others away from the fire area if doing so will ventilate other contaminated areas without enhancing extension. When opening windows, remove blinds, curtains, and shades to ensure that they won't restrict the removal of contaminants. Screens should always be removed because they can restrict

air movement by at least 50%. It isn't necessary to open double-hung or casement windows two-thirds down from the top and one-third up from the bottom. Maximum effectiveness will be accomplished by fully opening these windows at the top. Pivoting windows must be opened according to their design. If a window can't be opened, you may need to break it. Use as little force as necessary so that broken glass won't fly into the structure. If windows need to be broken above the grade floor and they cannot be broken from the ground or inside the building, they can be broken from floors above or the roof (whichever is more appropriate) by doing the following:

1. Tie a rope or cord around a tool such as a Halligan

2. Lower the tool down to the window you need to break

3. Secure the rope or cord either by tying it off or standing on the rope/cord

4. Pull the tool back up and toss it out away from the building.

This operation ensures that the tool will travel downwards and swing inwards to break the window that has been pre-selected. Always remember to clear all glass from a broken window (if possible). See chapter 6 for more information about breaking windows.

Horizontal ventilation can be enhanced by mechanical means. If multiple areas need to be ventilated, ventilate each area sequentially by using doors as partitions to reduce the size of each area. This maximizes the flow of fresh air through each ventilated area. When opening doors for ventilation operations, consider blocking or wedging them so that they won't close inadvertently.

Vertical ventilation. Depending on conditions, vertical ventilation can be the preferred method. One reason is because the heated products of combustion tend to rise, horizontal ventilation doesn't provide a natural avenue for rising contaminants. Also, within a confined area in a building, the highest concentrations of heat, smoke, and fire gases are in the upper regions. Ventilating directly overhead of a contaminated area in concert with natural air currents (figure 17–2) can exhaust the most dangerous by-products first. However, as personnel are working above ground, the elements of timing and the inherent risk factor must be constantly evaluated.

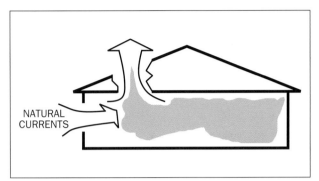

Figure 17–2. Vertical ventilation takes advantage of natural convection currents.

Roof venting operations can often begin by using available natural construction openings. Using building features such as skylights and roof scuttles can speed ventilation while causing minimal damage to the structure. However, natural construction openings should be used only when they are in close proximity to the fire or are strategically located in an area to be vented; such an opening can draw fire to its location. Cutting an opening in a roof to ventilate contaminants can be done in conjunction with natural openings or as a separate operation. Although cutting a roof requires special considerations and techniques, it can be the most effective method of ventilation if conditions warrant using it.

Method of movement

There are two basic methods used to ventilate structures to direct heat, smoke and fire gases to the exterior: natural and pressurized.

Natural ventilation. As a fire burns within a structure, the fire floor fills with hot smoke and gases. These products of combustion rise to fill all of the available spaces. A simple way to ventilate these by-products is to open the structure horizontally or vertically. The accumulated contaminants will then be relieved by natural convection. Before using horizontal or vertical openings in this way, consider the following.

Limited effectiveness. Natural ventilation depends on natural or self-generated movement. It can be enhanced or limited by prevailing wind currents and atmospheric conditions such as humidity and temperature differentials.

Specific openings. Depending on the type of structure, natural ventilation options may be limited to available openings such as windows and doors. These should lead directly from the contaminated area to the exterior of the structure.

Pressurized ventilation. Pressurized ventilation can be used to assist or replace natural ventilation.

Portable blowers. Although natural ventilation can be a viable method that is effective within certain limitations, it can be assisted or replaced with portable blowers. These can force by-products out of a contaminated area more rapidly (horsepower). They also allow you to use openings that are remote from heat, smoke, and fire, directing such contaminants through preselected, controlled openings or openings not normally viable by natural means. Portable blowers can overcome interior and exterior temperature differentials, as well as the effects of humidity. Overall, they can reduce the time necessary to ventilate a building as compared to natural ventilation.

HVAC systems. Many multi-story buildings use heating, ventilation, and air conditioning systems to control the internal environment. Such systems draw in outside air, change it to the desired temperature, circulate it, and exhaust it to the exterior. It is possible to use an HVAC system to rid a building of contaminants; however, doing so depends on the capabilities of the system and the expertise of the personnel operating it. It is imperative that the capabilities and operational techniques of a particular system be thoroughly understood because an HVAC system is also capable of spreading contaminants and fire to uninvolved portions of a building. Therefore, it is necessary to become familiar with HVAC systems during pre-fire planning inspections. If suppression personnel are unfamiliar with a particular HVAC system, operations should be handled by the appropriate building engineer or other similar responsible person.

Fog streams. Fog streams can be used for ventilation purposes. Simply direct a wide fog pattern out of a window so that the resultant venturi action will draw out contaminated contents. You can obtain maximum effectiveness by moving the nozzle in a clockwise motion 4 to 10 feet back from a window.

When this method has been tested, it has resulted in an air movement of over 2,000 cubic feet per minute (cfm). The effectiveness of this depends on the size of the opening, the area to be ventilated, the size of the nozzle and hose, and the hose discharge pressure. However, several factors should be evaluated before implementing this technique. Assess the other ventilation options; in virtually all cases, pressurized ventilation is more effective and safer. Ask where the water will go and whether it will cause additional damage or disrupt other operations. Also, determine the environment to which personnel will be subjected when using this technique. And determine if personnel and hose lines can be better used elsewhere.

Note: See the article "Nozzle Tests Prove Fireground Results" in Training Resources.

Ventilation openings

Four basic types of openings are used for ventilation operations: natural, natural construction, heat, and directional.

Natural openings. Ventilation isn't entirely dependent on suppression personnel. Some common examples of natural openings are the holes that burn through roofs, skylights that fail or burn through, automatic smoke vents that activate, and ventilators. Even if any of these are present, personnel must evaluate their effectiveness and consider the need for additional ventilation.

Natural construction openings. Natural construction openings such as windows, doors, skylights, and other similar openings can provide sufficient ventilation if they are of sufficient size and are in the right location relative to the fire.

Heat openings. Heat openings are opened in close proximity to a fire. They may be horizontal or vertical and are specifically designed to vent fire, heat, and smoke. Such openings are often referred to as offensive ventilation and should be a *first* priority.

Directional openings. Directional openings are opened ahead of a fire. Also known as strip ventilation, they are designed to change the horizontal direction or extension of fire, heat, and smoke to a vertical direction, reducing or curtailing horizontal spread. Such openings are referred to as defensive ventilation and should be a *second* priority.

Basic Safety Rules

Ventilation operations, whether horizontal or vertical, should be preceded by a set of SOPs and the ability to constantly follow appropriate safety rules that are applicable to each specific ventilation operation. As horizontal ventilation is relatively straightforward and discussed in the first portion of this chapter, let's focus on applicable safety rules for vertical ventilation, as it can be considerably more demanding and dangerous than horizontal ventilation. The following safety rules are designed to not only result in a timely vertical ventilation operation, but to also enhance the safety of this operation.

Ladders

Consider two ladders per roof to be the minimum for roof operations. If only one ladder is used, there is only one way off. Two ladders provide a primary and alternative means of escape. Before laddering the roof of a building, consider which areas offer the strongest support and the best routes of travel. Some areas of construction that offer good locations for ladders are corners, hips, ridges, valleys, and pilasters. Determine where your means of egress is located; know where and how to exit the roof, and whether or not you have an alternative means of escape. Extend ladders far enough above a roof so that roof personnel can easily see them.

Basic equipment

The following tools and equipment are recommended as a minimum for performing vertical ventilation:

- Full protective clothing
- Breathing apparatus
- Pickhead axe
- Pike pole
- Rubbish hook (or other suitable tools such as a Halligan, hook, and so on)

- Power saw (chain or rotary saw)
- Communications (walkie-talkie)

When using these tools and equipment, specific care must be applied to power equipment. When taking power equipment aloft, work in teams of two as a minimum, and emphasize safety. Power saws used for roof ventilation require an operator plus another member to act as a spotter. A power saw can be a dangerous tool, especially when it is used on a roof that is being weakened by fire and when cutting into an area below the decking that cannot be seen. A recent survey by the National Electronic Injury Surveillance System Program provided a few surprises regarding general power saw injuries. Most injuries occurred under ideal conditions. The operators tended not to be fatigued and had over six years of experience. Dress a firefighter in full protective equipment, put on breathing apparatus, run a power saw up to 5,000 rpm, add some smoke or darkness to the excitement being generated by the incident, and you have a prescription for an accident. It is essential to operate as a team. Without a safety member as an integral part of a roof team, what will keep a saw operator from tripping over a vent pipe or backing off a roof? The safety member is responsible for controlling the saw operator and watching the roof conditions near and along the path of the cut.

First member to the roof

Roof ventilation operations are usually performed in an area that isn't in close proximity to other firefighting personnel. If a solitary firefighter engaged in ventilation operations has a problem, who would assist the firefighter? Ventilation operations are simplified and safety is increased when a minimum of two personnel are deployed. The first ventilation member to a roof should have a sounding tool. Determine if the roof is strong enough to support the weight of personnel, or if the decking has been burned away under the roofing material. The strength of a roof can easily be determined by sounding it with an axe, pike pole, rubbish hook, hook, or other such tool. Always evaluate the integrity of a roof before stepping onto it. When the first ventilation member sounds a roof to determine its safety, that member also determines the path of

travel for other firefighters. Inexperienced members shouldn't determine the path of travel across a roof.

Read the roof

Before venturing off the ladder, personnel must take the appropriate time to read the roof. You need to determine the location of the fire, determine if it is burning in a specific location, or if it is extending from its original location. Ascertain how long the fire has been burning, if the fire is showing through the roof, and if a portion of the roof is sagging. Determine if the roof has ventilators, and if they are issuing smoke or fire. Determine if the roof is finished with tile/slate, which adds to the dead load of a roof. Verify if division walls are visible, and consider evaluating the building (and the roof) as a building under demolition.

Determine the type of roof

Before personnel venture away from their ladder, it is essential to know what the roof (that is supporting the ventilation team) is composed of. Some items of consideration are the type of roof, if it is conventional construction or lightweight construction, if the roof is covered with slate or tile that can make it difficult to traverse, and so on. To underscore an important point, consider the following: A flat roof is a common roof. It may be found in many areas, usually covered by composition-type materials. However, unless personnel are familiar with a particular building and roof, the structural members that will support it are an unknown element. The roof may be of lightweight concrete, 2-by-12-inch joists and 1-by-6-inch sheathing, 2-by-3-inch wooden I-beams and half-inch plywood/OSB, or open-web construction covered by corrugated metal. All of these roofs may look similar from the top, yet each will react differently when exposed to fire. Before committing personnel on an unknown roof, determine if it is structurally sound and whether it will provide adequate time to perform the intended operations.

This can be easily accomplished by taking a *plug* out of the roof to help you determine the type of roof construction. A plug is a small triangular piece of composition (only) that is cut with an axe or power saw and then removed to reveal the type of roof decking below. As an example, if the plug reveals corrugated metal decking (figure 17–3), the roof is probably a metal deck, built-up roof, which in reality is a metal lightweight roof. Notice the multiple layers of composition, which is an indicator of an older roof.

Figure 17–3. Removing the roof covering only can reveal the type and age of a roof.

If the plug reveals 1-by-6-inch sheathing, the roof is probably of conventional construction as 1-by-6-inch sheathing has not been used for over 50 years. Additionally, *multiple* layers of composition material should also be visible on older roofs.

If the plug reveals ½-inch plywood or OSB, the roof is likely a lightweight roof. Additionally, expect to see *minimal* layers of composition material on newer roofs.

Determine the location and extension of fire

Prior to initiating ventilation operations, you must determine the location and extension of fire. What did you see in your initial size-up? What areas of the roof are issuing fire or smoke? Ventilators, vent pipes, skylights, heat blisters and melting snow are excellent indicators. Consider the color, temperature, and pressure of any visible smoke. If necessary, an indicator opening can be made by using a power saw or the pick of an axe. Make a small (small is defined as not large enough for a firefighter to inadvertently step into and break an ankle, foot, and so on) hole in the roof decking to see whether smoke

or fire are below. Such an opening is often referred to as a kerf cut, an indicator hole, or an inspection hole. If any smoke is present, consider its characteristics. As an example:

- A lack of smoke is an indicator that fire is not in the immediate area.
- Cold, white smoke indicates that fire isn't in the immediate vicinity of the opening.
- Hot, black smoke under pressure, is a positive indicator that the fire is near (figure 17–4).
- Fire is an indicator that you may be in the wrong place at the wrong time.

Figure 17–4. Smoke from an indicator opening can be an indicator of the location of fire. (LAFD file photo)

Determine ventilation feasibility

Prior to committing personnel on a roof for ventilation operations, personnel must determine whether or not ventilation operations are feasible. This entails knowing the type of roof construction and the location of fire in relation to the roof structural members and roof personnel. When these two factors are known (as delineated in the two previous safety rules), then roof personnel should have a basic idea of the structural stability and amount of time that are available for a roof ventilation operation. Because of the introduction of lightweight construction, the days of randomly walking out on a roof and cutting a ventilation opening are over!

Walk the strong areas of a roof

Ventilation personnel should always concentrate on walking the strong areas of a roof, whether during fireground operations or training evolutions. The strong area of a roof can be easily summarized by the following considerations:

- The strongest area of a flat roof is the perimeter of the roof.
- The strongest areas of a gable roof are ridges, valleys, and over exterior bearing walls.
- The strongest areas of hip roofs are ridges, valleys, hips, and over exterior bearing walls.

Sound your path of travel

It is possible for sheathing, corrugated metal, or plywood/OSB decking to weaken or burn away without burning through the composition covering. Therefore, although a roof may look normal from the top, it may not support any weight. Sound it with an axe, pike pole, rubbish hook, hook, or other suitable tool to verify the roof's integrity. Sound the roof along your entire path of travel. Keep the sounding tool in front of your body to provide maximum stability in case you encounter a weak portion of the roof (figure 17–5).

Figure 17–5. When traversing a roof, sound the path of travel to indicate the strong and weak areas.

Roof personnel should also use their feet in conjunction with the sounding tool. Being sensitive with your feet can reveal a roof's bounce, flex, or sponginess (this concept is marginal at best on lightweight roofs). Think with your feet! Additionally, simply by sounding a roof, you can often determine its supported and unsupported sections, particularly on lightweight roofs.

Work from the weak area to the strong area/means of egress

Start ventilation cuts in the weakest portion of the roof (near the fire) and work away toward the strongest portion (unburned section). Spend as little time as possible on the weak portion. As an example, if a roof operation is commenced on a flat roof, start the cut in the middle of the roof (weakest portion) and cut towards the exterior wall (strongest portion). As a bonus, you will also be able to walk the strong portion of the roof back to your means of egress). Also, work away from the fire and toward the means of egress.

Keep the wind at your back

When possible, plan ventilation cuts so that the wind is at your back, with smoke and heat moving away as shown in figure 17–6. During the 1930s and 1940s, personnel were exposed to the by-products of wood, cotton, and paper. Today, personnel are plagued with the by-products of plastics, which have been proved carcinogenic. Consider using breathing apparatus when ventilating roofs.

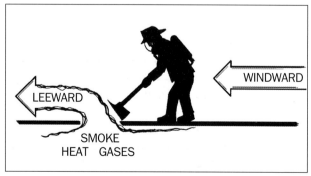

Figure 17–6. When using power saws on a roof, keep the wind at your back.

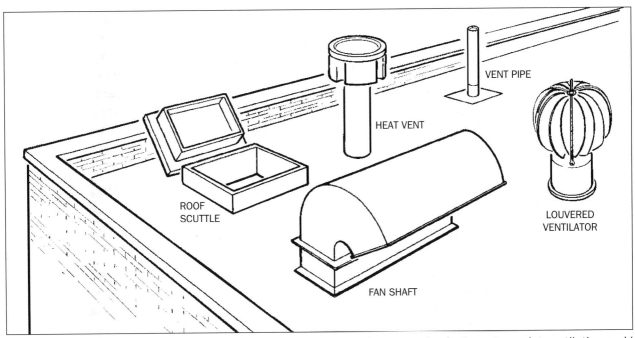

Figure 17–7. Be able to recognize and use various types of natural construction features to assist ventilation and/or detect the extension of fire.

Utilize natural construction features

Why cut an opening in a roof when natural construction features might be used to promote ventilation? Features such as scuttle covers, penthouse doors, and skylights can be fast and efficient means of opening a hole. A variety of features (figure 17–7) can be used for roof ventilation.

Air shafts. Air shafts or light wells may be found in older multi-story residentials, such as hotels. They are intended to provide air and light to inner bathrooms. An air shaft can be used for ventilation purposes by opening windows within the shaft to rooms that need ventilation. Remember that bathroom windows are usually small. They provide minimal ventilation. Close windows within the shaft as appropriate to prevent extension of heat and smoke. If present, you will need to remove the skylight at the top of the air shaft for this avenue to be effective.

Skylights. Skylights are good indicators of building floor plans. Skylights in residentials are located over hallways. In commercial occupancies, skylights are often placed over manufacturing areas. Depending on the type of structure, individual skylights may also be placed over stairways, light shafts, and air shafts. If a building has an attic, the area below the skylight is normally boxed off from the attic space. Therefore, when the skylight is opened, the interior of the building will be ventilated but the attic space won't be. Skylights may be opened for ventilation purposes by removing the entire assembly or by removing the glass panels. If the glass panels are removed individually, additional time may be required, and broken glass may fall into the building. The most effective method is to remove the skylight assembly completely or tip it onto the roof. This takes minimal time and reveals the entire opening. It also allows you to replace the skylight later. Skylights can be opened as follows.

Glass panel skylights. If a skylight is directly attached to the roof, cut the cricket along three sides. A cricket is the junction of the skylight (vertical line), roof (horizontal line), and composition covering (curved line). Tip the skylight over using the fourth, uncut cricket as a hinge. Use a pickhead axe or power saw to make the cuts.

Even if a skylight is mounted on 4- to 6-inch wood risers, as shown in figure 17–8, it can still be tipped over. Use an axe to remove the nails at two corners and lift up, using the opposing side as a hinge. Notice in the illustration that there is tar around the edges of the glass panels. This would make removal of the panels difficult to impossible. If it appears that tar hasn't been used around the edges of the glass and the panels can be quickly and easily removed, then remove the metal tabs or metal stripping along the bottom edge of the panels. This will allow them to slide out. If tar has been used as a sealant and the glass panels can't be removed easily, then remove the entire skylight.

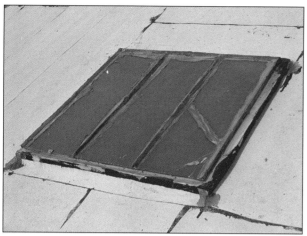

Figure 17–8. Older glass skylights are often sealed with tar, making the glass panels difficult to remove.

Remember, if you break the glass panels, the resulting shards of glass within the building will be a hazard to attack and search operations.

Plastic skylights. Newer buildings have plastic skylights, usually a 4-by-8-foot plastic bubble in a metal frame attached to a metal riser. First, try to pry the skylight away from the roof. If this can't be done easily, you may have to cut the plastic around the junction of the metal and plastic. Use an axe or power saw to make the cut. Colored and frosted plastic as well as fiberglass panels often serve as skylights, usually in roofs of corrugated metal. These panels can easily blend in with the existing roof, but they won't support any significant weight. If appropriate, they can easily be removed by raising or lifting the roof covering along one edge and then lifting the panel. If no personnel are below, you may opt to push the panel downward.

Penthouse. Opening the door to a penthouse can provide significant ventilation to the interior stair shaft and any hallways that are open to it. Due to the importance and simplicity of this, opening a penthouse should be a primary consideration. If a penthouse has a lower door at the bottom of the penthouse stairs, you must verify the conditions on the other side before opening it.

Monitor. A monitor is normally a rectangular structure on the ridge of a roof, is usually found on older commercial buildings, and was used to provide natural ventilation to the interior of a building. Ventilation can be increased from these structures by removing the slats, wire mesh, and/or glass (depending on how the monitor was constructed).

Roof scuttle. A roof scuttle is a small, covered opening providing access to the roof from the interior of a building. Some scuttles consist of a wood cover over wooden risers on the roof. The cover is normally protected by a composition covering. Unless the cover has been attached to the risers, it can be removed by prying it up with an axe or similar tool. Newer scuttles consist of metal risers with a spring-loaded metal door that is fastened and locked from the inside. These are difficult to open and should be left alone unless access from the roof to the interior is necessary. An opened roof scuttle can be a good indicator of the location of fire. However, if the open scuttle will draw fire, replace the cover.

Elevator house. Structures with elevators may have an equipment house on the roof. Elevator houses vary in size according to the number of elevators that they service. They are above the elevator shafts and open, by varying degrees, to the shafts. Opening the door and skylight (if present) can provide ventilation to the elevator shafts, reducing concentrations of heat and smoke within. Remember there is a shaft/smoke vent in the floor of these structures that likely will be unable to support the weight of a firefighter. Additionally, the equipment in an elevator house supports the full weight of the attached elevators. Therefore, any fire that has extended to this area may weaken the structural integrity of the elevator equipment supports, possibly causing collapse of the equipment and elevator cars. The resultant hazards and damage that this might cause cannot be overemphasized.

Ducting. Heating, air conditioning, and other ducts of various sizes are often found on roofs. Because these passageways may lead to the interior of a building, they need to be checked for heat and smoke. If heat and smoke are present, open the duct to provide some ventilation to the interior of the building, as well as to check for extension.

Vent pipes. Plumbing vent pipes extend through a given roof at various locations. Although they aren't considered to be ventilation avenues, they do travel horizontally and vertically within a building. Therefore, they are capable of spreading heat and fire. Smoke issuing from ABS plastic vent pipes is an excellent indicator that the pipe is burning somewhere within the building. ABS plastic pipes can readily burn and serve as extension avenues wherever the pipe travels. If the roof around any type of vent pipe shows signs of heat or smoke, it must be opened and checked. Additionally, a vent pipe that appears to be growing upward is an excellent indication that the roof is sagging and possibly about to fail.

Ventilators. Many types, sizes, and shapes of ventilators may be found on roofs. Ventilators can be classified as either venting a particular area within a building or providing some type of ventilation for a device or appliance. The size and shape will usually tell you which type it is. Ventilators for heat-producing appliances generally have a serrated top suspended over the pipe (refer to figure 17–7), whereas those that vent the attic or top floor usually have a capped top, and eyebrow type, or the familiar turbine. Smoke or heat coming from a ventilator indicates that the area it serves is affected by smoke or fire. If a ventilator is equipped with a screen, the screen should be removed because it will restrict the efficiency of escaping smoke. Don't remove the top of either a stationary or a rotating ventilator, however, because the ventilator stack can easily be damaged and restricted. Rotating ventilators are 30% more efficient when the turbine is operational as compared to when the turbine has been taken off. Ventilators are designed to ventilate. Let them do their job.

Cut only as deep as necessary

Personnel using power saws for ventilation purposes must control the depth of the cut and know what is needed to accomplish specific ventilation operations. Unless otherwise necessary, make ventilation cuts through the roof decking only. Cuts deeper than the roof decking increase the possibility of severing structural members and electrical conduit. As shown in figure 17–9, a cut deeper than 2 inches will compromise the structural members of a typical lightweight roof.

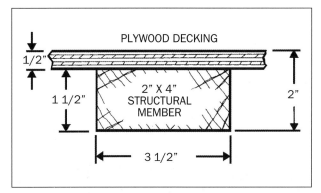

Figure 17-9. Cuts made deeper than the roof decking increase the chance of severing structural members.

Use the principle of clearspan

Clearspan is the uninterrupted or unsupported span of a roof or its structural members. In figure 17–10, the strong portion of the roof is the area where the roof ties into the walls (A), and the weakest portion is the center of the span (B). If this roof were to collapse as a result of fire, the structural members would most likely fail at the center and can remain attached at the walls.

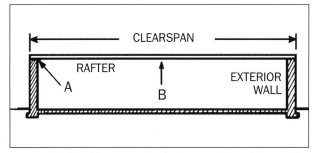

Figure 17-10. Clearspan is the unsupported span of a roof. Plan ventilation cuts to start in the weakest portion (B) and finish in the strongest portion (A).

Power equipment considerations

A power saw makes quick work of cutting through roofs, but the same quality that makes it so valuable also makes it potentially dangerous. Kickback, broken chains, and flying debris are hazards that can cause serious injury if personnel are unwary. To be safe, position yourself out of line with the saw (figure 17–11). Stand at a point 90 degrees away so that you will face the work rather than straddle it.

Figure 17–11. For safety, position yourself out of line with power saws.

Hose lines on the roof

If used correctly, a charged hose line with a spray nozzle can help in roof operations. Depending on the speed and direction of the wind, a spray stream can effectively deflect smoke and heat away from personnel. Also, if a ventilation opening begins to emit fire, combustible roofing material may ignite. Having a charged hose line at the ready helps you extinguish such surface fires before they become a serious hazard. However, hose lines on roofs must *not* be directed into ventilation openings to extinguish attic fires, particularly when an interior attack is underway and interior hose lines will be directed

into the attic. Under these circumstances, roof personnel could cause severe steam burns to interior personnel and force them out of the immediate area. Directing a hose line into an attic can also drive heat, smoke, and fire back into the building.

Size of ventilation openings

To adequately ventilate any building, make the opening commensurate with the amount of heat and smoke inside the building. Observe the pressure of the venting smoke. If the contaminants are coming out lazily, the opening probably doesn't have to be enlarged. However, if the contaminants are venting under pressure, then the opening is too small. Keep increasing the size of it or create additional openings as necessary. As large openings may be difficult to completely open, remember that several smaller openings that have been completely uncovered may be superior to one large opening that has only been partially opened. Additionally, the type of ventilation opening will usually affect its size. A cut that is made over a fire is usually square or rectangular, whereas strip openings cut ahead of a fire may be long and narrow.

Location of ventilation openings

Although the proper location for roof ventilation openings is dependent on the type of structure and incident, they can generally be divided into operations that consist of utilizing natural construction features and operations that cut openings in a roof.

Normally, natural construction features are opened *first* due to their capability to provide effective ventilation in a minimal amount of time. As an example, opening a penthouse door will ventilate a vertical stairshaft, opening skylights will ventilate hallways, and opening doors and/or windows can provide horizontal ventilation. Due to the minimal time constraints of these operations, it must be remembered that these types of openings must be in the right location as an opening has the ability to draw fire to its location.

Cutting openings in roofs can also provide effective ventilation but at the expense of time and the necessity of using power equipment. These openings are normally opened second. When ventilation openings are cut in a roof, the type of construction can determine the location of a ventilation opening. For example, conventional construction obtains its strength from the size of its structural members and can allow ventilation openings to be made over or near the fire if the roof is deemed safe. The key ingredient is the time available between impingement by flame and structural collapse. Conversely, lightweight construction is subject to fast failure rates when exposed to fire and will not allow adequate time for these operations. Therefore, any ventilation operations on lightweight roofs should focus on uninvolved portions.

To expand on the previous statement, let's consider a simple example. Assume the building in figure 17–12 is 100 feet long and the roof structural members are lightweight trusses.

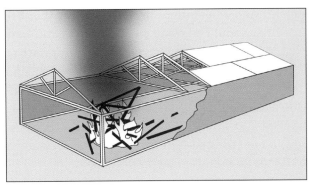

Figure 17–13. The fire has extended into the roof structural members with a resultant collapse of the exposed trusses.

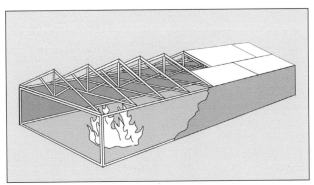

Figure 17–12. The fire in this building has not extended into the roof structural members.

Assume a fire is within the building (between the four walls, floor and ceiling only) as depicted. Because the fire is not directly exposing the roof structural members, the chance of these members collapsing is minimal at best. Therefore, the risk to firefighters on the roof is nominal, and roof operations can be more aggressive. However, assume the fire extends into the attic area and exposes the lightweight structural members as depicted in figure 17–13.

The structural members are prone to collapse in a short period of time (average 5 to 7 minutes) and the risk to firefighters on the roof is exceptionally high. In this case, roof operations with power saws (near or over the fire area) is not a safe or recommended operation. Let's summarize roof operations using figures 17–12 and 17–13:

If fire is *not* within an attic/cockloft area composed of lightweight construction, roof ventilation operations should be able to be conducted with minimal risk to personnel. If fire is within an attic/cockloft area comprised of lightweight construction, roof ventilation operations (near or over the fire area) can be categorized as an unsafe operation (see Training Resources, "Precious Faith Temple Church Fire"). Specifically using figure 17–13, let's ask an important question: When the portion of the burned roof structural members collapses as depicted, will the unburned portion also collapse or remain as depicted? If the unburned portion is not exposed to fire, it will not collapse into the building. This is a result of the roof and structural members being attached to the perimeter of the building, and the connecting member between the last truss that collapses and the first adjacent truss that does not collapse likely being ½-inch OSB, plywood, or other similar materials that are not strong lateral members. Therefore, roof ventilation operations on lightweight construction can be safely conducted only on those portions that are *not* exposed to fire and will likely consist of defensive operations (strip ventilation) only.

Use the roof construction to your advantage

When possible, make ventilation cuts parallel to structural members, yielding sections of decking nailed to a single rafter. This ensures ease of removal. Unlike horizontal openings, which are relatively easy to make, openings made in roof decking involve specific considerations if they are to be made quickly and efficiently. This doesn't imply that the openings should be small and made in haste. They should be easy to open and of a size appropriate to the needs of the incident. Therefore, when cutting and removing roof decking, don't cut a ventilation opening that cannot be easily and completely opened. A common recommendation for roof ventilation openings is 4 by 4 foot for dwellings and 8 by 8 foot for commercial properties.

Figure 17–14. This 8-by-8-foot opening couldn't be completed because the decking is nailed to three rafters. In this case, maximum efficiency was not obtained. (LAFD file photo)

Interestingly, there is often a lack of information on how to effectively open such openings. The 8-by-8-foot opening in figure 17–14 wasn't effectively opened due to a combination of reasons. The decking was nailed to three rafters, and it was necessary to reach out over the opening with a tool to pull it up while smoke and heat were issuing from below. If, in addition to the four perimeter cuts, an additional cut had been made between the three rafters, the opening would have been modified from one large section to three smaller sections, each attached to one rafter. These smaller sections could have easily been removed, exposing the entire opening. It is easier to remove decking nailed to one rafter than to two or three. The concept of center rafter operations is explained in chapter 18.

Don't be a roof shepherd

Ventilation personnel have a tendency to remain on roofs after the cuts have been made even though other fireground operations still need to be completed. In some cases, however, it may be necessary for a ventilation team to monitor the openings to verify possible changes in the fire's behavior When roof ventilation operations are finished, personnel should not spend unnecessary time *admiring their handiwork* (figure 17–15) because there are stronger and safer areas to rest in addition to other fireground priorities that likely need to be completed.

When roof operations are completed, exit the roof in a safe and timely manner.

Figure 17–15. When roof operations are completed, do not spend additional time admiring your handiwork.

Path of egress

Always know how to exit a roof safely. Generally, leave the roof the same way that you climbed onto it. Unless conditions have changed, consider the original path to be the best way off. If your path to the roof was safe, it should still be safe for your exit from the roof.

Ventilation Equipment

Although a wide range of tools and equipment may be used for ventilation, the following are some of the more common. When comparing products, remember that they will vary depending on a particular manufacturer. Therefore, the following comments are general in nature.

Pike pole

The venerable pike pole is found in a variety of lengths. It is used for various ventilation purposes such as breaking windows, removing tile or shingles from roofs, pulling up portions of roof decking, and opening ceilings. A useful part of the pike pole is the pointed end of the head, which allows you to start an opening in tough materials such as compressed lath and plaster. However, pike poles do not readily pull up cut sections of plywood/OSB type materials.

Rubbish hook

The rubbish hook was originally designed as a companion to the pike pole to move debris during overhaul and to strip shingles from roofs. Although the pike pole is versatile, the rubbish hook has proved superior in several ways. As a sounding tool, the heel of the head offers a strong, wide area to strike the roof. Also, the width of a rubbish hook simplifies pulling up cut portions of a roof. The width of the rubbish hook also makes it effective at opening plaster walls. This is graphically demonstrated when an opening has been made in a roof and it then becomes necessary to remove the ceiling to complete the ventilation avenue. Using the point of a pike pole is time-consuming compared to using the rubbish hook.

The increasing use of fiberglass has resulted in significant improvements over tools with wooden handles. Fiberglass offers a number of advantages, not the least of which is strength. Fiberglass is stronger than wood, virtually eliminating broken handles in pike poles, rubbish hooks, and axes. Fiberglass is also weatherproof, impervious to ordinary chemicals, and it won't conduct electricity. By its nature, it requires minimal maintenance, and it can be made in bright colors for easy identification even under

adverse conditions. D-grip handles provide additional flexibility, pulling power, and grip. As shown in figure 17–16, it is possible to purchase pike poles and rubbish hooks in three parts: a three-foot handle, a head, and an extension of any specified length.

This allows the tools to be broken down into three-foot lengths, making them interchangeable as well as convenient to store and carry.

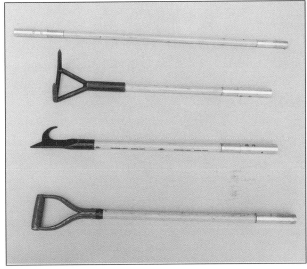

Figure 17–16. Some roof ventilation tools have interchangeable parts, increasing their versatility.

Axe

The axe is one of the most useful and dependable ventilation tools available. For ventilation operations, the pickhead axe is more versatile than its flathead counterpart since the pick can be used for a variety of tasks. Due to the grain of the wood in wooden handles, the strongest axis when prying is in line with the grain of the wood. Use care when prying against the grain. Most training materials regarding axes suggest cutting 60 degrees to the grain of the wood. Although this can provide acceptable results, it has several drawbacks. First, not all of the energy expended for cutting is effectively used. Also, it is an inefficient way of cutting plywood/OSB materials. Finally, when cutting, the axe head may plunge beneath the surface being cut, thereby becoming trapped. These problems can be overcome. Instead of cutting 60 degrees to the grain of the wood, bring the axe downward in a smooth arc so that the cutting edge is 90 degrees to the work surface at contact (figure 17–17). This results in more cutting power with less effort. It's important to position your feet properly so that the axe doesn't accidentally strike your leg or foot.

Figure 17–17. When cutting a roof with an axe, strike the work at 90 degrees instead of 60 degrees.

Therefore, stand with your feet toward the cut. This is also effective when cutting plywood/OSB. If the head of an axe becomes trapped in or below the work surface, simply use the toe end of a boot (steel toe only) as a fulcrum at the junction of the axe head and the handle to lever the axe out again (figure 17–18).

When cutting sheathing or plywood with an axe, determine the location of the rafters by sounding the roof with the head of the axe. Then, make the cut as close to the rafter as possible. This provides a firmer foundation for the cut. Axes can also be used with power saws on rock or gravel roofs. When inserting a power saw into a rock or gravel roof, the cutting teeth are damaged by initially coming down through the gravel surface when starting the cut. However, if a small cut is made with an axe, the power saw can then be inserted into the opening. This allows the teeth to pass through the opening to the underside of the roof, cutting through the wood and causing the rock or gravel surface to be thrown aside as the teeth move upward. The standard configuration for a pickhead axe is a six-pound head mated to a 32- or 33-inch handle. A handle length of 36 inches can be more effective for cutting. Wooden handles shouldn't be varnished. Keep them clean and rub them periodically with boiled linseed oil to enhance their grip and make them easier to use. Also, keep the cutting edge of an axe head sharp. When sharpening, the sides of the blade should curve toward the cutting edge rather than taper toward it. Curving these surfaces makes them more resistant to chipping when cutting hard materials. Use a scabbard when carrying an axe (figure 17–19). This allows you to use both hands for other purposes while keeping the axe available for use. Scabbards can be made of either leather or Cordura.

Figure 17–18. An axe head that is trapped below roof decking can be levered out by placing the toe of a boot between the roof and the junction of the head and handle.

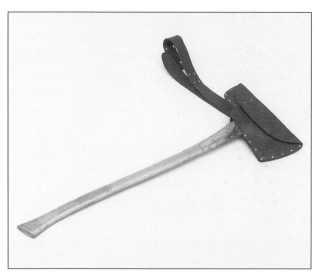

Figure 17–19. Carrying an axe in a scabbard allows both hands to be available for other purposes.

When traversing a pitched roof, keep the pickhead axe at the ready. The pick can be used to pierce the roof and act as an anchor in case you fall.

Halligan and hook

Some departments use a Halligan and a hook as standard roof ventilation tools and *marry* the tools together to simplify carrying. A hook can be used for punching an inspection hole, sounding, and pulling up roof decking although it is not as effective on plywood/OSB and sounding as a rubbish hook. A Halligan is a great multi-purpose tool, particularly for forcible entry, although its use on a roof is limited compared to a rubbish hook. This is primarily a result of its length, which limits its capabilities.

Thermal imaging camera (TIC)

More departments are successfully using thermal imagining cameras (TICs) for roof operations because they can locate an area under a roof that is exposed to fire and locate some structural members (larger members retain more heat than their surroundings). The value of this tool, however, is based on the expertise of the operator.

Rotary saw

Rotary and chain saws can both benefit from the following considerations. If possible, power

equipment from the same manufacturer can simplify maintenance, operational considerations, and purchasing parts; always use a compression release if so equipped; and when starting, pull the cord in short pulls to avoid early failure of the cord and/or jam the recoil mechanism. For proper starting techniques, see Training Resources, "Simple Saw Starting."

The rotary saw is a versatile tool that can simplify ventilation operations. It is adept at cutting thick or heavy roof composition usually found on older roofs, and it is most efficient when used on metal deck and lightweight concrete roofs. As with any power saw, the rotary saw can be dangerous, and it exhibits two unique tendencies that must be borne in mind. The blade will continue to spin at high rpms after the throttle is released, so never leave a saw unattended while it is still spinning. The blade should be stopped in the work material or monitored until it comes to a halt. Also, rotary saws exhibit a gyroscopic effect caused by the spinning blade. This can make the saw somewhat unwieldy when changing its cutting angle. When operating a rotary saw, use approved safety apparel that includes a helmet, gloves, protective clothing, and a face shield or goggles. If you suspect an electrical hazard, wear gauntlet gloves.

Rotary saws require the saw operator to move backward while cutting. Therefore, it is necessary to have an additional person on hand to act as a safety. This person will monitor the roof conditions, watch the path of the cut, and control the operator if necessary. Some rotary saws feature a reversible blade and guard, allowing them to be mounted either in the outboard or inboard position (figure 17–20). Mounting a blade and guard in the outboard position allows you to make cuts in close proximity to walls and other vertical surfaces.

This is best when used for forcible entry operations. The inboard position is preferred for ventilation because this position tends to balance the saw, reducing the gyroscopic effect. To prevent the blade from binding, never operate the saw without a blade guard. Adjust the guard so that the rear section is close to the workpiece. Particles, dust, and sparks will then be collected by the guard and directed away from the operator. Also, adjust the guard to control the proper depth of the cut. Enter all cuts at full rpm, and regulate the speed of the saw by increasing or decreasing the feed pressure.

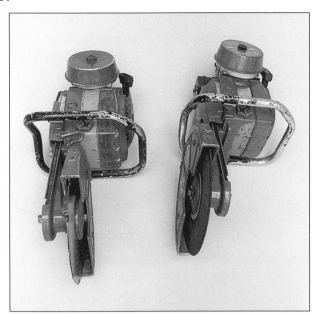

Figure 17–20. Most rotary saws allow the blade to be mounted in either an outboard or inboard position.

While cutting, personnel have a tendency to hold the saw off the roof. This places unnecessary weight on the spine and promotes binding. To reduce binding, make certain that the blade is kept at a right angle to the work surface by setting the saw on the roof and pulling it through the material (figure 17–21). With the blade and guard centered, and by using the guard as a depth gauge with the lower portion of the carrying handle as a pivot and skid, pulling the saw through the material will be simplified and the weight on your back will be reduced.

Additionally, always cut in a straight line, and don't twist the saw to either side. If binding occurs, release the throttle and immediately remove the blade from the cutting surface to avoid belt damage. Also, always use the lower portion of the blade for cutting purposes. If the upper portion of the blade is used for cutting, it may start to climb the cut. This will throw the saw up and back toward the operator.

Generally, carbide-tipped wood-cutting blades are the most effective types for most ventilation operations. These are very effective for cutting wood, light sheet metals, and acrylic. After each use, check for carbide tips that are missing or that have a visible radius. Remember that a rotary saw has a limited depth of cut. This is usually 3½ to 5 inches, depending on the type and model of saw. With a rotary saw,

it is difficult, at best, to feel the construction below when cutting a roof or attempting to determine the direction and spacing of the rafters. You should carry an appropriate tool to facilitate adjustment of the belt or to change the blade, if necessary. A belt can quickly be rendered useless if it is improperly adjusted.

Multi-use blades are becoming more popular because they cut a wide range of materials without having to change blades.

Figure 17–21. When cutting with a rotary saw, set the saw on the work surface to reduce stress on the back and keep the blade at a correct angle to the work surface.

Chain saw

As with a rotary saw, a chain saw can simplify numerous ventilation operations to a significant degree. It can be effective on roofs with wood decking. Depending on the size of the saw, it will cut the heavy layers of composition usually found on older structures, as well as the built-up layers of insulation and composition on metal-deck roofs. Additionally, chain saws have excellent reach and balance, which is beneficial during cutting operations. They are also lighter than rotary saws. Unlike other power tools, a chain saw allows you to feel or read the construction below to determine the location of the rafters and joists. At the same time, a chain saw presents a variety of hazards. The chain and cutting teeth are exposed, thus requires constant attention by the operator. Chain guards are available that can be adjusted to reveal only that portion of chain that is needed for cutting (figure 17–22). These guards can also be adjusted for a specific decking thickness, and thus be used as a depth gauge.

Figure 17–22. A chain guard increases safety and can serve as a depth gauge. A muffler guard reduces maintenance considerations. (Photo courtesy of Cutters Edge)

Although most chain saws are equipped with a centrifugal clutch and/or chain brake, the chain may continue to spin at idle. Therefore, turning around with a running saw must be done with extreme caution. Chains are capable of dislodging objects and throwing them with significant force. Wear appropriate eye protection to guard against nails, rocks, and splinters. Chain saws are capable of cutting to a depth governed by the length of the guide bar. This can vary from 12 to 24 inches, depending on the size of the bar. Saw operators *must* determine the proper depth of the cut to minimize the chance of cutting through structural members and electrical wires. The following guidelines are offered as general criteria when purchasing a chain saw for use on the fireground.

Engine size. Consider about four cubic inches (65 cc) to be the minimum displacement for fire service applications. An engine of this size provides adequate power for both residential and most commercial operations. For high altitudes or older roofs, 72 cubic inches (or larger) can be beneficial.

Guide bar. In most cases, the guide bar length should be about 18 to 20 inches. Shorter guide bars can cause the saw operator to put additional stress on

a back. Efficiency can be dramatically increased if a guide bar is equipped with a sprocket tip instead of the conventional hard nose bar. A sprocket tip reduces friction, allowing the chain to run at a higher speed and at a significantly lower temperature. This increases cutting speed. Tar won't tend to melt and clog the chain, and the chain will maintain proper tension for a longer period of time.

Chain. A carbide-tipped chain is superior to a conventional chain for ventilation operations. A chain pitch of .404 inches has proven to be durable. If possible, test before buying unless a particular chain has proved satisfactory. A particular carbide chain currently available is capable of effectively cutting thick roof deckings quickly. It can also cut up to 14-gauge steel without immediately rendering the chain and cutters ineffective.

Miscellaneous. A muffler guard is a piece of steel or aluminum that is attached to the front of the muffler to deflect dirt, tar, and other waste material away from the engine. This simple modification significantly reduces maintenance. Spraying commercially available vegetable shortening on a *clean* engine and bar also reduces maintenance and clean-up operations by minimizing the buildup of waste material. The chain adjustment tool can easily be mounted to a saw by attaching a short piece of ¼- or ⅜-inch fuel line hose to the hand bar and then inserting the tool into the hose. To make carrying easier, attach a nylon shoulder sling to the saw. This allows you to have both hands free for climbing ladders or carrying other equipment (this also applies to rotary saws).

As with a rotary saw, proper protective apparel is recommended as is the use of a safety person while cutting. A running power saw can create a significant communication problem, so the safety person may have to use gestures to communicate effectively. An effective method for a safety person to maintain control of the saw operator is to use an *underhanded* grip on the axe belt or SCBA frame. The following recommendations will improve the effectiveness and safety of chain saw operations.

Make a decision as to the area to be cut based upon the criteria given earlier in this chapter. An additional reason to work upwind of the fire is that a gasoline engine will produce maximum horsepower in clean air.

A safe depth of cut is defined as being deep enough to create a ventilation opening without impairing the structural integrity of the roof. Excessively deep cuts endanger ventilation members and subject the operator to electrical hazards. Striking foreign objects below the decking can also cause chain damage, rendering the saw inoperative.

If you cannot properly use a power saw under simulated conditions (training scenarios), you do not belong on a roof under fire conditions with one!

Chain lubrication. A lack of proper lubrication will cause premature wear to a chain and guide bar. Saws with automatic oilers should be adjusted for maximum oiling. Do not use common engine oils for guide bar oils or add kerosene, diesel fuel, or other such solvents to guide bar oils. Guide bar oils are specifically designed to provide the proper viscosity and anti-sling properties that common oils do not possess. Additionally, use light-viscosity chain bar oils in cold climates and heavier viscosity chain bar oils in warmer climates.

Chain tension and use. Proper chain tension is important to the longevity of a guide bar and chain. Tension should be checked before, during, and after each use. Tension is correct when the chain travels freely around the guide bar as it is pulled by hand and the bottoms of all tie straps and cutters just touch the underside of the guide bar rails. Avoid injury by using a rag or gloves when pulling a chain through its circuit of travel. Chains that have been adjusted while hot must be loosened afterward to allow for contraction. After use, inspect a carbide chain for three damaged teeth in a row, six damaged teeth total, any cracked or broken tie straps, and dull teeth, as indicated by a visible radius on their edges. If you find any of these conditions, replace the chain. If you have a bullet chain, and if 50 percent of the carbide is missing on a cutter, replace the cutter. If 30 to 40 percent of the cutters are damaged or broken, replace the cutters or the chain.

Blowers

Currently, blowers are powered by water, hydraulics, electricity, or gasoline. Depending on the manufacturer, they may be purchased in numerous sizes and configurations (figure 17–23). When selecting a blower, evaluating the sizes, power ratings, and features can be challenging at best.

Performance. Performance, or the amount of air that is moved by a particular blower, is measured in cubic feet per minute (cfm). This rating is an estimate, and different rating methods will yield different cfm ratings for the same blower. Therefore, be sure that comparable blowers have been rated by the same method because advertising claims can be very misleading.

Size. Consider the cfm necessary for a specific application, and factor in the exterior dimension and weight of the blower. Remember that a blower must be stored in your apparatus and hand-carried to the appropriate location. Generally, larger blowers offer greater versatility. However, as the size of a blower increases, so do the problems of storing it. Blowers of less than 18 inches generally do not offer high cfm ratings because they are limited in power and fan size. Blowers of 18 inches offer good cfm ratings and power choices. They perform satisfactorily in single-family dwellings but are limited in larger structures. Blowers of 21 inches are increasingly popular for their power output versus size. Twenty-four-inch blowers come in a wide range of power choices, and their size offers high cfm ratings, making them viable in most buildings.

Power. The power source for a blower is a major consideration due to setup time and effect on cfm ratings. Their setup time, however, can be a limiting factor. In the past, electric blowers did not produce the cfm of comparable water and gasoline-powered blowers. However, electric blowers are now powered with electronic variable speed motors that produce good cfm ratings when compared to an equivalent size of blower powered with a gasoline engine, these blowers can be lighter than their gasoline powered counterparts, and they do not produce carbon monoxide. Electric blowers can also be purchased with explosion-proof motors, which may be necessary in some instances.

For gasoline powered blowers, a 21-inch blower with a five horsepower gasoline engine has slightly less output than a 24-inch blower with a five horsepower gasoline engine, is about the same size as

Figure 17-23. Blowers are available in numerous sizes, power ratings, and styles.

an 18-inch blower with a five horsepower gasoline engine, is adequate for most single-family dwellings and commercial structures of moderate size, and can easily be handled by one person. Gasoline-powered blowers require little in the way of setup. However, one principal drawback of them is that they emit carbon monoxide. Tests have indicated that four-stroke engines produce about 140–170 parts per million (ppm) of carbon monoxide and can add that amount into a confined area that is being properly ventilated. However, exhaust tube extensions will move carbon monoxide emissions away from the blower, virtually eliminating the problem. Gasoline-powered blowers also require more maintenance than electric blowers, and the normal safety considerations of gasoline engines apply.

Weight. Consider weight when selecting a blower. Because it is a piece of equipment that must be moved into position, can it be deployed by one person or does it require two?

Gasoline

Chemicals, special additives, and ethanol have significantly changed the characteristics of modern gasoline as compared to a few short years ago. It is extremely important to use fresh gas in power saws. That means changing the gas at least once a month in your power equipment *and* supply cans. In most cases, it can be advantageous to add a fuel

stabilizer to gasoline, either for two-stroke or four-stroke power equipment as it has the capability to increase the *freshness* of gasoline. Additionally, gasoline should be stored in metal cans instead of plastic cans as the ethers in gasoline can penetrate the plastic and begin to degrade gasoline in several weeks, and should not be stored in the same compartment as composite blades for circular saws.

Training Resources

The following resources can provide additional information and/or another viewpoint for the subjects discussed in this chapter:

- Roof Operations: Listen, Look, and Cut, Michael A. Terpak, *Fire Engineering*, May 2000.

- Precious Faith Temple Church Fire, Lake Worth, Texas, Stuart Grant and Les Stephens, *Fire Engineering*, April 2006.

- Rotary Saw Use and Maintenance, Bill Gustin, *Fire Engineering*, June 2006.

- Small Town Truck Operations: Power Saws, Jamie C. Morelock, *Fire Engineering*, March 2007.

- Simple Saw Starting, Michael N. Ciampo, *Fire Engineering*, July 1998.

- Nozzle Tests Prove Fireground Realities, Jerry Knapp, Tim Pillsworth, and Sean P. White, *Fire Engineering*, February 2003.

- Saw Operations: The Tap Method, Michael N. Ciampo, *Fire Engineering*, April 2000.

- Venting Windows, Ray McCormack, *Fire Engineering*, March 1997.

- Using Power Tools Efficiently and Safely, Steve Shupert, *Fire Engineering*, February 2009.

- Truck Company Tools Across the Country, Michael N. Ciampo, *Fire Engineering*, March 2010.

Roof Ventilation Techniques for Conventional and Lightweight Construction

Introduction

Ventilation operations consist of systematic procedures to redirect and remove fire, smoke, heat, and gases from involved structures. Within this definition are two phrases that deserve additional emphasis. The term "systematic procedures" suggests that ventilation operations are most effective when they are based on a specific purpose or intent, and that they shouldn't be randomly used. They should be predicated on SOPs that form a foundation for consistent, effective, and safe operations. The second point of emphasis concerns the redirection and removal of heat, gases, and other contaminants from the fire area. Doing so decreases the possibility of flashover, reduces temperatures within the structure, improves visibility, and increases safety for anyone operating inside.

Properly executed, ventilation entails opening a building in a precise, calculated way to achieve a specific strategic purpose. Roof ventilation operations are most effective when they are designed to achieve a specific goal rather than as the result of a haphazard operation. Fire within a confined area can be channeled in a horizontal direction, reducing interior temperatures. Fire can also be redirected upward, thereby curbing horizontal extension. Proper ventilation openings can also be used to control extension by channeling fire or redirecting its travel. In short, a fire can be controlled to some degree by coaxing it to take a less dangerous path, thereby benefiting the search for occupants, enhancing suppression operations, and minimizing damage to the building. As with all other types of operations, safe and effective roof ventilation depends on training, knowledge, and the ability to implement basic principles and procedures. This can be expressed as an equation: *training + experience + pre-fire planning = SOPs*.

Practice

Look at the buildings in your district and become familiar with the different types of construction that you might encounter. Determine basic operations before a fire, not during it, and practice those operations on a frequent basis. An excellent method of practice is to use chalk to diagram routes of travel, features of construction, and where and how to place a ventilation opening. It is important to practice correctly. Human beings are creatures of habit. Chances are that firefighters will fight a fire exactly the same way that they practice for it. If they adhere to basic safety rules and correct ventilation procedures, then these tasks will be automatic at an incident, giving personnel a greater opportunity to focus on what needs to be accomplished.

Implementation

A working knowledge of various ventilation methods is necessary to meet the various types of roof construction that you will encounter. Just as there are similarities and differences between conventional and lightweight roof construction, the operations to ventilate them will likewise exhibit equivalent traits. Prior to performing ventilation on any roof, you must evaluate a number of variables, including the age and type of roof, the location of the fire, and the likely benefit of the operation. Always identify the type of roof and its construction before you initiate ventilation operations. This can often help you determine related strengths, hazards, and operational techniques. Also, the approximate age of a roof can indicate its characteristics, such as the thickness of the decking. For example, you can expect one layer of composition asphalt shingles for every 10 to 15 years of age, and older roofs were often constructed with larger and heavier structural members as compared to newer, lightweight roofs. This can translate into how much time a roof can withstand the effects of fire before failing. An additional consideration is the implementation of newer roof coverings such as membrane roofs, gypsum roofs, stone-coated steel roofing, Tectum roofs, and the list goes on. You should be familiar with appropriate ventilation techniques for these roofs if they are in your area of responsibility.

Terminology

The following terms, diagnostic methods, and specific cutting techniques will be used throughout this chapter and are presented here alphabetically.

Axe bounce. An axe can be used to determine the direction of the wood *sheathing* on a roof. Strike the roof decking perpendicular or parallel to any exterior wall. The axe will stick when it is parallel to the grain and bounce when it is at right angles to it. Determining the direction of the wood sheathing will also tell you the direction of the rafters underneath the sheathing, since the rafters normally run perpendicular to the grain of wood sheathing. The success of this test can be nullified by the presence of plywood-type materials, diagonal sheathing, or thick layers of composition.

Center rafter. Sections of roof decking to be removed for ventilation purposes should be cut so that they are nailed to one rafter or structural member only. Positioning parallel cuts on either side of a single rafter will ensure that the severed decking material can easily be removed. As illustrated in figure 18–1, two parallel cuts on either side of a single rafter or joist (1) should be sufficient for decking composed of wood sheathing or space sheathing that is covered with shingles, shakes, gravel, and several layers of composition. However, in the case of corrugated metal, plywood-type materials, or numerous layers of composition, two additional cuts (2) are required to remove the section easily.

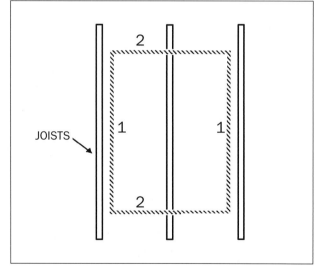

Figure 18–1. Sections of roof decking should be cut so they are nailed to one structural member. This ensures easy removal.

Coffin cut. This cut is created by cutting four sides (rectangular pattern) in wood decking material and a knockout at one of the corners by the head cut. This allows the top boards to be initially pulled allowing easy access to the remaining boards. Normally, the cut section of decking is pulled manually and not louvered. Also known as expandable cut.

Decking. Roof decking is composed of a base and a covering. The base is whatever material is attached to the structural members. It provides a foundation for the outer, visible roof. The base material may consist of wood sheathing, plywood-type materials, corrugated metal, or other materials. The covering consists of whatever material is designed to provide protection against the elements. Usually it consists of tile, shingles, layers of composition, and the like.

Dicing. Dicing is most effectively used on roofs with sheathing or space sheathing. Initially, you must determine whether a head cut is necessary. A head cut is used to breach thick or multiple layers of a roof covering, but it is normally unnecessary when the roof covering is minimal. If necessary, make a head cut perpendicular to the rafters (marked as optional, top of opening in figure 18–2). This can also help you determine the spacing of the rafters when you roll the saw over them. The head cut should be as long as the anticipated opening. Then make multiple dice cuts between and parallel to the rafters, intersecting the head cut.

The length of the dice cuts (A) shouldn't exceed the length of the tool (B) that will used to remove the decking material (figure 18–3). This ensures that personnel will stand in uncut or virgin portions of the roof when they remove the sections of decking. When done properly, the result will be sections of decking material nailed to a single rafter that can be removed by the J-hook motion.

If the roof decking is covered with numerous layers of material, use a base cut similar to the head cut (marked as optional, bottom of opening in figure 18–2) to facilitate removal of the sections. By louvering, you can easily enlarge the opening, if necessary, by making additional cuts between and parallel to the rafters.

Head cut. A cut through roof decking that is made perpendicular to the rafters (figure 18–4) and at the top of an opening.

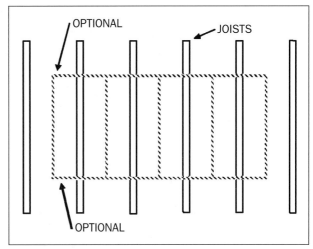

Figure 18-2. Dicing consists of making multiple cuts between joists on roofs with sheathing or space sheathing. A large opening can then easily be opened.

Figure 18-3. The length of the dice cuts should not exceed the length of the tool that will be used to remove the decking material.

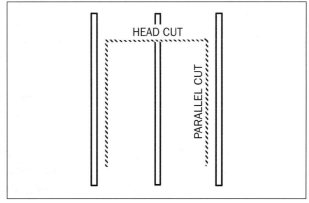

Figure 18-4. A head cut is made at the top of an opening and perpendicular to the structural members. Parallel cuts are made parallel to the structural members.

Indicator/inspection opening. A small opening made with an axe or power saw in roof decking and used to determine conditions within the structure. When placed in an area where personnel are working, it can also give a good indication of changing conditions, which may be vital to safety. For example, consider the following indications and how they would relate to operations: fire; hot-black-pressurized smoke; smoke with some pressure and color; cold-white smoke with minimal or no pressure; nothing showing. You can also monitor extension by placing indicator openings in the path of the fire; for example, in an area between a heat hole and where strip ventilation will be used. Placing an indicator opening in a roof on the *uninvolved* side of a division wall will help you determine whether there is a potential exposure problem. In all cases, clear away any insulation that may be present under the decking. You can remove it by pushing through the indicator hole with an axe handle, hook, or other similar tool. This opening can be created by either making a small hole with a hook, pickhead axe, or other similar tool, or making three small cuts to form a triangle shaped opening. The important point to remember is to make the opening small enough so that a firefighter cannot inadvertently step into the opening and break an ankle, foot, and so on.

45-degree inspection cut. An inspection cut made through the roof decking only and at an angle of 45-degrees to any exterior wall. Because roof structural members are either parallel or perpendicular to exterior walls, a 45-degree cut will ensure that the power saw (used at a 90-degree angle to the roof) intercepts a structural member. When the saw makes contact with a rafter, roll the rafter and continue the cut for approximately 1 foot (figure 18–5). Complete the operation by making two small cuts over the rafter as illustrated by the dotted lines. This will allow you to remove a small triangular portion of decking directly over the rafter. If necessary, the small section of roof that has been removed can be enlarged slightly to help you identify the construction.

When a small section of roof has been removed in this manner, the following factors can also be verified: the type of roof decking, the direction of the rafters, the depth of cut necessary to penetrate the decking, and the conditions inside. After you have learned the direction of the rafters, you must determine their spacing. Make a cut perpendicular to the rafters, and stop cutting when the saw encounters the adjacent rafter (2), thereby telling you the measurement.

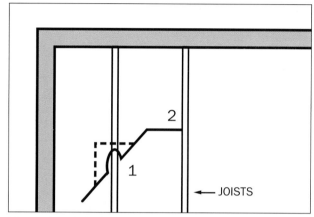

Figure 18-5. A 45-degree inspection cut can be used to determine the direction and spacing of rafters.

J-hook. The removal of sheathing is enhanced by using a J-hook motion with an appropriate tool as illustrated in figure 18–6. The J-hook motion brings the tool under and up to the decking in a smooth, forceful motion that will separate the decking materials from the rafter. It is simplified if the tool is used next to the rafter as illustrated.

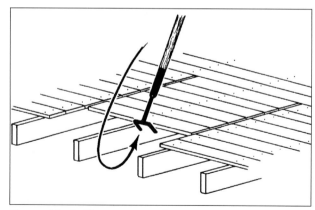

Figure 18-6. The removal of sheathing is enhanced by using a J-hook motion with an appropriate tool.

Kerf cut. A small insertion cut that is made with a power saw (or axe) in the decking and used as a quick indicator opening.

Louvering. The louver method can be used on most decking materials. It consists of making two parallel, longitudinal cuts on either side of a single rafter, then two parallel cuts that intersect them (figure 18–7). Ensure that the corners of the intersecting cuts overlap by several inches. The cut section of decking can then easily be louvered by pushing down on the near side of the panel and pulling up on the far side, using the rafter as a fulcrum.

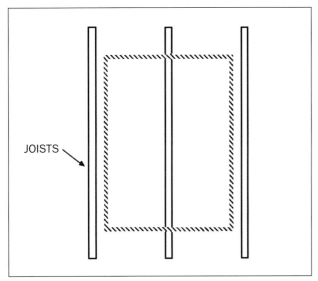

JOISTS

Figure 18–7. Roof decking can be louvered if, after four cuts have been made, the severed section is attached to one structural member as illustrated.

If you will be cutting additional panels, cut the second panel before louvering the first one, then cut the third panel prior to louvering the second one, and so on. This will reduce the exposure of personnel to heat and smoke. This operation is also known as hinging, butterfly, and other similar names.

Parallel cut. A cut through roof decking made parallel to the rafters (figure 18–4).

Personnel. It is recommended that a *minimum* of two personnel be used for roof ventilation operations. One person cuts the roof, and the other person acts as a safety and assists in opening the roof when appropriate. Commercial buildings may require a minimum of three personnel or more depending on the size of the building.

Plug. A small, triangular piece cut from a roof covering to reveal the type of roof decking and possibly the direction of the rafters. For example, metal joists are at right angles to the corrugations in metal deck roofs, and rafters are at right angles to the grain of sheathing. Additionally, if it is determined that the decking is composed of wood sheathing, the roof will likely be of conventional construction, whereas plywood/OSB decking materials indicate conventional or, most likely, lightweight construction. Corrugated metal usually denotes lightweight construction (see figure 17-3 in chapter 17).

Plywood-type materials. Plywood, particle board, oriented strand board (OSB), fiberboard, chip board, and other similar materials are commonly used as roof decking.

Pullback method. Use the pullback method only on decking composed of wood sheathing and space sheathing. Use two parallel cuts on either side of a single rafter to prepare the decking for removal (figure 18–4). If you encounter numerous layers of composition, use a score or head cut (marked as head cut) to simplify removal of the composition and the first few boards of sheathing. You can then easily remove the cut sheathing by striking the end of the top board with an axe or a rubbish hook. This will loosen or break the board and allow you to insert a tool underneath it to pry it from the center rafter. Use a J-hook motion to pull up and back on the remaining boards to remove the decking. As an alternative method, some departments use a cut out in one of the upper corners by the head cut to assist in removing the initial boards.

Rafters. Any of the structural members that slope from the ridge of a roof to the eaves and serve to support the roof. Can also be known as joists. Rafters are also used in flat roofs.

Rolling a rafter. When you encounter a rafter while cutting through roof decking with a power saw, lift the saw over the rafter and resume the cut (figure 18–8). Modern power saws are powerful and can easily cut through most decking and structural members. Therefore, whenever you insert a power saw into a roof covering, don't cut any deeper than necessary to remove the decking.

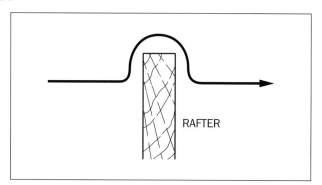

Figure 18-8. If you inadvertently encounter a rafter with a power saw, lift the saw over the rafter and resume the cut.

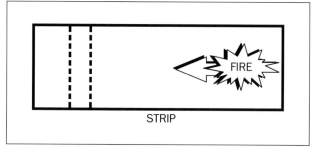

Figure 18-9. Strip ventilation openings are created ahead of a horizontally traveling fire to minimize extension.

Saw insertion. When inserting a chain saw into a roof to cut the decking and/or to feel a rafter, there are two positions that must be maintained for proper operation. If a chain saw is used to feel for a rafter, it must be used 90-degrees to the roof decking with about 8 inches of the bar inserted into the roof. This will maximize the feel of the saw on a rafter. Also, when a chain saw is used for other cutting operations, the saw should operate at a 45-degree angle to the roof decking with just enough insertion of the bar/chain to cut only the roof decking. This will maximize the ability of the saw to ride over rafters and other objects without inadvertently cutting through them. Both of these operations must be understood and practiced for proper utilization of a chain saw in roof ventilation operations.

Score cut. A light cut of the deck covering (usually composition) only. It is used to facilitate the removal of heavy or multiple layers of roofing material.

Strip. A long, narrow section of roof that is removed ahead of a horizontally traveling fire to minimize extension. Strip ventilation operations are normally performed across the width of a roof as illustrated in figure 18–9. To be effective, a strip should be about 3 feet wide. Strip operations are also known as trench ventilation.

Wet/dry. When cutting decking with a chain or rotary saw, the cut will either issue smoke and/or fire, or will not issue smoke and/or fire. This can be used to evaluate what is under the decking that is being cut during the cutting operation. From a simple perspective, visible smoke/fire is cutting wet, and no visible smoke/fire is cutting dry.

Methods

The following roof ventilation methods are not intended to limit the initiative and resourcefulness of ventilation personnel. However, there are numerous variations of these methods, some of which focus on altering a particular method so it is slightly larger, varying the order of cuts, louvering with or against the wind, changing the name of an operation, and so on. This section does not present all of the variations, but rather focuses on the basics that make these operations simple, yet effective. They are presented to stimulate thought and increase available options for when conditions favor their use.

When placing personnel on a roof for ventilation, follow these basic procedures as a minimum: if there is sufficient time, safety, and structural integrity, then roof ventilation will be a viable operation. Conversely, if time is not sufficient, safety is questionable, or if the roof is unstable, then you should immediately exit that roof. In conventional construction, the size of the structural members and their ability to withstand the effects of fire for reasonable lengths of time may mean that you can create ventilation openings directly over the fire. Compared to lightweight construction, conventional construction can resist the effects of fire for a greater period of time, potentially increasing the safety of roof operations. Conversely, due to fast failure rates and potential collapse hazards, ventilation openings should be placed as close to the fire as possible but on uninvolved portions of a lightweight roof. Ladder and approach the involved area from the uninvolved area; ladder the strong areas of the roof. Determine the location and extension of the fire, the type of roof, and the direction and spacing of the rafters. When possible, cut with the construction and, if possible, use the center-rafter principle. Use a minimum of two personnel. Be able to use a power saw either left or right handed (as you may not always be able to stand in the correct location on a roof when cutting the decking). If there is a ceiling within the building and a contaminated atmosphere below the ceiling, roof ventilation operations aren't complete until the ceiling has been opened to create a pathway from the contaminated area to the vent in the roof. Slash marks on head cuts denote the location of rafters underneath the roof decking and are only used for illustrative purposes. The following methods can be accomplished with a power saw or an axe.

Ventilation techniques for conventional construction

Conventional construction obtains much of its strength from the size of its structural members. This can allow you to open ventilation holes over the fire if the roof is deemed safe. Conventional construction can often allow the time and safety necessary to open a ventilation opening while the roof is in the process of being weakened by fire. When evaluating the type and construction of a conventional roof, it is important to determine the specific type of decking and its affect on ventilation operations. Diagonal sheathing can require longer cuts than straight sheathing to produce an opening of the same size. The presence of plywood/OSB-type materials on diagonal or straight sheathing (earthquake modifications) will require a deeper cut and four cuts around the perimeter of the opening to allow you to remove the decking material by louvering. Certain building modifications, such as earthquake protection, increase the potential for damaging power saws on metal straps, tie bars, and the like.

Pullback method

This method uses three cuts to prepare sheathing (usually found in older roofs) for removal. It isn't recommended for sheathing covered with plywood or plywood decking:

1. Make a head cut perpendicular to the rafters and across three of them. If

on a pitched roof, make the head cut approximately 1 foot below the ridge. (In this example, the direction of rafters is known as rafters on a pitched roof run from the ridge to the bottom of the roof. If this process is used on a flat roof, use a 45-degree inspection cut, if necessary, to determine the direction of the rafters.) When the saw makes contact with a rafter, roll over the rafter. This cut, depending on the thickness of the decking, may either be a score cut or one that is made through all of the layers.

2. Make two parallel cuts with the direction of the rafters between the center and outside rafters, starting at the head cut (figure 18–10).

Figure 18–10. To prepare wood sheathing for removal by the pullback method, make a head cut and two perpendicular cuts.

3. If necessary, loosen the top board next to the head cut by striking one end of the board with an axe, rubbish hook, hook, or other similar tool. This will allow you to insert a tool under the sheathing, where it is nailed to the rafter, and to remove the sheathing. Remove the top board using

a J-hook motion that will separate the board from the center rafter.

4. Continue to remove the decking material with a steady J-hook motion adjacent to the center rafter (figure 18–11). If necessary, score the composition covering at the bottom portion of the opening to ensure a clean and easy break. Using two personnel will simplify removal of the decking material. It is important that teams of personnel work together when removing decking.

Figure 18–11. The pullback method results in a cut section of roof decking nailed to one structural member, and it requires manual removal by ventilation personnel.

The results of using this method are:

- a head cut that can reveal the location and spacing of the rafters

- an opening that can be initiated near the ridge and opened toward the point of egress

- a section of decking to be removed that is only nailed to a single rafter

- decking material that doesn't drop into the structure.

You can use this method with or without a roof ladder on pitched roofs, depending on the pitch.

Dicing

This method consists of making multiple cuts between and with the direction of rafters, and is normally used on sheathing. A head cut may or may not be required. Similar to the pullback example, the direction of rafters is known on a pitched roof as they run from the ridge to the bottom of the roof. If this process is used on a flat roof, use a 45-degree inspection cut if necessary to determine the direction of the rafters. Note that Steps 1 and 2 may not be necessary if the spacing of the rafters is known (visible rafter tails under the eaves are an excellent indicator) or if the covering on the sheathing consists of shingles, shakes, or only several layers of composition:

1. Make a head or score cut perpendicular to and across two rafters when the rafter spacing is unknown (to determine rafter spacing) or when there are multiple layers of composition. The head or score cut should be located near the ridge if it is performed on a pitched roof.

2. If necessary, continue the head or score cut the approximate length of the intended ventilation opening, depending on the thickness of the layers of decking.

3. Make two cuts on either side of a single rafter. The length of these cuts should be less than the length of the tool that will be used to remove the severed section. This will ensure that personnel stand on uncut portions of the roof while they remove the decking.

4. Continue as necessary to achieve the desired length of the ventilation opening (figure 18–12).

5. Remove cut sections of the decking, beginning near the start of the head cut. Loosen the first section of sheathing to be removed by hitting the end of the first-cut board with an appropriate tool. Then, using the same tool with a J-hook motion, work down and across the slope of the roof until all the cut portions of the roof have been removed (figure 18–13).

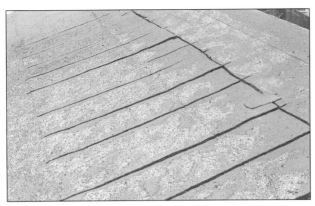

Figure 18-12. After locating two rafters for a dicing operation, make a head cut and the appropriate number of longitudinal cuts to achieve the desired length of the ventilation opening.

Figure 18-13. After the appropriate dicing cuts have been made, remove all cut portions of the decking material.

The results of this method are:

- a head cut that can reveal the location of the rafters and their spacing

- decking that can be moved away from the fire and toward egress

- removed decking that is nailed only to a single rafter

- a large section of roof that can easily be opened with minimal cutting operations

- decking material that doesn't drop into the attic.

Due to the area within some commercial occupancies, large openings may be necessary to provide adequate ventilation for the interior. Remember that layers of composition normally accumulate with the age of a building, and it is common to find numerous layers of composition on older roofs, which will significantly increase the thickness of the decking to be removed. This method will provide large ventilation openings in a relatively short period of time. This opening can easily be enlarged by the following operations:

1. Remove the cut portions of decking (from the previous operation) on the roof where additional cuts will be made.

2. Make additional cuts between the exposed rafters as necessary, and cut them according to the length of the tool that you will use to remove the severed portion (figure 18–14).

Figure 18-14. A dicing opening can be enlarged by making additional cuts between the exposed rafters.

3. Score the composition covering at the bottom of the opening (optional) and remove the cut portions of decking as necessary to achieve the desired opening (figure 18–15).

Figure 18–15. After additional cuts have been completed between the exposed rafters, remove the cut portions of decking as necessary to achieve the desired opening.

Louvering

This method can be effectively performed on any material (wood, metal, and so on), pitched or flat, that can be louvered as long as the center rafter principle is adhered to. A power saw and the principle of leverage do most of the work. If the direction of rafters are not known, use a 45-degree inspection cut if necessary:

1. Make a head cut perpendicular to and across three rafters. When you make contact with a rafter, roll the rafter with the saw.

2. Make two cuts between the center and outside rafters. The length of these cuts is determined by the size of the ventilation opening; however, 6 feet is an approximate rule of thumb.

3. Make a cut parallel and similar to the cut in Step 1 (figure 18–16).

4. When cuts intersect at the corners, continue several inches past the intersections to ensure that the decking has been completely severed at the corners. Remember to avoid compromising structural members.

5. The cut section of decking can now be vertically louvered on the rafter to complete the opening. Use an appropriate tool to push down on one side of the

panel, and if necessary, another tool to pull up on the opposite side (figure 18–17).

Figure 18–16. Louvering operations are accomplished by making four cuts that result in a section of roof decking attached to a single rafter.

Figure 18–17. A cut section of decking attached to a single rafter can be louvered by using the principle of leverage to simplify the operation.

6. You can enlarge the opening by making three additional cuts and louvering the additional panel as in Steps 1 through 5 (figure 18–18). This process can be repeated as necessary. Remember to enlarge the opening away from the fire and toward the point of egress.

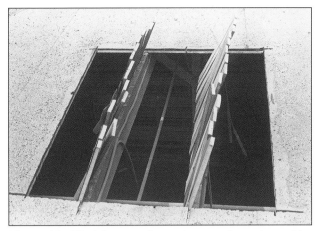

Figure 18–18. A louvered opening can be enlarged by making three additional cuts and louvering the cut panel. This process can be repeated as necessary.

This operation is simplified because personnel can use the weight and leverage of the panel against itself. The results of this method are:

- a head cut that reveals the location and spacing of the rafters

- louvered decking that is nailed only to a single rafter

- decking that can be louvered away from the fire and toward point of egress

- decking material that isn't usually dropped into the structure

This fourth bullet point depends on how the covering is secured to the base material, and it may result in composition sliding off the decking and into the structure. If composition material slides into an attic, it can be removed from the decking and placed on the roof.

If a cut section of decking cannot be louvered due to contact with sprinkler pipes, bracing, or other obstacles, you can use additional pressure on the panel to louver the decking material, or remove the louvered section to fully open the ventilation opening. This requires additional pressure on the panel, but it is possible. This method is faster and requires less effort than the pullback, dicing, or coffin cut methods.

Coffin cut

A coffin cut is based on the formula of a 7-, 9-, and 8-cut sequence and is best used on sheathing or space sheathing (figure 18–19). Prior to cutting, it is advantageous to determine the direction of rafters, which, if necessary, can be determined with a 45-degree inspection cut. When the direction of rafters is known, the long leg of the cuts (3) runs perpendicular to the rafters.

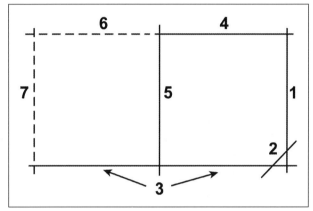

Figure 18–19. A coffin cut is based on the principle of the numbers 7, 9, and 8.

The sequence of cuts is:

1. Cut #1 is the top of the seven, approximately 3 to 4 feet long.

2. Cut #3 is approximately 6 to 8 feet long.

3. Cut #2 is a knockout cut, which allows for tool insertion to pull the roof boards.

4. Cut #4 and #5 are approximately 3 to 4 feet long, making the number 9.

5. If the opening needs to be enlarged, cuts #6 and #7 are approximately 3 to 4 feet long, making the number 8.

6. If a larger opening is necessary, the opening can be enlarged in either direction of the rectangle.

The results of this method are:

- a knockout that facilitates the initial removal of boards

- decking that personnel can remove away from the fire and toward egress

- a large section of roof that can be opened with minimal cutting operations

- decking material that doesn't drop into the attic.

Ventilation techniques for lightweight roofs

Determining the location for a ventilation opening in lightweight construction should be a primary concern due to fast failure rates, an increased use of glue to replace gang-nail plates, and a potential lack of safety margins. As a result, you should place roof ventilation openings as close to the fire as possible but on an uninvolved portion of a lightweight roof. If you cannot adhere to these stipulations, consider other options. Identification of lightweight construction begins with pre-fire planning, a basic knowledge of a particular district, and your ability to recognize certain structural characteristics. In dwellings, look for newer-style structures, those that have been constructed since the 1960s, and visible 2-by-4-inch rafter tails under the eaves indicate the presence of lightweight trusses. In commercial structures, look for concrete tilt-up walls, newer-styled buildings (particularly those with fascias), flat roofs, and buildings constructed since the 1960s.

Panelized roof

Because this distinctive type of roof is very popular in western states and is constructed with a unique pattern of structural members, let's review a ventilation assignment from the ground up and discuss why its construction can be used to augment ventilation operations. Ventilation operations begin with proper ladder placement. The strong areas of a panelized roof are the beams, purlins, and the perimeter of the building. Where a beam ties into an exterior wall would be an ideal place to position an initial ladder. However, due to numerous types of construction that have panelized roofs, locating this area can prove difficult. The four major exterior styles of buildings that use panelized roofs are:

Exterior pilaster. Exterior pilasters on concrete walls are used to strengthen the walls and to support beams. Beams will tie into the pilasters on two opposing walls of a structure. However, if you find pilasters on all four sides of the building (which is

common), it won't be possible to determine from the outside which of them support the beams.

Interior pilaster. Buildings with concrete walls that have interior pilasters are recognizable by two parallel lines on the exterior walls. These lines indicate the location of interior pilasters. Similar to exterior pilasters, interior pilasters can be found on only two or all four sides of a building.

Frame stucco and masonry. These buildings use interior posts or saddles to support the beams. From the exterior of the building, it isn't possible to determine where the beams tie into the exterior walls.

Concrete tilt-up wall. Vertical lines on exterior tilt-up concrete walls are expansion joints between the slabs of concrete. Pilasters or posts are anchored to the walls inside the building and support the beams. The beams may or may not tie into the wall at the expansion joints; therefore, expansion joints may not indicate the location of the beams.

When evaluating the four previous construction types, is it possible to know where the beams tie into the exterior walls of a building with a panelized roof? Unless personnel are familiar with a particular building or can rely on visible features, it is virtually impossible to tell. One area of a panelized roof that can be used for laddering is 8 feet from any corner. Because the spacing of the beams can vary from 12 to over 40 feet, the spacing of purlins is consistent at 8 feet due to the size of plywood-type materials. If a ladder is placed 8 feet from any corner, there is a 50% chance of laddering a purlin (Ladder 1, figure 18–20). By taking this approach, personnel will be thinking about the strong areas before accessing a panelized roof. Based on this idea, even if a ladder is placed 16 feet from any corner, this will still provide a 50% chance of laddering a purlin, and it will enhance the possibility of laddering a beam. Placing a second ladder to the building will provide an alternative means of egress (Ladder 2, figure 18–20).

Whenever possible, place a ladder on a known structural member. For the schematics in figures 18–20, 18–21, 18–23, 18–25, and 18–27, B-1 through B-3 are beams, and P-1 through P-3 are purlins.

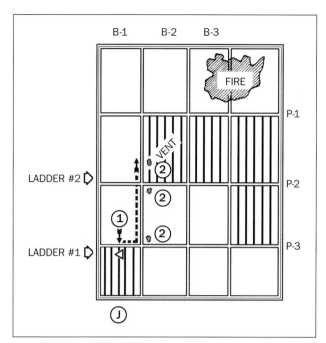

Figure 18–20. This schematic diagram for a panelized roof illustrates ladder placement and initial roof operations by a ventilation team.

Initially, ventilation personnel should determine the location of the beams and purlins to identify the strong areas of the roof. If the building has 4-by-8-foot plastic-bubble skylights, they have probably been placed lengthwise between the purlins. Therefore, the purlins run perpendicular to the length of 4-by-8-foot skylights and the beams run parallel to them. By using Ladder 1 and sounding the roof, you would determine that P-3 is a structural member and that the area to either side is unsupported plywood/OSB. If you don't locate a structural member, assume that it is parallel to and 8 feet from the wall where Ladder 1 has been placed. If necessary to make a 45-degree diagonal inspection cut, sound out a short distance from Ladder 1. Make the inspection cut out of the direction of travel (1). This can accomplish the following.

Verify the type of roof. In lightweight construction, joists and trusses are generally 2 feet on center. A diagonal inspection cut will cross a joist or rafter within 3 feet. A single 2-by-4 will indicate a panelized roof. A 2-by-4 with gusset plates, steel-tube webbing, or other hardware will indicate another type of lightweight roof. A joist of 2-by-8 inches or larger will indicate that this may not be a lightweight roof.

Determine the direction of the beams. 2-by-4 joists (J) run parallel to the beams. By determining the direction of the joists, you also discover the direction of the beams and purlins.

Depth of cut. The depth of the cut should be equal to the thickness of the decking and composition. Cut no deeper except for a specific reason.

Extension of fire. By piercing the foil insulation under the decking, you can determine the progress and location of the fire. What sort of smoke is venting out of the opening?

Feasibility of ventilation. By determining the type of roof and the location of the fire, you can normally determine the practicality of conducting ventilation operations on this type of roof. If the diagonal cut was venting hot black smoke or fire, it may not be safe to continue (remember that personnel should be able to operate on an uninvolved section of a lightweight roof). Additionally, you can quickly analyze the preceding factors without leaving the immediate area of egress (Ladder 1). After confirming the feasibility of a ventilation operation, the lead person would sound to a beam and then along it, toward the area of the fire. To check the extent of the fire, use an axe or power saw to make small indicator openings/kerf cuts (2) through the decking and foil insulation. Depending on the size of the building, these openings may be spaced about 10 feet apart. This will give a good indication as to the location of ventilation personnel in relation to the fire. Start ventilation operations when you reach an area that is both safe and over an uninvolved area of the building.

Although there are various ways to cut ventilation openings in a panelized roof, four methods are considered efficient, safe, and can be accomplished by personnel operating from the beams and purlins. It is recommended that a *minimum* of four personnel be used; two to cut and one to open the roof (five personnel are normally required for the fold-back method). An officer or other member supervises the ventilation efforts and observes the roof conditions, which includes watching the indicator openings.

Louver method

The louver method can be completed in four cuts (figure 18–21):

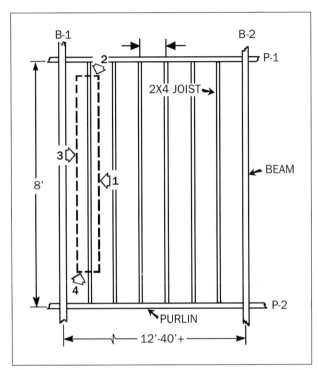

Figure **18–21.** A louver operation can be used by completing four cuts as illustrated.

1. Standing on B-1, make cut 1 about 3½ feet from and parallel to B-1. This cut is between the 2-by-4 joist and can be a maximum of 8 feet long.

2. Make Cut 2 about 3½ feet long.

3. Parallel to cut 1, make cut 3.

4. Similar to step 2, make cut 4.

By following this method, the results are:

• The cuts are made away from the fire and toward the means of egress.

• The opening is between personnel and the fire.

• A severed section of roof decking is nailed to a single 2-by-4 joist and may be hinged open.

• Material shouldn't drop into the building.

• Additional louvers may be used for extra openings or to create a strip across the building (figure 18–22).

Figure **18–22.** Additional louvers can be combined to create a strip opening.

Drop-panel method

The drop panel method can be quickly completed by two personnel and opens a large section of roof (figure 18–23):

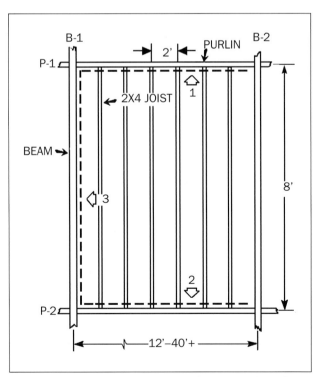

Figure **18–23.** A large section of panelized roof can be dropped into the building by completing the cuts as illustrated.

1. Standing on P-1, make Cut 1 from B-2 toward B-1 and about 6 inches from P-1. By making this cut 6 inches away from the purlins (P-1 and P-2), the power saw won't cut through the metal joist hangers, eliminating unnecessary damage to the saw. Make this cut deep enough to cut through the 2-by-4-inch joists.

2. Standing on P-2, make Cut 2 as in Step 1, and cut from B-2 toward B-1.

3. Standing on B-1, make Cut 3 parallel to B-1, and cut from P-1 toward P-2; or standing on P-1, make Cut 1 from B-l toward B-2 and about 6 inches from P-1.

4. Standing on P-2, make Cut 2 from B-2 toward B-1 (similar to Step 3).

5. Standing on B-1, make Cut 3 parallel to B-1, and cut from P-1 toward P-2.

These methods will accomplish several things. When Cut 3 is approximately halfway between P-1 and P-2, the entire panel will tear loose from the roof and fall into the building (figure 18–24). If the panel is restricted from falling by sprinkler pipes, it can easily be broken in half by striking the middle of it with pike poles, hooks, or rubbish hooks. Doing so probably won't damage the sprinkler pipes and heads. Also, the cuts will be away from the fire and toward the means of egress. The opening will be between personnel and the fire, and a large ventilation opening will have been opened in a short period of time.

A person with one chain saw can complete this operation in approximately 2 minutes. Ensure that no one is working below. You can verify this by delaying attack operations until ventilation has been completed or by using portable radios.

Panel-louver method

The panel louver method can open a large section of roof but requires numerous cuts to do so (figure 18–25):

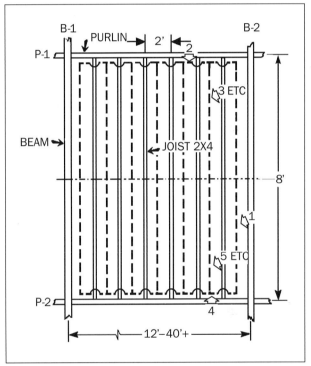

Figure 18–25. Numerous panels can be louvered by completing the cuts as illustrated.

1. Standing on B-2, make Cut 1 from P-2 to P-1 and parallel to B-2.

2. Standing on P-1, make Cut 2 from B-2 toward B-1 and about 6 inches from P-1. This cut should roll over the 2-by-4-inch joists. Ensure that the saw doesn't sever the joists.

3. Standing on P-1, make the cuts labeled 3, and so on, between the joists as indicated. These cuts should only go halfway across the panel.

Figure 18–24. When the appropriate cuts have been completed, the cut section of a panelized roof will tear loose and fall into the building.

4. Standing on P-2, make Cut 4 from B-2 toward B-1 and similar to Step 2.

5. Standing on P-2, make the cuts labeled 5, and so on, between the joists as indicated and similar to Step 3.

6. Louver the cut panels of decking to open all of them (figure 18–26).

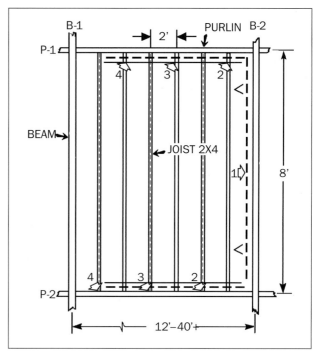

Figure 18–27. Sections of a panelized roof can be cut as illustrated and folded back to create a large opening. This is a time and personnel intensive operation.

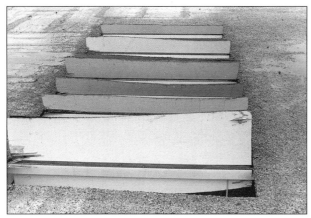

Figure 18–26. Louver the cut panels of decking to open all of the cut panels.

By following this method, the results are:

- The cut is away from the fire and toward the means of egress.

- The openings are between personnel and the fire.

- No material should fall into the building.

Although this method is effective, it does require more time than the preceding methods to complete the necessary cuts and louver the individual panels.

Fold-back method

The fold-back method requires at least four personnel to sequentially open sections of a roof to create the desired opening (figure 18–27).

This method is best accomplished with five personnel (however, four personnel is the minimum for this operation)—two with a pike pole, rubbish hook, and/or a hook on P-1, two with a pike pole, rubbish hook, and a power saw on P-2, and a safety person. At least one power saw is required:

1. Standing on B-2, make Cut 1 parallel to B-2, and cut from P-1 toward P-2 or P-2 toward P-1. Approximately 3 feet from P-1 and P-2, cut two small triangular holes that are large enough to accept a rubbish hook, pike pole, or hook.

2. Standing on P-1, make Cut 2 from B-2 toward B-1 and about 6 inches from P-1. This should cut through the first 2-by-4-inch joist and up to the second 2-by-4-inch joist.

3. Standing on P-2, make Cut 2 from B-2 toward B-1 and about 6 inches from P-2 and similar to Step 2.

4. While completing Cuts 2, it is necessary for the persons on P-1 and P-2 with a pike pole, rubbish hook, or hook to keep the panel from falling into the building by inserting the tools into the two small triangular holes along Cut 1 and pulling upward.

5. As Cuts 2 are completed, the panel can be pulled upward, which will break and separate the panel from the roof

Figure 18–28. To create the opening, pull the cut panel up and then backwards. (Photo courtesy of Kurt Zingheim)

at the second 2-by-4-inch joist (figure 18–28). The cut panel can then be pushed off to the side of the roof by the ventilation opening.

6. Standing on P-1, make Cut 3 from B-2 toward B-1 and about 6 inches from P-1. This cut should go through the second and third 2-by-4-inch joist and up to the fourth 2-by-4-inch joist.

7. Standing on P-2, make Cut 3 from B-2 toward B-1 and about 6 inches from P-2. This cut should go through the second and third 2-by-4-inch joist and up to the fourth 2-by-4-inch joist.

8. While completing Cuts 3, it is necessary for the persons on P-1 and P-2 with pike poles, rubbish hooks, or hooks to keep the panel from falling into the building by inserting the tools under the leading edge and pulling upward.

9. As Cut 3 is completed, the panel can be pulled upward and discarded as in Step 5. This process can be repeated as necessary to achieve the desired ventilation opening.

The results of this method are:

- The cut is away from the fire and toward the means of egress.

- The opening is made between personnel and the fire.

- No material will fall into the building.

This method is also time consuming, and it requires at least five personnel who are trained to perform it. Additionally, you won't be able to pull the cut panels upward if air-conditioning ducts and other systems are suspended from them, which is common. If you encounter this, either use another method, or repeat this procedure on another area of the roof. During these ventilation operations, the safety person should watch the small indicator holes/ kerf cuts (2) along B-1. This will give an indication of the extent of fire in relation to personnel on the roof. Personnel should exit the roof the same way that they approached it. If the approach was initially safe, it should still be considered safe.

Wooden I-beams and metal gusset plate construction

Wooden I-beams and metal gusset plate construction requires different operations than panelized roofs (figure 18–29).

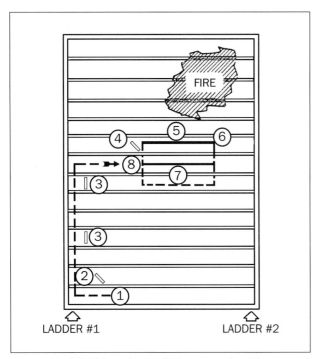

Figure 18–29. This schematic for a wooden I-beam and metal gusset plate roof illustrates ladder placement and initial roof operations by a ventilation team.

This type of construction is used on pitched and flat roofs alike, although flat roofs are more common with this construction. This operation focuses on flat roofs, although the same principles can apply to pitched roofs. On these types of roofs, the initial ladder (Ladder 1) should be placed to a corner. Consider the corners to be prime locations for ladders except when you encounter the overhang of a fascia. Ladder 2 will provide an alternative means of egress. The two major strong areas of these roofs are the corners and the perimeter. As with panelized roof operations, the first person on a roof should determine its safety by sounding it. Many of these roofs are flat, so the next concern should be to determine or verify the type of roof because the construction under the covering is anyone's guess. Does the roof have two-by-tens in a conventional configuration,

or 2-by-4 joists held together by metal gusset plates or glue? There is a vast difference between these types of construction. To determine the type and construction of a roof, sound out a short distance from the ladder. Make a 45-degree diagonal inspection cut out of the direction of travel (1). This allows you to verify the type and construction of the roof. When you open the 45-degree inspection cut, is the joist a two-by-eight (conventional construction) or a two-by-four (lightweight construction)? Additionally, if you encounter 1-by-6-inch sheathing instead of plywood-type materials, anticipate conventional construction. You can also easily determine this by removing a plug of composition roofing material, which will allow you to observe the decking material. A 45-degree inspection cut can also help you determine the direction of the joists or trusses, the depth of the cut, the extension of fire, and the feasibility of ventilation. If the preceding factors have been analyzed, and if the area under the roof is involved or soon to be involved with fire, reconsider your intended ventilation operation. Remember—consider working only on an *uninvolved* area of a lightweight roof. After the 45-degree inspection cut has been completed and ventilation operations are considered feasible, sound along the outside wall (dotted line 2) toward the location of the fire. To check the extent of fire, cut small inspection openings/kerf cuts (3) approximately every 10 feet or less. Stop when an area of the roof provides a good location to initiate a ventilation opening and is over an uninvolved portion of the building. A ventilation opening can be completed as follows:

1. If necessary, use a diagonal inspection cut (4) to verify the direction and spacing of the trusses.

2. Make the first cut (5) between and parallel to the trusses. Experience and the desired size of the opening will determine the length of Cuts 5 and 7. The long cuts of a ventilation opening should parallel the joists or trusses to reduce the chance of cutting through the structural members.

3. Make the second cut (6). Do not cut through the truss.

4. Make the third cut (7) between the trusses and parallel to Cut 5.

5. Make the fourth cut (8), connecting Cuts 5 and 7. Do not cut through the truss.

The results of this method are:

- The cut is made away from the fire and toward the means of egress.

- The ventilation opening is between the fire and personnel.

- The cut section of decking material is nailed to a single truss and can be hinged or louvered vertically to open the ventilation opening (figure 18–30).

Figure 18–30. A cut section of decking material can be louvered to create the opening.

Remember that it is difficult to remove plywood-type materials that are nailed to multiple joists or trusses.

The preceding ventilation opening may be enlarged by three additional cuts, designated by the dotted lines. This will easily enable a ventilation team to double the size of the ventilation opening and continue to enlarge it as necessary.

Fascias are common on commercial and some residential occupancies. Remember that they can be most effectively opened for ventilation or inspection by removing the rear vertical portion of the fascia from the roof.

The egress path from the roof should retrace the approach path.

Open-web bar joist

Open-web bar-joist construction frequently supports a metal deck roof or a metal built-up roof that may be covered with multiple layers of insulation material, tar, and composition (figure 18–31). Although these roofs are common, they can often be a paradox of efficiency and frustration. The building industry considers the metal built-up roof to be an efficient form of construction because the materials are widely available, the cost is less than that of a comparable wood roof, and it is an easy roof to install.

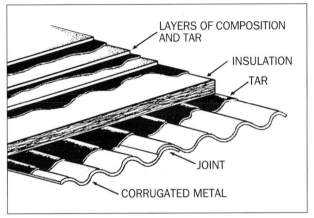

Figure 18–31. The decking of an open-web bar-joist roof normally consists of corrugated metal covered with multiple layers of insulation, tar, and composition.

Unfortunately, the metal deck roof can be difficult and dangerous when exposed to fire. Steel loses its strength at 800°F to 1,000°F, and such roofs have a quick failure rate with minimal warning. These roofs, therefore, demand specific ventilation techniques.

Although any roof that is exposed to fire can be hazardous to ventilation personnel, metal deck roofs present a unique hazard due to their ability to propagate and conceal fire, which can travel rapidly between the metal corrugations and composition coverings. The key is the first layer of tar in a metal built-up roof. When corrugated metal is heated, the tar liquefies and then gasifies. The gases cannot escape upward, so they travel between the corrugations in the steel decking and burn (figure 18–32). As the gases burn, they generate more fuel. Thus, the fire can be self-sustaining and independent of the original source of ignition.

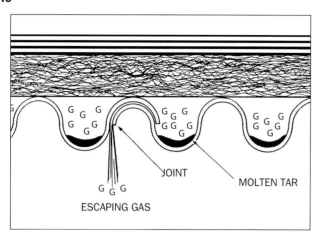

Figure 18-32. When corrugated metal is heated, the tar liquefies and then gasifies.

This type of fire can exhibit several hazards. For one, the insulation becomes a source of fuel. The fire is difficult to access, and it can cause or contribute to roof collapse. Such a fire also creates large volumes of dense smoke, which may hamper suppression efforts. Although this roof is capable of presenting specific hazards that aren't indigenous to other types of roofs, ventilation operations can still be conducted efficiently. Place the initial ladders to the corners of the building (refer to figure 18–29) because the strong areas of such buildings are at the corners and along the perimeter. Ladder 2 will provide an alternative means of egress:

1. Similar to other lightweight roof construction, ventilation personnel must first determine the type of construction. Sound out a short distance from the ladder. Out of the direction of travel (which is around the perimeter of the building), remove a small plug of composition until the base of the decking is visible (1). If corrugated metal is visible (refer to figure 17-3 in chapter 17), you have identified the roof and have also learned the direction of the bar joists, which run perpendicular to the corrugations.

2. Make a small indicator opening in the roof with the pick of an axe to determine the extension of fire and the feasibility of ventilation. Remember, if the roof is involved or soon to be involved under the area where personnel are committed, reconsider your operation. Work only on an uninvolved portion of this type of roof.

3. After you have completed a small indicator opening and consider ventilation to be feasible, sound along the outside wall (dotted line 2) toward the location of the fire. To check the extent of fire, use small indicator openings (3) approximately every 10 feet. Stop in an area that provides a good location to cut a ventilation opening and that is over an uninvolved area of the building.

4. Opening a metal built-up roof often entails a two-step process that first consists of removing the insulation/composition covering and then removing the decking from the metal bar joists. The reason for this is that rotary saw blades used to cut through corrugated metal decking won't cut completely through the composition covering, insulation and corrugated metal which can be around a total of 6 inches or more.

5. Using a chain saw or a rotary saw with a wood-cutting blade, make four cuts through the layers of insulation and composition only. Let the teeth of the power saw ride on top of the metal corrugations. The four cuts should at least form the approximate outline of the intended ventilation opening. An initial starting point is 8-by-8 feet.

6. Depending on the length of the intended opening, make a cross cut through the layers of the insulation/composition and four feet between the longitudinal cuts. This will ensure that the cut sections of insulation and composition are easily removed in one piece.

7. If the metal decking under the cut sections is cold, strike the composition sections to be removed with an axe, rubbish hook, hook, or similar tool. This will loosen the adhesive bond of the tar between the metal corrugations and the layers of insulation and composition,

facilitating removal. If the metal decking is warm, the cut sections are normally easily removed.

8. Remove the cut sections of insulation and composition and place them on the roof away from the fire (figure 18–33).

Figure 18–33. When opening a metal deck roof for ventilation, first remove the insulation-composition materials. The metal decking can then be cut and removed. (Photo courtesy of Kurt Zingheim)

The second step of this process consists of removing the metal decking with a rotary saw equipped with a metal-cutting blade or multi-use blade:

9. Determine the location of the metal joists. This is easily done by sounding with an appropriate tool. Remember, the bar joists run perpendicular to the corrugations.

10. Make cross cuts between the metal joists. This will ensure that each of the metal sections removed are attached to one metal joist only and that they can be removed with minimal effort.

11. Make perimeter cuts through the metal decking, similar to Step 1.

12. The cut panels can easily be louvered and separated from the metal joists and placed on the roof away from the fire.

Gypsum Roof Decking and Membrane Roofs

Gypsum roof decking is prefabricated from calcite concrete and wood chips and then molded into planks that are two inches thick, two feet wide, and eight feet long. These planks are then supported by structural members and then grouted together at all of their ends. Regrettably, the butted ends of these planks are often in between the roof structural members, are unsupported, are weakened by moisture, and vulnerable to early collapse under fire conditions. When cutting roof decking and a white or gray powder residue emits from the saw, a gypsum roof is present. In this case, some fire departments require their personnel to inform the incident commander and exit the roof.

Membrane roofs (or single-ply roofs) are unique in that they can be new or retrofitted old existing roofs. When new, multiple layers of insulation (such as polystyrene) are fastened to a roof and then a single sheet of membrane material for waterproofing is put on top. This material can be made from fiber resins or polyester fabric, which is then coated with asphalt compounds to provide a durable and lasting waterproof seal over a roof. This type of roof can also be applied to older existing roofs and is a superior material to the common tar-paper composition roofs (hence its popularity). This type of roof covering exhibits numerous significant hazards. It makes for a tight building and can hold in heat and increase flashovers, can burn with a significant intensity that can require a hose line on the roof to extinguish a rapidly extending fire, and forms a rigid surface that may not show any signs of a weakened decking. When exposed to heat or during combustion, it will emit numerous toxic gases; and after a relatively short period of time, it looks no different than an older conventionally covered roof. When conducting ventilation operations on this type of roof, it is generally recommended that the membrane must first be cut with a knife and peeled back (if not glued to the insulation). However, it can be cut with a chain saw operating at full rpm and with a sharp carbide chain that has a narrow gullet (which will cut rather than hook the membrane. The membrane can be seen to the side of the ventilation operation in figure 18–34). Be aware that when using a chain saw correctly on this roof that the membrane can quickly build-up on the interior of a chain guard if the saw is so equipped, and it is still possible for the membrane to begin to burn from the cutting operation due to its highly flammable nature.

Figure 18–34. It is possible to successfully cut a membrane roof with the proper equipment. (Photo courtesy of Kurt Zingheim)

Strip Ventilation Operations

Before we analyze the strategy and tactics of strip (trench) ventilation, let's review its basic concept. The term *strip ventilation* is perhaps more accurate because it connotes the removal of an outer covering as opposed to creating a deep furrow, which might sever structural members. By definition, strip ventilation operations are designed to redirect contaminants upward by creating a long, narrow opening in roof decking ahead of a horizontally extending fire (figure 18–35).

To accomplish these goals, you must adhere to three important operational and safety considerations. The first is a matter of timing. Because these operations can be time and resource intensive, and because a strip can actually accelerate the travel of fire, conduct strip ventilation as two distinct operations: one to cut the strip and one to open it. Also, always remember the time factor inherent in distance. If strip ventilation is necessary, allow enough distance between the extending fire and the cut zone to allow you to complete the opening before the fire has an opportunity to travel past it.

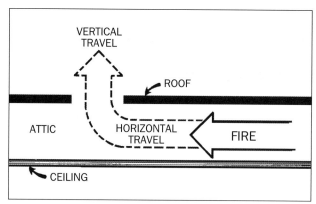

Figure 18–35. Strip ventilation is used to redirect and channel fire, heat, smoke, and fire gases from a horizontal direction to a vertical direction.

And always coordinate interior attack operations with strip ventilation. To be successful, strip operations require that the ceiling below (or as close as possible) be removed to allow access for a hose line to extinguish the attic fire. This takes coordination and communication between roof and interior personnel.

When the preceding considerations have been thoroughly evaluated, strip ventilation operations can be completed as follows.

Wood roofs

Strip ventilation openings that are cut against the construction require additional cuts and time to complete compared to those that are cut with the construction. Additionally, it is necessary to know the direction and spacing of rafters to effectively use the following methods.

Against the construction. Make two parallel cuts about 3 feet apart and across the rafters and section of roof to be ventilated. Depending on manpower and equipment, these cuts may be made consecutively or simultaneously. Then make cuts between the rafters, resulting in small sections that are easily hinged or louvered (figure 18–36).

The results of this method are

- an approximately 3-foot-wide strip across the roof of the building

- cut sections of decking that are nailed to a single rafter and easily louvered to open

- no material will fall into the building

Figure 18–36. Strip ventilation that is completed against the construction can be designed to result in small sections of roof decking that are easily louvered.

With the construction (center rafter). Make two parallel cuts on either side of a rafter. These cuts should be near the outside rafters. Make cross cuts between the parallel cuts about every 4 to 6 feet. This enhances removing or louvering plywood-type materials and multiple layers of roofing material. Remove the cut panels of decking. The results of this operation are

- a single large section of decking divided into 4-to-6-foot panels that are nailed to a single rafter and easily removed, hinged, or louvered (figure 18–37)

- no material will fall into the building.

Figure 18–37. Strip ventilation completed with the construction and in conjunction with a center rafter will result in panels that can be removed or louvered.

With the construction (between the rafters). Make a single cut between and next to one of two rafters. Make a parallel cut similar to the preceding cut and next to an opposing rafter. As a result of this operation, the cut section of decking will fall into the building (figure 18–38). Although this method drops material inward, the strip is completed quickly and no personnel are needed to remove or louver the cut sections manually. If no attic is present, ensure that no one is working below this operation.

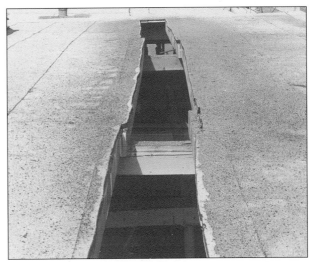

Figure 18–38. Strip ventilation completed with the construction and between two rafters is that the cut section of decking falls into the building or attic.

Metal deck roofs

Although the methods necessary to perform strip ventilation in metal deck roofs are similar to those appropriate for wood roofs, implementation is enhanced by using a two-step process that consists of removing the insulation and composition covering, then removing the metal decking from the bar joists. For example, a strip ventilation opening that is cut against the construction can be accomplished as follows:

1. Using a chain saw or a rotary saw with a wood-cutting blade, make two parallel cuts about 3 feet apart. It is only necessary to cut through the layers of composition and insulation. Let the teeth ride on top of the metal corrugations.

2. Make cross cuts every 4 feet between the parallel cuts (figure 18–39). This will ensure that the cut sections of composition/insulation can easily be removed.

Figure 18-39. Strip ventilation operations for metal deck roofs are enhanced by a two-step process that consists of cutting the insulation and composition covering for removal prior to cutting and removing the metal decking.

3. If the metal decking under the cut sections is cold, strike the sections to be removed with an axe or similar tool. This will loosen the adhesive bond of the tar between the metal corrugations and the layers of composition/insulation, facilitating removal. If the metal decking is warm, the cut sections should be easy to remove.

4. Remove the cut sections and place them on the roof away from the fire.

5. Using a rotary saw with a multi-use blade or metal-cutting blade, make two parallel cuts through the metal decking similar to those made in the layers of insulation/composition.

6. Make cross cuts between the metal bar joists. This process ensures that each of the metal sections are attached to a single bar joist only and can be louvered or removed with minimal effort.

7. Although this process is easily accomplished, it is a time-consuming operation that will destroy composite metal-cutting blades. The cuts necessary to produce the 40-foot strip in figure 18–40 consumed three composite metal blades. However, one multi-use blade could easily accomplish the same operation with one blade and in less time! *Note: If a multi-use blade is used it will simultaneously cut the insulation material and the metal decking.*

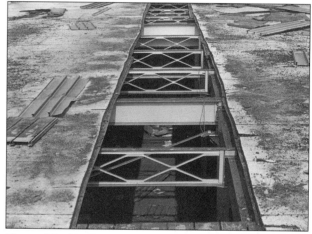

Figure 18-40. Strip ventilation in metal deck roofs can be a lengthy operation.

A derivative of this operation is based on the fact that gases from the liquefied tar between the metal decking and layers of composition cannot escape upward and are thus forced down between the joints in the steel decking. To prevent this, cut a strip through the layers of composition and insulation only to vent gases vertically from the liquefied tar and to remove the source of fuel for the self-sustaining fire. This tactic can quickly and easily be accomplished as follows:

8. As in Steps 1 through 4, cut a strip through the insulation material only (approximately three to six feet wide) and ahead of the rapidly moving fire.

9. Consider that after the strip has been completed, a hose line on the roof could extinguish any fire that crosses it.

Additionally, a hose line could be used to cool the exposed metal decking from the top.

Directional control of fire

An interesting application of strip ventilation is illustrated in figure 18–41. The left wing of this E-type hotel was well involved with fire, exposing the center wing. A strip was cut in the roof of the exposed side of the center wing. As fire from the exposed side began to extend into the attic of the center wing, the strip directed the fire up and away, allowing personnel appropriate time to save both the center wing and the building.

Each of the strip ventilation methods considered has its advantages and disadvantages. Choosing a particular method will depend on the type of incident and roof, staffing, individual preference, and your ability, which will develop through a combination of training and experience.

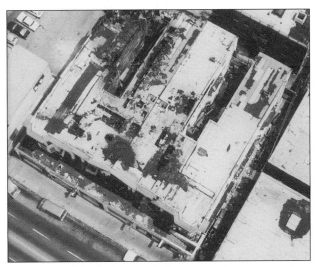

Figure 18–41. Strip ventilation can also be used to redirect the travel of fire as illustrated by this incident.

Pitched Roofs

When personnel conduct ventilation operations on roofs that are flat or slightly sloped, walking across them and maintaining a solid footing while working normally aren't major concerns. However, as the pitch of a roof increases and/or the presence of

ice, snow, tile, and other similar hazards increases the difficulty of safely operating on a pitched roof, so should the importance placed on maintaining stability during ventilation. When properly used, roof ladders can provide the stability necessary to conduct a safe and efficient roof ventilation operation. After you have decided to use a roof ladder, place it next to the section of roof to be ventilated. At this point, several problems usually surface. Personnel are hesitant to work away from a roof ladder. Also, personnel are usually unsure of their operation. This is usually due to a lack of training and experience with roof ladders. The greater the slope of a roof, the more obvious these two problems become.

Currently, there are several popular roof ladder methods that are used for ventilation operations on sloped roofs. Personnel can place a roof ladder next to an area that will be ventilated and try to work exclusively from the ladder. Depending on the expertise of personnel, this may prove cumbersome, particularly with a rotary saw. Foot holes placed in the roof will allow personnel to work away from the roof ladder; however, foot holes can prove dangerous when an attic is involved with fire or pressurized with hot gases. And personnel can use two roof ladders on either side of an area of roof that will be ventilated. Although this is a safe and proven method, its effectiveness is minimized by the number of personnel required and the time necessary to raise two roof ladders, which may be long enough to negate the affects of timely ventilation. Additionally, one of the roof ladders may be in the escaping heat and smoke, depending on the direction of the wind.

The following method is presented as a simple, effective alternative. It requires a power saw, a roof ladder, and two personnel:

1. Initially, raise a ground ladder to the roof and on the windward side of the section to be ventilated; then place an appropriate roof ladder on the roof where ventilation operations are anticipated, with the base of the roof ladder near the ground ladder. This provides easy access from the ground ladder to the roof ladder and places both ladders on the windward side of the anticipated ventilation opening. In all cases, be sure the hooks are set on the ridge of the roof.

Also, be aware that while the ladder is being positioned, the hooks may retract on older roof ladders with weak hook springs. If they do, you will need to reset them.

2. Position two personnel on the roof ladder, one with a power saw near the ridge of the roof, and one with an axe behind the top person. The axe will be used for footing, and this lower person will act as a safety.

3. To facilitate working away from the roof ladder, insert the pick end of the axe into the roof about 2 feet from the ladder. This offers several advantages. Using an axe in conjunction with a roof ladder will facilitate the stability of personnel. Also, the person with the power saw can be securely positioned on the roof ladder and the axe (figure 18–42). The second person acts as a safety, enabling the person with the power saw to cut up to 6 feet away from the ladder and still remain stable. It is vital to ensure that the axe is securely embedded in the roof. The weight of a person on the head of the axe will force the pick into the roof and prevent it from dislodging. The blade can present a potential safety hazard, however, because it will be pointing upward. Simple covers may be made out of plastic, hose, or other similar materials and fitted over the blade when appropriate.

Using an axe this way means that personnel won't have to rely on foot holes. Obviously, you shouldn't consider foot holes to be safe when the attic is involved with fire or superheated smoke and gases. Other tools can be used in place of an axe. As an example, the pick-end of a Halligan can be inserted into a roof with the handle of the Halligan parallel to the beam of the ladder, and a rubbish hook can also be used if the tangs are driven into a roof, the handle perpendicular to and across the ladder, and the second firefighter stands on the junction of the handle and ladder beam thereby supporting the handle. This allows the firefighter with the saw to step out on the handle.

The following procedure may be used to eliminate the need for multiple roof ladders and additional personnel:

1. For a louver operation, make a head cut near the ridge to determine the location and spacing of the rafters.

2. Make a cut parallel to and between the center rafter and outside rafter (figure 18–42).

3. Remove the axe from the roof. The person with the power saw can stand on the roof ladder.

4. Make a cut parallel to and between the center rafter and inside rafter or roof ladder.

5. The ventilation personnel exchange the power saw and the axe.

6. The person with the axe louvers the cut roof decking (figure 18–43).

7. To enlarge the opening, move lower on the roof ladder and repeat Steps 2 through 6.

Figure 18-42. When using a roof ladder on a pitched roof, make a cut parallel to and between the center rafter and outside rafter while positioned on the ladder and a tool that is properly imbedded in the roof.

Figure 18–43. After exchanging tools, the person with the axe louvers the cut section of roof decking.

The results of using this method are:

- stability and efficiency for personnel on roof ladders

- a head cut that will reveal the location and spacing of the rafters

- decking that is nailed only to a single rafter and can easily be louvered

- decking material that shouldn't drop into the building

- a ventilation opening that can easily be enlarged as necessary.

When using this method on tile or slate roofs, remember that you may have to remove ridge tile to increase the stability of the roof ladder and to ensure positive engagement of ladder hooks into the ridge. You must remove tile to embed an axe and before using power saws or axes to cut the decking. Tile and similar materials can be heavy and may reduce the structural integrity of a roof under fire conditions. Although tile or slate is easily removed with an axe, it can be a time-consuming operation. Tile can be removed with the pick end of an axe or by hand. This latter method can be faster than with an axe, and results in fewer small pieces to remove. You can also break the tile with the flat portion of the axe head. This results in smaller pieces of tile falling from the roof. Ensure that personnel in the immediate area are aware that tile will be falling. Aerial ladder devices require fewer personnel and less time to place into operation compared to roof ladders.

Lightweight Steel-Tile Roofs

Lightweight steel-tile roofs have become popular because they can be installed over tar and gravel, wood shakes, shingles or composition, fiberglass shingles, plywood-type materials, or open rafters. Installation is often composed of 4-foot-by-15-inch stone-coated interlocking panels of 26-gauge steel nailed to 1-by-4-inch wood battens that have been nailed to an existing or new roof. Although these panels result in an attractive roof, they can be difficult to remove by hand and can cause severe cuts to the hands, even when wearing gloves. When you encounter this type of roof and ventilation operations are necessary, an opening is best created with an axe or carbide-tipped power saw, although the latter is the most effective. Cut the steel panels, the batten substructure, and the original roof decking simultaneously and similarly to the louver procedure. This will require cutting deep enough to sever the various layers of materials. When the four cuts have been completed, louver the panel (figure 18–44).

Figure 18–44. When cutting a lightweight steel-tile roof for a louver operation, cut deeply enough to sever the steel panels, batten substructure, and roof decking.

Alternative Roofs

The title of *alternative roofs* is representative of the fact that older roofs were primarily composed of metal decking, wood sheathing, or plywood and OSB materials. However, as the building industry continues to develop and use different materials that are cost effective to replace the aforementioned materials, firefighters are being confronted with roofs that can be classified as "I haven't seen that before and I wonder how long that will last in a fire." A prime example is a new material called TECTUM (figure 18–45) panels that are two layers of fiberboard separated by foam and installed over roof structural members to form a roof decking. The material is then covered with composition, etc. This (as well as other examples) is a prime reason why firefighters engaged in roof ventilation operations must not only know what they are standing on and where the fire is in relation to them, but also sound their path of travel while traversing a roof.

Figure 18-45. These roofing panels are made from fiberboard and foam. (Photo courtesy of Erin Sawall)

Training Resources

The following resources can provide additional information and/or another viewpoint for the subjects discussed in this chapter:

- Using a Chain Saw in Peaked-Roof Ventilation, Peter, F. Kertzie, *Fire Engineering*, April 2001.

- Ventilation in Wood-Frame Structures, Peter F. Kertzie, *Fire Engineering*, April 2005.

- Challenges in Pre-Engineered Metal Roof Decks of Polyiso Foam, John Novak, *Fire Engineering*, October 2004.

- Venting Single-Ply Roofs, Tony Papoutsis and Scott Reichenbach, *Fire Engineering*, February 1992.

- Safe Roof Operations, Stuart Grant and Les Stephens, *Fire Engineering*, June 2006.

- Peaked Roof Ventilation: Cutting Through Asphalt Shingles, Kai W. Rieger, *Fire Engineering*, November 2005.

- Operating Safely on Peaked Roofs, Doug Leihbacher, *Fire Engineering*, January 2005.

- Roof Ventilation at Attic Fires, Doug Leihbacher, *Fire Engineering*, September 2003.

- Primary Roof Operations at Multiple-Dwelling Fires, John Flynn, *Fire Engineering*, July 2003.

- Steel Roofs: A View from the Top, James F. McMullen, *American Fire Journal*, September 1996.

- Ventilation Tactics for Pitched Wood Roofs, Mark Waters, *Fire Engineering*, July 2004.

- Creating a Roof Cutting Prop, Daniel P. Sheridan, *Fire Engineering*, August 2007.

- Ventilating Lightweight Roof Assemblies: The Inverse Trench Cut, John "Skip" Coleman, *Fire Engineering*, March 2010.

Index

S

T